ACOUSTIC WAVE PROBLEMS

Viktor T. Grinchenko

Igor V. Vovk

Vladimir T. Matsypura

begell house, inc
New York • Connecticut

Acoustic Wave Problems

Viktor T. Grinchenko, Igor V. Vovk, Vladimir T. Matsypura

ISBN: 978-1-56700-459-5 Print
ISBN: 978-1-56700-460-1 eBook

PRINTED IN THE UNITED STATES OF AMERICA 1 2 3 4 5 6 7 8 9 0

Library of Congress Cataloging-in-Publication Data

Names: Grinchenko, Viktor Timofeevich, author. | Vovk, I. V. (Igor' Vladimirovich), author. | Matsypura, Vladimir T., author.
Title: Acoustic wave problems / Viktor T. Grinchenko, Igor V. Vovk, Vladimir T. Matsypura.
Description: Danbury, Connecticut : Begell House, Inc., [2018] | Includes bibliographical references and index.
Identifiers: LCCN 2018060793 (print) | LCCN 2019000408 (ebook) | ISBN 9781567004601 (electronic) | ISBN 9781567004595 (hardcover : alk. paper)
Subjects: LCSH: Sound-waves--Scattering. | Sound-waves. | Sound-waves--Damping | Sound-waves--Diffraction. | Elastic waves--Diffraction. | Fluid mechanics.
Classification: LCC QC243.3.S3 (ebook) | LCC QC243.3.S3 G745 2018 (print) | DDC 534/.2--dc23
LC record available at https://lccn.loc.gov/2018060793

Direct Inquires To Begell House, Inc., 50 North Street, Danbury, Connecticut 06810

TABLE OF CONTENTS

Foreword

Don't worry. We will soon like these equations.
Once at a lecture.

From the MFTI newspaper "For science" as of March 27th, 1987

After such an epigraph, a popular warning that Stephen Hawking had once given, springs to mind: "For every equation in the book, the readership would be halved." The authors understand a certain validity of such a statement. However, the subject of this book, given the nature of the problems considered, and the traditions of their presentation in the scientific literature, excludes any possibility to forego most of the equations.

The book considers a wide range of classical acoustic wave problems using the ideal compressible fluid model. The large majority of the presented problems belong to the harmonic, or monochromatic, waves. However, the book also presents the problems, the solution of which practically realizes the potential for the construction of unsteady wave fields using effective solutions for the harmonic case.

The major idea that inspired the authors in their work has been discussed for a long time and has been used in one form or another in mathematical physics. To date, a rather large set of solutions of boundary problems for vector and scalar wave problems in so-called canonical domains has been accumulated. Such canonical domains are for instance domains with boundaries defined by the same family coordinate surfaces in spherical, Cartesian, cylindrical, spheroidal coordinate systems. The essence of the proposed idea is to build analytical forms for solving boundary problems in rather complex, non-canonical domains, based on the results of the analytical solution of the boundary problems for canonical domains.

In the book, the focus is on construction an impeccable analytical solution for the boundary problem in such complex non-canonical domains. In this case, the concept of *the general solution of the boundary problem* formulated based on the analysis of classical problems plays a crucial role. An important part of constructing solutions of classical problems lies in using expression of desired functions that easily fit the boundary conditions. These expressions contain integrals and infinite series with unknown coefficients. Using the expressions in boundary conditions give rise to functional equations. These equations are easily solved with the use of the properties of completeness and orthogonality of the functions in the expressions. It is important that the shape of such expressions is independent of the boundary conditions. Changing them only leads to a change in the values of series coefficients or of the densities of integral representations. An expression that has such properties for a given non-canonical domain we call the general solution of the boundary problem.

The expressions contain integrals and infinite series with unknown coefficients. With their use of the boundary conditions give rise to functional equations. These equations

ISBN: 978–1–56700–459–5

are easily solved with the use of the properties of completeness and orthogonality of the functions in the expressions. It is important that the shape of such expressions is independent of the boundary conditions. Changing the boundary conditions leads only to changes in the values of the unknown coefficients. An expression that has such properties for a given non-canonical domain we call the general solution of the boundary problem.

The components of the general solutions of problems presented in the book are particular solutions of the Helmholtz equation in Cartesian, cylindrical and spherical coordinates. It turned out that by using these solutions it is possible to obtain general solutions for a great number of new boundary problems in the areas of complex shapes. The book presents only a part of them.

Of course, the implementation of such an approach makes it no longer possible to obtain both complete and orthogonal system of functions at the boundary. However, the completeness of the system is preserved and provides assurance that reliable quantitative estimates of field characteristics can be obtained. In order to obtain such characteristics we need to solve infinite systems of algebraic equations. Using modern computers provides a possibility to get the shape quantitative estimates of wave fields by solving some finite systems without much difficulty.

The range of problems considered under this approach is rather broad, due to the possibility to construct general solutions in the domains whose boundaries are approximated by the parts of coordinate surfaces of different coordinate systems. The entire region of the existence of the field is divided into subdomains, in each of which the general solution of the boundary problem is built. The actual realization of this possibility has led to the need to substantially generalise approaches to the formulation of the boundary and conjugation conditions at the border of the subdomains.

Finally, in anticipation of the main content, you can go back to the epigraph. To enjoy the use of the equations the reader wants to see an interesting physical result flowing from the analysis of its solutions. When setting each problem, discussed in the book, the authors understand that "It is important to realize that the mathematical solution of a set of equations is not the answer to a physical problem" (Philip M. Morse). The physical analysis of the solutions is the focus of the book. These solutions are analyzed quantitatively in wide range of the basic parameters. The results are presented in a manner that ensures an understanding of the calculation data.

The results presented in this book were accumulated over many years, and the accumulation process was stimulated by a variety of practical needs. Collaboration with colleagues and students was crucial for the authors. We are grateful to all of them.

CHAPTER 1

Model of Medium: Statement of Boundary Problems and Solution Method

"...it was customary for wealthy pirates to start with a proper wardrobe"

Mark Twain, "Adventures of Tom Sawyer"

1.1 THE MODEL OF MEDIUM

The solution of many problems of applied acoustics is based on the use of different, often quite complex, model of a continuous medium. However, in many cases, important knowledge about the processes of generation, propagation of sound, its interaction with obstacles can be obtained using a simple model of media—ideal compressible fluid.

All the problems considered in this book, are defined on the basis of this model of medium. When the disturbances are propagating, the change in the state of the medium, is characterized by the following values: particle velocity $\upsilon(\mathbf{r}, t)$, pressure $p(\mathbf{r}, t)$ and acoustical compression $s(\mathbf{r}, t) = [\rho(\mathbf{r}, t) - \rho_0]/\rho_0$. Here ρ and ρ_0 are the densities of disturbed and undisturbed medium, respectively. The value ρ_0 is one of the fundamental physical characteristics of medium. The second characteristic is the adiabatic modulus of volume compression χ. It relates the change of pressure to the acoustical compression:

$$p = \chi s. \tag{1.1}$$

When medium is undisturbed, it is uniform, i.e., ρ_0 and χ—are constant. The equation of state of acoustic medium is used in form (1.1). It means that the assumption of smallness of disturbances is accepted. The information about the limits of applicability of linear model in acoustics is in works [1–3].

In many practically interesting cases the interaction of such ideal medium with the objects, which radiate and scatter the sound, can be described using the indicated above characteristics. The study on sound field characteristics is conducted on the basis of solutions of corresponding boundary problems. But in some cases, it is impossible to describe the sound field without including the deformation properties of radiator or scatterer. In such cases, it is necessary to solve the corresponding matching problems. To set such problems, we use geometrically linear relationships of continuum mechanics and the linear state Eq. (1.1).

We assume that the medium deformation is physically and geometrically linear when sound is propagating. It allows simplifying the system of equations of hydrodynamics. The problem of determining the motion in medium reduces to finding one scalar function

ISBN: 978–1–56700–459–5 © 2018 by Begell House Inc.

of the point position $\mathbf{r}$ and time t. This function is usually the velocity potential $\Phi(\mathbf{r}, t)$, or pressure $p(\mathbf{r}, t)$. These functions satisfy the wave equation

$$\Delta\Phi = \frac{1}{c_0^2}\frac{\partial^2\Phi}{\partial t^2}, \tag{1.2}$$

Where sound velocity c_0 is determined by medium properties:

$$c_0^2 = \chi/\rho_0. \tag{1.3}$$

Functions $\Phi(\mathbf{r}, t)$ and $p(\mathbf{r}, t)$ are related as follows

$$p(\mathbf{r}, t) = \rho_0\frac{\partial\Phi}{\partial t}. \tag{1.4}$$

If we know potential $\Phi(\mathbf{r}, t)$, it allows us to find the velocity vector of particles motion in medium

$$\boldsymbol{\upsilon} = -\text{grad}\Phi. \tag{1.5}$$

The differential operators, which are in Eqs. (1.2) and (1.5) for the arbitrary orthogonal coordinate system ξ_1, ξ_2, ξ_3 with scale Lame multipliers H_1, H_2, H_3 are as follows [4]:

$$\begin{gathered}\Delta\Phi = \text{div}\,\text{grad}\Phi;\\ (\text{grad}\Phi)_{\xi_i} = \frac{1}{H_i}\frac{\partial\Phi}{\partial\xi_i}, \quad i = 1, 2, 3;\\ \text{div}\mathbf{u} = \frac{1}{H_1H_2H_3}\left[\frac{\partial}{\partial\xi_1}(H_2H_3u_1) + \frac{\partial}{\partial\xi_1}(H_3H_1u_2) + \frac{\partial}{\partial\xi_1}(H_1H_2u_3)\right].\end{gathered} \tag{1.6}$$

The parameters for the Cartesian cylindrical and spherical coordinate systems are as follows:

$$\begin{gathered}\xi_1 = x, \quad \xi_2 = y, \quad \xi_3 = z, \quad H_1 = H_2 = H_3 = 1;\\ \xi_1 = z, \quad \xi_2 = r, \quad \xi_3 = \phi, \quad H_1 = 1, \quad H_2 = 1, \quad H_3 = r;\\ \xi_1 = r, \quad \xi_2 = \theta, \quad \xi_3 = \phi, \quad H_1 = 1, \quad H_2 = r, \quad H_3 = r\sin\theta.\end{gathered} \tag{1.7}$$

The system of relations given above describes the general regularities of small disturbances propagation in the ideal compressible liquid. These regularities do not depend on the way of appearance of these disturbances, on the obstacles that sound encounters when propagating, and on domain of space occupied by medium. The indicated relations are the mathematical model of acoustical medium, and they are not the mathematical model of some concrete acoustical situation. The concretization of situation is reached by a priori setting the form of the boundary of sound field existence domain, the physical properties of boundary, the sound sources location, and initial state of medium [5–7].

Below we will consider the concrete acoustical problems within the framework of assumption of special time dependency of field characteristics. It is assumed that all the characteristics of sound field depend on time by harmonic law, like the functions

$\exp(i\omega t)$, $\exp(-i\omega t)$ or $\cos(\omega t)$, where ω is circular frequency. Thus we simplify the mathematical formulation of problem and there is no need to consider the initial conditions. The solutions for simple harmonic motion can be used to consider more complicated non-stationary problems.

The using of complex time function allows us to write the condition at infinity [8,9] in the simplest form, when considering the sound field in unbounded medium. This condition is physically equivalent to the situation when there are no sound sources at infinity. Concerning the pressure function, for example, this condition can be written as the following asymptotic relation $p_{r\to\infty} \to A(\phi, \theta)\exp(\pm ikr)/r$, where the sign is opposite to the sign in time exponents given above.

For harmonic waves the main equation for the velocity potential (1.2) turns into Helmholtz equation

$$\Delta\Phi(\mathbf{r}) + k^2\Phi(\mathbf{r}) = 0, \quad k = \omega/c_0. \tag{1.8}$$

Equation (1.4) leads to linear relation between pressure and velocity potential:

$$p = -i\omega\rho_0\Phi. \tag{1.9}$$

Consequently, it does not matter, whether the initial problem is formulated relative to pressure function or potential function. Both these functions satisfy the Helmholtz equation.

Wave number k, added to expression (1.8), is connected with the most important characteristics of harmonic wave motion, wave length λ, by equation $k = 2\pi/\lambda$. The presence of specific linear dimension in acoustical problems is rather important for setting up the acoustical problems, for developing the solution methods, and for estimating accuracy of the quantitative characteristics of sound field. This circumstance is very important for characteristics of the problem. The matter is that the non-stationary problem, as well as the problem of harmonic wave motions, belongs to the correctly formulated boundary problems of mathematical physics. It means that, in the general case, some small changes in the initial statement of problem lead to the small changes in results. Such general characteristics of the correctness of statement of problem are almost content-free, because the specific scales to estimate the values of disturbances are not indicated. The wave length is one of important scale in acoustics.

The definition of such scale values according to the physics of problem is important for understanding of peculiarities of considered problems. In some cases it allows us to simplify their statement of boundary problems (the examples are given below). The classical example of the importance of using the immanent scales is connected with Newton assumption that process of sound propagation is isothermal. This assumption leads to a significant error in the estimation of sound velocity in gases in spite of small (with respect to everyday experience) the temperature fluctuations in sound waves.

Thus, using the harmonic waves allows us to indicate the natural linear parameter of problem—wave length. We will talk about small or large deviations in linear dimensions of radiating or scattering objects in relation to this parameter.

The complex wave number is used, if it is necessary to take into account the losses in medium connected with sound absorption in the specific acoustical problems. If the temporal multiplier is $\exp(-i\omega t)$, complex wave number is $\tilde{k} = k(1 + i\alpha), \alpha > 0$. Then the function of sound field potential for the plane wave should be represented in the form $\exp(ikx)\exp(-\alpha kx)$. If the temporal factor is $\exp(+i\omega t)$, it is necessary to take $\tilde{k} = k(1 + i\alpha)$, $\alpha < 0$ to describe the decaying wave correctly [10].

1.2 THE STATEMENT OF ACOUSTICAL BOUNDARY PROBLEMS

The statement of problems of sound radiation and dissipation includes the constructing of physical model of the object under consideration. The mathematical model is formed on this base. The constructing of physical model is mainly intuitive. It is based on the experience and qualification of researcher. For example, it is necessary to find out, which elements of the structure of the object are important (in the context of their influence on the field, radiated or scattered by object), and which are not important. As a rule, such analysis allows us to simplify the form of the object. It makes the problem solution simpler. Then, it is necessary to conduct the analysis of the acoustical (mechanical) properties of material of the surfaces of the object, which are in contact with the medium. For example, if the wave impedance of the material $\rho_1 c_1$ of the object surface exceeds the wave impedance of medium $\rho_0 c_0$, we can consider the object surface to be acoustically rigid. The normal component of the particle velocity on such surface is equal to zero

$$\upsilon_n = 0|_S\,, \tag{1.10}$$

where υ_n—normal component of the particle velocity of the medium on the object surface, S-object surface.

When $\rho_1 c_1 \ll \rho_0 c_0$, the acoustical pressure in the medium on such surface is equal to zero

$$p = 0|_S\,. \tag{1.11}$$

In the cases when the wave impedance of medium and wave impedance of material of the object surface are comparable, the acoustical properties of the object surface can be described with the help of local impedance

$$Z = -p/\upsilon_n|_S, \tag{1.12}$$

the sign "$-$" here is connected with the fact that the positive sound pressure (medium compression) on the surface S should lead to the particle velocity appearance, directed inside the object. In this situation we need the quantitative data about value Z.

Impedance condition (1.12) means that particle velocity in any point of the object surface, depends only on pressure in this point, and it does not depend on pressure in other points of surface. In other words, it is assumed that the elastic waves do not propagate along the object surface. Note that the properties of local impedance are typical for the surfaces of rubber-like and some other synthetic materials [11].

If the propagation of elastic modes of the object surface cannot be ignored, the particle velocity in some point of the object surface depends not only on pressure in this point but also on pressure in other points of the surface. The investigation of the sound field characteristics, in this case, should be performed on the basis of matching problem solution. The full system of matching conditions, which includes kinematic (the equality of normal object velocities and acoustic medium) and dynamic conditions should be fulfilled on the contact surface. Dynamic conditions determine: 1) the normal component of stress tensor σ_n is equal to acoustical pressure (with minus), 2) shearing stress vector $\vec{\tau}_s$ is equal to zero on the contact surface

$$\sigma_n = -p|_S, \ \vec{\tau}_s = 0, \tag{1.13}$$

where $p|_S$—sound pressure on the object surface.

If the part of the surface S_1 performs induced oscillations under the action of external forces, the object becomes the sound radiator. It is necessary to coordinate the normal component of particle velocity υ_n^* on this part of the object surface with particle velocity υ_n of medium on the same surface. The evident continuity condition is usually used for it

$$\upsilon_n^* = \frac{1}{i\omega\rho} \frac{\partial p}{\partial n}\bigg|_{S_1}. \tag{1.14}$$

Let us consider now some variants of particle velocity

$$\upsilon_n^* = \upsilon_0 f(S_1), \tag{1.15}$$

where $f(S_1)$—is known function of particle velocity distribution over the surface S_1, and υ_0—particle velocity in some point of the surface [10,12]. The way of supplying the energy to sound radiator, the process of transformation of energy into mechanical energy of the surface S_1 oscillations, and the property of material of the radiating object are not important under this approach.

The situation is more complex and realistic, when velocity υ_n^* is not known. But we know the external forces acting on the surface S_1. Then besides the condition (1.14) we should also use the differential equation, describing the mechanical oscillations of the surface S_1 of the following form

$$L^* \vec{\upsilon}_1 = \vec{F}, \tag{1.16}$$

where L^* is differential operator, determining the inner mechanical properties of surface S_1, $\vec{\upsilon}_1$—particle velocity vector, and $\vec{F}$—external forces vector. It is necessary to have the quantitative data about the external forces.

The situation is even more complex and realistic if the external forces are unknown for us, but we know the electrical voltages. This situation appears when it is necessary to estimate the acoustic properties of electro-acoustic sound radiators. The radiators transform the electrical energy to mechanical energy, and then to acoustical energy. In this case, we should use not only condition (1.14), but also the differential equations, determining the type of electromechanical transformation. For example, when considering piezoceramic electro-acoustical radiator it is necessary to use the equations of piezoeffect.

It is important to choose the coordinate system (or several coordinate systems) according to object geometry. In this case, the solution of problem of sound scattering by objects or sound radiation by objects can be simpler.

Consequently, mathematical statement of radiation and sound scattering problems includes Eq. (1.8) and condition (1.14). Some of the conditions (1.10)–(1.13), (1.15), (1.16) should be included, depending on the inner structural complexity and on the depth of investigation. The other conditions can also be used (for example, the conditions on the sharp edges), allowing to specify the problem solution and to make it exclusive. Almost all the variants of mathematical and physical models construction described above will be used in the following chapters.

We note finally that the methods of acoustical and electromagnetic wave's description are similar. That is why a lot of results (but not all of them), obtained for acoustical phenomena, are true for electromagnetic waves (see [13]).

1.3 CONDITIONS ON THE EDGE

The representations of domains boundaries used in acoustics are the idealization. We consider the boundary within the context of Euclidean geometry. As a result, the boundary conditions are often set on the surfaces, which include corner points and lines. When ideal liquid or gas flow around such parts of boundary the pressure jumps appear. Consequently, the local singularities in the field of velocities appear. The velocity of the medium particles tends to infinity when approaching such angular point. We should take this fact into account to set the boundary acoustical problems correctly [8,14,15].

The nature of the problem of setting the conditions on the edge can be easily understood from the following reasoning. Let us consider the neighborhood of wedge point (Fig. 1.1). The wedge is infinite in the direction, perpendicular to the figure plane. The position of the arbitrary point M in medium (in the vicinity of wedge) is determined by coordinates r and ϕ. The sides of wedge $\phi = 0$ and $\phi = \phi_0$ are ideal—acoustically soft or acoustically rigid. There is a sound field in the domain outside the wedge and it is necessary to define its structure in the neighborhood of wedge point.

The line of our reasoning is quite simple. We consider the wedge point as a sound source. Let us indicate some domain with radius r_0, in the neighborhood of wedge point, which is less than the wave length $r_0 \ll \lambda$. It is known that in the small neighborhood of such corner points the medium can be considered as incompressible [8,9,16]. The motion of medium inside the domain $r < r_0$ can be examined on the basis of analysis of Laplace equation solutions (in the polar coordinate system) for velocity potential [8,17]

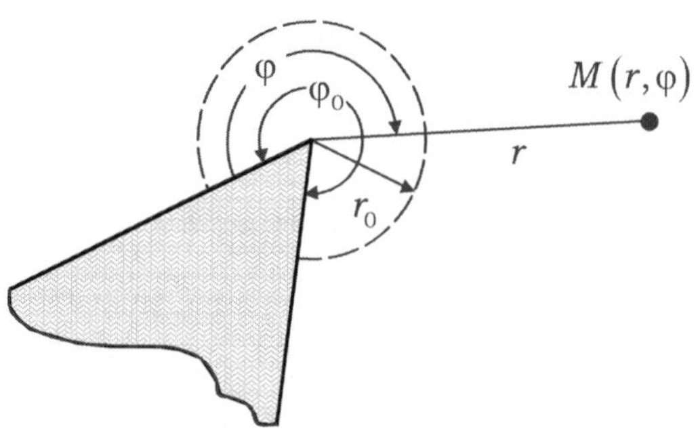

FIG. 1.1: Domain geometry in the vicinity of wedge point

$$\Delta\Phi(r,\phi) = 0, \tag{1.17}$$

$$\Phi(r,0) = \Phi(r,\phi_0) = 0, \tag{1.18}$$

$$\frac{1}{r}\frac{\partial\Phi(r,0)}{\partial\phi} = \frac{1}{r}\frac{\partial\Phi(r,\phi_0)}{\partial\phi} = 0. \tag{1.19}$$

Boundary conditions (1.18) correspond to the acoustically soft wedge surfaces, and (1.19) —acoustically rigid surface. We can consider the situation, when one face of the wedge is acoustically soft, and the other is acoustically rigid.

The Laplace equation solution (1.17) for potential Φ should satisfy some additional condition to describe the motion character in domain $r < r_0$. Its physics is quite simple— representation for Φ should ensure the sound fields matching in domains $r < r_0$ and $r > r_0$ when sound field in domain $r > r_0$ is arbitrary. It is simple to construct such solution.

It is easily checked that Laplace equation has the following solutions in polar coordinates r, ϕ

$$\Phi_\nu(r,\phi) = r^{\pm\nu}\begin{Bmatrix}\cos(\nu\phi)\\ \sin(\nu\phi)\end{Bmatrix}, \tag{1.20}$$

where ν—arbitrary.

The domain contains point $r = 0$, and that is why only the solutions with positive degrees should be used ν. We consider such solution in more details

$$\Phi_\nu(r,\phi) = r^\nu\left[A\cos(\nu\phi) + B\sin(\nu\phi)\right]. \tag{1.21}$$

We use arbitrariness (ν, A, B) to satisfy boundary conditions on the wedge sides. Let us consider the case when the boundaries are soft. It follows from the condition on face $\phi = 0$ that $A = 0$. The condition on face $\phi = \phi_0$ leads to the following possible values $\nu_n = n\pi/\phi_0$, $n = 1, 2, \ldots$. The constant B_n in each of such particular solutions remains arbitrary.

These computations allow us to find the final solution of initial problem—to find the full representation for the field in domain $r < r_0$. The system of functions $\sin(n\pi\phi/\phi_0)$, $n = 1, 2, \ldots$ is complete on interval $0 \le \phi \le \phi_0$, then the required solution of Laplace equation can be represented in the form of infinite series

$$\Phi(r,\phi) = \sum_{n=1}^{\infty} B_n r^{\nu_n}\sin(\nu_n\phi). \tag{1.22}$$

This series is a complete Fourier series on the boundary of domain $r = r_0$. We can fulfill the matching conditions with domain $r > r_0$, when selecting the values of constant B_n.

We consider that the representation of (1.22) for potential velocities is complete. It means that we should not add any additional solutions to describe the motion of medium inside the domain $r \le r_0$. But it also means that not a single summand can be excluded from (1.22). It is impossible to fulfill the arbitrary conditions on the boundary $r = r_0$ after such exclusion. Consequently, the conclusions about the structure of sound field, received on the basis of (1.22) correspond to the accepted models of medium and boundary.

Whatever is the solution method of problem of sound field radiation in the domain with angular point, the obtained presentation should give the same character of sound field in the vicinity of edge, as presentation (1.22).

Keeping in mind Eq. (1.22) we consider the expressions for the particle velocities of the particles of medium:

$$\begin{aligned}\upsilon_\phi &= -\frac{\partial \Phi}{r\partial \phi} = -\sum_{n=1}^{\infty} B_n \nu_n r^{\nu_n - 1} \cos(\nu_n \phi),\\ \upsilon_r &= -\frac{\partial \Phi}{\partial r} = -\sum_{n=1}^{\infty} B_n \nu_n r^{\nu_n - 1} \sin(\nu_n \phi).\end{aligned} \tag{1.23}$$

All the summands are finite when $\phi_0 \leq \pi$, in expressions (1.23), when approaching the edge. If ϕ_0 is larger than π, the first summands in these expressions contain multiplier r with negative degree. Then particle velocity tends to infinity when approaching the edge. The speed is maximum in the case when $\phi_0 = 2\pi$, i.e., in the vicinity of semi-plane boundary. In this case, the speed increases, when approaching the edge, proportional to $r^{-1/2}$. When $\phi_0 = 3\pi/2$, the increase in speed is determined by multiplier $r^{-1/3}$.

The presentation for potential velocities, similar to (1.22), can be obtained in the case of acoustically rigid boundaries of wedge. It is as follows:

$$\Phi(r, \phi) = \sum_{n=1}^{\infty} B_n r^{\nu_n} \cos(\nu_n \phi), \quad \nu_n = \frac{n\pi}{\phi_0}. \tag{1.24}$$

As we can see, the character of peculiarities in the field of velocities in this case is the same as in the case of acoustically soft boundaries of wedge.

The obtained results allow us to formulate the conditions on edge in the problems of sound radiation and scattering. The idea is that the sound field should have local peculiarities in the field of velocities in the vicinity of sharp wedges with opening angles $\phi_0 > \pi$ in the context of model of ideal compressible liquid. The angle ϕ_0 determines velocity with which the components of particle velocity tend to infinity and their angular distribution in the vicinity of point. These peculiarities can be determined independently from the boundary problem solution for the domain with angles. These peculiarities can be considered as known in advance. It should be noted that when we set the conditions on edge, we set not only the character of peculiarity but also the corner distribution of the field near the wedge point. The a-priory knowledge of angle distribution of field characteristics is substantial for construction of effective algorithms of boundary problems solution for the domains with angle points [17–20]. This statement is proved in the following chapters.

1.4 PARTIAL DOMAINS METHOD

The exact description of sound fields, radiated or scattered by elastic objects of complex form, is a rather difficult task. A lot of analytical and numerical methods of boundary problems solution appeared. Sometimes the process of development of new solution methods of acoustical problems is considered as creating the analytical and numerical methods independently of one another. Such point of view simplifies the real situation. A lot of acoustical problems were solved over the years. The analysis of all the obtained results shows that the progress is provided by the simultaneous development of analytical and numerical approaches.

Among the great number of different methods of boundary problems solution, only the method of finite differences can be called pure numerical. The differential formulation of problem here is used only when we write the main equations. The partial derivatives then are replaced with finite differences, and initial continuum is replaced with the system with finite number of degrees of freedom.

In the context of applied problems solutions, the analytical solutions have only one flaw—the number of such solutions is limited. Consequently, we use the methods, which allow us to broaden the range of problems, for which we can construct the effective and useful solutions for quantitative estimation of physical field's characteristics.

Some analytical solutions are used in one form or another in these approaches to construct the general solution. The typical example is method of boundary integral equations in different forms [21]. The presentation of Green function in the infinite space is used as analytical solution when constructing the method.

An interesting idea is taken as a basis T-Matrix method. An infinite set of particular solutions of wave equation in the form of spherical harmonics is used as analytic base for the solutions construction of boundary problems of scattering by finite bodies of different form [22]. The using of such method enlarges the scope of problems, having analytical solution, but its practical application is limited. The fundamental defect of this method is the absence of a-priori estimates of limitations and the difficulties in accuracy estimation of the obtained quantitative estimates.

The ideology of the proposed approach, which is widely used in this monograph, was formed long time ago in mathematical physics. We should point out its close connection with partial domains method [19,23], Riemann-Hilbert method [24] and Schwartz-Neumann method [25]. The partial solutions and other types of equations are used in method of partial domains in canonical coordinate systems for analytical solution construction in non-canonical domains. Let us consider this method in more details.

1.4.1 Main Provisions of the Method of Partial Domains

Consider the elementary problem on sound propagation in the composite wave guide filled with an ideal liquid of density ρ_0 and sonic velocity c_0. The problem geometry is depicted in Fig. 1.2. For the sake of definiteness, the wave guide walls are considered to be acoustically rigid. All characteristics of the harmonic sonic field within this domain can be derived via the function of potential of velocities Φ, which satisfies the Helmholtz equation

$$\frac{\partial^2\Phi}{\partial x^2} + \frac{\partial^2\Phi}{\partial y^2} + k^2\Phi = 0, \quad k = \frac{\omega}{c_0}. \tag{1.25}$$

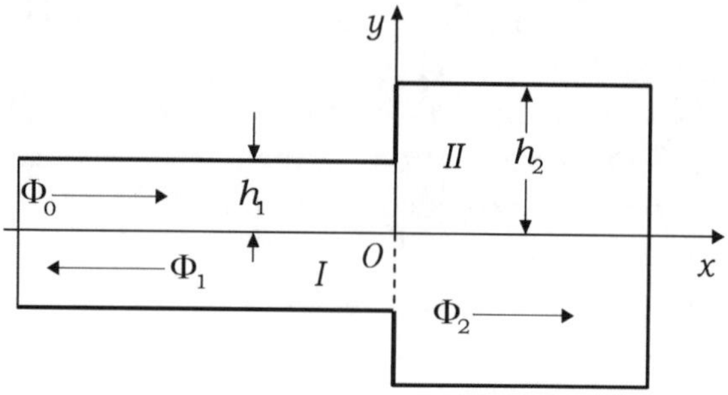

FIG. 1.2: Geometry of a composite wave guide

Here ω is frequency of a harmonic wave, while a temporary multiplier is set in the form of $\exp(-i\omega t)$.

It is postulated that an incident wave propagates in the left part of the wave guide, which wave is controlled by the following potential

$$\Phi_0 = A_0 \exp(ikx). \tag{1.26}$$

The fact that the lower propagation mode is considered here is not critical. The domain, where the sonic field exists, can be subdivided into two partial domains (subdomains) I and II, which are separated by line segment $x = 0$, $|y| \le h_1$ (see Fig. 1.2), and such subdivision is quite natural from the purely geometrical viewpoint. More critical is the consideration that such subdivision is lucrative for construction of the problem solution in general. The point is that the expression for the potential function in subdomain II has the following form:

$$\Phi_{II}(x, y) = \sum_{n=0}^{\infty} B_n \exp(i\xi_n x) \cos\left(\frac{n\pi y}{h_2}\right), \tag{1.27}$$

where

$$\xi_n = \begin{cases} \sqrt{k^2 - (n\pi/h_2)^2}, & k > n\pi/h_2, \\ i\sqrt{(n\pi/h_2)^2 - k^2}, & k < n\pi/h_2, \end{cases} \tag{1.28}$$

is the general representation of the sonic field in a semi-infinite wave guide $x \ge 0$. The generality concept implies that the representation form of a potential remains unchanged irrelevant of the boundary conditions at boundary $x = 0$. A similar construction can be made for the left part of the wave guide (subdomain I). The incident wave is considered to be preset, while the general representation for the reflected field is easily derived as

$$\Phi_I(x, y) = \sum_{m=0}^{\infty} D_m \exp(i\eta_m x) \cos\left(\frac{m\pi y}{h_1}\right), \tag{1.29}$$

where

$$\eta_m = \begin{cases} \sqrt{k^2 - (m\pi/h_1)^2}, & k > n\pi/h_1, \\ i\sqrt{(m\pi/h_1)^2 - k^2}, & k < n\pi/h_1. \end{cases} \tag{1.30}$$

Thus, the general representation of the sonic field within the total domain of the wave guide has been constructed. The above expressions contain two sequences of arbitrary magnitudes A_n and B_m. By selecting their values, it is possible to satisfy precisely the conjugation condition at boundary $x = 0$ in the wave guide. The validity of this statement follows from the general properties of the Fourier series.

The above example vividly illustrates the essence of one of the concepts, which is critical for the method of partial domains, i.e., the concept of the boundary problem general solution. This concept is quite simple: the given set of particular solutions for a system of differential equations in partial derivatives is referred to as the general solution of a boundary problem for the given domain, provided representation of this solution allows one to satisfy to arbitrary boundary conditions at its boundary.

As a rule, the conclusion on the solution generality can be made based on properties of the Sturm-Liouville function systems. Expressions (1.27) and (1.29) are the general solutions of boundary problems for semi-infinite wave guides with rigid walls.

Within the framework of such an approach, it is easy to construct the general solutions, for example, for the domains limited by segments of straight lines (parts of co-ordinate surfaces of the Cartesian co-ordinate systems). Figure 1.3 depicts a polygon with boundary surfaces, all of which are described in a uniform co-ordinate system. For simplicity we will consider that nonzero values of sonic velocity (normal derivative of the potential) are set only within polygon sides $y = 0$, $y = b$, and $y = d$. The choice of partial domains here is quite ambiguous, whereas either a vertical, or a horizontal dashed line can be used as a border between them. In the second case, the general solutions in domains I $(0 \le x \le a)$ $0 \le y \le b$ and II $(0 \le x \le l)$ $d - b \le y \le d$ are reduced to the following form

$$\Phi_I(x, y) = \sum_{m=0}^{\infty} \cos(\alpha_m x) \left[a_m f_1^{(m)}(y) + b_m f_2^{(m)}(y)\right],$$

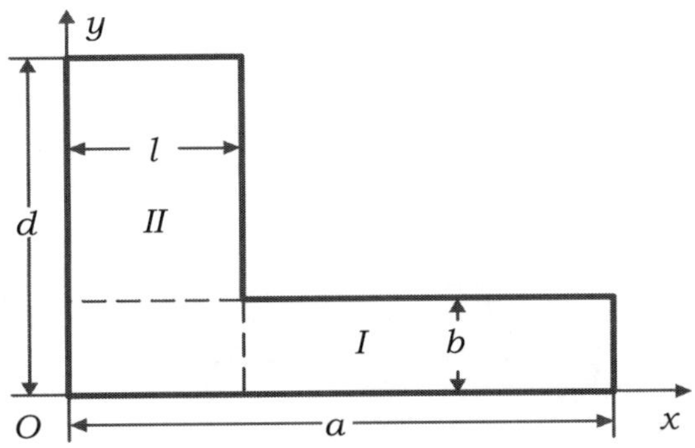

FIG. 1.3: Polygonal domain geometry

$$\Phi_{II}(x,y) = \sum_{n=0}^{\infty} \cos(\beta_n x)\left[c_n g_1^{(n)}(y) + d_n g_2^{(n)}(y)\right]. \tag{1.31}$$

Here $\alpha_m = m\pi/a$, $\beta_n = n\pi/l$,

$$f_1^{(m)} = \begin{cases} \cos\left(\sqrt{k^2-\alpha_m^2}\,y\right), & k>\alpha_m, \\ \exp\left(-\sqrt{\alpha_m^2-k^2}\,y\right), & k<\alpha_m, \end{cases}$$

$$f_2^{(m)} = \begin{cases} \sin\left(\sqrt{k^2-\alpha_m^2}\,y\right), & k>\alpha_m, \\ \exp\left(-\sqrt{\alpha_m^2-k^2}\,(y-b)\right), & k<\alpha_m, \end{cases} \tag{1.32}$$

$$g_1^{(n)} = \begin{cases} \cos\left(\sqrt{k^2-\beta_n^2}\,y\right), & k>\beta_n, \\ \exp\left(-\sqrt{\beta_n^2-k^2}\,(y-d)\right), & k<\beta_n, \end{cases}$$

$$g_2^{(n)} = \begin{cases} \sin\left(\sqrt{k^2-\beta_n^2}\,y\right), & k>\beta_n, \\ \exp\left(-\sqrt{\beta_n^2-k^2}\,(b-y)\right), & k<\beta_n. \end{cases}$$

The problem of completeness of these solutions is resolved quite simply. Each of the expressions for potentials contains two Fourier series with arbitrary coefficients, which allows one to satisfy the arbitrary boundary conditions within horizontal segments of the boundary, as well as the conjugation condition within line $y = b$. Zero boundary conditions for sonic velocity are identically satisfied in the vertical segments of the boundary.

1.4.2 Solution of an Interior Problem for a Triangular Domain

The above discussion illustrates the procedure of constructing the general solution of a boundary problem in partial derivatives. Generalization of the constructed solutions is provided by representation of the general solution for an interior of, e.g., a rectangular triangle with rigid walls (Fig. 1.4). The rectangular type of triangle is considered here

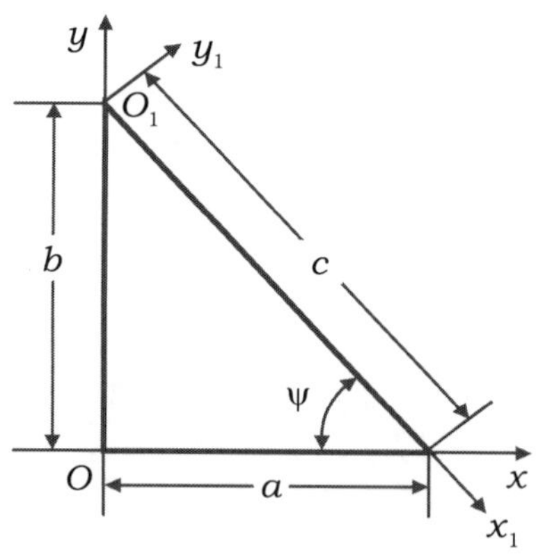

FIG. 1.4: Geometry of a triangular domain

only for simplification of the respective calculations. Generally, no restrictions on the triangle form are imposed. Boundaries of the triangle under study are parts of co-ordinate surfaces of two Cartesian co-ordinate systems xOy and x_1Oy_1.

The general solution of a boundary problem in eigenvalues is reduced to the following form

$$\Phi = \sum_{m=0}^{\infty} A_m \cos(\alpha_m x)\, f_m^{(1)}(y) + \sum_{n=0}^{\infty} B_n \cos(\beta_n y)\, f_n^{(2)}(x) + \sum_{l=0}^{\infty} D_l \cos(\gamma_l x_1)\, f_l^{(3)}(y_1), \tag{1.33}$$

where $\alpha_m = m\pi/a$; $\beta_n = n\pi/b$; $\gamma_l = l\pi/c$.

Here the Fourier series corresponding to each of the triangle sides are explicitly expressed, and their presence in the solution representation for function Φ ensures generality of the boundary problem solution. In regard to functions $f_i^{(j)}$, proceeding from the domain geometry, the following expressions are applicable for them:

$$f_m^{(1)}(y) = \begin{cases} \cos\left(\sqrt{k^2 - \alpha_m^2 y}\right), & k > \alpha_m, \\ \exp\left(-\sqrt{\alpha_m^2 - k^2 y}\right), & k < \alpha_m, \end{cases}$$

$$f_n^{(2)}(x) = \begin{cases} \cos\left(\sqrt{k^2 - \beta_n^2 x}\right), & k > \beta_n, \\ \exp\left(-\sqrt{\beta_n^2 - k^2 x}\right), & k < \beta_n, \end{cases} \tag{1.34}$$

$$f_l^{(3)}(y_1) = \begin{cases} \cos\left(\sqrt{k^2 - \gamma_l^2 y_1}\right), & k > \gamma_l, \\ \exp\left(\sqrt{\gamma_l^2 - k^2 y_1}\right), & k < \gamma_l. \end{cases}$$

The assumption of this solution generality is based on the Fourier series properties and requires no detailed consideration. The pattern of the relations connecting values of arbitrary constants A_m, B_n, and D_l in representation of the potential (1.33) can be described upon consideration of the following example, i.e., the boundary condition at the rigid wall $x = 0$. This condition yields the functional equation

$$\sum_{n=0}^{\infty} B_n \cos(\beta_n y) \frac{df_n^{(2)}(0)}{dx} + \sum_{l=0}^{\infty} D_l \Big\{ \sin\psi \cos[\gamma_l (y-b)\sin\psi] \times \frac{df_l^{(3)}[-(y_1-b)\cos\psi]}{dy_1} - \gamma_l \cos\psi \sin[\gamma_l (y_1-b)\sin\psi] \times f_l^{(3)}[-(y_1-b)\cos\psi] \Big\} = 0. \tag{1.35}$$

In view of the properties of the system of functions $\cos(\beta_n y)$, $n = 0, 1, \ldots$ within segment $0 \leq y \leq b$, this functional equation is transformed into the following infinite system of linear algebraic equations

$$B_n = \sum_{l=0}^{\infty} b_{nl} D_l, \quad n = 0, 1, ... \tag{1.36}$$

Noteworthy is that coefficients of this system can be derived in the explicit form.

Equation of type (1.36) also holds for boundary conditions at side $y = 0$. If boundary conditions are met at the triangle hypotenuse, this corresponds to an infinite system containing all unknowns A_m and B_n in each equation. The set of three homogeneous bound infinite systems controls the requirements imposed by the boundary conditions on factors of the Fourier series in (1.33). By equating the determinant of this system to zero, we get the equation controlling the eigenfrequencies of a finite domain filled with an acoustic medium.

The above construction makes it possible to consider one more aspect of the considered method. The line of argument leading to representation of the required sonic field potential in the form of (1.33) implies an ambiguous form of the solution. In particular, if $\cos(\alpha_m x)$ in the first term of Eq. (1.33) is substituted by $\sin(\alpha_m x)$, the property of completeness of this term will still hold. Both systems of functions are full/broad and orthogonal within segment $0 \leq x \leq a$. It is easy to see that the first system yields simpler relations if the boundary conditions are met in rigid walls, whereas, in case of soft walls (with zero pressure in them), the preference should be given to the second system.

An option of varying the solution with preservation of its completeness will be considered further in Section 1.4.4. Now we consider noteworthy an important distinction between problems formulated for a wave guide and a composite rectangle, on the one hand, and for triangular domain—on the other hand. While, in first two problems, the boundary surface of the field existence domain is described by parts of co-ordinate surfaces of one system, this is not the case in the problem for a triangular domain. A possibility of simultaneous operation in various co-ordinate systems essentially expands the circle of problems of mathematical physics, for which the construction of an analytical solution is possible, and this issue should be considered in more details.

1.4.3 Domains Limited by the Co-Ordinate Surfaces of Different Families

Apparently, no general scheme is available, in what concerns the approach to the specified class of problems. Therefore, we'll illustrate the basic features of such an approach by a particular example. To some extent, this is be related to generalization of the above problem on a sonic field in conjugated flat wave guides (see Fig. 1.2). Consider a problem on sound emission in a wave guide shown in Fig. 1.5. A flat semi-infinite wave guide $x \leq 0$, $|y| \leq h$ is conjugated with a wedge-type wave guide with an opening angle $2\theta_0$. For the sake of definiteness, walls of this wave guide are considered to be rigid. A wave, which falls from the negative infinity, can be described by any set of propagation modes. For definiteness we will consider the case of an individual normal wave

$$\Phi_0 = Q_n \cos(\alpha_n y) \exp\left(i\sqrt{k^2 - \alpha_n^2}\, x\right). \tag{1.37}$$

The unknown quantities here are characteristics of the passed and reflected sonic fields.

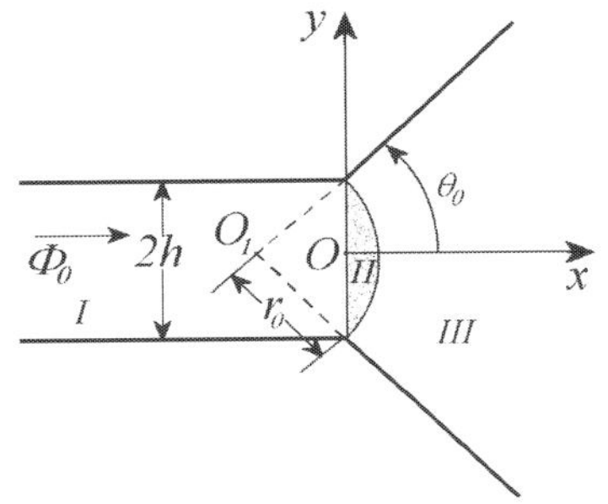

FIG. 1.5: Geometry of a complicated wave guide in the form of conjugated flat and wedge-type wave guides

For construction of the general solution of the problem the total domain of existence of a sonic field is naturally subdivided into three subdomains as is shown in Fig. 1.5.

For each of them it is possible to construct a general representation for the potential of velocities which makes it possible to satisfy boundary conditions on the walls of the flat and wedge-type wave guides, as well as the conjugation conditions between these subdomains. For domain I this representation is easily derived and it includes an incident wave and a full set of the incident and heterogeneous reflected waves:

$$\Phi_I = \Phi_0 + \sum_{n=0}^{\infty} A_n \cos(\alpha_n y) \begin{cases} \exp\left(-i\sqrt{k^2 - \alpha_n^2}x\right), & k > \alpha_n, \\ \exp\left(\sqrt{\alpha_n^2 - k^2}x\right), & k < \alpha_n. \end{cases} \tag{1.38}$$

If we specify that $\alpha_n = n\pi/h$, expression (1.38) will satisfy the boundary conditions at the wave guide walls. A set of arbitrary factors A_n allows one to satisfy the conjugation conditions at the boundary with subdomain II.

Representation of the sonic velocity field potential in subdomain III has also quite a simple form. It is expedient to represent it in polar co-ordinates

$$\Phi_{III} = \sum_{m=0}^{\infty} B_m \cos(\beta_m \theta) H^{(1)}_{\beta_m}(kr). \tag{1.39}$$

Here $\beta_m = m\pi/\theta_0$. The choice of angular functions in the solution ensures that boundary conditions are satisfied at the wedge-type wave guide walls, while radial functions are chosen according to the emission conditions. The sequence of arbitrary factors should be determined from the conditions of conjugation with subdomain II.

For construction of the general solution of the boundary problem in domain II, taking into account the character of its boundary, it is necessary to use particular solutions of the Helmholtz equation in the Cartesian and polar co-ordinate systems. At the boundary of subdomain II for $x = 0$, the full and orthogonal system of functions is represented by $\cos(\alpha_n y)$, $n = 0, 1, 2, \ldots$. At the second boundary for $r = r_0$, $|\theta| \leq \theta_0$, the system of functions $\cos(\beta_m \theta)$, $m = 0, 1, 2, \ldots$ possesses the same properties. In view of this, the sonic field potential in subdomain II can be reduced to the following form

$$\Phi_{II} = \sum_{n=0}^{\infty} D_n \cos(\alpha_n y) \begin{cases} \cos\left(\sqrt{k^2 - \alpha_n^2}\, x\right), & k > \alpha_n \\ \exp\left(-\sqrt{\alpha_n^2 - k^2}\, x\right), & k < \alpha_n \end{cases} + \sum_{m=0}^{\infty} C_m \cos(\beta_m \theta) J_{\beta_m}(kr). \tag{1.40}$$

The structure of this solution coincides with that of the interior problem solution for a triangular domain. However, the physical essence of a problem in this case is somewhat different. If in the first case it was the issue of deriving the characteristic forms of oscillations in a final volume, in the second case, subdomain II is intermediate for energy transfer from the flat wave guide in the wedge-type one. The energy flow through this domain should be distinct from zero. These are absolutely natural physical requirements, which have the same character, as emission conditions. At first sight, the solution in the form of (1.40) fails to satisfy the specified power requirements. Thus, in the first sum by n, the solution contains standing waves by x co-ordinate, which waves transfer no energy. Similar standing waves by the radial co-ordinate are described by function $J_{\beta_m}(kr)$.

Thus, when taken separately, each term in expression (1.40), does not transfer any energy. However, taking advantage of the above-mentioned ambiguity of a choice of particular solutions within the framework of the considered method, one can eliminate the noted formal difficulties by way of substituting the solutions by the radial co-ordinate and by x co-ordinate. Thus, instead of the Bessel functions $J_{\beta_m}(kr)$ it is possible to apply the solution in the form of Hankel functions $H^{(2)}_{\beta_m}(kr)$, which describe the waves running to the origin of co-ordinates in a polar co-ordinate system. Such waves should represent the perturbations reflected from a heterogeneity. However, such substitution is not even necessary. The analysis shows that the full solution in the form of (1.40) controls a nonzero energy flow through subdomain II. Physically, the situation here is identical to that occurring during the perturbation passage through a layer/stratum of a finite thickness, when the glancing angle of an incident wave is below the critical value. Here the energy transfer occurs as a result of interaction of the heterogeneous waves “adhered“ to the opposite boundaries. The particular calculations made for various forms of the potential representation in subdomain II confirm that the energy-based considerations also require no changes in the form of (1.40).

Noteworthy are also some computing aspects for the method characterization. It is clear that, in the case under study and other similar cases, factors of infinite systems cannot be calculated explicitly and, therefore, numerical determination of integrals is required. However, the number of terms of the expansion series obtained after reduction of infinite systems, depends on many factors. Thus, for the wave dimensions of a wave guide are $2h/\lambda \leq 1.5$ it would be sufficient to reduce the number of terms in each of series in (1.38)–(1.40) to five, which would limit the energy-based solution error to only few fractions of percent. The problem of sound transmission through the domain of plane and wedge waveguides matching is considered in chapter seven.

The proposed general approach to construction of solutions of boundary problems with use of particular solutions in various co-ordinate systems often appears useful even in those cases where, apparently, solutions from a single co-ordinate system could be used. We will illustrate it by an example of a classical problem of sound emission by a finite cylinder. We will focus only on the problem of subdividing the domain under study into partial domains (subdomains).

The geometry of a cylindrical emitter (cross section) and its basic geometrical parameters are shown in Fig. 1.6. It is assumed that the oscillating velocity is set at the emitter surface. Construction of the general solution for a field outside the finite cylinder can be performed with use of various systems of subdomains. Thus, as one of variants in the total domain of the field existence it is possible to select the outer part of a cylindrical concavity $R \geq R_0$ and two semi-infinite cylinders $z \geq h$, $R \leq R_0$, and $z \leq -hR \leq R_0$. Alternatively, one can use as subdomains two half-spaces $z \leq -h$, $z \geq h$ and a stratum with a cylindrical concavity $|z| \leq hR \geq R_0$. For each of such subdomains one can construct the general solution of the boundary problem. However, practical application of such solutions is problematic due to certain difficulties of a quite obvious physical nature. These are related to the fact that it is very hard to satisfy the conditions of emission by the boundaries of partial subdomains (depicted by dashed lines in Fig. 1.6) by the solutions constructed for the above subdomains. The problem of taking into account the conditions of emission does not arise if one uses another system of partial subdomains. Boundaries of these subdomains are parts of a spherical surface $r = r_0$ and of a cylinder surface. These domains (I–IV) are shown in Fig. 1.6. The general solution for the outer part of a spherical surface $r = r_0$ as a part of the general solution of the problem automatically accounts for the emission conditions. Solutions in domains I–III are easily constructed with use of particular solutions of the Helmholtz equation in cylindrical and spherical co-ordinate systems. Within the framework of such an approach, the solutions of some new problems of sonic emission have been derived [26,27].

1.4.4 Supplementation of the Boundary Conditions in the Potential Theory Problem

In some cases, even application of notorious full solutions of the boundary problem does not exclude some difficulties encountered during transformation of the functional equations of the boundary conditions into an algebraic form. This implies a specific problem

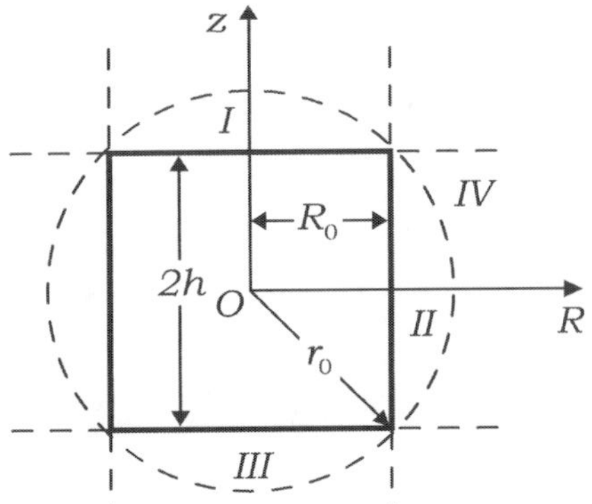

FIG. 1.6: Cross section of a cylindrical emitter

of expanding the boundary conditions from physically set boundaries to some nonexistent subdomain. We will discuss a technique for overcoming the specified difficulties. At first, in order to simplify our calculations, we will consider the elementary problem of the potential theory, which permits a strict derivation of all necessary proof.

The inner problem is being solved for a harmonic function $\varphi(x, y)$ in a rectangle $|x| \le a$ $|y| \le b$ depicted by solid lines in Fig. 1.7. The boundary conditions at the rectangle boundaries are set as follows

$$\varphi(\pm a, y) = 0, \quad \varphi(x, \pm b) = \cos\left(\frac{\pi x}{2a}\right). \tag{1.41}$$

The given problem is so simple that the required solution can be presented with no additional comments. This solution looks like

$$\varphi(\pm a, y) = \cos\left(\frac{\pi x}{2a}\right)\frac{\cosh(\pi y/2a)}{\cosh(\pi b/2a)}. \tag{1.42}$$

Consider now the following sets of particular solutions of the Laplace equation:

$$\varphi(x, y) = \sum_{m=0}^{\infty} B_m \cos(\gamma_m y)\frac{\cosh(\gamma_m x)}{\cosh(\gamma_m a)} + \sum_{k=0}^{\infty} A_k \cos(\delta_k x)\frac{\cosh(\delta_k y)}{\cosh(\delta_k b)}. \tag{1.43}$$

Here the following designations are used

$$\gamma_m = \frac{(2m+1)\pi}{2b}, \quad \delta_k = \frac{(2k+1)\pi}{2l}. \tag{1.44}$$

The introduced magnitudes contain a new geometrical parameter l. This parameter controls the size of the rectangle depicted by dashed lines in Fig. 1.7.

It makes sense to consider Eq. (1.43) only within domain $|x| \le a$ $|y| \le b$. For $|x| > a$ the first series will be dispersing unless a special choice of factors B_m is provided, whereas within the specified domain it is possible to treat it as the general solution of the problem for a rectangle. Such statement is based on the fact that expression (1.43) contains two Fourier series, and by choosing the required values of factors A_k and B_m, one can to satisfy the arbitrary boundary conditions at the rectangle sides.

Certainly, expression (1.43), which yields the solution of the initial boundary problem (1.41), is much more complicated than the solution in the form of (1.42), and

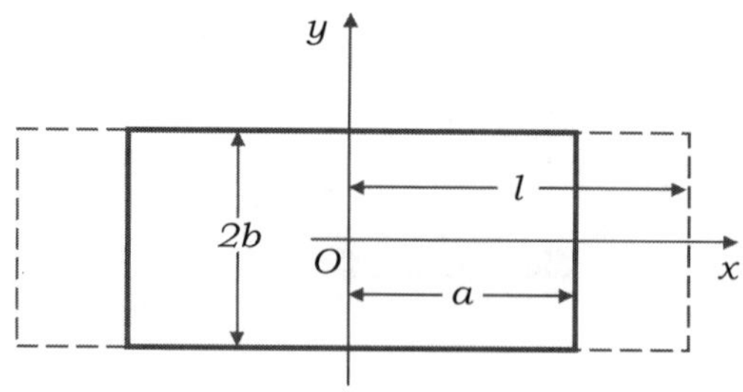

FIG. 1.7: Geometry of a rectangular domain

application of such a complicated representation is expedient only from the standpoint of deriving some general conclusions about the used method. Therefore, derivation of such conclusion is our main purpose.

Use of expression (1.43) for meeting the boundary conditions at surfaces $x = \pm a$ leads to derivation of the relations between values of constants A_k and B_m. Therefore, we will firstly consider the boundary conditions at surfaces $y = \pm b$. These conditions imply the following functional equation

$$\sum_{k=0}^{\infty} A_k \cos(\delta_k x) = \cos(\alpha_0 x), \quad \alpha_0 = \frac{\pi}{2a}, \quad |x| \le a. \tag{1.45}$$

The system of trigonometric functions $\cos(\delta_k x)$ is not orthogonal within interval $|x| \le a$, while it is possible to specify various ways of deriving constants A_k from this equation. The most simple one is based on replacement of Eq. (1.45) by the following relation:

$$\sum_{k=0}^{\infty} A_k \cos(\delta_k x) = \begin{cases} \cos(\alpha_0 x), & |x| \le a, \\ \xi(x), & a < |x| \le l. \end{cases} \tag{1.46}$$

Here function $\xi(x)$ is arbitrary. Based on the completeness and orthogonality properties of the system of functions $\cos(\delta_k x)$ within interval $|x| \le l$, one can easily derive factors

$$A_k = \frac{2\alpha_0 \cos(\delta_k a)}{l\left(\alpha_0^2 - \delta_k^2\right)} + \xi_k, \tag{1.47}$$

where

$$\xi_k = \frac{2}{l} \int_0^l \xi(x) \cos(\delta_k x)\, dx. \tag{1.48}$$

Using the expansion of hyperbolic functions into the Fourier series

$$\frac{\cosh(\delta_k y)}{\cosh(\delta_k b)} = \sum_{m=0}^{\infty} \frac{2(-1)^m \gamma_m}{b\left(\gamma_m^2 + \delta_k^2\right)} \cos(\gamma_m y), \tag{1.49}$$

and the boundary conditions at surface $x = \pm a$ it, we get

$$B_m = -\sum_{k=0}^{\infty} A_k \frac{2(-1)^m \gamma_m}{b\left(\gamma_m^2 + \delta_k^2\right)} \cos(\delta_k a). \tag{1.50}$$

Thus, for the required potential function $\varphi(x, y)$ there are two different representations: (1.42) and (1.43). Both of them satisfy the harmonic equation and boundary conditions (1.41). Hence, these expressions should be identically equal in each point inside the considered rectangle and at its boundaries. This is strictly proved by expansion of both expressions into the double Fourier series within a rectangle. Moreover, this identity is valid for any function $\xi(x)$, which controls the supplementation of the boundary conditions.

The above conclusion can be taken as a principle in further consideration of a special class of problems for which construction of the general solutions within the limits of the method of partial domains is applicable in the absence of physically set boundary conditions at all co-ordinate surfaces linked to the complete set of functions. Now will consider a particular example of such problem.

1.4.5 The Problem of Sonic Emission by Intersected Cylinders

The problem of emission or sound dissipation by the system of unjoining circular cylinders with parallel axes is well-known. Formation of infinite systems for series expansions by wave functions is carried out on the basis of the summation theorem [28]. The respective formulas of re-expansion are also available for the case where the distance between centers of these cylinders is less than sum of their radii. The respective boundary of domain, outside of which the sonic field is considered, is shown in Fig. 1.8. The emitting surface is formed by circular arcs $r_1 = a_1$, $\theta_0^{(1)} \leq \theta_1 \leq -\theta_0^{(1)}$ and $r_2 = a_2$, $\theta_0^{(2)} \leq \theta_2 \leq 2\pi - \theta_0^{(2)}$. At these surfaces the velocity normal component is set for the ideal compressible medium particles

$$\upsilon_r(a_1, \theta_1) = F_1(\theta_1), \quad \upsilon_r(a_2, \theta_2) = F_2(\theta_2). \tag{1.51}$$

The general representation of the potential function, which satisfies the Helmholtz equation, is quite obvious for the case under study, and for a temporary characteristic of type $\exp(-i\omega t)$ it has the following form

$$\Phi = \sum_{m=0}^{\infty} A_m H_m^{(1)}(kr_1)\cos(m\theta_1) + \sum_{n=0}^{\infty} B_n H_n^{(1)}(kr_2)\cos(m\theta_2). \tag{1.52}$$

Such representation satisfies term by term the equation of Helmholtz, conditions of emission and conditions of periodicity of the field outside the emitting body. Boundary conditions (1.51) should be satisfied by choosing specific values of arbitrary factors A_m and B_n.

In case of unjoining cylinders, where the boundary conditions are set at full circles, the procedure of meeting boundary conditions is simple and yields an infinite system of linear algebraic equations. Meanwhile, application of expression (1.52) to the case under

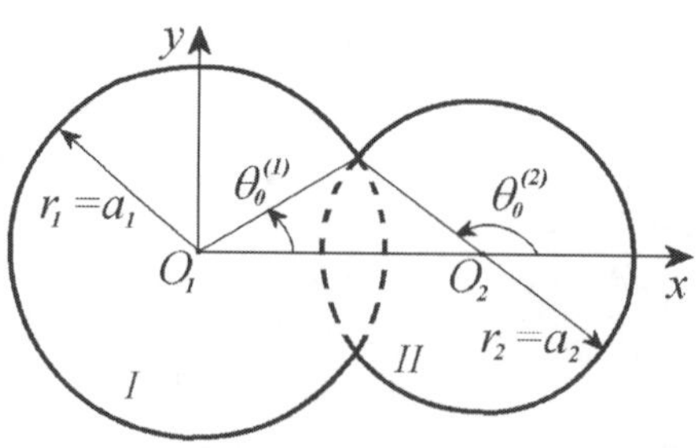

FIG. 1.8: System of intersected circular cylinders

study produces a similar situation to that earlier discussed for a rectangular domain. The boundary conditions at circular arcs depicted by dashed lines in Fig. 1.8 are not specified, since these arcs are not physical boundaries of the domain. The results obtained for the potential theory problem in Section 1.4.4 allow one to assume that, at these non-physical segments of the boundaries, conditions (1.51) can be supplemented by arbitrary functions. Here the quantitative characteristics of the sonic field should be independent on the form of these functions.

Such general conclusion is quite natural. However, since practical application of this approach it related to solving infinite systems of equations, it is expedient to get some particular quantitative estimations for the problems of this type. Some results obtained for the emitting system formed by two identical cylinders with dimensions $a_1/\lambda = a_2/\lambda = 1.2$ are given below. The wave distance between axes of cylinders (Fig. 1.8) is equal to $O_1O_2/\lambda = d/\lambda = 1.8$, thus $\theta_0^{(1)} = \theta_0^{(2)} = 41.4°$. At the emitting surfaces the uniform distribution of the oscillating velocity ($F_1(\theta_1) = F_2(\theta_2) = F_0$) is preset. In the further discussion, symbols N and M correspond to the number of the factors, to which the infinite series are truncated during calculations. The results are provided for two limiting cases where the velocity preset at the dashed sections of cylindrical surfaces is either zero, or the same as at the active part of the boundary, i.e., equal to F_0.

Continuous expansion of the boundary conditions to the total surface significantly improves the reduction method convergence while solving infinite systems and evaluating the field characteristics. In this case, at $N = M = 20$ the problem solution obtained is practically exact one. If zero values of the oscillating velocity are preset at the expanded boundaries, the convergence deteriorates, as it always occurs during summation of the Fourier series. However, the quantitative estimations obtained explicitly indicate that with the increasing number of factors being taken into consideration results in convergence of estimated characteristics of close and distant fields, thus confirming a possibility of generalizing the conclusion drawn in the previous section to the general case.

Figures 1.9 depict the calculated results on the oscillating velocity and pressure at the emitter surfaces, normalized accordingly by F_0 and $\rho c F_0$, for two above-mentioned ways of supplementation of the boundary conditions. The distinctions between these curves are observed only within a narrow zone close to intersection points of the cylinders at

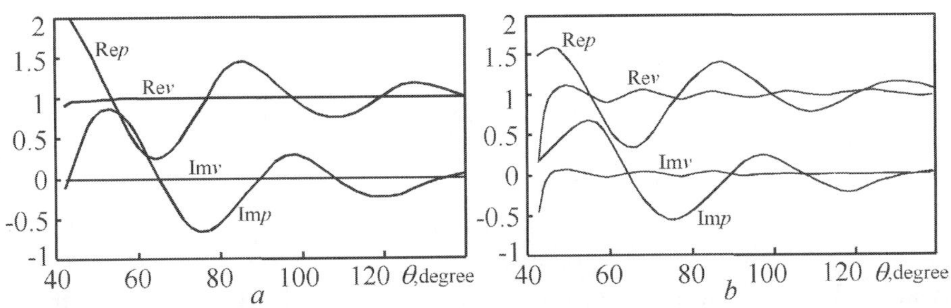

FIG. 1.9: Normalized values of real and imaginary parts of particle velocity and pressure on the surface of radiator, when velocity is set on the dashed (nonphysical) surfaces of cylinder in the form: (a) $\upsilon_r(a_1, \theta_1) = \upsilon_r(a_2, \theta_2) = F_0$, (b) $\upsilon_r(a_1, \theta_1) = \upsilon_r(a_2, \theta_2) = 0$

$\theta_0 = 41.4°$. Here the Gibbs effect is observed, which is intrinsic for the Fourier series of discontinuous functions.

Additional data on the influence of types of boundary conditions can be obtained by considering quantitative estimations of the total emitted energy. Values of the total emitted energy calculated via the distant field characteristics are convenient to be normalized by magnitude $\rho c F_0^2 a$. Since, in practical calculations, the infinite series in the general solution are truncated, the emitted energy is some function of the number of remaining terms of a series. A minor alteration of this magnitude with increasing number of terms can be used as one of the indications characterizing the accuracy of the boundary problem solution.

In case of continuous expansion of the boundary conditions, we get the following values of the emitted energy, if the number of remaining terms of the truncated series in (1.52) is fifteen and twenty, respectively: $W(15, 15) = 115.3$ and $W(20, 20) = 115.4$. These values practically coincide. If zero velocity values are preset at the expanded boundaries, the following estimations are obtained: $W(10, 10) = 109.2$; $W(15, 15) = 111.1$ and $W(20, 20) = 112.0$. In the comparative analysis of the results, of interest is not only the distinction between two types of boundary conditions, but also the dynamics of estimate variation with increasing number of terms of truncated series. It is obvious that, in the limiting case, solutions for two various situations will practically coincide.

1.4.6 Specifics of the Method Application to the Case of Overlapping Partial Domains

In the previous section, we discussed the exterior problem of sound radiation by the system of intersecting cylinders. Solid lines indicate the boundary of the field domain. It is the parts of circumferences in Fig. 1.8. The sound field is formed in the areas $r_1 > a_1$ and $r_2 > a_2$. It is clear that sound field characteristics are periodic. Therefore, the form of the general solution of the boundary problem for the potential selected as the expression (1.52).

Dashed sections of the circumferences do not belong to the sound field domain. Therefore, we can formally set boundary conditions there. This allows one to use the properties of completeness and orthogonality of function sets in presenting solutions for partial domains. Since the dashed sections of the circumferences do not belong to the area of the field, the boundary parameters there can be set arbitrarily. Thus, the problem of sound radiation receives a unique solution. Calculations presented in Section 1.4.5 confirm that such approach can be used to solve specific problems.

To assess possibilities of the method more completely we should also consider the problem of determining characteristics of the sound field within the region shown in Fig. 1.8. This will illustrate another aspect of the method of partial domains, which will allow us to broaden the range of considered problems. Let us look at Fig. 1.8. Now the dashed areas of the circumferences are inside the field area and the value of the potential or its normal derivative on them cannot be set arbitrarily. In addition, the requirement that the sound field has to be periodic remains in force. To construct the general solution of the boundary problem for the inner domain, we should replace Hankel functions with

Bessel functions in Eq. (1.52). Therefore, for the internal problem, expression for the sound field potential has to be presented as

$$\Phi = \begin{cases} \Phi_1 = \sum\limits_{m=0}^{\infty} A_m J_m(kr_1)\cos(m\theta_1), & 0 \le r_1 \le a_1, \quad 0 \le \theta_1 \le 2\pi, \\ \Phi_2 = \sum\limits_{m=0}^{\infty} B_m J_m(kr_2)\cos(m\theta_2), & 0 \le r_2 \le a_2, \quad 0 \le \theta_2 \le 2\pi. \end{cases} \tag{1.53}$$

Here A_m and B_n are arbitrary coefficients. Their value must be derived from the boundary conditions on those parts of the border, which are marked with solid circular arc lines.

When analysing the forced problem for a bounded domain, filled with an ideal fluid, it should be noted that for certain values of the frequency there is no bounded solution of the problem. In what follows we assume that the frequency of forced oscillations do not coincide with a natural frequency of the specified domain. It should be understood, that the values of these natural frequencies could be found using the general solution (1.53).

To obtain algebraic expressions to determine unknown coefficients in (1.53), we have to set either potential or its normal derivative on full circles with centres in points O_1 and O_2. For example, conditions for potential on a solid area of the circumference O_2 is defined. Then for the dashed area of the circumference, O_2 we write the condition $\Phi_1 = \Phi_2$. We use similar approach for boundary conditions on dashed area that is inside circumference O_2. Such approach of using the method of partial domains is used in [29]. Its efficiency illustrates when considering a particular problem in Section 7.9.

This procedure allows you to create an infinite system of algebraic equations for the unknown coefficients A_m and B_n:

$$\begin{aligned} A_m &= \sum_{n=0}^{\infty} \alpha_{mn} B_n + F_m, \\ B_n &= \sum_{m=0}^{\infty} \beta_{nm} A_m + Q_n. \end{aligned} \tag{1.54}$$

The values of coefficients A_m and B_n define expression (1.53) for the sound field potential. Each term in (1.53) satisfies Helmholtz equation, while the expression itself as a whole satisfies boundary conditions on the domain boundary. Thus, we built a complete solution of the boundary problem. When calculating value of characteristics of the field within the area bounded by the dashed portions of circumferences, one can use either component of the potential in (1.53). Note, that near circumference intersection points (see Fig. 1.8) vibrational velocity values have local singularity. Therefore, when solving system (1.54), we can use the technique, outlined below in Section 2.1.2.

1.5 CONCLUSIONS

The main benchmarks for elaboration of analytical methods for solving the boundary problems in non-canonical domains are discussed. Although the argumentation is based on the examples of boundary problems for the Helmholtz and Laplace equations, the developed approaches are quite applicable for other types of equations, as well, including those related to solving the electro-dynamical problems.

In certain cases, calculations required for construction of solutions and subsequent transformations of the functional equations to the algebraic form seem to be quite cumbersome. However, in estimation the above method capabilities it is necessary to focus on its advantages. First of all, the most important is that each solution is truly general and, hence, this solution form is the most suitable. The general properties of the solution imply that augmentation of the number of particular solutions being involved in the analysis will always lead to improvement of the accuracy of calculated results. Functions, application of which ensures fullness of solutions, are the most lucrative for meeting the boundary conditions from the mean squared approach standpoint. Such way of meeting the boundary conditions is the most adequate and compatible with the physical sense of the majority of problems of mathematical physics.

It is noteworthy that application of the simple reduction method to basic infinite systems of linear algebraic equations ensures derivation of correct quantitative estimations of considered fields using a small number of particular solutions involved in the analysis. Moreover, systems containing several tens equations guarantee that the accuracy of calculated results will exceed practical needs. Within the framework of the proposed approach, the problem of accuracy of the required results is resolved simply enough. For this purpose, it is enough to check whether conjugation conditions of fields are satisfied at the boundaries of partial subdomains.

Such direct verification of the solution accuracy can involve certain difficulties if the boundary of the field existence domain has salient/corner points similar to those at the order of flat and wedge wave-guides (see Fig. 1.5). In this case, the field of velocities has a singularity near these corner points, which excludes the possibility of direct verification of the velocity matching near these points. Some additional information on the solution quality can be provided by verification of the energy conservation conditions.

Occurrence of singularities in a field is typical for domains with a sharp variation of the boundary shape. In acoustics, the singularity behavior in the field of velocities near such points is known in advance, and this can be used for an essential improvement of the efficiency of algorithms of deriving quantitative estimations [17]. Some other problems are also encountered during the method application. First of all, this is a problem of the solution unambiguity. Availability of several options in a choice of the solution form definitely allows one to achieve a certain simplification of calculating procedures, but no particular recommendations on optimizing this choice are yet available. It is important that the general solution of a boundary problem is not only special mathematical expression. This expression gives possibility to describe all physical peculiarities in considered sound field.

REFERENCES

1. Zarembo, L.K. and Timoshenko, V.I., *Nonlinear Acoustics*, Moscow: Moscow University Printing House, 1984. [in Russian]

2. Gurbatov, S.N., Rudenko, O.V., and Saichev, A.I., *Waves and Structures in Nonlinear Nondispersive Media,* Berlin: Springer, 2011.
3. Whitham, G.B., *Linear and Nonlinear Waves,* New York: Wiley, 1999.
4. Morse, P.M. and Feshbach, H., *Methods of Theoretical Physics,* Vol. 2, New York: McGraw Hill, 1953.
5. Richardson, I.D., *Some Methods of Applied Acoustics*, Moscow: Voenizdat, 1962. [in Russian]
6. Rayleigh, J.W.S., *The Theory of Sound, Two Volumes in One,* New York: Dover Publications, 1945.
7. Tyurin, A.M., Stashkevich, A.P., and Taranov, E.S., *Basic Hydro-Acoustics*, Leningrad: Sudostroenie, 1966. [in Russian]
8. Hönl, H., Maue, A., and Westpfahl, K., *Theorie der Beugung*, Berlin: Springer, 1961.
9. Isacovich, M.A., *General Acoustics*, Moscow: Nauka, 1973. [in Russian]
10. Shenderov, E.L., *Sound Radiation and Scattering*, Leningrad: Sudostroenie, 1989. [in Russian]
11. Glazanov, V.E., *Some Problems of Sound Propagation in Elastic Mediums*, Taganrog: Printing House TRTI, 1973. [in Russian]
12. Smaryshev, M.D., *Directivity of Hydro-Acoustic Antennas*, Leningrad: Sudostroenie, 1973. [in Russian]
13. Shenderov, E.L., *Wave Problems of Hydroacoustics*, Leningrad: Sudostroenie, 1972. [in Russian]
14. Vainshtein, L.A., *Diffraction Theory and the Factorization Method*, Moscow: Sovetskoe Radio, 1966. [in Russian]
15. Noble, B., *Method Based on the Winer-Hopfe Techique for the Solution of Partial Differential Equations,* London: Pergamon Press, 1958.
16. Ufimtsev, P.Ya., *Fundamentals of the Physical Theory of Diffraction*, Hoboken: John Wiley & Sons, 2007.
17. Grinchenko, V.T. and Vovk, I.V., *Wave Problems of the Sound Scattering on Elastic Shells,* Kiev: Naukova Dumka, 1986.
18. Grinchenko, V.T., Axisymmetric problems of elastic theory for semi-infinite round cylinder, *Appl. Mech.*, vol. **1**, no. 1, pp. 109–118, 1965. [in Russian]
19. Grinchenko, V.T., *Equilibrium and the Established Oscillations of Elastic Bodies of Finite Dimensions*, Kiev: Naukova Dumka, 1978. [in Russian]
20. Grinchenko, V.T. and Ulitko, A.F., Characteristic Features of the Stress State of Rigidly Fixed Round Plate, *Report of Academy of Science USSR*, Series A, no. 2, pp. 162–164, 1970.
21. Pathak, A.G. and Stepanishen, P.R., Acoustic harmonic radiation from fluid-loaded infinite elastic plates used elasticity theory, *J. Acoust. Soc. Am.*, vol. **94**, no. 1, pp. 1700–1709, 1993.
22. Waterman, P.C., Matrix theory of elastic wave scattering, *J. Acoust. Soc. Am.*, vol. **60**, pp. 567–580, 1976.
23. Mittra, R. and Lee, S.W., *Analytical Techniques in the Theory of Guided Waves,* New York: The McMillan Company, 1971.

24. Shestopalov, V.P., *Method of the Ryman-Hilbert Problem in the Theory of Diffraction and Propagation of Electromagnetic Waves*, Kharkov: Kharkov University Printing House, 1971. [in Russian]

25. Kantorovich, L.V. and Krylov, V.I., *Approximate Methods of Higher Analysis,* New York: Intersience Publisher, 1958.

26. Vovk, I.V. and Grinchenko, V.T., Sound radiation by a number of coaxial piezoceramic shells, *Sov. Phys. Acoust.,* vol. **37**, pp. 130–135, 1991.

27. Vovk, I.V. and Grinchenko, V.T., Expanding the capabilities of the method of partial domains in application to wave radiation and scattering problem, *Sov. Phys. Acoust.*, vol. **35**, pp. 17–21, 1989.

28. Bateman, H., *Higher Transcendental Functions*, Vol. 2, New York: McGraw-Hill, 1953.

29. Vovk, I.V., Gomilko, A.M., and Gorodetskaya, N.S., Concerning the application features of partial domains method in the wave problems, *J. Acoust.*, vol. **41**, no. 3, pp. 399–404, 1995. [in Russian]

CHAPTER 2

Sound Radiation

"It's as simple as tit-tat-toe, three-in-a-row, and as easy as playing hooky"

Mark Twain, "Adventures of Huckleberry Finn"

2.1 SOUND RADIATION BY CYLINDRICAL RADIATORS

The problem of sound radiation by cylinder of infinite length holds a special place in theoretic and applied acoustics. The reason for this is that the corresponding boundary problems admit a simple and rigorous solution when the distribution of particle velocity on radiating surface is arbitrary [1–4]. At the same time, cylindrical radiators become widely used in applied acoustics and, first of all, in hydroacoustics [5–10]. All the analysis data obtained in solving the problem of sound radiation by cylinder of infinite length are applicable to the cylinders of finite length in the cases when the length of a real cylindrical radiator is much more than its diameter and radiating wave length. The multi-functionality of the modern hydroacoustic stations necessitates the control of the parameters of their radiators. Such parameters as directional properties, resonance frequency, radiation impedance and so on. The methods of control of these parameters can be divided into passive and active. Among the passive methods are, for example, the applications of different screens. The principal defect of these methods, along with their high efficiency, is that the parameters of the radiators cannot be changed while the radiators are in operation. Among the active methods are the methods of control of the radiators parameters by changing electrical voltage. Their primary advantage is the possibility of operating control while the radiators are in operation. Active methods are widely used for the control of directional properties of the hydroacoustic phased antenna arrays [7–9,17].

We will consider some passive and active methods of control of the cylindrical radiators parameters.

2.1.1 Sound Radiation by Cylinder through a Closed Annular Layer

The problem of sound radiation from an infinite cylinder through a closed annular layer (plane problem) was considered in a series of works. The plane problem of radiation from a cylinder through a closed cylindrical elastic shell filled with medium, when its thickness is considerably less than the length of the wave in the material of the shell, was considered in work [11]. The plane problem of radiation of cylindrical transformer

ISBN: 978–1–56700–459–5

through one and two closed annular layers of arbitrary thickness was investigated in works [12,13]. The aim of investigation was to determine the band broadening capability of piezoceramic transformer. We will consider such problem in spite of the fact that many properties of sound transmission through the closed cylindrical layer were studied out in the indicated works. We can gain a better understanding of fundamental difference between the sound transmission through the plane and curved layers, when considering this problem.

We assume that sound is radiated from a cylinder of infinite height with radius r_0 in liquid or gas medium with wave impedance ρc through the closed annular layer with thickness $r_1 - r_0$, Fig. 2.1. The acoustic properties of the layer material are characterized by wave impedance $\rho_1 c_1$. In accordance with the fundamental principles of partial domains method, we partition all the sound field existence domain into two partial domains: domain I $(r_0 \leq r \leq r_1)$ and domain II $(r > r_1)$. The velocity potentials in the indicated partial domains are represented as follows

$$\Phi_1 = \sum_{n=-\infty}^{\infty} \left[A_n J_n(k_1 r) + B_n N_n(k_1 r)\right] e^{jn\phi},$$
$$\Phi_2 = \sum_{n=-\infty}^{\infty} E_n H_n^{(1)}(kr)\, e^{jn\phi}, \tag{2.1}$$

where $k_1 = \omega/c_1$, $k = \omega/c$, $n = 0, \pm 1, \pm 2, ...$; $J_n(\,)$, $N_n(v)$ and $H_n^{(1)}(\,)$—Bessel, Neumann, and Hankel functions.

The unknown complex coefficients A_n, B_n, and E_n are determined from the system of functional equations, which contains the matching conditions of sound field components at the interface of partial domains $r = r_1$ and boundary condition on the surface $r = r_0$:

$$\Phi_1 = \Phi_2, \ r = r_1,$$

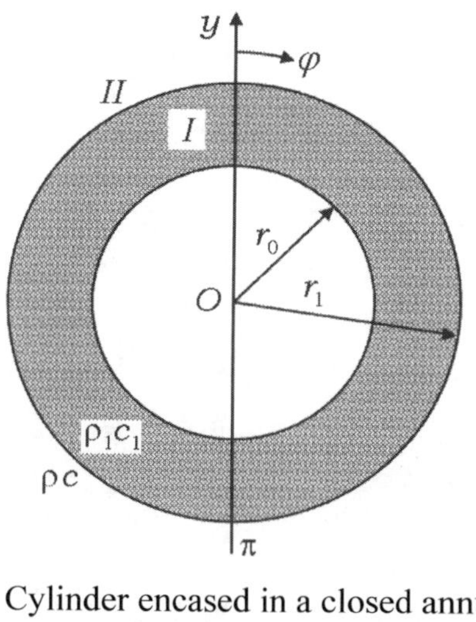

FIG. 2.1: Cylinder encased in a closed annular layer

$$\frac{\partial \Phi_1}{\partial r} = \frac{\partial \Phi_2}{\partial r}, \; r = r_1,$$
$$-\frac{\partial \Phi_1}{\partial r} = \upsilon_0 f(\phi), \; r = r_0, \tag{2.2}$$

where υ_0—reduced amplitude of particle velocity on the surface of the cylinder; $f(\phi)$—particle velocity distribution function.

We represent the function $f(\phi)$ as Fourier series

$$f(\phi) = \sum_{n=-\infty}^{\infty} a_n e^{jn\phi}, \quad a_n = \frac{1}{2\pi}\int_0^{2\pi} f(\phi)\, e^{-jn\phi} d\phi. \tag{2.3}$$

Taking into consideration expression (2.3), we transform functional Eqs. (2.2) into the system of linear algebraic equations in unknown complex coefficients

$$\rho_1 A_n \left[J_n(k_1 r_1) + B_n(k_1 r_1)\right] = \rho E_n H_n^{(1)}(k r_1),$$
$$k_1 A_n \left[J_n'(k_1 r_1) + B_n N_n'(k_1 r_1)\right] = k E_n H_n^{(1)'}(k r_1), \tag{2.4}$$
$$-k_1 A_n \left[J_n'(k_1 r_0) + B_n N_n'(k_1 r_0)\right] = \upsilon_0 a_n.$$

The prime on the radial functions in system (2.4) and below means that it is a derivative of argument kr (or $k_1 r$) with corresponding value of r.

Two parameters are interesting for us in the considered problem: dimensionless radiation impedance of the cylinder Z and the coefficient, which can be named as sound transmission coefficient through an annular layer,

$$k_{ins} = \left|\Phi_1/\Phi_0\right|, \tag{2.5}$$

where Φ_0—potential velocity, formed by the cylinder when the layer is absent on the surface $r = r_1$, and Φ_1—potential velocity, formed by the cylinder with the layer. In general case, each of these parameters can be defined for any harmonic component in Eqs. (2.1), i.e., for each concrete value of n. These characteristics are the most interesting for the case of zero harmonic, when $f(\phi) = \text{const}$.

Figure 2.2 shows frequency dependencies of the cylinder impedance components for two ratios of the wave impedances of the layer material and surrounding acoustic medium $m = \rho_1 c_1/\rho c$. The layer thickness in both cases is determined from the equation $2r_0/(r_1 - r_0) = 2.8c/c_1$. The components of the cylinder impedance without an annular layer (curve 3) and for the case of an infinite plane layer (curves 4) are given here for comparison. Figure 2.2 shows the values of the wave thickness of the layer along the horizontal axis $\tilde{d} = (r_1 - r_0)/\lambda_1$, where λ_1—wave length in the material of the layer.

As it follows from the data in Fig. 2.2, the values of radiation impedance components for the annular and plane layers almost coincide with each other when the values of kr_0 are large (in fact when $kr_0 > 2$). The situation is much different when $kr_0 < 2$, and when the annular layer $kr_0 \to 0$, when $R \to 1$. It follows from the fact that the annular layer is finite and the plane layer is infinite. The value of X has elastic nature in the domain $\tilde{d} < 0.25$ for the case of a plane layer. In the case of annular layer, the value of X has at first the nature of mass, then it follows the resonance ($X = 0$), and only then X has the nature of elasticity. It should be noted that an abrupt increase in the value of R is observed at this resonance frequency.

We can give simple physical interpretation to the indicated regularities. For the case of a plane layer, the character of X is determined only by the wave thickness of the layer, because the impedance of infinite medium behind the layer (for plane incident wave) is of active character and is equal ρc. For the case of an annular layer (for cylindrical wave), the character of X is determined not only by the wave thickness of the layer but also by the impedance of medium behind the surface $r = r_1$. As it is known, the reactive component of this impedance has the nature of mass [curve 3, Fig. 2.2(b)]. Thus it is obvious that the resonance in the domain $kr_0 \approx 0.4$ is caused by interaction between the elasticity of the annular layer and additional mass of medium surrounding the layer. The resonance vibrations lead to an abrupt increase of the active component of radiation impedance of the cylinder at these frequencies.

It is reasonable that the indicated phenomenon affects favorably the efficiency of sound energy radiation from the cylinder. This is proved by the data shown in Fig. 2.3. The curves 1 and 2 here correspond to two values of m in Fig. 2.2. There is an abrupt increase of k_{ins} at the frequencies corresponding to the maximum of R. The cylinder with a layer (particle velocity on its surface being the same) produces higher pressure in the external medium than the cylinder without a layer. Note that with an increase of m, the indicated resonance frequency rises, and the amplitude of the maximum of R and k_{ins} decreases.

The peculiarity in the behavior of the value of k_{ins}, which was indicated above, is absent in the case of a plane layer. Here k_{ins} is always less than or equal to unity. That is why the pressure produced by plane radiator without a layer is always higher than pressure of radiator with a layer (curve 3).

We take note of one more circumstance. While analyzing the peculiarities of the behavior of the value of X, we do not pay special attention to the character of its resonances. In the case of a plane layer, [curve 4, Fig. 2.2(b)] the first value $X = 0$ when

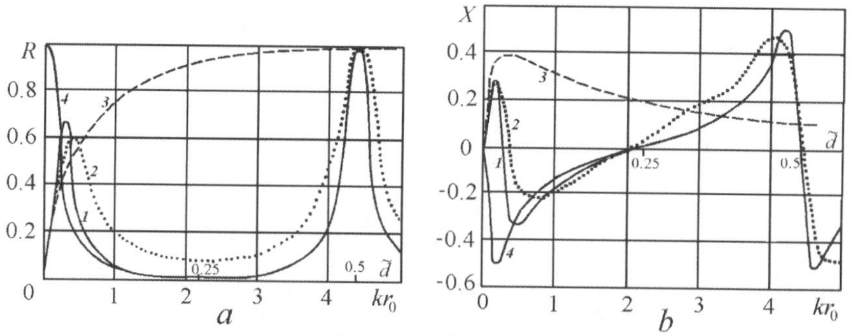

FIG. 2.2: Frequency dependencies of the cylinder radiation impedance; (a) real part, (b) imaginary part; 1, 2—correspond to $m = 0.15$ and $m = 0.3$; 3—annular layer is absent, 4—plane layer

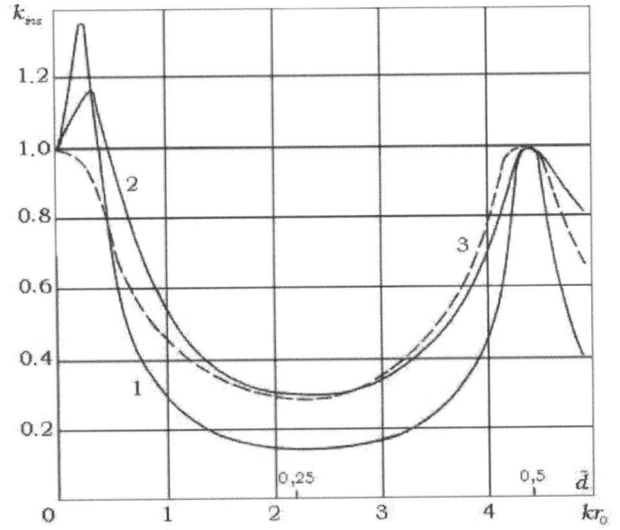

FIG. 2.3: Frequency dependence of sound transmission coefficient; 1, 2—correspond to $m = 0.15$, and $m = 0.3$; 3—plane layer

$\tilde{d} = 0$ corresponds to antiresonance, and values $X = 0$ when $\tilde{d} = 0.25$ and $\tilde{d} = 0.5$—to resonance and antiresonance respectively. It follows from the analysis of a plane wave passing through a plane layer when $m < 1$ [3,4,14] and Forster theorem [15]. The picture is somewhat different in the case of an annular layer. The value of $X = 0$ when $\tilde{d} = 0$ corresponds to resonance. Then we have antiresonance in the vicinity of $\tilde{d} \approx 0.04$, and from here on the values and the sequence of the resonances and antiresonances coincide with a plane layer.

Thus, we can effectively control the radiation impedance of the cylindrical radiator and increase its sound pressure in the far field in some frequency range, using a close annular layer of material with a certain wave impedance.

2.1.2 Sound Radiation from a Cylinder through an Open Annular Layer

Now we consider a plane problem of sound radiation from a cylinder when a cylinder is incased in an annular layer (Fig. 2.4). Sound energy can be radiated only through the clearance formed by removing the part of a layer. The width of the clearance is defined by the angles $\phi = \pm\, \phi_0$. The radiators of this type are widely used in hydroacoustics for making different electroacoustic transformers and arrays of different application [5,7,9,10,16,17]. We turn first to the case when the layer is absolutely nontransparent and its surfaces are acoustically compliant (In the following paragraphs we will consider finite transmissivity of a layer). The boundary conditions are as follows:

$$\begin{aligned} &\Phi = 0, \quad r_0 \le r \le r_1, \quad |\phi| = \phi_0, \\ &\Phi = 0, \quad r = r_1, \quad \phi_0 \le |\phi| \le \pi. \end{aligned} \tag{2.6}$$

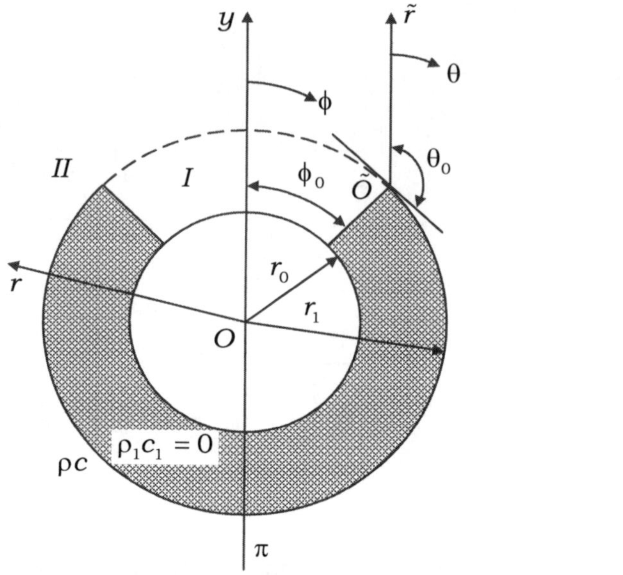

FIG. 2.4: Cylinder encased in an open annular layer

We assume that the distribution of particle velocity $\upsilon_0 f(\phi)$ is given on the exposed part of the surface of cylindrical radiator. In order to simplify the problem, we assume that this distribution is symmetric with respect to Oy-axis. Then

$$-\frac{\partial \Phi}{\partial (kr)} = \upsilon_0 f(\phi), \quad r = r_0, \quad 0 \leq |\phi| \leq \phi_0. \tag{2.7}$$

We partition the whole domain into two partial domains: $r_0 \leq r \leq r_1$, $|\phi| \leq \phi_0$ (domain I) and $r \geq r_1$, $0 \leq |\phi| \leq \pi$ (domain II).

Taking into account the boundary conditions indicated above, we represent the field in each partial domain as follows:

$$\begin{aligned} \Phi_1 &= \sum_{q=0}^{\infty} \left[A_q J_{\nu_q}(kr) + B_q N_{\nu_q}(kr)\right] \cos \nu_q \phi, \\ \Phi_2 &= \sum_{n=0}^{\infty} S_n H_n^{(1)}(kr) \cos n\phi. \end{aligned} \tag{2.8}$$

The expression for Φ_1 satisfies Helmholtz equation with arbitrary ν_q. Taking

$$\nu_q = \frac{(2q+1)\pi}{2\phi_0}, \tag{2.9}$$

we satisfy the first condition in (2.6) at the ends of the layer $\phi = \pm\, \phi_0$. For such ν_q, the solution for the first domain contains two functions, on the surfaces $r = r_0$ and $r = r_1$, $|\phi| \leq \phi_0$, represented by Fourier series with undetermined coefficients. The solution

Φ_2 for the outer domain $r \geq r_1$ satisfies Sommerfeld condition [18] and enables us to satisfy the boundary condition on the surface $r = r_1$, $0 \leq \phi \leq 2\pi$. These properties of the expressions for Φ_1 and Φ_2 allow us to fulfill the boundary conditions (2.6) and (2.7) by choosing the arbitrary constants.

Condition (2.7) allows us to relate the values of the coefficients A_q and B_q to each other. Substituting the expression for Φ_1 in (2.7), we have

$$B_q = \frac{L_q - A_q J'_{\nu_q}(kr_0)}{N'_{\nu_q}(kr_0)}, \tag{2.10}$$

where $L_q = -(2\upsilon_0/k\phi_0)\int_0^{\phi_0} f(\phi)\cos\nu_q\phi d\phi$.

Bearing in mind formula (2.10), we can transform the expression for Φ_1 into the following form

$$\Phi_1 = \sum_{q=0}^{\infty}\left[A_q\Delta_q(kr) + L_q\frac{N_{\nu_q}(kr)}{N'_{\nu_q}(kr_0)}\right]\cos\nu_q\phi, \tag{2.11}$$

where $\Delta_q(kr) = J_{\nu_q}(kr) - J'_{\nu_q}(kr_0)N_{\nu_q}(kr)/N'_{\nu_q}(kr_0)$.

It is necessary to use functional equations, containing the continuity condition of the sound field at the interfaces of the partial domains and boundary condition on the surface $r = r_1$, to determine the unknowns in expressions (2.8)

$$\Phi_2 = \begin{cases} \Phi_1, & 0 \leq |\phi| \leq \phi_0, \\ 0, & \phi_0 \leq |\phi| \leq \pi, \end{cases}$$
$$\frac{\partial\Phi_2}{\partial r} = \frac{\partial\Phi_1}{\partial r}, \quad 0 \leq |\phi| \leq \phi_0. \tag{2.12}$$

In the result of algebraization of functional Eq. (2.12), we obtain the following infinite system of linear algebraic equations of the second kind in the unknown coefficients:

$$S_n H_n^{(1)}(kr_1) - \frac{1}{\delta_n}\sum_{q=0}^{\infty} A_q\Delta_q(kr_1)a_{nq} = \frac{1}{\delta_n}\sum_{q=0}^{\infty} a_{nq}L_q N_{\nu_q}(kr_1)/N'_{\nu_q}(kr_0),$$
$$A_q\Delta'_q(kr_1) - \frac{1}{\phi_0}\sum_{n=0}^{\infty} S_n H_n'^{(1)}(kr_1)a_{nq} = -L_q N'_{\nu_q}(kr_1)/N'_{\nu_q}(kr_0), \tag{2.13}$$

where

$$\delta_n = \begin{cases} 2\pi, & n = 0, \\ \pi, & n \neq 0, \end{cases} \qquad a_{nq} = \begin{cases} \dfrac{2\nu_q(-1)^{q+1}\cos n\phi_0}{n^2 - \nu_q^2}, & n \neq \nu_q, \\ \phi_0, & n = \nu_q, \end{cases}$$

$\Delta'_q(kr_1)$—derivative of $\Delta_q(kr)$ with respect to kr when $r = r_1$.

To prove the existence of a solution for the considered boundary problem, we have to prove the solution existence for system (2.13), which provides convergence of infinite series (2.8) for the field components in all considered domain.

Usual approach to analysis of solvability of the system of such kind is to prove its regularity [19]. We introduce new unknowns to examine system (2.13) for regularity:

$$\tilde{S}_n = S_n H_n'^{(1)}(kr_1), \quad \tilde{A}_q = A_q \Delta_q(kr_1). \tag{2.14}$$

Initial system (2.13) is rearranged in the form

$$\begin{gathered}
\tilde{S}_n \frac{H_n^{(1)}(kr_1)}{H_n'^{(1)}(kr_1)} - \frac{1}{\delta_n} \sum_{q=0}^{\infty} \tilde{A}_q a_{nq} = \frac{1}{\delta_n} \sum_{q=0}^{\infty} L_q a_{nq} N_{\nu_q}(kr_1)/N'_{\nu_q}(kr_0), \\
\tilde{A}_q \frac{\Delta'_q(kr_1)}{\Delta_q(kr_1)} - \frac{1}{\phi_0} \sum_{n=0}^{\infty} \tilde{S}_n a_{nq} = -L_q N'_{\nu_q}(kr_1)/N'_{\nu_q}(kr_0).
\end{gathered} \tag{2.15}$$

Complex functional dependence of system (2.15) coefficients on coefficient number makes regularity analysis difficult. Consequently we focus on the proof of quasiregularity of this system. We can simplify the elements of matrix by using the known asymptotic representations of Bessel, Neumann and Hankel functions, which are considered as functions of n and ν_q orders. Basing on these representations [20] we obtain

$$\frac{H_n^{(1)}(kr_1)}{H_n'^{(1)}(kr_1)} = \frac{kr_1}{n}, \quad \frac{\Delta'_q(kr_1)}{\Delta_q(kr_1)} = \frac{\nu_q}{kr_1}, \quad \frac{N'_{\nu_q}(kr_1)}{N'_{\nu_q}(kr_0)} = \left(\frac{r_0}{r_1}\right)^{\nu_q+1}. \tag{2.16}$$

Using expression (2.16), we find the following infinite system, which is equivalent to initial system (2.15):

$$\begin{gathered}
\tilde{S}_n^{(1)} - \frac{2n\cos^2 n\phi_0}{\pi} \sum_{q=0}^{\infty} A_q^{(1)} \frac{1}{n^2 - \nu_q^2} = \frac{2n\cos^2 n\phi_0}{\pi k r_1} \sum_{q=0}^{\infty} L_q a_{nq} N_{\nu_q}(kr_1)/N'_{\nu_q}(kr_0), \\
\tilde{A}_q^{(1)} - \frac{2\nu_q}{\phi_0} \sum_{n=0}^{\infty} \tilde{S}_n^{(1)} \frac{1}{n^2 - \nu_q^2} = -L_q \left(\frac{r_0}{r_1}\right)^{\nu_q+1},
\end{gathered} \tag{2.17}$$

where $\tilde{A}_q^{(1)} = \tilde{A}_q(-1)^{q+1}\nu_q/kr_1$, $\tilde{S}_n^{(1)} = \tilde{S}_n \cos n\phi_0$.

The coefficients $1/\left(n^2 - \nu_q^2\right)$ change the sign in each row of system (2.17), and that is why the following analysis is difficult. We substitute the value of $\tilde{A}_q^{(1)}$, obtained from the second equation of system (2.17), into the first and we find $\tilde{S}_n^{(1)}$

$$\tilde{S}_n^{(1)} = \frac{4n\cos^2 n\phi_0}{\pi} \sum_{g=1}^{\infty} \tilde{S}_g^{(1)} a_{ng} + b_n, \tag{2.18}$$

where $a_{ng} = (1/\phi_0) \sum\limits_{q=0}^{\infty} \nu_q/\left(n^2 - \nu_q^2\right)\left(g^2 - \nu_q^2\right)$.

It is difficult to verified weather the quasiregularity conditions of system (2.18) are fulfilled for the case of the arbitrary value of angle ϕ_0. That is why we confine ourselves to consideration of the specific and the most unfavorable case, when $\phi_0 = \pi$. When the coefficients a_{ng} take the following form

$$a_{ng} = \frac{1}{\pi}\sum_{q=0}^{\infty} \frac{q+1/2}{[(q+1/2)^2 - n^2]\,[(q+1/2)^2 - g^2]}. \tag{2.19}$$

The general term of infinite series (2.19) can be represented by the following integral:

$$\frac{q+1/2}{[(q+1/2)^2 - n^2][(q+1/2)^2 - g^2]} = \frac{1}{n^2 - g^2}\int_0^{\infty}[\mathrm{ch}(nt) - \mathrm{ch}(gt)]e^{-(q+1/2)t}\,dt. \tag{2.20}$$

The infinite series in the expression for a_{ng} can be transformed, and the coefficients of infinite system (2.18) can be represented as follows:

$$\begin{aligned} a_{ng} &= \frac{1}{\pi(n^2 - g^2)}\int_0^{\infty}[\mathrm{ch}(nt) - \mathrm{ch}(gt)]\sum_{q=0}^{\infty} e^{-(q+1/2)t}\,dt \\ &= \frac{1}{2\pi(n^2 - g^2)}\int_0^{\infty}\mathrm{sh}^{-1}(t/2)\,[\mathrm{ch}(nt) - \mathrm{ch}(gt)]\,dt. \end{aligned} \tag{2.21}$$

In improper integrals in this expression can be represented by special functions [21]. Then we obtain:

$$a_{ng} = \frac{1}{\pi(n^2 - g^2)}[\psi(n+1/2) - \psi(g+1/2)], \tag{2.22}$$

where $\psi(n+1/2)$ and $\psi(g+1/2)$—psi-functions.

The properties of psi-functions [21,22] enable us to find out an important peculiarity, namely the constancy of their sign. This circumstance simplifies the calculation of the sum of coefficients moduli in system (2.18). Bearing in mind the quasiregularity analysis, we can use the asymptotic expression for a_{ng}

$$a_{ng} = \frac{1}{\pi}\left[\frac{\ln(n/g)}{n^2 - g^2} - \frac{B_2}{4n^2 g^2}\right], \tag{2.23}$$

which gives their value with an accuracy to $1/n^4$ compared with unity. Here B_2—Bernoulli number.

The representations of psi-functions in the form of finite series and the corresponding asymptotic expressions [21] were used to obtain expression (2.23).

Bearing in mind expression (2.23), we have the following expression to examine the quasiregularity of system (2.18):

$$F_n = \frac{4n}{\pi^2}\sum_{g=G}^{\infty} a_{ng} \leq \frac{4n}{\pi^2}\sum_{g=G}^{\infty}\left[\frac{\ln(n/g)}{n^2-g^2} - \frac{B_2}{4n^2g^2}\right] + O\left(\frac{1}{n^4}\right), \tag{2.24}$$

where G—Hankel function index in system (2.13). These functions can be replaced with their asymptotic representations starting with this index.

For the large values of n the sum of series (2.24) can be determined using the methods described in [23]. Taking into account that

$$\sum_{g=G}^{\infty}\frac{\ln(n/g)}{n^2-g^2} < \frac{1}{n}\int_0^{\infty}\frac{\ln(n/g)}{1-(n/g)^2}d(n/g),$$

we have the estimation for the large values of g

$$F_n \leq 1 - \frac{1}{36n} = 1 - Q(n). \tag{2.25}$$

Thus system (2.18) is quasiregular. We can find out whether the solution exists considering the behavior of $Q(n)$ function and the absolute terms in expression (2.18). The absolute terms behavior analysis in system (2.18) shows that they decrease as $1/n^2$, i.e., more rapidly, than function $Q(n)$ in expression (2.25). This points to the possibility of using the reduction method when finding of approximate solution of system (2.18) and thus of system (2.15).

Taking into consideration that solution of system (2.15) exists, we can show that the infinite series in expressions (2.8), and also the corresponding series for particle velocity converge at all points of partial domains. This goes to prove that the solution of the considered boundary problem exists. The regularity of infinite system plays almost no part for finding of solution and for the following quantitative estimation of the physical quantities characterizing the field.

More information about the possibilities of the method of solving the boundary problems in complex domains can be obtained on the basis of the analysis of peculiarities of considered physical fields. In this connection we briefly analyze the stages of the solution of problem of radiation from the cylinder through an open layer, preceding the derivation of infinite system (2.15).

A single-valued solution of the considered problem can be obtained, if we fulfill the following conditions: 1) boundary conditions on the surface of the cylinder and the layer; 2) continuity conditions of the field components at the interface of the partial domains; 3) radiation conditions at infinity; 4) if we indicate the specific type of peculiarity of particle velocity when approaching the corner points (edges).

Fulfillment of the boundary conditions allows us to determine ν_q, and also to indicate the relation between A_q and B_q. The form of the radial functions in the expression for Φ_2 is determined by the radiation conditions at infinity. We obtain the infinite system of algebraic equations relative to the coefficients of infinite series for Φ_1 and Φ_2 due to the fulfillment of the continuity conditions.

As we can see, the fourth condition was not used.

When we are using the indicated above way of sound field representing, we can have different solutions of infinite system (2.15) due to the existence of the solutions with different peculiarities of particle velocity in the corner points. In this case, the algorithm of converting infinite system into finite system must provide the adequate solution, which corresponds to the peculiarities of the velocities field in the corner points.

The question of peculiarity-definition in the field of the velocities of particles of acoustic medium in the vicinity of the edges was discussed in Chapter 1. The algorithm that allows us to use the data about known in advance peculiarities, when performing the calculations, is also given in Chapter 1. In this case (Fig. 2.4), particle velocity has the peculiarity $1/\tilde{r}^{1/3}$, where $\tilde{r}$—the distance from the summit of an angle, in the vicinity of right angle, formed by the surface $r = r_1$ and the plane $\phi = \phi_0$. Consequently, the distribution of particle velocity radial component on the surface $r = r_1$, can be represented in the following form

$$\upsilon_r(\phi) = \begin{cases} \dfrac{-ks}{(\cos\phi - \cos\phi_0)^{1/3}} + \upsilon_1(\phi), & 0 \le \phi \le \phi_0, \\ -\dfrac{ks}{(\cos\phi_0 - \cos\phi)^{1/3}} + \upsilon_2(\phi), & \phi_0 \le \phi \le \pi, \end{cases} \tag{2.26}$$

where s—certain constant, $\upsilon_1(\phi)$ and $\upsilon_2(\phi)$—regular functions.

The main components in velocity representation (2.26), considered on the full interval $0 \le \phi \le \pi$ can be represented by Fourier series

$$\upsilon_r(\phi) = \sum_{n=0}^{\infty} a_n \cos n\phi. \tag{2.27}$$

The most slowly decreasing part of a_n coefficients is defined in this case by expression

$$\tilde{a}_n = \frac{2}{\pi}\int_0^{\phi_0} \frac{ks\cos n\phi}{(\cos\phi - \cos\phi_0)^{1/3}} d\phi + \frac{2}{\pi}\int_{\phi_0}^{\pi} \frac{ks\cos n\phi}{(\cos\phi_0 - \cos\phi)^{1/3}} d\phi. \tag{2.28}$$

When we use the integral representations for Legendre function of the first and the second kind [24], it is not difficult to show that

$$\tilde{a}_n = ks\left[P_{n-1/2}^{-1/6}(\cos\phi_0) + \frac{2}{\pi}\frac{\Gamma(n+1/3)}{\Gamma(n+2/3)} Q_{n-1/2}^{1/6}(\cos\phi_0)\right]. \tag{2.29}$$

Asymptotic representations of Legendre function for large values of indexes [24,25] allow us to obtain the following asymptotic estimations for the coefficients of series (2.27):

$$\tilde{a}_n = A\frac{\cos n\phi_0}{n^{2/3}}, \quad n \gg 1. \tag{2.30}$$

Note that expression (2.30) determines the behavior of unknown values $\tilde{S}_n$ in system (2.15). If we use only the first part of expression for $\upsilon_r(\phi)$ in (2.26) and expend it into series in complete functions system $\cos \nu_q \phi$ on the interval $0 \le \phi \le \phi_0$, we can obtain the estimation of asymptotic behavior of unknowns $\tilde{A}_q$ in system (2.15):

$$\tilde{A}_q^{(0)} = B\frac{(-1)^{q+1}}{\nu_q^{5/3}}, \quad q \gg 1. \tag{2.31}$$

Equations (2.30) and (2.31) show the assumed asymptotic properties of unknowns in system (2.15). They are obtained on the basis of analysis of the conditions at the edge, which are necessary for existence of a unique solution of boundary problem. We can verify the consistency of asymptotic estimations (2.30) and (2.31) to prove the existence of such solution for infinite system (2.15).

Bearing in mind the properties of unknowns $\tilde{A}_q$, we define asymptotic properties of $\tilde{S}_n$. When the indexes n are large, we have the following expression from the first equation in system (2.15)

$$\tilde{S}_n^{(0)}\frac{kr_1}{n} = \frac{1}{\pi}\sum_{q=0}^{\infty}\tilde{A}_q a_{nq} + O\left(n^{-2}\right), \quad n \gg 1. \tag{2.32}$$

We make an estimation of sum in the last expression:

$$\begin{aligned} J &= \frac{1}{\pi}\sum_{q=0}^{\infty}\tilde{A}_q a_{nq} = \frac{1}{\pi}\sum_{q=0}^{\infty}\tilde{A}_q^{(0)} a_{nq} + \frac{1}{\pi}\left\{\sum_{q=0}^{Q-1}\tilde{A}_q a_{nq} - \sum_{q=0}^{Q-1}\tilde{A}_q^{(0)} a_{nq}\right\} \\ &= \frac{1}{\pi}\sum_{q=0}^{\infty}\tilde{A}_q^{(0)} a_{nq} + O\left(n^{-2}\right), \end{aligned} \tag{2.33}$$

where Q is the value of index q. Bessel and Neumann functions can be replaced with their asymptotic representations starting from this index; $\tilde{A}_q^{(0)}$—asymptotic value of the coefficients $\tilde{A}_q$ when $q \gg 1$.

Taking into account expression (2.31) we obtain

$$J = B\cos n\phi_0 \sum_{q=0}^{\infty}\frac{1}{\nu_q^{2/3}\left(n^2 - \nu_q^2\right)}. \tag{2.34}$$

It is not difficult to make an estimation of sum in this expression using the method of summability of infinite series [23].

$$\sum_{q=0}^{\infty}\frac{1}{\nu_q^{2/3}\left(n^2 - \nu_q^2\right)} \le \frac{\phi_0}{\pi n^{5/3}}\int_0^{\infty}\frac{dx}{x^{2/3}\left(1 - x^2\right)} = \frac{\phi_0\sqrt{3}}{2n^{5/3}}. \tag{2.35}$$

Thus, with an accuracy to $O\left(n^{-5/3}\right)$, we obtain

$$J = B\frac{\phi_0\sqrt{3}}{2}\frac{\cos n\phi_0}{n^{5/3}}. \tag{2.36}$$

Bearing in mind the estimations (2.36) from (2.32) we find

$$\tilde{S}_n^{(0)} = B\frac{\phi_0\sqrt{3}}{2kr_1}\frac{\cos n\phi_0}{n^{2/3}}. \tag{2.37}$$

We can assure that the estimations (2.30) and (2.31), are in agreement with the properties of the coefficients of system (2.15), using the second equation in system (2.15) and performing similar calculations.

The algorithm of determining solution for system (2.15) must provide the solution, the asymptotic properties of which are defined by Eqs. (2.30) and (2.31). Consequently, when passing from infinite to finite system, (when N and Q are unknowns) we set

$$\begin{aligned} \tilde{S}_n &= \tilde{S}_N\frac{\cos n\phi_0}{\cos N\phi_0}, \quad n \gg N, \\ \tilde{A}_q &= \tilde{A}_N\left(\frac{\nu_Q}{\nu_q}\right)^{5/3}(-1)^{q-Q}: \ q \gg Q. \end{aligned} \tag{2.38}$$

Expressions (2.38) enable us to bring system (2.15) to the form

$$\begin{aligned} &\tilde{S}_n\frac{H_n^{(1)}(kr_1)}{\tilde{H}_n'^{(1)}(kr_1)} - \frac{1}{\delta_n}\sum_{q=0}^{Q-1}\tilde{A}_q a_{nq} - \frac{\tilde{A}_Q}{\delta_n}\sum_{q=Q}^{\infty}\left(\frac{\nu_Q}{\nu_q}\right)^{5/3}a_{nq} = \frac{1}{\delta_n}\sum_{q=0}^{\infty}L_q a_{nq}\frac{N_{\nu_q}(kr_1)}{N'_{\nu_q}(kr_1)}, \\ &\tilde{A}_q\frac{\Delta_q'(kr_1)}{\Delta_q(kr_1)} - \frac{1}{\phi_0}\sum_{n=0}^{N-1}\tilde{S}_n a_{nq} - \frac{\tilde{S}_N}{\phi_0}\sum_{n=N}^{\infty}\left(\frac{N}{n}\right)^{2/3}\frac{\cos n\phi_0}{\cos N\phi_0}a_{nq} = L_q\frac{N'_{\nu_q}(kr_1)}{N'_{\nu_q}(kr_0)}. \end{aligned} \tag{2.39}$$

The received above data on decreasing nature of coefficients of infinite series, which represent the sound fields in partial domains, allow us to estimate the degree of convergence of these series. Thus, we are basing on the following representation, when studying the far field characteristics:

$$\Phi_2 \approx \Phi_0\sum_{n=0}^{\infty}\bar{S}_n\cos n\phi, \quad \Phi_0 = \left(\frac{2}{\pi kr_1}\right)^{1/2}e^{i(kr-\pi/4)}. \tag{2.40}$$

Bearing in mind expression (2.30), for $\bar{S}_n$ we find

$$\bar{S}_n \approx \text{const}\frac{(kr_1)^n\cos n\phi_0 e^{in\pi/2}}{n^{2/3}(n-1)!}, \quad n \gg 1. \tag{2.41}$$

Fourier series (2.40) with the coefficients $\bar{S}_n$ for the cylinder with a layer with not large wave diameters converges very rapidly. This means that it is enough to know the values of some first coefficients $\bar{S}_n$ to make quantitative description of the far field. A similar estimation can be made for the near-field. Bearing in mind expressions (2.11) and (2.14), the potential velocity in domain I can be represented by the following series:

$$\Phi_1 = \sum_{q=0}^{\infty} \left\{ A_q \frac{\Delta_q(kr)}{\Delta_q(kr_1)} + L_q \frac{N_{\nu_q}(kr)}{N'_{\nu_q}(kr_0)} \right\} \cos \nu_q \phi. \tag{2.42}$$

Taking into account the asymptotic properties of the coefficients $\tilde{A}_q$ (2.31) we can see, that the convergence of series (2.42) when $r = r_0$ is the same, as of Fourier series with coefficients

$$\frac{2}{\pi} \frac{(-1)^{q+1}}{\nu_q^{5/3}} \left(\frac{r_0}{r_1} \right)^{\nu_q} + \frac{(-1)^{q+1} kr_0}{\nu_q^2}. \tag{2.43}$$

The analysis of convergence of infinite series shows that it is enough to know several first coefficients in the corresponding series to obtain quantitative estimations of sound field characteristics. This means that we can solve system (2.15) by simple reduction method and define these characteristics. The possibilities of such simplifying approach were discussed in the first chapter when we were considering the simplest mixed problems. We will give additional quantitative data here, to prove the possibilities of such approach. It is necessary to take into account that reduction method can be used if the following conditions are fulfilled:

1. when the number of unknowns in system (2.15) increases, its solution approaches the solution of a system with desired asymptotic properties;

2. the values of unknowns with small numbers will be defined with sufficient degree of accuracy when the number of equations in system obtained by reduction of system (2.15) is comparatively small.

The check of these conditions can be realized by comparing the results of system (2.15) solution, obtained by simple reduction, and of system (2.39), which was obtained with consideration of asymptotic properties of unknowns.

The values of $\tilde{S}_n$, obtained by simple reduction of system (2.15) when $N = 6$ and $Q = 6$, are given in the left column of Table 2.1. It was assumed that $kr_0 = 1.6$, $kr_1 = 2.89$, $\phi_0 = \pi/2$ and $f(\phi) = 1$. The data in Table 2.1 allows us to answer positively both questions posed above. It is enough to consider system containing twice as much of unknowns than it is necessary for obtaining quantitative estimates of the field characteristics, to find the values of several first unknowns. The last statement is also proved by the results of calculations for the different geometric and wave parameters of a cylindrical radiator with an open layer, the results are given in the first chapter, and in works [26,27].

The performed analysis shows, that the far field and radiation impedance with good approximation can be defined if we solve infinite system (2.15) by simple reduction method. But when we study the local characteristics of the field, especially on the surface $r = r_1$, where the series defining pressure and particle velocity have the slowest convergence. It is necessary to use system (2.39) here, applying the known methods of improving the convergence of Fourier series [19].

TABLE 2.1: The values of the coefficients calculated by the two methods

n	$\tilde{S}_n$	
	Simple reduction method	**Asymptotic properties method**
0	$-1.397148\times10^{-1}+i4.193470\times10^{-2}$	$-1.379932\times10^{-1}+i4.343100\times10^{-2}$
1	$-3.615388\times10^{-2}+i2.376839\times10^{-1}$	$-3.345683\times10^{-2}-i2.346520\times10^{-1}$
2	$-1.397027\times10^{-1}-i3.035603\times10^{-2}$	$-1.410888\times10^{-1}-i2.456649\times10^{-2}$
3	$-5.670435\times10^{-4}+i2.611376\times10^{-2}$	$-5.770254\times10^{-4}+i2.671194\times10^{-2}$
4	$-2.328268\times10^{-2}+i4.271593\times10^{-3}$	$-2.000019\times10^{-2}+i4.205960\times10^{-3}$
5	$-3.206226\times10^{-5}-i3.395070\times10^{-3}$	$-3.404527\times10^{-6}-i3.342761\times10^{-3}$
6	$-1.540533\times10^{-2}+i3.235455\times10^{-3}$	$-1.650261\times10^{-2}+i7.595264\times10^{-3}$
7	$-7.250446\times10^{-8}+i7.387453\times10^{-4}$	$-6.120918\times10^{-7}+i7.417587\times10^{-4}$
8	$-1.115143\times10^{-2}-i6.101197\times10^{-3}$	$6.327824\times10^{-3}-i8.38423\times10^{-4}$
9	$-2.862725\times10^{-7}-i1.719243\times10^{-4}$	$-1.085472\times10^{-6}-i1.76887\times10^{-4}$
10	$-8.814035\times10^{-3}+i7.677106\times10^{-3}$	$-1.392445\times10^{-2}+i2.575785\times10^{-2}$
11	$-1.021961\times10^{-7}+i4.224494\times10^{-5}$	$-5.933418\times10^{-8}+i4.245868\times10^{-5}$

2.1.2.1 Directional Properties and Radiation Impedance of a Cylinder Encased in an Open Annular Layer

We can proceed with the investigation of quantitative characteristics of the field of a cylindrical radiator encased in an open annular layer, basing on the analysis performed in the previous paragraph.

Our main interest is the far field of a radiator, which will be characterized by directivity pattern and concentration factor [8]. The normalized directivity pattern $R(\phi)$ can easily be obtained from the expression defining the external field of radiator (2.8), if we use the asymptotic representations of Hankel functions for the large values of argument:

$$R(\phi) = \frac{\sum_{0}^{N} S_n e^{in\pi/2} \cos n\phi}{\sum_{0}^{N} S_n e^{in\pi/2} \cos n\phi_1}, \tag{2.44}$$

where N is determined by the order of the finite system of equations; pressure amplitude is normalized to pressure in the direction ϕ_1 ($\phi_1 = 0$ from here on).

Concentration factor cannot be determined for plane problem. We can use, however, the value

$$K = \pi \left[\int_0^{\pi} R^2(\phi) d\phi \right]^{-1}, \tag{2.45}$$

which is interpreted as concentration factor per unit height of cylindrical radiator in [8].

The characteristics of the far field and near field will be examined basing on the numerical solution of system (2.15). The system reduction order was chosen on the basis of the analysis of series convergence performed above. For the data given below, the number of unknowns S_n was 12, and the number of unknowns A_q was determined from the condition $\nu_q \approx 12$. Particle velocity uniform distribution over the surface of a cylinder was taken in all calculations, i.e., $f(\phi) = \text{const}$.

Characterizing angular distribution of sound energy in the far field, it should be noted that it is nonuniform. Figure 2.5 shows the directivity patterns for some values of angle ϕ_0. As we can see, the main part of sound energy is radiated in the front semi-space $|\phi| \leq \pi/2$. Small value of sound energy is radiated into the back semi-space.

We proceed now to more detailed analysis of the directional properties of the radiator. Figure 2.6 shows the dependency of the width of main directional lobe $\phi_{0.7}$ (at the level 0.7 of maximum of $R(\phi)$) and the concentration factor K on angle $2\phi_0$. It follows from these data, that when the wave thickness of a layer is relatively large, there is some angle ϕ_0, at which the width of the main lobe is minimum and concentration factor is maximum. This circumstance is due to the following reasons. The wave dimension

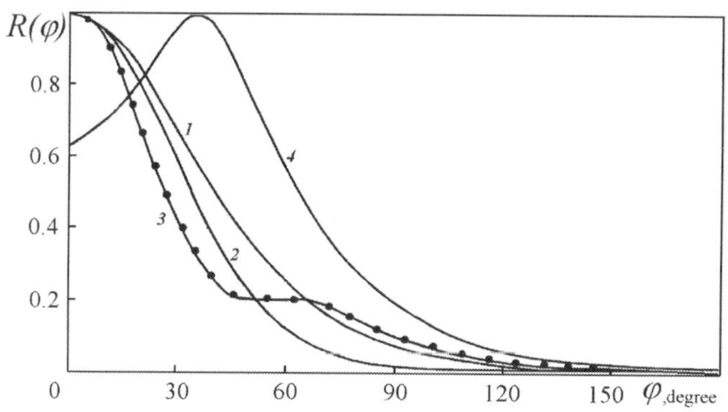

FIG. 2.5: Directivity patterns for $kr_0 = 4.52$ and $kr_1 = 6.9$; 1, 2, 3, 4—correspond to $\phi_0 = \pi/6, \pi/4, \pi/3, \pi/2$

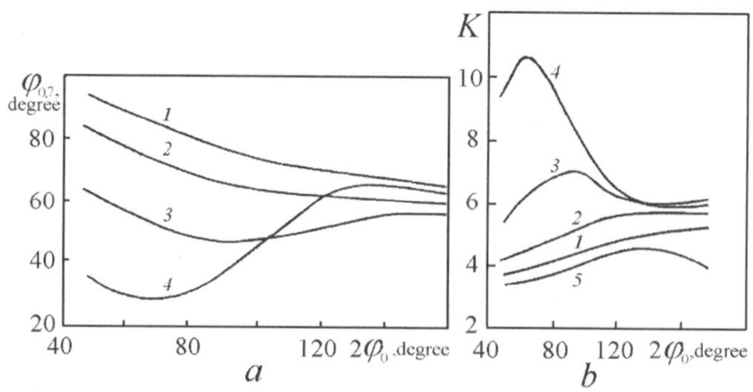

FIG. 2.6: The dependencies of the main directional lobe width (a) and concentration factor (b) on angle $2\phi_0$ for $kr_0 = 2.26$; 1, 2, 3, 4—correspond $kr_1 = 2.89, 4.14, 6.18, 14.82$, curve 5—corresponds to the case, when $(r_1 - r_0) \leq 0.1$

of the radiator aperture (by aperture we mean the distance between the points $r = r_1$, $\phi = \pm\, \phi_0$) is small for small angles ϕ_0, and that is why the values of $\phi_{0.7}$ are large and the value of K is small. With an increase of ϕ_0, the wave dimension of aperture and the radiator directivity increases, this is reflected in the decrease of $\phi_{0.7}$ and an increase of K. This process is going on until the phase distortions, caused by cylindricality of the wave front in aperture, reach substantial value. It is because of this circumstance, that the main lobe narrowing slows down and, when the values are large enough, the main lobe begins to expand again and, in some cases, it splits into two parts (see, for example, curve 4, Fig. 2.5). Some stabilization of the directional properties of the radiator may be expected when the wave thickness of a layer decreases (when the value of kr_0 is fixed). According the calculations, directivity pattern and concentration factor change scarcely in the domain $(r_1 - r_0)/\lambda \leq 0.1$, Fig. 2.6(b), curve 5.

We turn our attention to the band properties of cylindrical radiator with an open annular layer, i.e., frequency dependencies of the main characteristics. Figure 2.7 shows the family of curves, characterizing frequency dependency of the values $\phi_{0.7}$ and K at the various values of angle ϕ_0. It follows from this data, that there are angles at which the values of $\phi_{0.7}$ and K almost do not depend on frequency. Thus, at certain geometrical parameters, the radiator has the same shape of directivity pattern in rather broad frequency band.

It is interesting to consider the limiting case, when $\phi_0 = \pi$ and annular layer degenerates into a band, the width of which tends to zero. Effective radiation of sound energy is observed in the wide range of angles in this case (approximately from 0 to 120°), and only in the domain of bigger angles, the level of the field decreases rapidly. Figure 2.8 shows some typical directivity patterns. These data illustrate the form of directivity pattern for this special case, and also the efficiency of the band in suppression of the cylinder back radiation. In hydroacoustics, the radiators with such directional properties are called the radiators with sector pattern and they are widely used in practice [8].

Now we proceed to the analysis of peculiarities of the cylinder radiation impedance in an open annular layer. Using general definition for radiation impedance and the expression for sound field in the first partial domain, we can represent radiation impedance per unite height of a cylinder as follows:

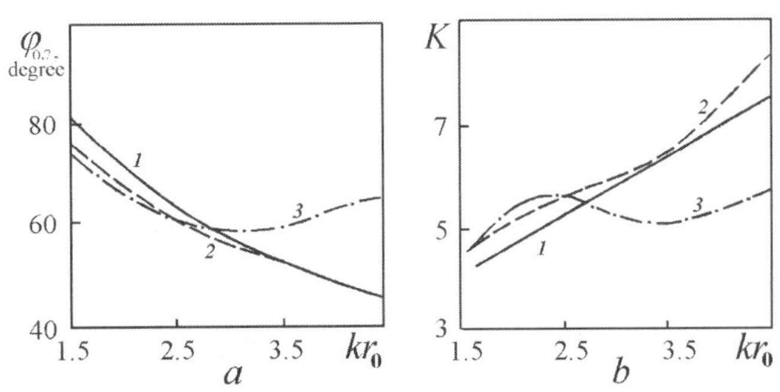

FIG. 2.7: Frequency dependence of $\phi_{0.7}$—(a) and value of K—(b) for $r_1/r_0 = 1.8$; the angles $\phi_0 = \pi/4, \pi/3, \pi/2$ correspond to the curves 1, 2, and 3

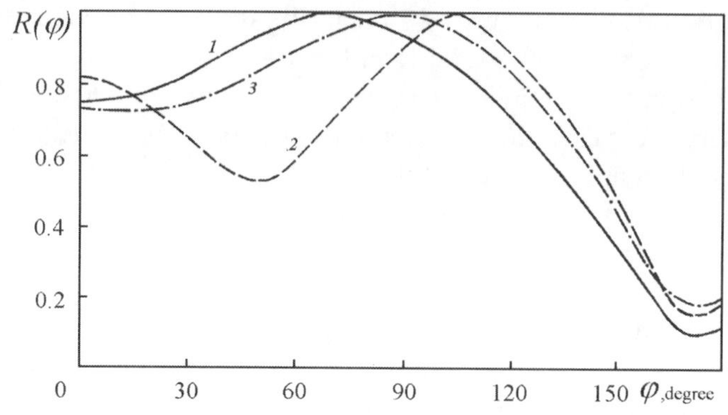

FIG. 2.8: Directivity pattern for $\phi_0 = \pi$ and $kr_1 = 6.18$; 1, 2, 3—correspond to $kr_0 = 1.57, 2.26, 3.14$

$$Z = \rho c\bar{S}(R + iX), \tag{2.46}$$

where $\bar{S} = 2r_0\phi_0$—radiating area of the source per unit height,

$$R = \frac{1}{\pi r_0}\sum_{q=0}^{\infty} \mathrm{Im}(A_q)\frac{L_q}{N'_{\nu_q}(kr_0)},$$

$$X = -\frac{1}{2r_0}\sum_{q=0}^{\infty}\left[\frac{2}{\pi}\mathrm{Re}A_q + r_0 L_q N_{\nu_q}(kr_0)\right]\frac{L_q}{N'_{\nu_q}(kr_0)}.$$

Figure 2.9 shows frequency dependencies of radiation impedance components for different values of angle ϕ_0. When we analyze the behavior of curves with a rise in

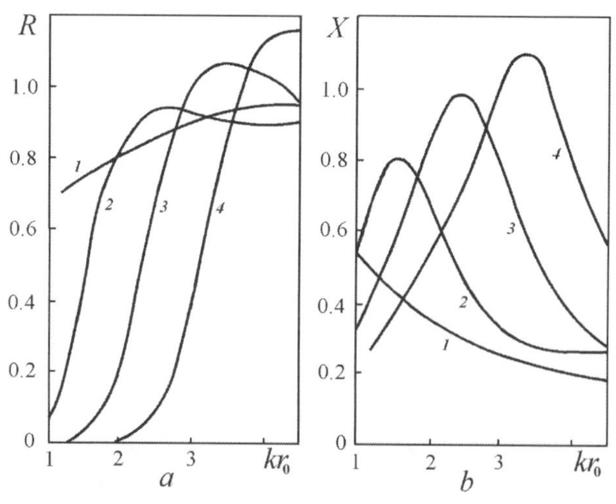

FIG. 2.9: Cylinder radiation impedance with an open layer for $r_1/r_0 = 2.5$; (a)—real part, (b)—imaginary part; 1, 2, 3, 4—correspond to $\phi_0 = \pi/2, \pi/4, \pi/6, \pi/8$

frequency, the following peculiarities can be indicated: when angle ϕ_0 decreases, the impedance components reach relatively large values when the values of kr_0 are large; the values R and X increase when the angle ϕ_0 decreases and the value of r_1/r_0 increases in the peak region of the curves. We can make a conclusion about the influence of ratio r_1/r_0 basing on the comparison of the data, shown in Fig. 2.9, with the calculation results for $r_1/r_0 = 1.1$. Slope of the curves, representing the change in R when kr_0 increases, decreases in this case, and the general level of the radiating system reactivity decreases.

Thus the cylinder in an open annular layer can effectively radiate acoustic energy, starting with some frequencies that are determined by the value of angle ϕ_0. Its efficiency decreases rapidly below these frequencies. The rate of this decrease is determined by the ratio r_1/r_0.

In conclusion we note the following. Taking $r_1 \to \infty$, we will have the case considered in [28]. The thickness of the open annular layer is infinite in this situation. Comparing the results, given in [28], with data shown in Fig. 2.9, we can conclude that the values of R and X coincide with each other with graphical accuracy. It allows for the conclusion: when $r_1/r_0 \geq 2.5$ diffraction effects, caused by finite thickness of a layer, do not influence the impedance.

Thus we can control the directional properties of cylindrical radiator and its concentration factor, using acoustically compliant open layer. And we can stabilize directivity pattern, in some frequency range, selecting the layer properties.

2.1.3 Sound Radiation from a Cylinder through an Open Annular Layer Having a Finite Transmissivity

In the previous paragraph an open annular layer was assumed to be acoustically compliant, that is absolutely nontransmissive. Now we consider a more complex situation, when the material of a layer has a certain transmissivity and is characterized by certain wave resistance $\rho_1 c_1$, Fig. 2.10. As in the case of acoustically soft layer, we also assume that particle velocity is given on the surface of the cylinder

$$\upsilon = \upsilon_0 f(\phi), \quad f(\phi) = f(-\phi), \tag{2.47}$$

and the following conditions are met at the outer ends of the layer

$$\Phi = 0, \quad \phi = \pm\,\phi_0, \quad r_1 \leq r \leq r_2. \tag{2.48}$$

When constructing the expressions for sound field inside the layer we consider the inner ends to be acoustically stiff:

$$-\partial\Phi/\partial r = 0, \quad \phi = \pm\,\phi_0, \quad r_1 \leq r \leq r_2. \tag{2.49}$$

Besides the accepted presumptions, we also assume the presence of clearance between the layer and the cylinder.

To solve the stated problem, we increase the number of partial domains comparing with the problem considered in the previous paragraph. We partition the entire domain of existence of the field into four partial domains: domain I ($r_0 \leq r \leq r_1$, $|\phi| \leq \pi$),

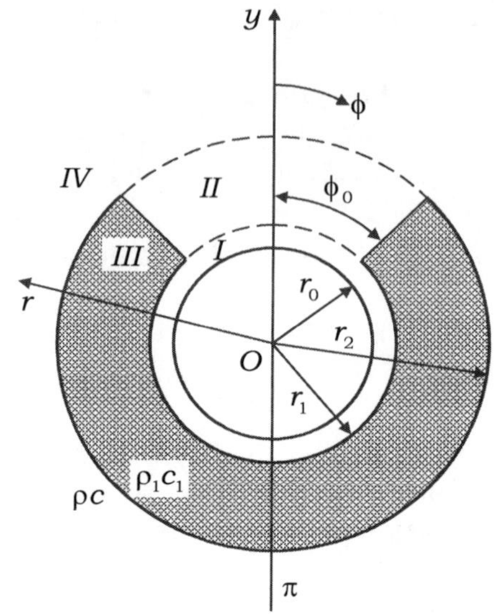

FIG. 2.10: Cylinder encased in an open annular layer of finite transmissivity

domain II ($r_1 \le r \le r_2$, $|\phi| \le \phi_0$), domain III ($r_1 \le r \le r_2$, $\phi_0 \le |\phi| \le \pi$), and domain IV ($r \ge r_2$, $|\phi| \le \pi$). We write the velocity potential in the partial domains in the form:

$$\begin{aligned}
\Phi_1 &= \sum_{n=0}^{\infty} [A_n J_n(kr) + B_n N_n(kr)] \cos n\phi, \\
\Phi_2 &= \sum_{q=0}^{\infty} \left[C_q J_{\nu_q}(kr) + D_q N_{\nu_q}(kr)\right] \cos \nu_q \phi, \\
\Phi_3 &= \sum_{s=0}^{\infty} [E_s J_{\nu_s}(k_1 r) + F_s N_{\nu_s}(k_1 r)] \cos \nu_s(\pi - \phi), \\
\Phi_4 &= \sum_{n=0}^{\infty} L_n H_n^{(1)} \cos n\phi.
\end{aligned} \tag{2.50}$$

Here the notation is the same as in the previous paragraph, $\nu_s = sn/(\pi - \phi_0)$ and is determined from the boundary condition (2.49), $k = \omega/c$, $k_1 = \omega/c_1$.

The complex coefficients of series (2.50) must be determined from the following system of functional equations. Besides the boundary conditions (2.47)—(2.49) these equations contain the matching conditions at the interfaces of the partial domains:

$$-\partial\Phi_1/\partial(kr) = \upsilon_0 f(\phi), \quad 0 \le [\phi] \le \pi, \quad r = r_0,$$

$$\Phi_1 = \begin{cases} \Phi_2, & |\phi| \le \phi_0, \quad r = r_1, \\ \Phi_3, & \phi_0 \le |\phi| \le \pi, \quad r = r_1, \end{cases}$$

$$\partial\Phi_1/\partial(kr) = \partial\Phi_2/\partial(kr), \quad |\phi| \le \phi_0, \quad r = r_1,$$

$$k\partial\Phi_1/\partial r = k_1\partial\Phi_3/\partial r, \quad \phi_0 \leq |\phi| \leq \pi, \quad r = r_1,$$

$$\Phi_4 = \begin{cases} \Phi_2, & |\phi| \leq \phi_0, \quad r = r_2, \\ \Phi_3, & \phi_0 \leq |\phi| \leq \pi, \quad r = r_2, \end{cases} \tag{2.51}$$

$$\partial\Phi_2/\partial(kr) = \partial\Phi_4/\partial(kr), \quad |\phi| \leq \phi_0, \quad r = r_2,$$

$$k\partial\Phi_4/\partial r = k_1\partial\Phi_3/\partial r, \quad \phi_0 \leq |\phi| \leq \pi, \quad r = r_2,$$

Algebraization of system (2.51), enables us to obtain an infinite system of linear algebraic equations of the second kind. We shall not write out this system below because its structure has not vital difference from the structure of system (2.13). We proceed with an analysis of quantitative data.

We first determine the influence of the transmissivity of the layer on the directional properties of the radiator. Figure 2.11 gives the directivity patterns of a radiator with a quarter-wave layer and different values of $m = \rho_1 c_1/\rho c$. The calculations were made for the following geometric and wave parameters values: $2r_0/\lambda = 0.7$, $r_1/r_0 = 1.01$, $\tilde{d} = (r_2 - r_1)/\lambda_1 = 0.25$. Here λ and λ_1—sound wavelength in surrounding medium and in the material of the layer respectively.

It follows from these data that an increase in the transmissivity of the layer has the effect of increasing the back radiation level. When $m \leq 10^{-3}$, directivity pattern almost coincides with the pattern calculated for acoustically compliant layer (compare curves 1 and 2). And starting with $m \approx 10^{-2}$, it is modified considerably and back radiation is almost doubled (curve 3). Further increase of m parameter leads to a greater increase of back radiation level. The given data apply to the case in which the acoustic transmissivity of the layer is close to the minimum, since $\tilde{d} = 0.25$ [3,14]. It is logical at this point to inquire into the role played by the thickness of the layer.

This question can be answered by referring to Fig. 2.12, in which the back radiation level $R(\phi = \pi)$ is given as a function of the wave thickness of the layer. The initial

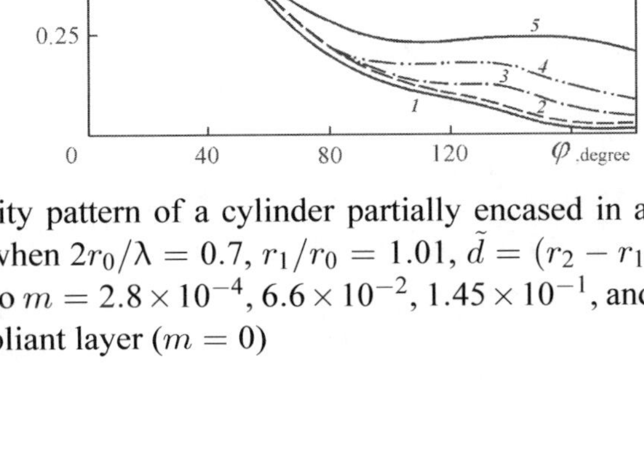

FIG. 2.11: Directivity pattern of a cylinder partially encased in a partially-transparent open annular layer when $2r_0/\lambda = 0.7$, $r_1/r_0 = 1.01$, $\tilde{d} = (r_2 - r_1)/\lambda_1 = 0.25$; 2, 3, 4, and 5—correspond to $m = 2.8 \times 10^{-4}$, 6.6×10^{-2}, 1.45×10^{-1}, and 0.3; 1—corresponds to acoustically compliant layer ($m = 0$)

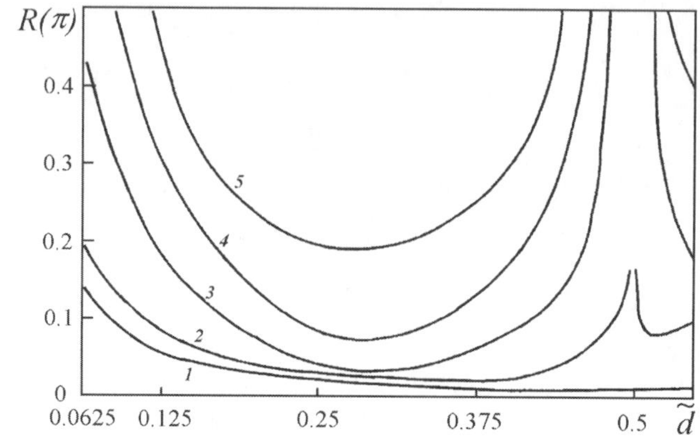

FIG. 2.12: Back radiation level versus wave thickness of the annular layer; 1—corresponds to acoustically compliant layer, 2, 3, 4, 5—correspond to the same values of m, as in Fig. 2.11

values of geometric parameters in this case are the same as for the curves in Fig. 2.11. It is seen that the total level of the back radiation value increases with increase of m for any value of the wave thickness of the layer.

For the small values of $\tilde{d}$ the value of back level of radiation increases due to the increase in the transmissivity of the layer and an increase in the level of the waves diffracted by the radiator. In the region $0.25 \leq \tilde{d} \leq 0.35$ we observe a minimum of back radiation because the transmissivity of the layer is minimum for such wave thickness [3,14]. Thereafter the value of $R(\pi)$ increases due to the increase of the transmissivity of the layer, and in the region $\tilde{d} = 0.5$, attains a maximum, whereupon it again decreases. Consequently, the dependence of the back radiation level of the cylinder on the wave thickness of the layer is qualitatively similar to the dependencies of the acoustic transmissivity of a closed annular or plane layer.

The foregoing data pertains only to the case in which the clearance between the cylinder and the layer is very small, being 1% of the radius of the cylinder. We now ascertain how this clearance affects the far-radiated field. Figure 2.13 gives a series of directivity patterns for various clearance widths. It is seen that an increase in clearance induces distortions of the directivity pattern mainly in the interval of angles $|\phi| \leq \pi/2$. The size of the clearance has practically no influence on the back-radiation level (when the value of m is very small). This characteristic is attributable to the fact that for large clearances and low transmissivity of the layer the acoustic energy "leaking out" from beneath the layer through the gap $\phi = \pm\, \phi_0$, $r_0 \leq r \leq r_1$ begins to play a significant part.

We now analyze the radiation impedance of the considered cylinder. Figure 2.14 gives frequency dependencies of the dimensionless components of the radiation impedance of the cylinder for various values of m. To simplify the analysis of the data, the values of wave thickness $\tilde{d}$ are plotted along the horizontal axis.

Characterizing the frequency dependence of the real part of radiation impedance R [Fig. 2.14(a), curves 1–3], we remark that it is much different from the frequency

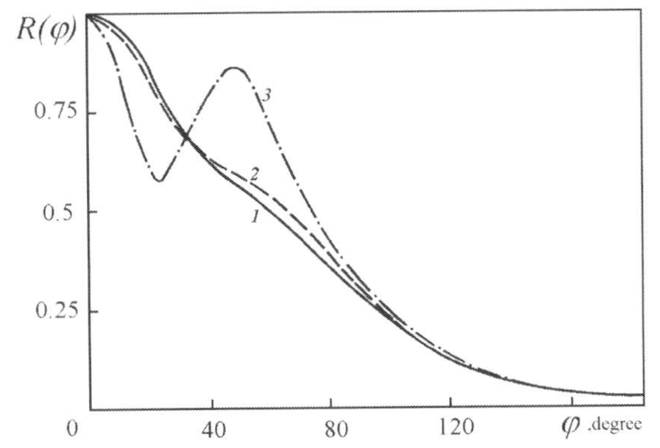

FIG. 2.13: Directivity pattern for $2r_0/\lambda = 1$, $\tilde{d} = 0.35$, $m = 2.8 \times 10^{-4}$; 1, 2, 3—correspond to $r_1/r_0 = 1.01, 1.2, 1.5$

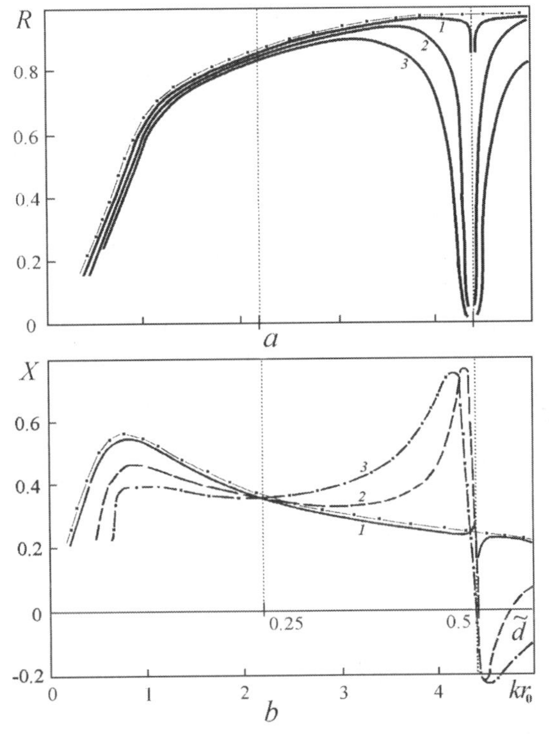

FIG. 2.14: Frequency dependence of radiation impedance for $2r_0/\lambda = 0.7$ and $r_1/r_0 = 1.01$; (a)—real part, (b)—imaginary part; 1, 2, 3—correspond to the values $m = 2.8 \times 10^{-4}$, 6.6×10^{-2}, 1.45×10^{-1}; curve, punctuated with dots,—for the cylinder, encased in an open, acoustically compliant layer without clearance

dependence of the value of R for the cylinder with acoustically compliant layer, which is immediately adjacent to its surface, [the curve is punctuated with dots in Fig. 2.14(a)]. First of all, we can observe an abrupt decrease (down to zero) in the frequency response in the vicinity of $\tilde{d} \approx 0.5$ and secondly, the general level of the value of R decreases with an increase of m and especially in the vicinity of $\tilde{d} \approx 0.5$. Frequency dependence of the imaginary part of the radiation impedance has the interesting peculiarities. It follows from Fig. 2.14(b) (curves 1–3), when $\tilde{d} < 0.25$ the level of value of X decreases with an increase of m, and when $\tilde{d} > 0.25$, on the other hand, it increases. In the vicinity of $\tilde{d} = 0.5$, the value of X attains a maximum, after which it decreases abruptly, changes sign, and attains a minimum. According the calculations, with further increase in frequency, we can observe the fluctuations of the value of X and a change of its sign. By contrast, the frequency dependence of the value of X, for the cylinder with acoustically compliant layer immediately adjacent to its surface [the curve is punctuated with dots in Fig. 2.14(b)], has not any peculiarities in the vicinity of $\tilde{d} \approx 0.5$. This behavior of imaginary part of radiation impedance can be explained as follows.

We assume that $X = X_1 + X_2$, here X_1 is defined on the exposed part of the surface of the cylinder $|\phi| \leq \phi_0$, and X_2 underneath the layer $\phi_0 \leq |\phi| \leq \pi$. As it was shown in the previous paragraphs, the value of X_1 always has the nature of the mass and is characterized by the curve punctuated with dots in Fig. 2.14(b). In this case, the value of X_2 is equal to the value of X, which was determined in paragraph 1, [Fig. 2.2(b)]. The value of X_2 depends on the wave thickness of the layer and it has by turns the nature of mass and of elasticity. If we add algebraically the ordinates of curve 1 in Fig. 2.2(b) with the ordinates of the curve punctuated with dotes in Fig. 2.14(b), at each frequency, we will obtain the curve close to curve 2 in Fig. 2.14(b).

Thus the behavior of the frequency dependence of radiation impedance imaginary component of the cylinder with an open annular layer of finite transmissivity is determined by resonance effects occurring in a singular mechanical circuit formed by the mechanical parameters of the layer, by the additional mass of the surrounding medium on the outer surface of the layer $r = r_2$, $\phi_0 \leq |\phi| \leq \pi$, and by additional mass of the medium on the exposed part of the surface of the cylinder $r = r_0$, $|\phi| \leq \phi_0$. On the basis of the foregoing analysis we can explain the abrupt decrease of the value of R in the vicinity of $\tilde{d} \approx 0.5$. Here we have antiresonance (parallel resonance) of the indicated mechanical circuit, and the average value of the sound pressure at the surface of the cylinder becomes a minimum. As a result, a paradoxical effect is observed: in the event of total transmissivity of the layer, the efficiency of radiation of acoustic energy by the cylinder is insignificant.

Next we consider the influence of the clearance between the layer and the cylinder on the radiation impedance. For small wave dimensions of the clearance, when $r_1/r_0 \leq 1.1$, the values of R and X practically coincide with those for $r_1/r_0 \leq 1.01$. The most significant variations for R and X occur for large values of r_1/r_0, when the wave width of the clearance $(r_1 - r_0)/\lambda$ lies in the vicinity of 0.25. In this case, the value of X can change sign. These attributes of the behavior of the radiation impedance of the cylinder are associated with the presence of a standing wave in the clearance due to the interaction of the direct wave radiated by the cylinder and the wave reflected from the layer. When the wave width of the clearance attains a value close to 0.25, the considerable increase of the pressure and, hence, of the radiation impedance is observed on the part of the cylinder surface underneath the layer. In this case the radiation impedance of the cylinder is almost completely determined by the processes taking place in the clearance.

Thus, the degree of the transmissivity of the layer can influence significantly the directional properties of the cylindrical radiator as well as its radiation impedance. In particular, with increasing transmissivity of the layer, the level of its back radiation increases and, under certain conditions, its efficiency as an acoustic source can decrease considerably.

2.2 SOUND RADIATION FROM THE CYLINDRICAL SHELLS

The problems of sound radiation from the cylindrical sources were considered above. It was assumed that normal component of particle velocity is set on the radiating surfaces. Such sources are the particle velocity generator with infinitely large internal impedance because particle velocity does not depend on surrounding medium reaction.

The real sound sources are characterized by finite value of their internal mechanical impedance, in contrast to such idealized sources. As a result of this, the particle velocity of their surfaces depends on internal impedance and on reaction of external medium. The sound fields can also be in the intra-cavities of the sources and this fact also influences their particle velocity.

That is why the sound fields levels, radiated by real sources can differ considerably from the levels predicted with the help of idealized schemes. The thin elastic shells will be considered below as the sound sources.

2.2.1 Sound Radiation from a Shell, Filled with Acoustic Medium

Consider the sound source in the form of an infinitely long elastic cylindrical shell. The shell executes pulsing vibrations induced by uniformly distributed periodic load p (Fig. 2.15).

Load p can be created by different ways. If the material of the shell is piezoelectric, the load p can be easily created with the help of electrical voltage applied to the electrodes of the shell [29]. We will consider this possibility in the following paragraphs.

We suppose that there is a vacuum inside of the shell, in domain I ($\rho_1 c_1 = 0$). The shell is surrounded by liquid with impedance ρc from outside (domain II). Sound

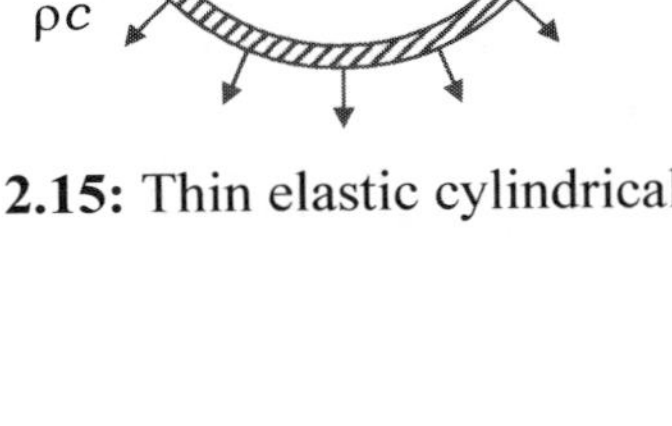

FIG. 2.15: Thin elastic cylindrical shell

pressure, created by pulsing shell in the surrounding medium, can be represented as follows [2,3,30]:

$$p_2 = -i\omega\rho A H_0^{(1)}(kr). \tag{2.52}$$

Unknown complex constant A is defined from the matching condition on the surface $r = r_1$

$$p = p_2 + \upsilon_2 Z, \quad \upsilon_2 = \upsilon_0, \tag{2.53}$$

where υ_0—radial particle velocity of the shell, Z—mechanical impedance of the shell, $\upsilon_2 = \partial p_2 / [i\rho c \partial(kr)]$.

When the constant A is defined, we set $p = 1$, and then we have

$$p_2 = \frac{i\rho H_0^{(1)}(kr)}{H_0'^{(1)}(kr_1)\tilde{Z}},$$
$$\tilde{Z} = Z_1 + Z_2, \quad Z_1 = Z/\rho c, \quad Z_2 = -i\frac{H_0^{(1)}(kr_1)}{H_1^{(1)}(kr_1)}. \tag{2.54}$$

The value Z_2 characterizes radiation impedance of pulsing cylinder, as it is known from [2,3,30]. Imaginary part of the cylinder radiation impedance defines the added mass of medium. Ignoring dissipative losses, the mechanical impedance of the shell is $Z = -i\omega m + is/\omega$ [29,31], where $m = \tilde{\rho}(r_1 - r_0)$ and $s = E(r_1 - r_0)/r_0^2$—mass and elasticity of the surface unit of the shell, and $\tilde{\rho}$ and E—the density and Young modulus of shell's material. It is obvious that the resonance frequency of the pulsing shell will decrease in the presence of the medium. This fact is well known in the literature (see, for example, [32]).

We complicate the task, assuming that liquid with wave impedance $\rho_1 c_1$ is present in the shell. Then, sound pressure inside the shell should be represented as follows:

$$p_1 = -i\omega\rho_1 B J(k_1 r), \quad k_1 = \omega/c_1. \tag{2.55}$$

The values A and B in expressions (2.52) and (2.55) are defined now from the conditions

$$p = p_2(r_1) - p_1(r_0) + \upsilon_2(r_1)Z, \quad \upsilon_1(r_0) = \upsilon_2(r_1) = \upsilon_0. \tag{2.56}$$

Note that in condition (2.56) the particle velocities of outer and inner surfaces of the thin shell are supposed to be equal. The smaller is ratio $(r_1 - r_0)/r_0$, the truer is this fact for not very high frequencies.

Using condition (2.56), and bearing in mind (2.52) and (2.55) we obtain

$$p_2 = \frac{i\rho H_0^{(1)}(kr)}{H_0'^{(1)}(kr_1)\left[Z_1 + Z_2 + Z_3\right]}, \tag{2.57}$$

where $Z_3 = i(\rho_1 c_1/\rho c)(J_0(k_1 r_0)/J_1(k_1 r_0))$.

Value Z_3 characterizes radiation impedance into the shell. This value has imaginary character, in contrast to Z_2. For $k_1 r_0 < 2.405$, the value of Z_3 has elastic nature. It has the nature of mass in the domain $2.405 < k_1 r_0 < 3.832$. It has elastic nature again when $3.832 < k_1 r_0 < 5.52$ and so on. The impedance Z_3 has the alternating nature, and Fig. 2.16 illustrates the change in Z_3 nature with an increase in frequency.

Thus, the presence of liquid inside the shell can lead to a decrease and to an increase in the resonance frequency of the shell. It depends on the wave dimensions of the shell at its resonance frequency. If the resonance frequency corresponds to the domain $k_1 r_0 < 2.405$, the presence of liquid inside the shell leads to an increase in resonance frequency.

2.2.2 Sound Radiation by a Shell with a Cylindrical Insertion

Consider the case when a cylinder, which is non-transparent for sound and coaxial with the shell, is inside of the shell (Fig. 2.17). The shell executes pulsing vibrations induced by uniformly distributed periodic load $p = 1$ (the same situation was considered in the previous paragraph). The field in the annular space (domain I) between the shell and this cylinder is described as follows:

$$p_\mathrm{I} = -i\omega\rho_1 \left[C J_0(k_1 r) + D N_0(k_1 r) \right]. \tag{2.58}$$

Using condition (2.56) and bearing in mind (2.58) and (2.52), we can easily obtain the expression for the outer field of the pulsing shell:

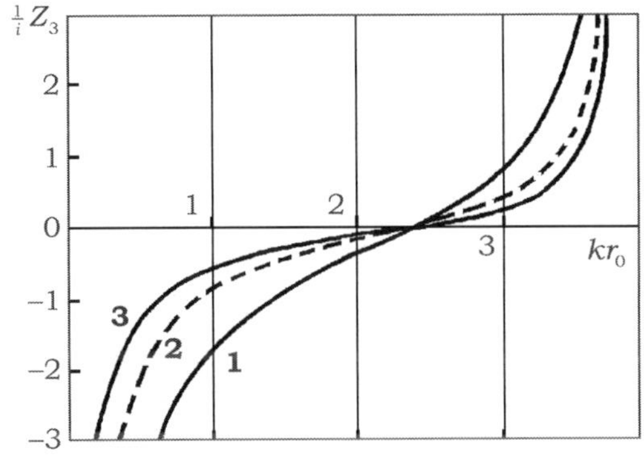

FIG. 2.16: Frequency dependence of impedance Z_3; curves 1, 2, 3—correspond to $\rho_1 c_1/\rho c = 1, 0.5, 0.3$

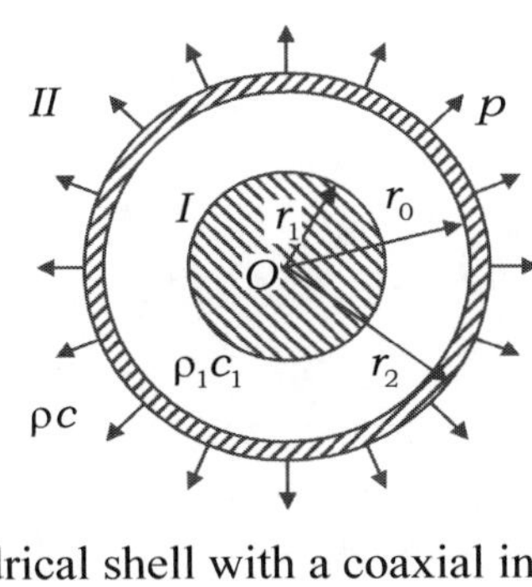

FIG. 2.17: Thin elastic cylindrical shell with a coaxial insertion in the form of a cylinder

$$p_2 = \frac{i\rho H_0^{(1)}(kr)}{H_0^{\prime(1)}(kr_1)\left[Z_1 + Z_2 + Z_4\right]}, \tag{2.59}$$

where

$$Z_4 = \begin{cases} i\dfrac{\rho_1 c_1}{\rho c}\left[\dfrac{J_0(k_1 r_0)N_1(k_1 r_2) - N_0(k_1 r_0)J_1(k_1 r_2)}{J_1(k_1 r_0)N_1(k_1 r_2) - N_1(k_1 r_0)J_1(k_1 r_2)}\right] \\ -i\dfrac{\rho_1 c_1}{\rho c}\left[\dfrac{J_0(k_1 r_0)N_0(k_1 r_2) - N_0(k_1 r_0)J_0(k_1 r_2)}{J_1(k_1 r_0)N_0(k_1 r_2) - N_1(k_1 r_0)J_0(k_1 r_2)}\right] \end{cases}.$$

The upper expression for Z_4 defines the impedance component for an acoustically hard cylinder, and the lower expression—for an acoustically soft cylinder.

The frequency dependency of the value of Z_4 for the acoustically hard internal cylinder is similar to the dependencies, shown in Fig. 2.16. That is why it is not very interesting to consider this case. The case when the internal cylinder is acoustically soft is much more important. Figure 2.18 shows frequency dependencies of Z_4 for this case. As we can see, the impedance Z_4 has the nature of mass at low frequencies. That is why the resonance frequency of the shell decreases. This property should be born in mind when it is necessary to decrease resonance frequency. For example, when it is necessary to decrease the frequency of pulsing electro-acoustic transducer in the form of piezoceramic cylindrical shell [5,7,10,29,32–35].

We consider the following example. Consider a piezoceramic shell with an internal acoustically soft cylinder: $r_0 = 0.027$ m, $r_1 = 0.03$ m, $r_2/r_0 = 0.35$, $\rho c = \rho_1 c_1 = 1.5 \times 10^6$ kg/(m^{2}s), $\tilde{\rho} = 7.2 \times 10^3$ kg/m^3, $E = 0.86 \times 10^{11}$ Pa. The resonance frequency of the shell, calculated with consideration of the data in Fig. 2.18, is about 12 kHz. The resonance frequency of the shell itself in vacuum is about 19.2 kHz. Thus, using internal

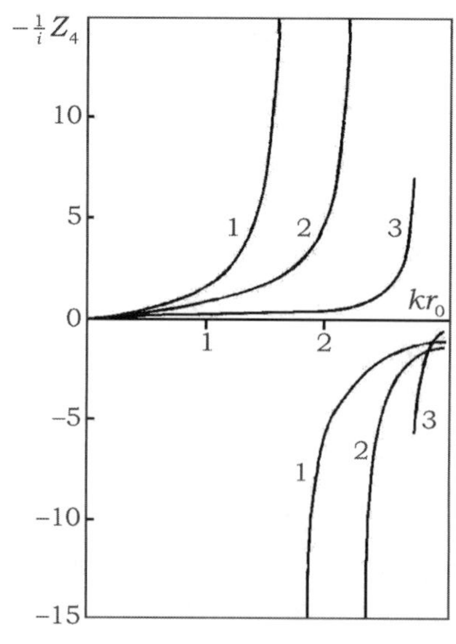

FIG. 2.18: Frequency dependence of the impedance Z_4 when $\rho_1 c_1/\rho c = 1$; curves 1, 2, 3—correspond to $r_2/r_0 = 0.28, 0.425, 0.72$

acoustically soft cylinder, the resonance frequency of the shell was decreased by 1.6 times.

The properties of the cylindrical shell with the insertion in the form of an ideal cylinder can be explained from physical standpoint. In the case of a hard cylinder, the change of volume between the insertion and the vibrating shell can happen only due to medium compression. That is why, at relatively low frequencies, the resistance of the liquid in the internal space has elastic character, and the eigenfrequency of the system increases. In the case of a soft cylinder, the elastic properties of liquid are of no importance. The change of volume between the shells at relatively low frequencies happens due to the shift of liquid layer almost as a single whole. That is why the reaction of the layer of liquid to the motion of the shell has the nature of mass and its eigenfrequency decreases. First of all we are interested in quantitative data in this example. The effect that was discussed is important when considering the vibrations of the piezoceramic shells in water.

2.2.3 Sound Radiation by Piezoceramic Shell with an Asymmetric Insertion

Only zero (pulse) mode of the shell can be excited in the radially symmetric system considered above. For this reason, sound radiation into exterior medium is non-directional. It is possible to affect not only the frequency characteristics but also the directed characteristics of such radiators by breaking the radial symmetry of the system (shifting the insertion with regard to the axis of the shell). The examples are considered below. These examples illustrate the methods, which permit:

1. to decrease resonance frequency of the radiator in the form of an electro-elastic shell without significant decrease in energetic efficiency of the system;

2. to ensure the high directivity of the radiation of acoustic energy into exterior medium.

Consider the problem of acoustic radiation by the infinite thin piezoelastic cylindrical shell with radius r_0 and thickness h (see Fig. 2.19). There is an insertion inside the shell in the form of the sector of a round cylinder with radius a_0. The axis O_1 of the sector is shifted with respect to the axis of the shell O on distance l. Summit opening angle of the sector is equal $2(\pi - \theta_0)$. We confine ourselves to the case, when the system is symmetric with respect to the direction $\phi = 0$. The domain between the shell and insertion is filled with liquid with density ρ_1 and sound velocity c_1. Consider that the surfaces of piezoceramic shell are coved with electrodes and harmonic electrical voltage applied to electrode coating is

$$\Psi = \Psi_0 e^{-i\omega t}, \quad r = r_{1,2}, \tag{2.60}$$

where Ψ—electrical potential, $r_{1,2} = r_0 \pm h/2$.

Assuming $h/r_0 \ll 1$, we will use Kirchhoff-Love hypothesis describing the movement of the shell [39–41]. Let u be the circumferential displacement of the middle surface of the shell and w be the radial one. Consider that piezoceramic shell is polarized in

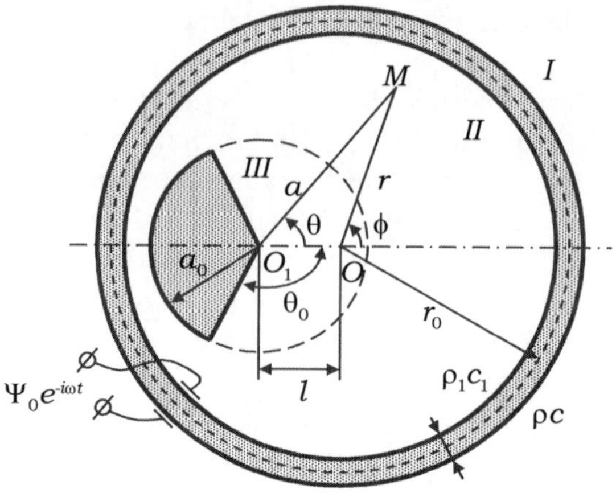

FIG. 2.19: Piezoceramic shell with nonaxisymmetric insertion in the form of a cylindrical sector

thickness. Hence, on the basis of the Kirchhoff-Love hypothesis and the relation of piezoceramic shells of revolution [36], it appears that the equation for the harmonic forced vibrations of the shell can be written as

$$\frac{d^2u}{d\phi^2} - \frac{dw}{d\phi} + \xi_p\beta^2\left(\frac{d^2u}{d\phi^2} + \frac{d^3w}{d\phi^3}\right) + k^2r_0^2w = 0,$$

$$\frac{du}{d\phi} - w - \xi_p\beta^2\left(\frac{d^3u}{d\phi^3} + \frac{d^4w}{d\phi^4}\right) + k^2r_0^2w$$
$$= -\frac{r_0}{h}(1+\nu)d_{31}\Psi_0 - \frac{i\omega r_0^2}{h\tilde{\rho}\tilde{c}^2} \times [\rho\Phi_1(r_1,\phi) - \rho_1\Phi_2(r_2,\phi)], \quad (2.61)$$

where $\tilde{k} = \omega/\tilde{c}$, $\beta^2 = h^2/12r_0^2$, $\xi_p = 1 + ((1+\nu)/2)(k_p^2/(1-k_p^2))$; Φ_1 and Φ_2—acoustic potentials for the liquids inside and outside the shell, $\tilde{\rho}$, $\tilde{c}$, ν—density, longitudinal wave velocity and Poisson's ratio for piezoceramics, d_{31}—piezomodulus, k_p—is the planar electromechanical coupling coefficient of the shell.

The displacement of the shell will be defined as

$$u(\phi) = \sum_{n=1}^{\infty} U_n \sin n\phi,$$
$$w(\phi) = \sum_{n=0}^{\infty} W_n \cos n\phi. \quad (2.62)$$

Acoustic velocity potential in outer domain I is represented as follows

$$\Phi_1(r,\phi) = \sum_{n=0}^{\infty} A_n H_n^{(1)}(kr)\cos n\phi, \quad k = \omega/c. \quad (2.63)$$

Partial domains method is used for consideration of acoustic fields in the inner domain of the shell. We partition the entire domain between the shell and insertion into two subdomains, as it is shown in Fig. 2.19. Subdomain II ($a > a_0, r < r_0$) and subdomain III ($0 < a \leq a_0, |\theta| \leq \theta_0$), and we write the expressions for velocity potential Φ_2 and Φ_3, respectively.

Fields matching condition must take place on the common bound of these subdomains

$$\Phi_2 = \Phi_3, \quad \frac{\partial \Phi_2}{\partial a} = \frac{\partial \Phi_3}{\partial a}, \quad a = a_0, \quad |\theta| \leq \theta_0. \tag{2.64}$$

Consider cylindrical part of the insertion surface to be acoustically hard. Then

$$d\Phi_2/da = 0, \quad a = a_0, \quad \theta_0 \leq \theta \leq 2\pi - \theta_0. \tag{2.65}$$

The expression for Φ_2 is defined as:

$$\Phi_2(a, \theta) = \sum_{n=0}^{\infty} [B_n J_n(k_1 a) + D_n N_n(k_1 a)] \cos n\theta, \tag{2.66}$$

and for Φ_3:

$$\Phi_3(a, \theta) = \sum_{n=0}^{\infty} F_n J_{\mu_n}(k_1 a) \cos \mu_n \theta, \quad k_1 = \omega/c_1. \tag{2.67}$$

Eigenvalues μ_n in (2.67) in the case of acoustically hard sector part of the surface

$$\partial\Phi_3/\partial\theta = 0, \quad 0 \leq a \leq a_0, \quad |\theta| = \theta_0, \tag{2.68}$$

are as follows

$$\mu_n = \pi n/\theta_0, \tag{2.69}$$

and for acoustically soft

$$\Phi_3 = 0, \quad 0 \leq a \leq a_0, \quad |\theta| = \theta_0, \tag{2.70}$$

here

$$\mu_n = \pi(2n + 1)/2\theta_0. \tag{2.71}$$

Equality conditions of normal velocities of liquid and shell must be fulfilled on the surfaces of piezoceramic shell:

$$\begin{aligned} -i\omega w &= -\partial\Phi_1/\partial r, \quad r = r_1, \\ -i\omega w &= -\partial\Phi_2/\partial r, \quad r = r_2. \end{aligned} \tag{2.72}$$

To differentiate the field Φ_2 with respect to r we write the field Φ_3 in an (r, ϕ) coordinate system. For this purpose we use addition theorem for cylindrical functions [21,22]:

$$\Phi_3(r, \phi) = \sum_{n=0}^{\infty} \sum_{m=0}^{\infty} \cos n\phi J_{m-n}(k_1 l) [B_m J_n(k_1 r) + D_m N_n(k_1 r)]. \tag{2.73}$$

We substitute expressions (2.62), (2.63), (2.66), (2.67) into (2.61), (2.64), (2.65), (2.72), taking into account μ_n. Thus we obtain the infinite system of linear algebraic equations for determining unknown coefficients A_n, B_n, D_n, F_n, U_n, W_n, due to completeness and orthogonality of eigenfunction. This is initial system for obtaining quantitative data about the acoustic properties of radiator.

Consider some insertion configuration variants, which are important on the practical side:

1. insertion in the form of a complete cylinder ($\theta_0 = 0$); partial domain III disappears in this case and there is no need to use the expression for the field Φ_3;

2. insertion in the form of a sector ($\theta_0 \neq 0$) for $l = 0$; coordinate systems (r, ϕ) and (a, θ) coincide with each other in this case, and expression (2.73) is transformed into identity.

Numerical analysis was made for the shell with $h/r_0 = 0.107$ when $r_0 = 0.075$ m. Shell material is piezoceramics CTBS-3 [38] with the following characteristics: $\tilde{\rho} = 7.2 \times 10^3$ kg/m^3, $\tilde{c} = 3.4 \times 10^3$ m/s, $\nu = 0.3$, $k_p^2 = 0.34$, $d_{31} = -1.6 \times 10^{-10}$ C/N. It was assumed in calculations, that exterior medium and interior medium are the same: $\rho_1 = \rho_2 = 10^3$ kg/m^3 and $c_1 = c_2 = 1.5 \times 10^3$ m/s. Ratio a_0/r_0 was set equal 2/3.

Frequency dependency of acoustic power per unit length of the shell's surface (Fig. 2.20) was calculated to estimate the efficiency of the considered radiating system. The frequencies related to the resonance frequency of the pulse mode of the shell in vacuum f_0 are plotted on the abscissa.

Consider first the simple case (curve1), when the insertion is complete coaxial with the shell hard cylinder, $\theta_0 = 0$ and $l = 0$. As we might expect, there is only one resonance in the system, which corresponds to the pulse mode, and peak power which corresponds to this resonance. Its displacement to a higher position from the resonance of the shell in

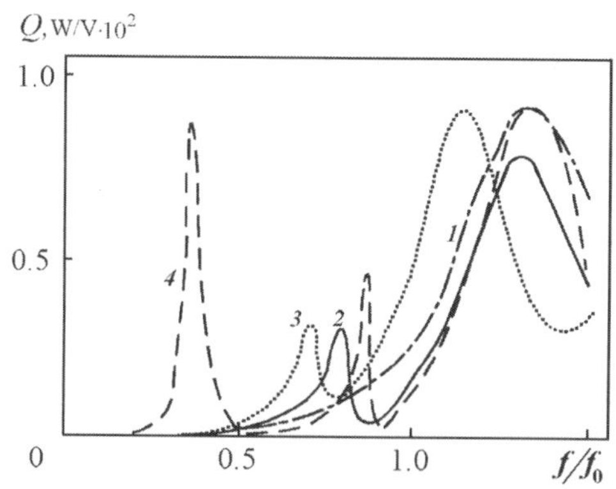

FIG. 2.20: Frequency dependence of radiated acoustic power (per unit length of the shell); 1—for cylindrical coaxial insertion $\theta_0 = 0$, $l = 0$; 2—for cylindrical insertion, shifted from the axis of the shell, $\theta_0 = 0$, $l/r_0 = 1/4$; 3—for sectoral insertion with acoustically hard edges $\theta_0 = \pi/2$, $l = 0$; 4—for sectoral insertion with acoustically soft edges $\theta_0 = \pi/2$, $l = 0$

vacuum is explained by influence of the impedance of inner layer of liquid. The last here has an elastic character at these frequencies (this effect was discussed in the previous paragraph).

For the systems with broken radial symmetry the situation changes significantly. When we shift the cylinder, another peak of radiating power appears in the vicinity of $f/f_0 \approx 0.78$ (curve 2). A similar effect is present in the case when there are sector insertions with acoustically hard and soft edges (curves 3 and 4 respectively). As the figure shows, new power peak is even more shifted into the low-frequency region in these cases. Sizable shift (to one and half octave in comparison with f_0) can be observed when there is a sector insertion with acoustically soft edges.

For better understanding of the behavior of the considered radiating system, we turn now to Fig. 2.21, where the frequency dependencies of the first three modes of radial velocity of the shell for the case of insertion in the form of the hard sector are shown (according the calculations, the level of the excitation of the higher vibrating modes is lower, and their amplitudes, in most cases, are small in comparison with the first three modes). In the radially symmetric system only one eigenmode is excited under the selected method of electrical loading. That is why only one resonance of the system and one corresponding to its peak of power exists. In the system with broken symmetry, on the contrary, the subsequent modes appear. The amplitude of these modes can be compared with the amplitude of pulse mode. It means that in this case energy contributed to the system is efficiently redistributed between the subsequent forms of vibrations. Particularly striking is the "oscillation build-up" of the shell on oscillatory mode in the region of low frequencies. The point is that, as it is known, two eigenfunctions exist for each shape of vibrations (starting with $n = 1$) for coupled longitudinal-and-flexural vibrations of cylindrical shell. These eigenfunctions correspond to fundamentally different stress and strain states of the shell and belong to two branches of the solution of the dispersion equation—the low frequency and high frequency branches. For the low frequency branch $w \gg u$, and circumferential stresses on the middle surface of the shell are insignificant (quasi-flexional vibrations). The high frequency branch corresponds to the conditions with

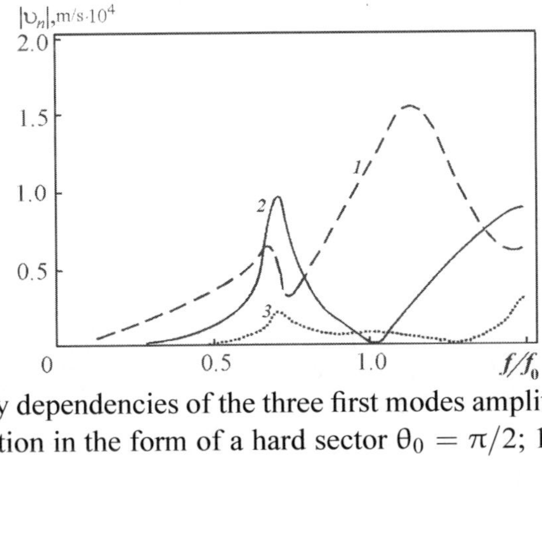

FIG. 2.21: Frequency dependencies of the three first modes amplitudes of the shell radial velocity for the insertion in the form of a hard sector $\theta_0 = \pi/2$; 1–3—correspond to the modes $n = 0, 1, 2$

significant circumferential stresses along the middle surface ($w \ll u$, quasi-longitudinal vibrations) [37]. The frequencies, which belong to the upper brunch, are well above the frequency of the pulsing resonance f_0. And the shapes of vibrations, corresponding to these frequencies, cannot have influence in the domain $f/f_0 < 1$. The several natural frequencies are in the range under consideration for the low frequency branch at the chosen value of h/r_0. A natural frequency of the shell's vibrations for $n = 1$ is formally equal to zero. From the physical standpoint it corresponds to the oscillations of the shell as a single whole, when it does not experience elastic deformations. In this case for all frequencies the mechanical impedance of the shell has the nature of mass. The interaction between the shell's mass, the added mass of outer liquid, and the liquid layer inside the shell, which has elastic impedance, gives the prerequisite for the appearance of relatively low-frequency resonance in the system, which is absent in the spectrum of the natural frequencies of the shell in a vacuum. It is clear that we can attain a greater decrease of the frequency of the system resonance indicated above, when increasing general elasticity of the internal layer of liquid. It can be attained for example by making the part of the insertion surface acoustically soft. This fact is illustrated by curve 4 in Fig. 2.20.

It should be noted for the insertion with combined acoustic characteristics of the surface that one more peak of radiation appears in the vicinity of $f/f_0 \approx 0.85$ (Fig. 2.20, curve 4). The appearance of this peak is determined by resonance phenomena in the liquid layer between the acoustically soft part of the insertion surface and the opposite part of the shell's surface, because at these frequencies the thickness of this layer is close to quarter-wavelength in liquid.

As it follows from the analysis of the frequency dependencies of the radial velocities modes of the shell for different sizes and properties of an insertion (Fig. 2.21), the peaks of power are defined by major influence of pulse mode of the shell vibrations in the frequency domain $1 < f/f_0 < 1.5$.

We turn now our attention to the consideration of the directional properties of the system in the vicinity of radiation peaks. Figure 2.22 shows directivity patterns of power for the shifted cylindrical insertion (curve 1), of acoustically hard semi-cylinder (curve 2), and semi-cylinder with combined properties of the surface. As it is seen from the graphs, the considered radiating system has strongly pronounced directivity. The main part of energy is radiated in the direction $\phi = 0$. It is interesting to note that the phases of pulsing and oscillating modes are close at these peaks. Curve 4 corresponds to the directivity pattern with semi-cylindrical insertion with combined properties of the surface. Here system has strongly pronounced directivity, but the main part of energy is radiated in the opposite direction ($\phi = \pi$). The reason of it is that pulsing and oscillating modes are excited in anti-phase in this case.

At the peaks of power, when $1 < f/f_0 < 1.5$, radiation does not have pronounced directivity, because the amplitudes of vibrating modes with $n \geq 1$ at these frequencies are small in comparison with the amplitude of pulse mode.

We can add to the foregoing that the numerical researches, conducted for a broad range of values of geometrical parameters of the insertion a_0, θ_0 and the shift l show, that the amplitude of low-frequency peak of radiating power increases with an increase of the insertion a_0 radius. For small θ_0 the behavior of the system in the domain of low

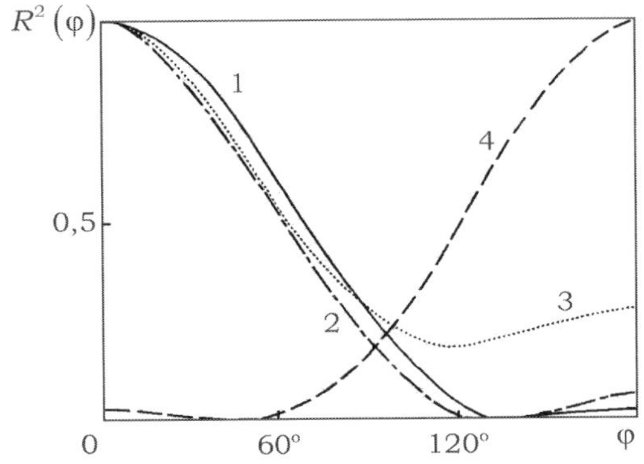

FIG. 2.22: Directivity patterns of power; 1—for the insertion with $\theta_0 = 0$, $l/r_0 = 1/4$ at frequency $f/f_0 \approx 0.78$; 2—for the insertion in the form of a sector with acoustically hard edges $\theta_0 = \pi/2$, $l = 0$ at frequency $f/f_0 \approx 0.69$; 3—for the insertion in the form of a sector with acoustically soft edges $\theta_0 = \pi/2$, $l = 0$ at frequency $f/f_0 \approx 0.37$; 4—for the insertion in the form of a sector with acoustically soft edges $\theta_0 = \pi/2$, $l = 0$ at frequency $f/f_0 \approx 0.85$

frequencies is close to the behavior of the system with a cylindrical insertion. If l/r_0 is close to zero, the amplitudes of low-frequency peaks are negligible and frequency dependence of radiating power is little different from the case of radially symmetrical system. The amplitudes of low-frequency peaks increase with an increase of l, due to the system asymmetry increase. The same tendency is observed when θ_0 increases to $3\pi/4$. With further increase of θ_0 the efficiency of the insertion influence decreases. For the insertion with hard surface, the case when $\theta_0 = \pi$ is the same as the case when there is no insertion.

Thus sector insertions with different acoustical properties of the surfaces inside the radiating shell have a great impact on acoustical parameters of the considered system. The first two or three vibrating modes of the shell play the major part in radiating (in the considered frequency range). New resonances, that appear at frequencies $f/f_0 < 1$, are of particular interest. These resonances are absent in eigenfrequency spectrum of the shell in vacuum. The comparatively high directivity of radiation at the frequencies of these resonances when the wave diameter of the shell is approximately 0.3–0.5 of wave length in liquid, is also of interest.

It should be noted in conclusion that it is known from the literature how to excite the shell on the first mode by separating the electrode coatings and applying on their parts electrical signals in antiphase [10]. In this case the circumferential mechanical stresses of opposite signs appear in the shell. These stresses lead to the shell's vibrations at upper eigenfrequency of the first vibration mode that is 1.41 times higher than the frequency of the pulsing vibrations resonance of the shell. In the considered case the excitation of the oscillatory mode was achieved without separation of electrode coatings but through trapping of energy contributing to the system through the zero mode by subsequent vibration forms. As this takes place new resonance emerge at the frequency that is lower than the pulsing resonance of the shell. This resonance emerge due to the interactions between the elements of the system. These elements have the impedances of different character in this frequency band.

2.2.4 Sound Radiation from a Piezoceramic Shell with an Internal Insertion in the Form of a Piezoceramic Shell

Passive methods of control of cylindrical radiators acoustic parameters were considered above. Consider now one of the active methods of these parameters control.

Consider two coaxial infinite piezoceramic shells, Fig. 2.23. The space between the shells and outer space are filed with liquid with wave impedance $\rho c = \rho_0 c_0$. Electrical voltages u_1 and u_2 are applied to the outer and inner piezoceramic shells. They differ in amplitude and in phase in general case. We suppose that the piezoceramic shells are radially polarized, the internal and external side surfaces are coved with electrodes, and the thickness of each shell δ is small in comparison with its radius ($\delta_1 \ll r_0$ and $\delta_2 \ll r_2$).

The potentials of sound field's velocities, created by the shells in outer space I, and in the space between the shells II, can be written in the form

$$\Phi_1 = AH_0^{(1)}(kr), \tag{2.74}$$

$$\Phi_2 = BJ_0(kr) + CN_0(kr), \tag{2.75}$$

where A, B, C—unknown complex coefficients.

Radial shifts of external and internal shells can be defined from the following differential equations:

$$\frac{\partial^2 \xi_1}{\partial t^2} + \omega_1^2 \xi_1 = \frac{d_{31}^{(1)}}{s_{11}^{E_1} \rho_1 r_1} E_1 + \frac{1}{\rho_1 \delta_1} \left[p_1(r)|_{r=r_0} - p_2(r)|_{r=r_1} \right], \tag{2.76}$$

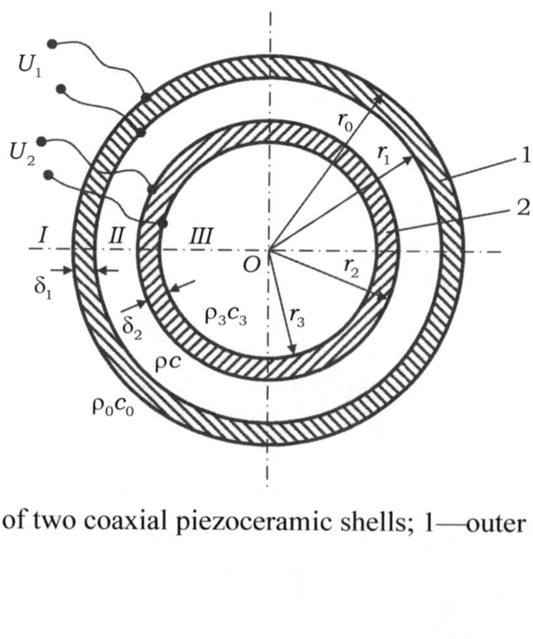

FIG. 2.23: System of two coaxial piezoceramic shells; 1—outer shell, 2—inner shell

$$\frac{\partial^2 \xi_2}{\partial t^2} + \omega_2^2 \xi_2 = \frac{d_{31}^{(2)}}{s_{11}^{E_2} \rho_2 r_2} E_2 + \frac{1}{\rho_2 \delta_2} \left[p_2 (r)|_{r=r_2} \right], \tag{2.77}$$

where ξ_1, ξ_2—the radial shifts of the shells; ω_1, ω_2—the eigenfrequencies of pulsing vibrations of the shells; $d_{31}^{(1)}$, $d_{31}^{(2)}$—the piezoelectric modulus of the shells materials; $s_{11}^{E_1}$, $s_{11}^{E_2}$—the elastic constants of the shells materials; ρ_1, ρ_2—the densities of the shells materials; $p_1 = -j\omega\rho_0\Phi_1$, $p_2 = -j\omega\rho\Phi_2$, E_1, E_2—the electric field intensity in the shells materials.

Note that indexes "1" and "2" are used in equations to denote the geometric and physical parameters of the external and internal shells respectively. Equation (2.77) was written with consideration of the assumption that there is no medium ($\rho_3 c_3 = 0$) in the inner space of the internal piezoceramic shell (domain III).

We transform differential Eqs. (2.76) and (2.77) into the following functional equations for the harmonic excitation of the piezoceramic shells $E_1 = \bar{E}_1 \exp(-j\omega t)$ and $E_1 = \bar{E}_1 \exp(-j\omega t - \phi)$:

$$\upsilon_1 z_1 = n_1 u_1 + p_1 (r)|_{r=r_0} - p_2 (r)|_{r=r_1}, \tag{2.78}$$

$$\upsilon_2 z_2 = n_2 u_2 + p_2 (r)|_{r=r_2}, \tag{2.79}$$

where υ_1, υ_2—the radial components of particle velocities of piezoceramic shells; u_1, u_2—exciting voltages; n_1, n_2—the coefficients of electromechanical transformation per unit surface of the piezoceramic shells: $n_1 = d_{31}^{(1)} 2\pi / s_{11}^{E_1} r_1$; $n_2 = d_{31}^{(2)} 2\pi / s_{11}^{E_2} r_2$; z_1, z_2—the mechanical impedances of the piezoceramic shells per unit surface: $z_1 = j(\omega m_1 - \bar{s}_1/\omega)$; $z_2 = j(\omega m_2 - \bar{s}_2/\omega)$; m_1 and m_2, $\bar{s}_1$ and $\bar{s}_2$—the mass and elasticity of piezoceramic shells respectively per unit surface: $m_1 = \rho_1\delta_1$; $\bar{s}_1 = \delta_1 / r_1^2 s_{11}^{E_1}$; $\bar{s}_2 = \delta_2 / r_2^2 s_{11}^{E_2}$.

We find the unknown complex coefficients A, B, C in expressions (2.74) and (2.75) by solving Eqs. (2.78) and (2.79) and evident equalities

$$-\left.\frac{\partial \Phi_1}{\partial r}\right|_{r=r_0} = \upsilon_1; \quad -\left.\frac{\partial \Phi_2}{\partial r}\right|_{r=r_1} = \upsilon_1; \quad -\left.\frac{\partial \Phi_1}{\partial r}\right|_{r=r_2} = \upsilon_2.$$

Omitting intermediate computations, we represent the expression defining outer sound field of coaxial piezoceramic shells as follows:

$$p_1 (r) = \frac{(u_1 n_1 + u_2 n_2 \Psi) H_0^{(1)} (kr)}{(z_H + z_1 + z_B) k H_0^{\prime(1)} (kr_0)}, \tag{2.80}$$

where

$$z_H = -j\rho c \frac{H_0^{(1)} (kr_0)}{H_0^{\prime(1)} (kr_0)}, \quad z_B = z_{B2} + z_2^*, \quad z_{B2} = j\rho c \frac{J_0 (kr_0)}{J_0' (kr_0)},$$

$$z_2^* = \left[J_0 (kr_2) - j \frac{z_2}{\rho c} J_0' (kr_2) \right] \Psi,$$

$$\Psi = 2j\rho c\left\{\pi k r_0 [J_0(kr_2) N_0'(kr_0) - J_0'(kr_0) N_0(kr_2)] - i\frac{z_2}{\rho c}[J_0'(kr_2)N'(kr_0) - J_0'(kr_0)N_0'(kr_2)]\right\}^{-1}.$$

It is apparent from expression (2.80) that the sum in the denominator in parentheses is the complex impedance of the system of two thin-wall coaxial shells in liquid. It consists of the radiation impedance z_H of the cylinder with radius r_0, the natural mechanical impedance z_1 of the outer shell, and the inner impedance z_B, which is the load of the outer shell from inside of the shell. Internal impedance z_B consists of the impedance of the layer z_{B2} between the shells and impedance z_2^*, produced by the internal shell with natural mechanical impedance z_2.

2.2.4.1 Acoustic Properties of Two Coaxial Piezoceramic Shells

Acoustic parameters of the system of two piezoceramic shells were calculated using expression (2.80). This system has the following characteristics: for the outer shell—outer radius 102 mm, inner 90 mm, the composition of piezoceramic TBC-3 [38]; for the inner shell—outer radius 74 mm, inner 65 mm, the composition of piezoceramic CTBS-3 [38]. Figure 2.24 illustrates the frequency dependencies of normalized acoustic pressure, produced by the system of shells in far field, for various values of amplitudes and phases of electric voltage, applied to the electrodes of the shells.

As we can see, the frequency dependencies of acoustic pressure have resonant behavior. When the voltage $u_2 = 0$, there are two approximately equal maximums, the value of which depends on the amplitudes of voltage ratio u_2/u_1, and the difference of their phases ϕ. In absence of phase shift (Fig. 2.25), the first pressure maximum increases

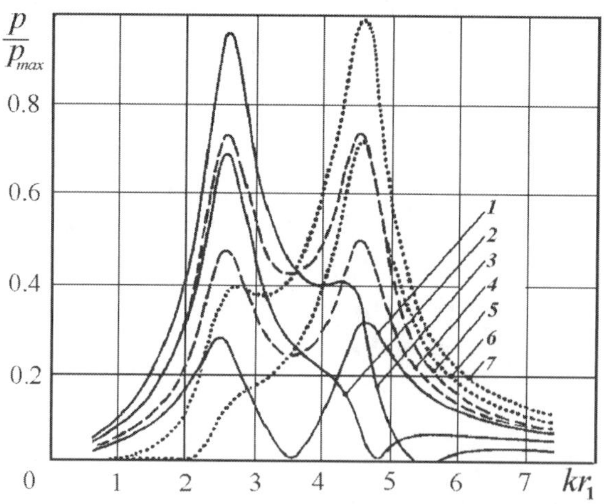

FIG. 2.24: Frequency dependencies of normalized acoustic pressure, produced by the system of the shells in far field: 1—$u_2/u_1 = 0$, $\phi = 0$; 2—$u_2/u_1 = 0.6$, $\phi = 0$; 3—$u_2/u_1 = 1.0$, $\phi = 0$; 4—$u_2/u_1 = 0.6$, $\phi = \pi/2$; 5—$u_2/u_1 = 1.0$, $\phi = \pi/2$; 6—$u_2/u_1 = 0.6$, $\phi = \pi$; 7—$u_2/u_1 = 1.0$, $\phi = \pi$

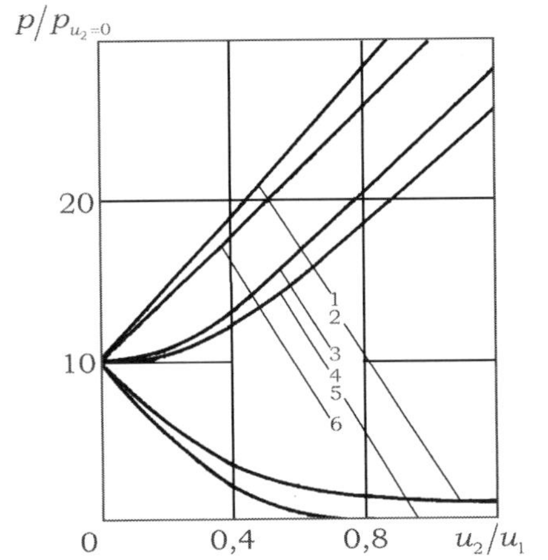

FIG. 2.25: The maximum values of acoustic pressure in exterior medium versus the ratio of electrical voltages, applied to the shells: 1, 3, 5—the first maximums respectively, when $\phi = 0; \pi/2; \pi$; 2, 4, 6—the second maximums respectively, when $\phi = 0; \pi/2; \pi$

proportionally with an increase of the amplitudes of voltage ratio u_2/u_1 (curve 1), and the second decreases, tending to zero (curve 2). General form of frequency dependence also changes (Fig. 2.24, curves 1–3).

The phase shift also greatly influences the behavior of maximums. When $\phi = \pi/2$, both maximums increase almost uniformly with an increase of voltages ratio u_2/u_1 (Fig. 2.25, curves 3 and 4). But frequency dependence almost does not change (Fig. 2.24, curves 4 and 5).

For the phase shift $\phi = \pi$ the frequency dependence (Fig. 2.24, curves 6 and 7) and the values of pressure maximums of voltages ratio u_2/u_1 (Fig. 2.25, curves 5 and 6) change in another way than in the case when $\phi = 0$. The first maximum decreases, tending to zero (Fig. 2.25, curve 5), and the second maximum increases (Fig. 2.25, curve 6).

There is a shift of the resonances of the system of piezoceramic shells on frequency scale (Fig. 2.26), when the value of annular space changes $r_1 - r_2$, and $\omega_1 = \omega_2$. The change in frequencies ratio ω_2/ω_1 when the thickness $r_1 - r_2$ remains unchanged (Fig. 2.27, curve 1) leads to the shift of the resonance frequency of the system of piezoceramic shells. The shift of resonances is followed by the change in maximum pressure

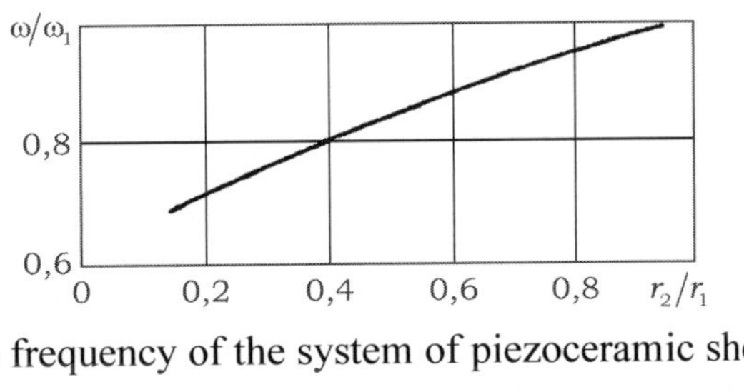

FIG. 2.26: Resonance frequency of the system of piezoceramic shells versus the ratio of their radii

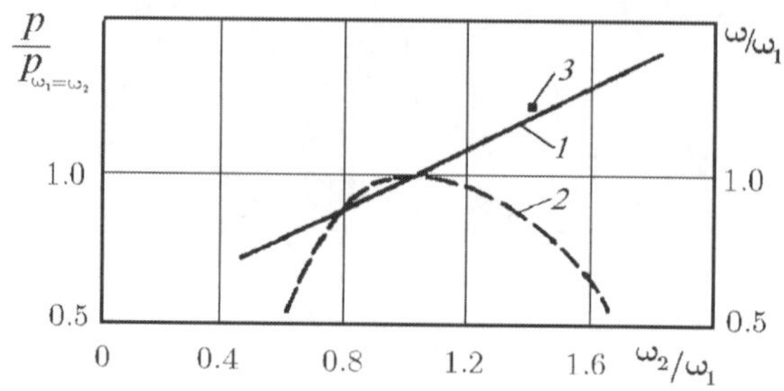

FIG. 2.27: Resonance frequency (curve 1) and acoustic pressure (curve 2) of the system of piezoceramic shells versus the eigenfrequencies of the shells ratio; 3—(point)—eigenfrequency experimental value

amplitude (Fig. 2.27, curve 2). The maximum value of this pressure is observed when $\omega_1 = \omega_2$.

This analysis shows that the vibrations of the coaxial piezoceramic shells in liquid depend on the ratio of the voltages applied to the shells, the difference of phases between them, the value of annular space, and the resonance frequencies of the shells. We can solve practical problems, choosing these parameters adequately.

The efficiency of this active method of control of radiator's acoustic parameters is demonstrated experimentally (Fig. 2.28) with the help of the laboratory model of a radiator. This model consists of two coaxial piezoceramic shells, composed of the piezoceramic rings with diameter 74 mm and 52 mm and composition CTBS-3.The voltages ratio is $u_2/u_1 = 1$, and the difference of phases $\phi = 0$.

As we can see, it is possible to increase the radiator efficiency without an increase of the dimensions and applied electric voltage with the help of the system of two piezoceramic shells. Thc value of the resonance frequency of the system of shells in water lies between the values of the resonance frequencies of the shells composing the system.

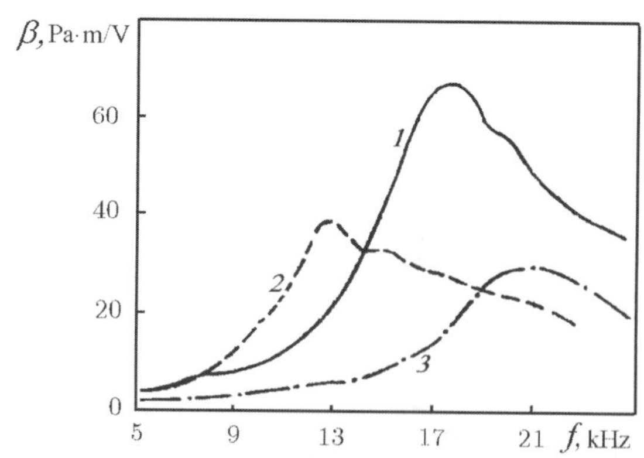

FIG. 2.28: Frequency dependence of the laboratory model sensitivity in sound radiation mode: 1—system of shells; 2—outer shell separately; 3—inner shell separately; electrical voltage applied to the shells was the same in all three experiments

Thus, the active method that was considered allows us to control the parameters of the system of coaxial piezoceramic shells in water. This method can be used in production of the multifunctional hydroacoustic cylindrical radiators.

2.3 SOUND RADIATION FROM A FINITE-HEIGHT CYLINDER

A number of the infinite-length cylindrical radiators were considered above. Such calculation model can be used to describe the acoustic properties of the real finite-height cylindrical radiators, when their height h is much more than the diameter $2r_0$ and the length of the radiated sound waves λ. Such model cannot be used in the cases, when the height of cylindrical radiators is finite and comparable with the diameter and length of the radiated sound waves, and particularly when it is less than these parameters, because this model does not include diffraction effects at the cylinder ends. These effects have a significant impact on radiators acoustic characteristics. There are some difficulties in obtaining strict and effective solution of the problem of sound radiation from the finite-height cylinder (from a viewpoint of sound field quantitative properties characterization). In this connection, the approximate solutions of such problem are of great practical importance. One of the approaches is to replace a cylinder of finite length with an infinite cylinder [11,49]. A part of the cylinder with the height h radiates sound waves, and the rest of the cylinder surface is acoustically hard. Such model allows us to obtain simple analytical expressions for defining the radiation field of real cylindrical radiators, when $h \gg \lambda$. For the cases when $h \leq \lambda$, this approach leads to errors in estimation of the quantitative characteristics of the field, because the diffraction of the waves at the ends of the cylinder cannot be taken into account. Below we will consider the method based on the substitution of cylindrical radiator with spherical radiator, when the conditions $h < 2r_0$ and $h < \lambda$ are fulfilled.

The supposed computational scheme is based on the approximation of cylindrical radiator of finite length [Fig. 2.29(a)] by equatorial ball layer. The cross section of the layer is illustrated in Fig. 2.29(b), (the contour of the cross-section of the cylinder is marked with dashed line). We will see below that the substitution of the finite-length

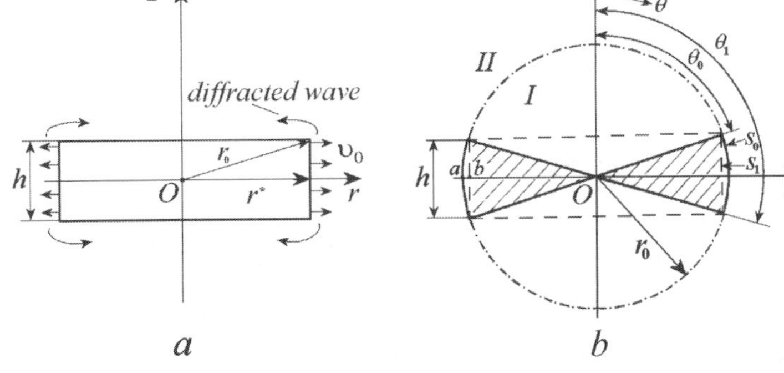

FIG. 2.29: Cross-section of a cylindrical radiator of finite length (a) its ball layer approximation (b)

cylinder with the ball layer allows us to obtain the good approximation of quantitative estimations of acoustic characteristics of real finite-height cylindrical radiators. Suppose that the distribution of the normal component of particle velocity is given on the spherical surface of the ball layer, and this distribution depends only on θ coordinate:

$$-k\frac{\partial \Phi}{\partial (kr)} = \upsilon_0 f(\theta), \quad r = r_0, \quad \theta_0 \le \theta \le \theta_1, \tag{2.81}$$

and the following conditions are fulfilled at the end surfaces of a ball layer:

$$\partial\Phi/\partial\theta = 0: \quad 0 \le r \le r_0, \quad \theta = \theta_0, \quad \theta = \theta_1. \tag{2.82}$$

Suppose that $\theta_1 = \pi - \theta_0$, and function $f(\theta)$ is symmetric with respect to the plane $z = 0$. We represent the sought velocity potential $\Phi(r, \theta) = \Phi(r, \pi - \theta)$ in terms of the velocity potentials in partial domains I and II:

$$\Phi_1 = \sum_{\nu=1}^{\infty} A_\nu i_{m_\nu}(kr) P_{m_\nu}(\cos\theta), \quad 0 \le \theta \le \theta_0, \quad 0 \le r \le r_0, \tag{2.83}$$

$$\Phi_2 = \sum_{n=0}^{\infty} B_n h_n^{(1)}(kr) P_n(\cos\theta), \quad 0 \le \theta \le \pi/2, \quad r \ge r_1, \tag{2.84}$$

where A_ν, B_ν—complex coefficients, m_ν—roots of an equation $\partial P_{m_\nu}(\cos\theta)/\partial\theta = 0$ for $\theta = \theta_0$, $n = 0, 2, 4, \ldots,$ $i_{m_\nu}(kr)$ and $h_n^{(1)}(kr)$—spherical Bessel and Hunkel functions, $P_{m_\nu}(\cos\theta)$ and $P_n(\cos\theta)$—Legendre functions.

In accordance with partial domains method, the unknowns A_ν, B_ν must be defined from the system of functional equations. The system combines the continuity conditions at the interface of boundary domains and boundary condition (2.81):

$$\Phi_2 = \Phi_1, \quad r = r_0, \quad 0 \le \theta \le \theta_0,$$

$$\frac{\partial \Phi_2}{\partial (kr)} = \begin{cases} \dfrac{\partial \Phi_1}{\partial (kr)}, & r = r_0, \quad 0 \le \theta \le \theta_0, \\ -\dfrac{\upsilon_0 f(\theta)}{k}, & r = r_0, \quad \theta_0 \le \theta \le \pi/2. \end{cases} \tag{2.85a}$$

The conversion of this system to an algebraic set of equations is based on the completeness and orthogonality of the functions $P_{m_\nu}(\cos\theta)$ on the interval $0 \le \theta \le \theta_0$ and $P_n(\cos\theta)$ on the interval $0 \le \theta \le \pi/2$. As a result we obtain the infinite system of linear algebraic equations of second kind:

$$\begin{gathered} A_\nu i_{m_\nu}(kr_0) - \frac{1}{N_\nu}\sum_{\nu=1}^{\infty} B_n h_n^{(1)}(kr_0) N_{\nu n} = 0, \\ B_n h_n'^{(1)}(kr_0) - (2n+1)\sum_{\nu=1}^{\infty} A_\nu i'_{m_\nu}(kr_0) N_{\nu m} = -\frac{\upsilon_0}{k}(2n+1)F_n, \end{gathered} \tag{2.85b}$$

where

$$N_\nu = \int\limits_{\cos\theta_0}^{1} [P_{m_\nu}(\cos\theta)]^2\, d(\cos\theta), \quad F_n = \int\limits_{0}^{\cos\theta_0} f(\theta) P_n(\cos\theta) d(\cos\theta),$$

$$N_{\nu n} = \int\limits_{\cos\theta_0}^{1} P_{m_\nu}(\cos\theta) P_n(\cos\theta) d(\cos\theta).$$

We proceed to the analysis of the ball layer acoustic properties. Analysis is based on the numerical calculations, obtained for a wide range of values kr_0. Computational and experimental data correspond to $\theta_0 = 75°$, $h/2r_0 = 0.2588$, $f(\theta) = \text{const}$.

Let us consider the directional properties of the ball layer. Figure 2.30(a) shows directivity patterns $R(\theta) = \Phi_2(\theta)/\Phi_2(\theta = \pi/2)$. Figure 2.30(b) shows directivity patterns of cylindrical radiator from the works [11,49]. These data comparison shows that the diffraction of sound waves by the end of the radiator has a great impact on the directional properties. Figure 2.31 illustrates dependency of the value of sound pressure relative level in the direction of axis of symmetry $\sigma = \Phi_2(\theta = 0)/\Phi_2(\theta = \pi/2)$ on kr_0. As we can see, frequency dependence of the value of σ, when the ends of radiator are free, has strongly pronounced oscillating character. The oscillation amplitude is maximum in the domain $1.5 \leq kr_0 \leq 8$. With further increase of kr_0 the oscillations disappear and the computed data begin to coincide with the data computed using methods [11,49].

The experimental investigation of the directivity of the laboratory model of cylindrical radiator with the flat end surfaces was conducted. The aim of this investigation was to check the effectiveness of the solution method of the considered problem and to prove that the finite-height cylinder approximation by the ball layer is possible. Figure 2.30(a) shows experimental directivity patterns (curves 6, 7, 8) and Fig. 2.31 shows experimental dependency of the value of σ (curve 3) on kr_0. Comparing the experimental and calculation data, we can see that they are in good agreement with each other.

We turn again to Fig. 2.31. The type of the dependence of the value of σ can be explained as follows. The edge wave, which appears at the edge line (Fig. 2.29), propagates

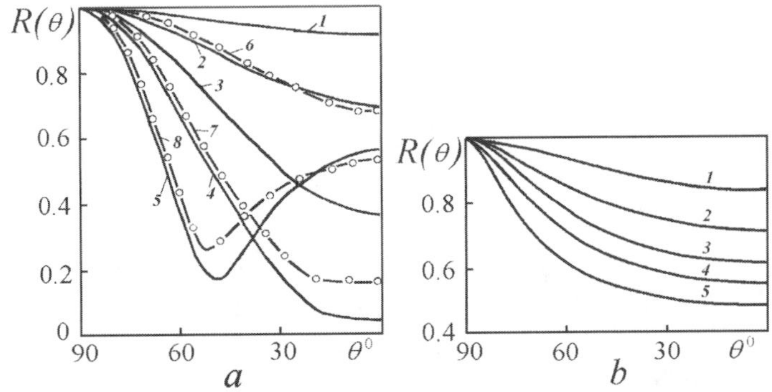

FIG. 2.30: Computational directivity patterns; (a)—ball layer, (b)—according [11,49]; curves 1, 2, 3, 4, 5—correspond to the values of kr_0 equal 0.5, 1, 1.5, 2, 3; curves 6, 7, 8—experimental directivity patterns, and corresponding to them values of kr_0 equal 1, 2, 3

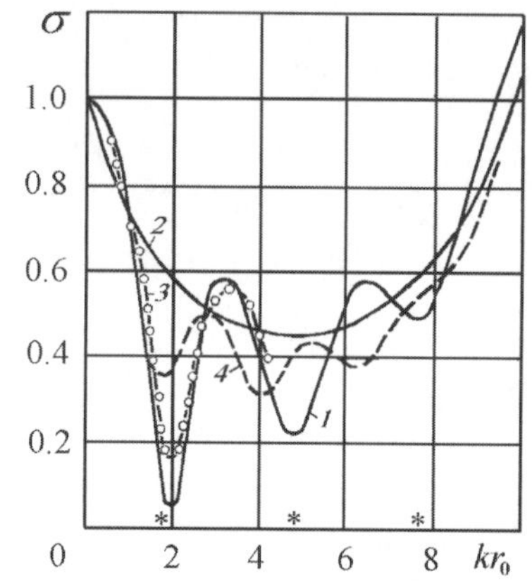

FIG. 2.31: Relative pressure level in the z direction; 1—acoustically hard ends (our calculation), 2—acoustically hard ends (calculation according [11,49]), 3—acoustically hard ends (experiment), 4—acoustically soft ends (our calculation)

along the end surface of the radiator, reaches diametrically opposite part of radiating surface. The edge wave is added to the direct wave, radiated by this part of the surface, or is deducted, depending on phase, which is defined by ratio $2r_0/\lambda$. It is obvious that the worst cases (pressure level decreases in $\theta = \pi/2$ direction) appear, when the radiator's diameter is close to $(2q+1)\lambda/2$, where $q = 0, 1, 2, \ldots$. This idea is easy to verify. We mark the values of kr_0 on the X axis. These values correspond to the values of $2r_0/\lambda$, equal 0.5, 1.5, and 2.5 (in Fig. 2.31 they are marked with asterisks). We can see that the minimums on curve 1 are in the vicinity of these points. It should be added that the amplitude of the curve 1 oscillation decreases with an increase of kr_0, and when $kr_0 > 8$, curves 1 and 2 begin to match each other. There are two reasons for it. With an increase of kr_0, the relative contribution of the edge wave decreases when it interferences with a direct wave. The wave height of the radiator increases and, as a result of it, the radiator directivity increases in $\theta = \pi/2$ direction. The tendency of curves 1 and 2 to match when $kr_0 > 8$ $(h/\lambda > 0.65)$ justifies assumptions made in [11]. When $h/\lambda > 1$ the sound field of the finite-height cylinder can be described adequately with a help of mathematical model used in this work.

We turn now to the dependencies, which characterize the impedance of the considered radiator (Fig. 2.32). The presented dependencies show that the diffraction at the ends of radiator has great impact on its impedance. The greatest impact is in the range of values of $kr_0 \approx 1 - 6$. This influence can be ignored when the values of kr_0 are large. Comparing curve 1 in Fig. 2.32 and curve 1 in Fig. 2.31, we can see that the active component of the ball layer impedance decreases, when the values of kr_0 are the same as the values at which the field decreases in $\theta = \pi/2$ direction. It is most pronounced when kr_0, is equal 2 and 4.7.

Consider the case when the ends of the cylinder are acoustically soft. It is enough to substitute the boundary condition at the end surfaces of the ball layer (2.82) with the condition

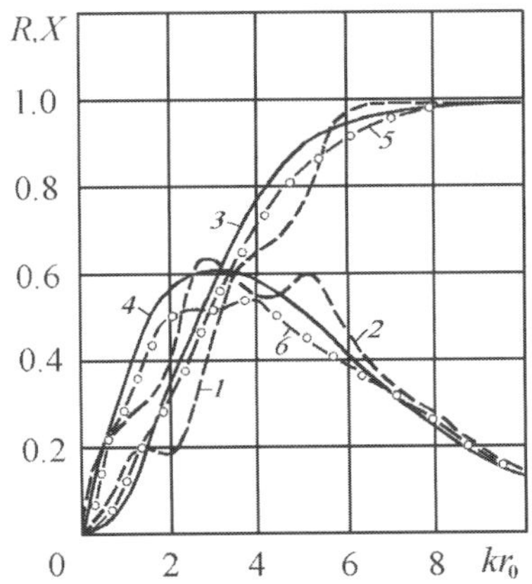

FIG. 2.32: Radiation impedance; 1, 3, 5—real part, 2, 4, 6—imaginary part; 1, 2—ball layer with acoustically hard ends, 3, 4—according to the method [11,49], 5, 6—ball layer with acoustically soft ends

$$\Phi = 0, \quad 0 \leq r \leq r_0, \quad \theta = \theta_0, \quad \theta = \theta_1, \tag{2.85c}$$

and to define the value of m_ν in expression (2.83) as the roots of transcendent equation $P_{m_\nu}(\cos\theta) = 0$ for $\theta = \theta_0$. The calculations were made including these conditions. The results of the calculations are given in Figs. 2.31 and 2.32 (curves 4 and 5, 6 respectively). If we compare the quantitative data for the cases of the acoustically hard and acoustically soft ends of cylindrical radiator, we can make a conclusion: acoustically soft ends cause the change in the acoustic properties of a radiator. Firstly, sound radiation level decreases along the symmetry axis of radiator. Secondly, its active part of radiation impedance increases, especially in the range of values of $kr_0 \approx 2.2$ and $kr_0 \approx 4.7$.

As stated above, the computational scheme proposed in this work can be used to estimate the field of radiation of a finite-height cylinder, if the following inequalities are valid $h < \lambda$ and $h < 2r_0$. But it is possible to give more accurate estimate. For this purpose, we specify a value, characterizing the deviation of spherical radiating surface of the ball layer from the cylindrical surface of the finite-height cylinder. In this case, it makes the most sense to choose the value of ab (Fig. 2.29). Suppose $ab \leq \lambda/16$. Then limit wave dimension of the height of cylindrical radiator, which can be approximated by the ball layer, is defined by the condition

$$h/\lambda \leq \cos\theta_0/(1 - \sin\theta_0). \tag{2.86}$$

In conclusion we note, that the research of a cylindrical radiator of finite height was conducted in work [50] in 1975, using presented here method. Works [51,52] appeared later, in 1976 and 1980. The acoustical properties of finite cylinder were studied in these works, but using the other methods. It should be noted that according to the data in work [51], there is a good agreement between the results obtained by different methods.

2.4 SOUND RADIATION BY A GRATING OF FINITE HEIGHT CONSISTING OF PIEZOCERAMIC SHELLS

In traditional formulation of problems involving acoustic radiation by structures made up by vibrating bodies, the usual starting point is the assumption that the normal component of the particle velocity is specified at the surface of the bodies [2–4,11,18,30]. Once this problem is solved, it is possible to calculate the various local and global field characteristics of interest and practical importance, e.g., space distribution of acoustic pressure, radiated power, radiation impedance, directivity diagram etc.

However, this formulation is seriously deficient in one respect: by prespecifying the values of the particle velocity, we exclude in principle the possibility of taking into account changes in these particle velocities due to the reaction of the surrounding medium and acoustic interaction between bodies.

This simplified approach to solving problems that involve acoustic radiation can in fact be justified to some extent if the bodies are situated at considerable distances from one another, and their intrinsic impedances greatly exceed the radiation impedance. However, in cases where the spacing between the bodies is not large, or when the intrinsic mechanical impedances are comparable to the radiation impedance, the assumption of a prespecified velocity leads unavoidably to considerable errors in estimating the local and global field characteristics. The real vibratory velocities of bodies are determined by the level of energy supplied to each body, the value of the frequency, the mutual positions of the bodies in space, their shapes, etc., and in general may differ greatly from one another. In particular, these statements apply to the case of acoustic radiation by several closely spaced electroacoustic transducers when their intrinsic frequencies fall within the range of radiated frequencies (e.g., acoustic radiation by resonant transducers used as components of an antenna array).

Linear antenna array consisting of a coaxial structure made up of several elastic piezoceramic shells will be considered below. We develop a method for determining the particle velocity of the shells and their characteristic radiation field, taking into account the reaction of the medium, the physical and geometric parameters of the shells, and acoustic interactions between them, as well as the amplitudes and phases of the electrical voltages applied to the shells electrodes. By solving the "wall-plug" problem, i.e., when the prespecified quantities are not the particle velocities but rather the input electrical voltages, we can make an effective choice of the set of signal generators needed to create a required distribution of particle velocities at the shell surfaces.

Setting up and solving the proposed problem, we will show how the method of partial domains can be used, when the boundaries between the partial domains coincide with the parts of the coordinate surfaces of different coordinate systems [53].

Let us consider the axisymmetric problem of acoustic radiation by a set of several (five in this case) thin elastic coaxial cylindrical piezoceramic shells, which can execute radial oscillations only (Fig. 2.33). We shall assume that the ends of the structure are closed by acoustically rigid dampers, that there is a vacuum in the interior volume, and that there are no gaps between the shells. We apply an electrical voltage, which in general case has its own amplitude and phase, to the electrodes of each shell. For concreteness we

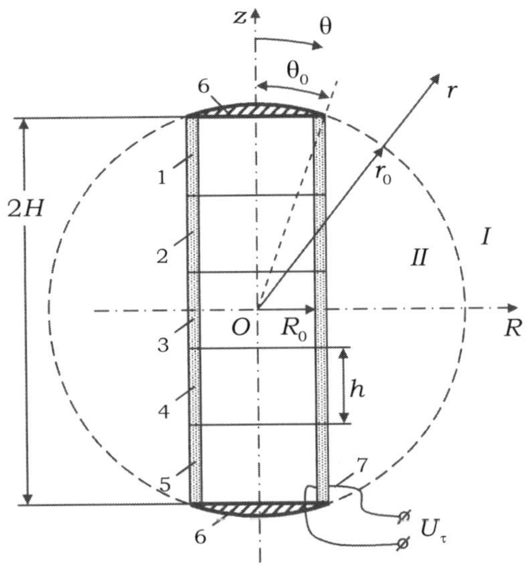

FIG. 2.33: Linear grating of coaxial piezoceramic shells; 1–5—shells, 6—dampers, 7—feed circuit

limit ourselves to the case of radially polarized shells, which implies that the electrodes are located on the cylindrical surfaces of the shells. The entire structure is immersed in an ideal compressible medium characterized by the wave impedance ρc.

The solution to this problem will be based on use of the modified method of partial domains [53]. In our formulation we will include the differential equations that describe induced vibrations of the piezoceramic shells under the action of electrical voltage applied to them. Incorporating the usual method of solution, we introduce a spherical (r, θ) and a cylindrical (R, z) systems of coordinates with a common origin at the point O, and the entire region where the field exists is divided up into the following two subregions (Fig. 2.33): region I $(r \geq r_0, 0 \leq \theta \leq \pi)$; region II $(r \leq r_0, \theta_0 \leq \theta \leq \pi - \theta_0, R \geq R_0, |z| \leq H)$. Let us express the velocity potential in these regions using the following sets of special solutions to the Helmholtz equation [53]:

$$\Phi_1 = \sum_{n=0}^{\infty} A_n h_n^{(1)}(kr) P_n(\cos\theta), \tag{2.87}$$

$$\Phi_2 = \sum_{m=0}^{\infty} B_m M_0^{(1)}(k_m R) \cos\frac{m\pi}{2H}(z+H) + \sum_{\nu=0}^{\infty} C_\nu i_{q_\nu}(kr) \Delta_{q_\nu}(\cos\theta), \tag{2.88}$$

where

$$k_m = \left[k^2 - (m\pi/2H)^2\right]^{1/2}, \quad M_0^{(1)}(k_m R) = \begin{cases} H_0^{(1)}(k_m R), & k \geq m\pi/2H, \\ K_0(k_m R), & k < m\pi/2H, \end{cases}$$

$$k = \omega/c, \quad \Delta_{q_\nu}(\cos\theta) = P_{q_\nu}(\cos\theta) + D_\nu Q_{q_\nu}(\cos\theta),$$

q_ν and D_ν, are determined from the system

$$\frac{\partial}{\partial \theta}\Delta_{q_\nu}(\cos\theta) = 0, \quad \theta = \theta_0, \quad \theta = \pi - \theta_0.$$

Basic notation, the same as in [53,54] is used in expressions (2.87) and (2.88).

We represent the field Φ_2 as a sum of the fields in cylindrical and spherical coordinate systems to meet the boundary conditions on the surface of piezoceramic shells and the matching conditions at the interface of the partial domains. It should be appreciated that in this case spherical coordinates should be transformed into cylindrical coordinates and vice versa. When it is necessary to meet the boundary conditions on the cylindrical surface of piezoceramic shells, the spherical coordinates r, θ in the second sum of expression (2.88) should be transformed into the cylindrical coordinates R, z. When it is necessary to meet the matching conditions on the spherical boundary of partial domains, the cylindrical coordinates in the first sum of expression (2.88) should be transformed into spherical coordinates.

In order to describe the radial oscillations of the shells we use the following differential equation [55,56]:

$$\rho_1\delta\frac{\partial^2\xi}{\partial t^2} + \frac{\delta}{R_0^2 S^E}\xi + \mu\frac{\partial\xi}{\partial t} = \frac{d_{31}E\delta}{R_0 S^E} + \Delta p, \tag{2.89}$$

where t and ξ—time and displacement, d_{31}, S^E, ρ_1, μ—the piezomodulus, the rigidity, the density of the shell material and the loss resistance, R_0, δ—the mid-radius and thickness of the shell, Δp—the pressure difference between the inner and outer surfaces of the shell, E—electric field strength in piezomaterial of the shells.

This way of formulating the equations of motion for the shell reflects the customary assumptions about the character of its deformation. Let us assume that within each shell the displacement of its mean surface is unchanged along the z coordinate. In the general case this change would have to be taken into account. However, the corresponding vibrational modes have a nearly zero efficiency coefficient for electromechanical coupling and their inclusion does not affect the conclusions we will arrive at below. The use of more general equations of motion for the shell and the inclusion of arbitrary boundary conditions at the ends can be implemented within the framework of the computational method we shell describe.

By taking the electric field intensity in the form $E_\tau = E_{0\tau}e^{-i\omega t}$ and including the fact that there is no field within the shell, we can easily transform Eq. (2.89) to the form

$$F_\tau + \upsilon_\tau Z_\tau = N_\tau U_\tau, \tag{2.90}$$

where τ—labels the shell (it varies from 1 to T), υ_τ and F_τ—particle velocity the force of reaction of medium on the external surface of the τth shell

$$U_\tau = E_\tau\delta, \quad Z_\tau = -i\omega M_\tau\left(1 - \omega_{0\tau}^2/\omega^2 + i\omega_{0\tau}/\omega Q_\tau\right), \quad N_\tau = d_{31}2\pi h_\tau/S^E,$$

where $M_\tau, N_\tau, h_\tau, Z_\tau, Q_\tau$—the mass of τth shell, its electromechanical conversion coefficient, its height, its intrinsic mechanical impedance, its mechanical Q-factor, $\omega_{0\tau}^2 = (\rho_1 R_0^2 S^E)^{-1}$—the eigenfrequency of the shell in a vacuum.

Here it is appropriate to pause for the following considerations. In accordance with the theory of electroacoustic transducers [10,29], Eq. (2.90) can be considered an equivalent electromechanical system of the sort shown in Fig. 2.34(a). Since the voltage source with emf ε_τ is ideal (i.e., it has no internal impedance), we have $U_\tau = \varepsilon_\tau$. However, for a practical source of electricity there always is a certain internal impedance $Z_{j\tau}$. Furthermore, there exists an electrical capacitance $C_{0\tau}$ between the electrodes of the shell, as well as dielectric loses, which will be neglected in view of their smallness. If we want to include these elements, we should replace the circuit in Fig. 2.34(a) by the circuit shown in Fig. 2.34(b). In order to simplify the subsequent contributions and calculations, it is expedient to replace the left-hand (from points 1 and 2) side of the circuit in Fig. 2.34(b), in accordance with the theorem of equivalent generator [10], by the circuit shown in Fig. 2.34(c). Making use of what have been said, it is not difficult to transform Eq. (2.90) into Eq. (2.91), which includes the internal impedance of the source and the capacitance between the electrodes of the shells:

$$F_\tau + \upsilon_\tau \left(Z_\tau + Z^*_{j\tau} N^2_\tau\right) = \varepsilon^*_\tau N_\tau, \tag{2.91}$$

where $Z^*_{j\tau} = Z_{j\tau} X_{0\tau} / \left(Z_{j\tau} + X_{0\tau}\right)$, $\varepsilon^*_\tau = \varepsilon_\tau X_{0\tau} / \left(Z_{j\tau} + X_{0\tau}\right)$, $X_{0\tau} = i/\omega C_{0\tau}$,

$$C_{0\tau} \approx \varepsilon^T_{33}(1 - k^2_{31}) 2\pi R_0 h_\tau / \delta,$$

ε^T_{33} and k_{31}—the dielectric permittivity of the shell material and the energy electromechanical coupling coefficient. When $Z_{j\tau} = 0$ Eq. (2.91) reduces to Eq. (2.90).

We now can formulate a system of functional equations, derived from the conditions of continuity of the field at the boundaries of the subregions and the boundary conditions

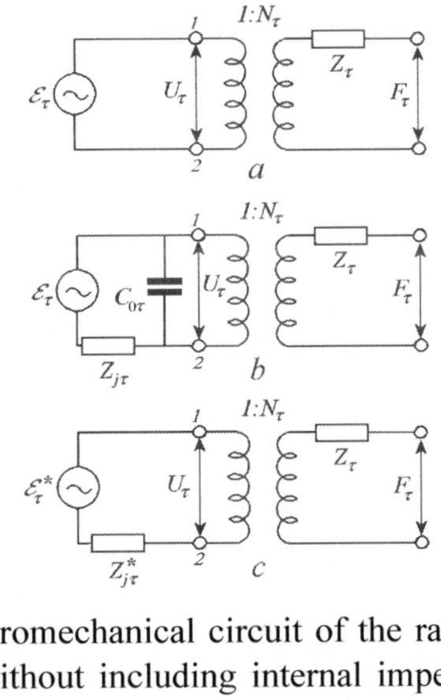

FIG. 2.34: Equivalent electromechanical circuit of the radially polarized shell with an electrical generator; (a)—without including internal impedance $Z_{j\tau}$ of the source and electrical capacitance $C_{0\tau}$; (b)—including the quantities $Z_{j\tau}$ and $C_{0\tau}$; (c)—with an equivalent electrical generator

at the outer surfaces of the shells. Taking into account the assumptions we have stated above and the properties of the dampers at the ends, we have

$$
-\frac{\partial \Phi_1}{\partial r} = \begin{cases} 0, & 0 \le \theta \le \theta_0, \quad r = r_0, \\ -\partial \Phi_2/\partial r, & \theta_0 \le \theta \le \pi - \theta_0, \quad r = r_0, \\ 0, & \pi - \theta_0 \le \theta \le \pi, \quad r = r_0, \end{cases}
$$

$$
\Phi_1 = \Phi_2, \quad \theta_0 \le \theta \le \pi - \theta_0, \quad r = r_0,
$$

$$
-\frac{\partial \Phi_2}{\partial R} = \begin{cases} \upsilon_1, & -H \le z \le -H + h_1, \quad R = R_0, \\ \upsilon_2, & -H + h_1 \le z \le -H + h_1 + h_2, \quad R = R_0, \\ \dots\dots\dots & \\ \upsilon_T, & H - h_T \le z \le H, \end{cases}
$$

$$
2\pi R_0 \int_{-H}^{-H+h_1} p_2 dz + \upsilon_1 (Z_1 + Z_{j1}^* N_1^2) = \varepsilon_1^* N_1, \tag{2.92}
$$

$$
2\pi R_0 \int_{-H+h_1}^{-H+h_1+h_2} p_2 dz + \upsilon_2 (Z_2 + Z_{j2}^* N_2^2) = \varepsilon_2^* N_2,
$$

$$
2\pi R_0 \int_{H-h_T}^{H} p_2 dz + \upsilon_T (Z_T + Z_{jT}^* N_T^2) = \varepsilon_T^* N_T,
$$

here $p_2 = -i\omega\rho\Phi_2$.

The conversion of system (2.92) to an algebraic set of equations, which is carried out by well-known methods, leads to an infinite system of linear algebraic equations with respect to the unknowns $A_n, B_m, C_\nu, \upsilon_\tau$. In view of the awkwardness of this system, and the fact that the systems very similar in structure were described above, we deem it advisable to omit it here.

2.4.1 Shells Excitation by Harmonic Electrical Signal

Let us analyze the shells vibrations and the field radiated by the shells. The frequency, the electrical sources parameters, and the shells parameters vary over a wide range. In all these calculations we assumed that there are five identical shells ($T = 5$). We also assumed that $H/R_0 = 1.67 N_\tau = 8$ N/V, $Q = 15$, $M_\tau = 1.5$ kg, $\omega_{0\tau} = \pi \times 1.4 \times 10^4$ s^{-1}, $C_{0\tau} = 0.085$ μF, $\rho c = 1.5 \times 10^6$ kg/m^{2}s. The number of complex unknowns in the algebraic system we actually solved was 44. When we checked the extent to which the boundary conditions and the conjugate conditions at the boundaries of the subregions were satisfied, we found that when this number of unknowns is retained the discrepancy in the field components (excluding points that were immediately next to the edges $r = r_0$, $\theta = \theta_0$, $\theta = \pi - \theta_0$) did not exceed 1–5%. The discrepancy in energy (we compared the radiated power at the surface of the shells and in the far-field zone) did not exceed 2%.

We turn now to the analysis of the data obtained from these calculations. Let us first consider the simplest case, for which the same electrical voltage, for example, $\varepsilon_\tau = U_\tau = 1B$ is applied to each shell from an ideal source $Z_{j\tau} = 0$. This situation is close to the encountered in practice, where the inequality $Z_{j\tau} \ll Z_\tau$ is usually valid. We turn now to Figs. 2.35 and 2.36, which show the frequency dependencies of the modulus and phase of the normalized particle velocities of the shells, their radiated powers and electroacoustic efficiencies. We note that the symmetry of the problem with respect to the plane $z = 0$ implies that $\upsilon_1 = \upsilon_5$ and $\upsilon_2 = \upsilon_4$. Therefore, quantitative data are

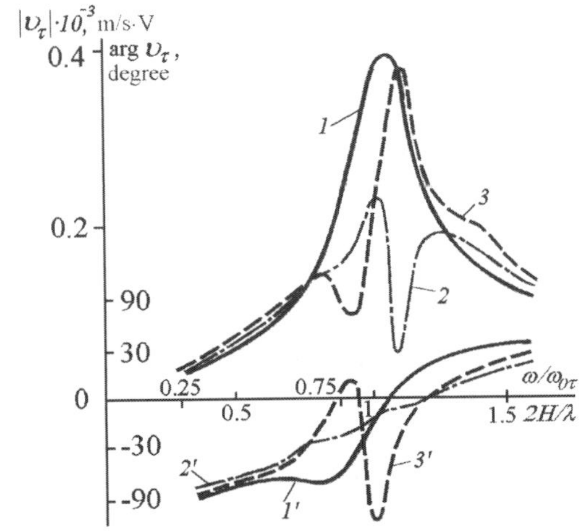

FIG. 2.35: Frequency dependencies of the modulus (curves 1–3) and phase (curves 1′–3′) particle velocities for the first, second and third shells, respectively

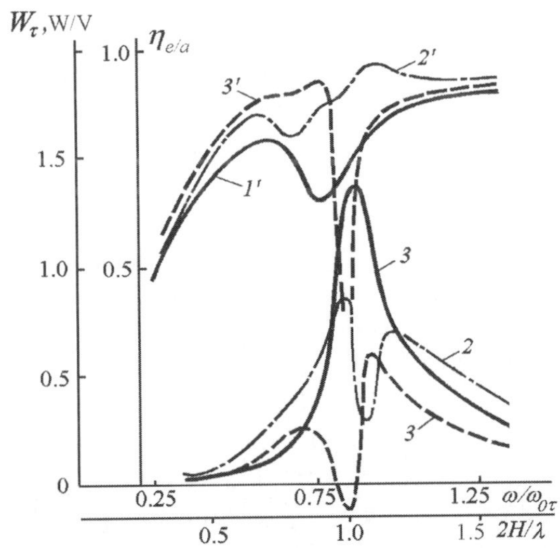

FIG. 2.36: Frequency dependencies for the radiated acoustic power (curves 1–3) and electroacoustic efficiency (curves 1′–3′) for the first, second and third shells, respectively

presented in Figs. 2.35 and 2.36, and also on the following figures, for the first tree shells (see Fig. 2.33).

It is natural to distinguish three zones in these dependencies: the first zone (low-frequency) where $\omega/\omega_{0\tau} < 0.6$, the second (resonance) zone where $0.6 \leq \omega/\omega_{0\tau} < 1.1$ and the third (high-frequency) zone where $\omega/\omega_{0\tau} \geq 1.1$. It is clear that in the first zone the particle velocities of the shells, their radiated powers and efficiencies fall rather rapidly with a decrease in frequency, and these parameters become the same for all the shells. It is connected with the fact that as the frequency decreases, the intrinsic mechanical impedance of the shells increases (in this zone it has an elastic character), while the radiation impedance decreases, because the wave radius of the shells decreases. Therefore, the role of interactions between the shells mediated by the field is quite insignificant, and particle velocities are determined entirely by intrinsic mechanical impedances of the shells. At high frequencies (in the third zone) the decrease in the particle velocities and powers (with an increase in frequency) is also caused by the increase in the intrinsic mechanical impedance of the shells, which in this zone has an inertial character. We also observe a tendency of the velocities and powers of the shell to become equal in this frequency range; however this occurs more slowly than in the first zone, because the real part of radiation impedance increases with an increase of frequency, and the acoustic interactions play more important role than in the first zone. An increase in the shell efficiency, which is caused by an increase in radiation impedance with an increase in frequency, is a characteristic feature of the third zone as well.

Now let us consider the resonance zone, in which the intrinsic mechanical impedances of the shells are comparable to the shell radiation impedances. The reaction of the medium and field-mediated energy exchange between the shells begin to play a significant role. In this zone the shell characteristics differ considerably from one shell to the other. For example, the moduli of the particle velocities of the shells may differ by factors of 5–8, while the value of frequency at which the phase of a given characteristic takes on the value zero is different for all the shells. Furthermore, the phase characteristic of the central shell crosses the frequency axis three times, indicating that the character of the total mechanical impedance of the shell changes several times from elastic to inertial and conversely. The frequencies where the phase characteristic intersects the abscissa are the eigenfrequencies of the system "shells—surrounding medium." Consequently, we can conclude that the acoustic interaction leads to a broadening and enriching of the spectrum of eigenfrequencies of the acoustic source. Our calculations show that an increase of Q-factor of the shells and a decrease of their thickness (while preserving the remaining parameters) lead to an increase in the number of the points, in which the phase characteristics go through zero. The same is true for the other shells.

Here we note an interesting effect which at first glance may seem paradoxical. It is clear that the velocity phase of the central shell becomes less than –90° in the resonance region, in the vicinity of the frequency $\omega/\omega_{0\tau} \approx 0.85$, i.e., the velocity vector passes into the left half of the complex plane. This implies that the real part of the total mechanical impedance becomes negative, and that the shell is converted from a radiator of energy to a consumer (absorber) of energy from the field. This fact is well confirmed by the behavior of the frequency dependence of the radiated power and the efficiency: at this frequency the power acquires a negative value, and the efficiency passes through zero. This effect is a consequence of the strong interaction between the end and central shells. It is easier to understand if we note that the wave height of the entire set of the shells $2H/\lambda \approx 1$, at the frequency $\omega/\omega_{0\text{T}} \approx 0.85$ (in Figs. 2.35 and 2.36, we have also plotted the values of $2H/\lambda$ on the abscissa in addition to the frequencies), and consequently the wave spacing between the central and edge shells is close to a half-wavelength in the surrounding medium. If we take into account that the total power radiated by two end shells exceeds the power radiated by the single central shell, it becomes apparent that it is possible not only to slow down the vibrations of the central shell but also to force the shell to execute vibrations with a phase less than –90°. The calculations show that as the Q-factor of the shells increases to 80, we begin to observe absorption of energy from the medium in the second and fourth shells as well; however this occurs at a somewhat higher frequency $\omega/\omega_{0\text{T}} \approx 0.93$.

The calculations of the near field of intensity vector $\vec{I}$ were carried out to visualize the wave processes in resonance frequency zone. The results of the calculations are shown in Fig. 2.37. As it is shown in Fig. 2.37(a), when the shells interaction effects are relatively small and their particle velocities and radiated powers do not differ greatly from each other, the radiated energy flow outgoes from all the shells in radial direction. The situation is totally different when the interaction between the shells is strong and the mean shell begins to absorb energy from the field, see Fig. 2.37(b). We can see here, that the sound energy flow outgoes from two first shells. Then this flow divides into two flows: one is propagating in the surrounding medium and the other is flowing into the central shell.

The case when $Z_{j\text{T}} = 0$ was considered above. While analyzing the radiation in the resonance zone, it is important to take into account the internal resistance of the generator. In particular, when the internal impedance of the source has a purely active character $R_{j\text{T}}$, we find the following picture. As the value of $R_{j\text{T}}$ increases, the overall radiated acoustic power drops, and the absorption of energy by the shells from the field ceases. In order to illustrate this, the frequency dependencies of the total power radiated by all the shells for the case $Z_{j\text{T}} = 0$ (curve 2) and $\text{Re}Z_{j\text{T}} = 250$ Ohm (curve 3), are showed in Fig. 2.38. The reasons for the decrease in acoustic power are obvious—a part of the power of the electrical source is dissipated because of its internal resistance.

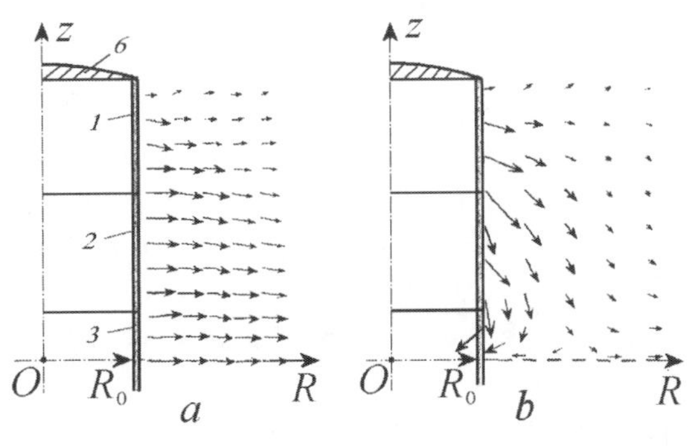

FIG. 2.37: Near field of intensity vector; (a)—$\omega/\omega_{0\text{T}} \approx 1$, (b)—$\omega/\omega_{0\text{T}} \approx 0.85$

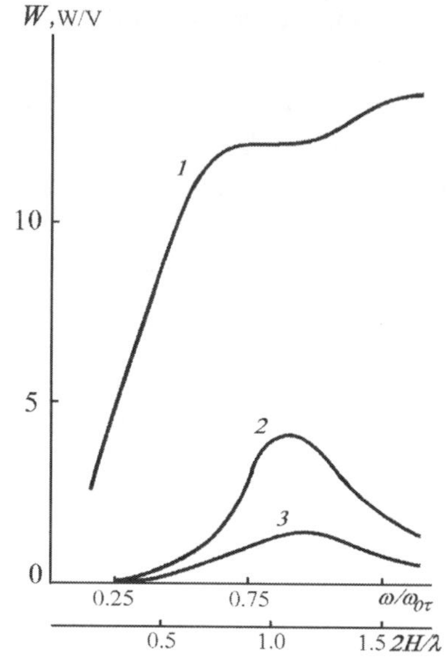

FIG. 2.38: Frequency dependencies of total power, radiated by all the shells; 1—$\upsilon_\tau = \upsilon_{\text{lim}} = 0.4 \times 10^{-3}$ m/s; 2—$Z_{j\tau} = 0$; 3—$\text{Re}Z_{j\tau} = 250$ Ohm

Let us sum up the analysis we have carried out. We have shown that when a system of five identical shells is excited by the same electrical voltage: a) the particle velocities, the radiated powers and the electroacoustic efficiencies in all of the shells are different; b) a particularly large difference in these parameters is observed in the frequency region close to the eigenfrequency of the shells, and this phenomenon is very interesting for the practical applications; c) under certain conditions, some shells can pass from the regime of energy radiation to the regime in which they absorb energy from the external medium.

It is obvious that the effects we have elucidated can be very detrimental to the energy efficiency of this type of antenna array, and also may significantly hinder matching with sources of electrical energy. In particular, the considerable difference in particle velocities of the shells prevents full exploitation of the possibilities of such arrays with regard to radiated power. It becomes obvious, if we take into account that the level of mechanical stress, arising in the material of the vibrating shells, is proportional to the magnitude of the particle velocity. Consequently, there always exists a limiting attainable particle velocity υ_{lim}, above which mechanical disruption of the shells takes place. If we refer to Fig. 2.35, and if we assume that $\upsilon_\tau = \upsilon_{\text{lim}} = 0.4 \times 10^{-3}$ m/s, it becomes clear that the end shells vibrate at the limit of their strength capabilities at a frequency $\omega/\omega_{0\tau} \approx 0.85$, while the remaining shells are in a very low-stress regime. In this case the second and fourth shell will radiate almost half the energy than the end shells, while the central shell in general radiates nothing at all (it absorbs energy from the field). Thus, the radiator has relatively low efficiency. Nevertheless, we can point some possible ways to control the acoustic characteristics of the radiator, which allow us to reduce detrimental effect of acoustic interaction between the shells and to prevent the particle velocities from exciding their limiting attainable value υ_{lim}.

2.4.1.1 The Methods of Control of the Grating Acoustical Characteristics

One of the possible methods is to force all of the shells to oscillate with the same velocity, equal to the maximum allowable velocity υ_{lim}. In this formulation of the problem the unknown is the electromotive force ε_τ of the electrical sources. The initial system (2.92) must be modified: we replace the previously unknown $\upsilon_1, \upsilon_2, \ldots, \upsilon_\tau, \ldots, \upsilon_T$ by the value υ_{lim} (which is determined by the allowable values of the mechanical stresses in piezoceramic [29,38,57]). Than we rearrange all of the equations of system (2.92) starting with the fourth by interchanging the positions of the second terms on their left sides with the terms on their right sides.

After making these changes in system (2.92) and setting $\upsilon_\tau = \upsilon_{\text{lim}} = 0.4 \times 10^{-3}$ m/s, we used them to calculate the results shown in Figs. 2.38–2.40. This value of the particle velocity was chosen for convenience of comparison with the data shown in Fig. 2.35. We can see that the end shells have the maximum values of the velocities that are 0.4×10^{-3} m/s. This value of the velocity is conventionally taken to be the limiting velocity. In the general case, it is necessary to determine υ_{lim}, taking into account the allowable mechanical voltages in the material of the shell, including the corresponding safety factor coefficients. [38,40,57,58]. Comparing the data shown in Figs. 2.40 and 2.36, and also curves 2 and 1 in Fig. 2.38, we can see that by maintaining equal particle velocities of all the shells, we increase the radiated acoustic power both of the individual shells and of the system as a whole, increase efficiency, and completely eliminate any tendency of the shells to change from the regime of radiating to the regime of absorbing energy from the field.

FIG. 2.39: Frequency dependencies of the modulus (curves 1–3) and phase (curves 1′–3′) of electrical voltage on the electrodes of the first, second, and third shells respectively when $\upsilon_\tau = \upsilon_{\text{lim}} = 0.4 \times 10^{-3}$ m/s and $Z_{j\tau} = 0$

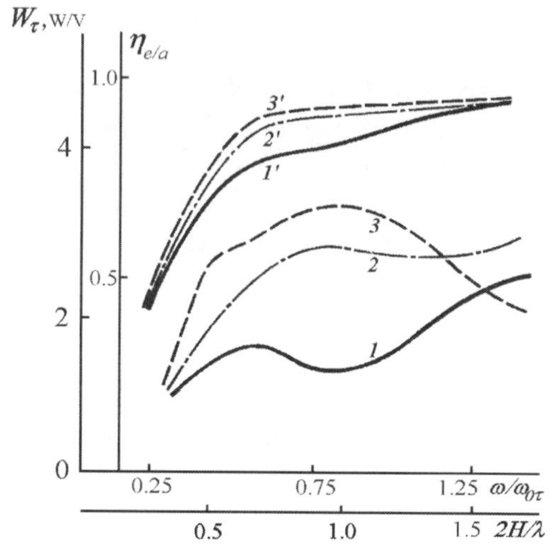

FIG. 2.40: Frequency dependencies of the radiated acoustic power (curves 1–3) and efficiency (curves $1'$–$3'$) for the first, second, and third shells respectively, for $\upsilon_\tau = \upsilon_{\text{lim}} = 0.4 \times 10^{-3}$ m/s and $Z_{j\tau} = 0$

The calculations show that in this case the directivity diagram of the system improves, in particular, the level of the lateral side-lobes decreases in the direction of the z axis, see Fig. 2.41. The band properties of the system improve, and the nonuniformity of the frequency characteristics of the radiation impedances of the shells decreases. However it should be emphasized that the improvement of the fundamental characteristics of this system is accomplished at the cost of increasing the electrical intensities applied to the shells (first of all in the frequency domain that is below and above the resonance frequency) and imposing specific frequency dependences on these intensities (see Fig. 2.39).

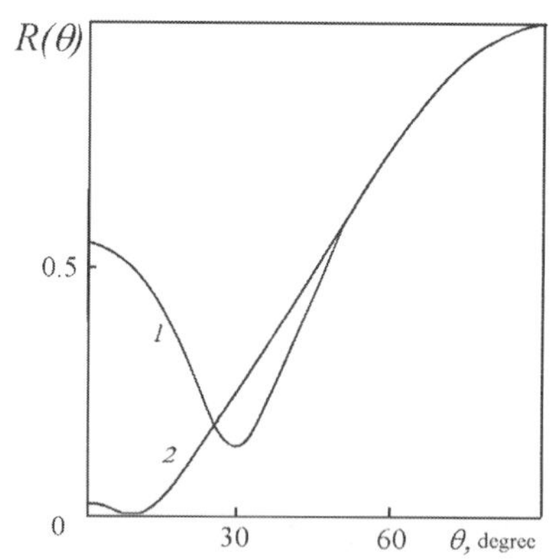

FIG. 2.41: Directivity pattern for $\omega/\omega_{0\tau} \approx 0.85$; 1—the shells are excited by the same electrical voltages, 2—the shells are excited by the different electrical voltages, in order to ensure the same particle velocity for all the shells

We can do it differently to avoid a considerable rise of voltage applied to the shells. Let us apply some electric voltage U^* to one of the shells. We add the condition of equality of particle velocities of all the shells to the system of functional Eq. (2.92). Then electrical voltages, applied on the other shells and the particle velocities become the unknown quantities. The numerical experiment was conducted for the case when the given electrical voltage was applied to the second shell (it is also applied on the fourth shell because of the symmetry of the problem about plane $z = 0$), i.e., $U_2 = U_4 = U^*$. The conditions $\upsilon_1 = \upsilon_2 = \upsilon_3 = \upsilon_4 = \upsilon_5$ are to be met. Figure 2.42 illustrates the results of these calculations, from which it follows that there is a 20% increase of electrical voltage in the considered frequency band on the central shell only. The velocities of all the shells are equal and their frequency dependence is determined by the character of the mechanical impedance of the shells and radiation impedance.

It should be said that, the proposed ways of the grating characteristics control by means of the due choice of the electrical voltages, applied to the shells, can be used

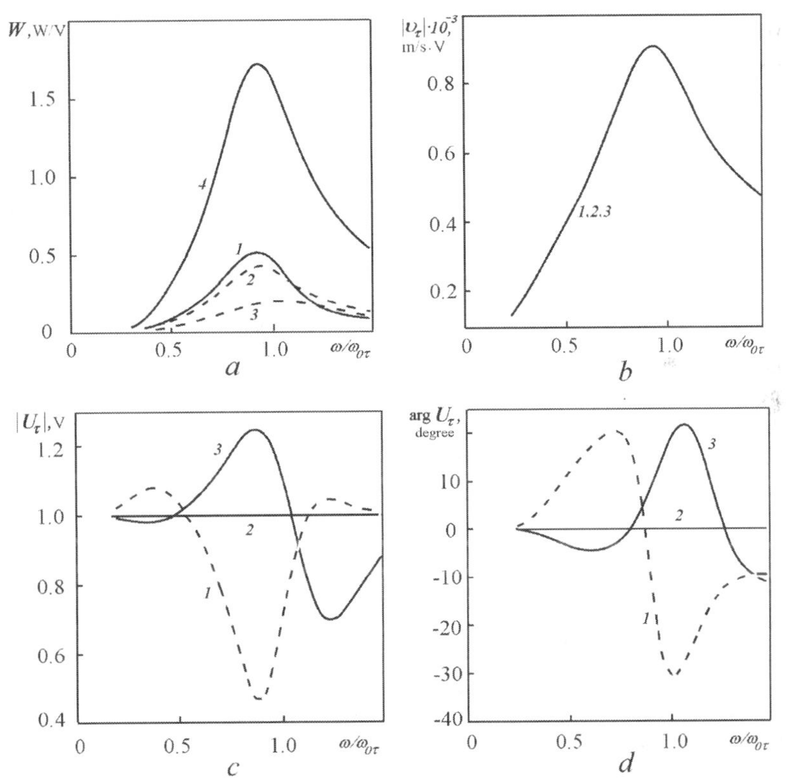

FIG. 2.42: Frequency dependencies of the main characteristics of the grating of shells when the voltage is fixed on the second and fourth shells and all particle velocities of the shells are equal; (a)—acoustic power radiated by the shells and total power of the grating, (b)—the amplitudes of the particle velocities of the shells, (c)—the amplitudes of electrical voltages, (d)—voltage phases; curves 1, 2, 3 correspond to the numbers of the shells, curve 4 shows total power radiated by the grating

effectively not only to protect the shells from the mechanical overloads but also to control the directivity pattern of the grating. For example, we can provide a certain inclination (or even "swinging") of the directivity pattern in a vertical plane [7–9] by choosing the phases of the electrical voltages applied to the shells. The consideration of these problems lies beyond the scope of this monograph.

The ways of decreasing the influence of the interaction effects between the shells in the grating, controlling the electrical voltages applied to them, were considered above. These ways can be called "active" methods. But we can also mention some "passive" methods, when the grating parameters control is carried out using the screens or making some changes of geometrical character. In particular, it is well-known [2,4,8,11] and physically evident, that an increase in distance between the sound radiators weakens the acoustic interaction mediated by the field. From the other side, it is known [7,8], that an expansion of the area of radiating surface of the grating elements also leads to a reduction of the acoustic interaction role. The reason of it is the averaging effect enhancement of the reaction force of the surrounding medium, affecting on the radiating surface of the grating elements. The two variants of the considered grating of piezoceramic shells are given here to show the effectiveness of these "passive" methods.

In the first variant, the second and fourth shells were replaced with the cylindrical acoustically rigid insertions of the same size that the shells. Thus, the distance between the neighboring shells centers was doubled. In the second variant, the height of the shells was increased by factor of 1.665. In both variants, the total height of the grating is $2H$, but the number of the shells is decreased to three. Figure 2.43 shows that both "passive"

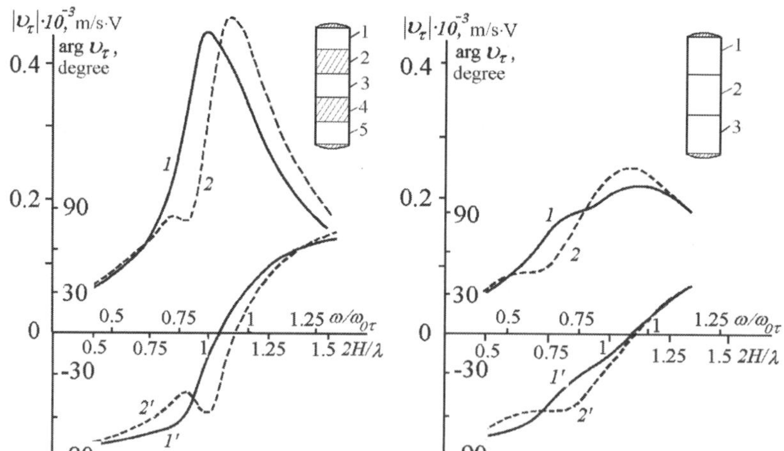

FIG. 2.43: Frequency dependencies of the amplitude and phase of the particle velocity for two variants of the grating of three shells; (a)—the first variant: acoustically rigid insertions are between the shells; (b)—the second variant: the height of the shells is increased by factor of 1.665; curves 1 and 2 correspond to the amplitudes of the particle velocities of the shells, and curves 1′ and 2′—to the phases of velocities; the gratings are illustrated at the top right of the both graphs for the first and second variants, respectively

methods effectively diminish influence of the acoustic interaction of the shells. The frequency dependencies of the amplitudes and phases of the particle velocities of the shells in each grating become similar in form. Each shell has one resonance frequency, and these frequencies are almost the same. Note, that in the second variant [Fig. 2.43(b)] the amplitude of particle velocity at the resonance frequency is two times smaller. This is a direct consequence of an increase of the shells radiation impedance due to an increase of the radiating surface of the shells.

2.4.2 Exciting of the Shells by Pulsed Electric Signal

Sound radiation by the linear grating of piezoceramic shells excited by monochromic harmonic electric signal was considered above. Using the terminology of applied hydroacoustics, we can say that such sound radiators work in continuous mode. But in practice such radiators usually work in pulsing mode. In particular, the arrays of the acoustic and hydroacoustic radars of different application [5,9,34,35,57]. Two questions arise: a) what are the changes that impulse excitation introduces in the shells vibrations and sound field? and b) whether the acoustic characteristics, estimated in pulsing and continuous modes of a radiator, differ (and how much)?

We will try to answer these questions in terms of consideration of the grating of piezoceramic shells. We assume that the same electrical voltage is applied to all the shells in the form of radio pulses sequencing

$$U(t) = \begin{cases} U_0 \sin \omega t, & 0 \le t \le \bar{\tau}, \\ 0, & \bar{\tau} \le t \le \bar{T}, \end{cases} \tag{2.93}$$

where U_0—the impulse amplitude, $\omega = 2\pi f = 2\pi/T$, here T—the carrier frequency cycle, $\bar{\tau}$ and $\bar{T}$—the duration and pulse-repetition cycle. Then the relative pulse duration can be defined as $q = \bar{T}/\bar{\tau}$, and the number of periods of carrier in impulse $N = \bar{\tau}/T$.

We expand function (2.93) in Fourier series

$$U(t) = \sum_{n=1}^{\infty} [a_n \cos(n\Omega t) + b_n \sin(n\Omega t)] = \sum_{n=1}^{\infty} U_n \cos(n\Omega t - \psi_n). \tag{2.94}$$

Here a_n, b_n—coefficients of Fourier series, $n = 1, 2, \ldots,$ $\Omega = 2\pi\bar{f}$, $\bar{f} = 1/\bar{T}$—pulse-repetition frequency, $U_n = \sqrt{a_n^2 + b_n^2}$, $\cos\psi_n = a_n/U_n$. When performing computation we break of Fourier series at some term n^*, which is chosen in such a way as to provide preassigned accuracy of approximation of function (2.93).

Thus, we have finite number n^* of monochromatic harmonic electrical signals U_n. We solve the problem of sound radiation by linear grating of piezoceramic shells for each signal taken separately and, as a result, we define the complex values of particle velocities $\upsilon_\tau^{(n)}$ of the shells and the other acoustical properties of the grating. It is obvious now that when the shells are excited by pulse signal (2.93), the particle velocity of the shells will be determined by the series $\upsilon_\tau = \sum_{n=1}^{n^*} \upsilon_\tau^{(n)}$.

Note that the calculations of acoustic properties, affected by pulse signal, can be simplified and unified. For this purpose we have to precalculate the transfer functions of particle velocity $H_\upsilon(\tau, \omega)$ of τth shell or the transfer function of pressure $H_p(r, \theta, \omega)$. For example, the velocity transfer functions can be defined as

$$H_\upsilon(\tau, \omega) = \frac{\upsilon_\tau(\omega)}{U_\tau}.$$

Then the reaction of τth shell on impulse action (2.93), including (2.94), can be represented as follows:

$$\upsilon_\tau(t) = \sum_{n=1}^{\infty} [a_n \cos(n\Omega t) + b_n \sin(n\Omega t)] H_\upsilon(\tau, \omega = n\Omega).$$

The calculations were made according to the algorithm given above, and the results were obtained. We are going to analyze these results. At first we consider the situation, when only one central shell is excited by pulsed electric signal (2.93), and the other shells are braked. We suppose that the number of periods of carrier in the impulse $N = 10$, the relative pulse duration $q = 3$, the number of the terms kept in series (2.94) is $n^* = 50$. Consider two values of carrier frequency: the carrier frequency coincides with the resonance frequency of the shell in the surrounding medium (in this case $\omega/\omega_{0\tau} \approx 0.875$), the carrier frequency is lower than resonance frequency (in this case $\omega/\omega_{0\tau} \approx 0.5$).

As it follows from the data in Fig. 2.44(a), the impulse calculated by (2.94), when there are 50 terms of series, is almost the same as the impulse calculated by Eq. (2.93).

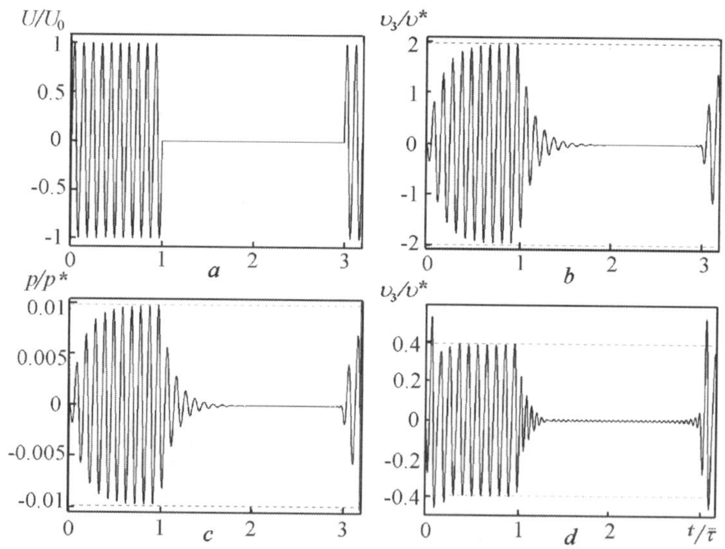

FIG. 2.44: Timing normalized diagrams (only central shell is excited, the rest of the shells are braked); (a)—impulse applied to the shell (calculation by formula (2.94) when $n^* = 50$); (b, c)—particle velocity of the shell and pressure in the far field when $\omega/\omega_{0\tau} \approx 0.875$; (d)—particle velocity of the shell when $\omega/\omega_{0\tau} \approx 0.5$; (time interval a little more than one impulse repetition period is shown on the diagrams)

The timing diagrams of particle velocity of the shell and the pressure in the far field in the region of the shell resonance [Figs. 2.44(b) and 2.44(c)] illustrate the typical distortions as transient processes. They are typical for an impulse, coming through a mechanical resonance system [15]. This is a gradual rise of the pulse leading edge and a decay of the trailing edge. Here and below the levels, calculated in continuous mode of the grating, are pointed with horizontal dashed lines.

It is interesting to note that transient processes take not much time at frequencies below the resonance of the shell, see Fig. 2.44(d). It is to be expected, because the transfer functions of the grating below the resonance frequency region are relatively weakly dependent on frequency. At the same time, at the beginning and at the end of impulse, short "bursts" of particle velocity amplitude can be observed, which can be detrimental for the real shells.

The situation is different when all the shells take part in the field formation and there is a strong acoustic interaction between them. Consider Figs. 2.45(a)–2.45(c), on which the diagrams of system response are represented for the situation which corresponds to maximum radiating power. Note that it is impossible to indicate the resonance

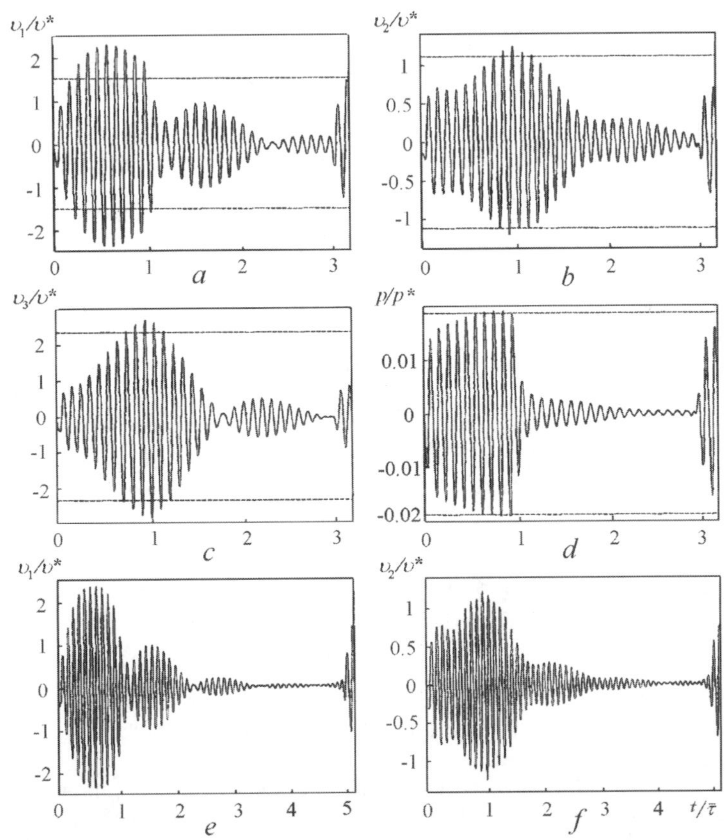

FIG. 2.45: Time diagrams of the system (all the shells are excited at frequency $\omega/\omega_{0\tau} \approx 0.875$): (a–c)—particle velocities of the shells when $q = 3$; (d)—pressure in the far field when $q = 3$; (e, f)—particle velocities of the first and second shells when $q = 5$

frequency of the grating because each shell has its own frequency in their combined action (Fig. 2.35). As we can see, in the result of acoustic interaction, the forms of impulses of the particle velocities of the shells differ from each other and from the form of initial electric impulse. Besides, the forms of the pulses differ markedly from the classical velocity pulse form, when one shell is radiating [see Fig. 2.44(b)]. The laws of pulse rise and pulse decay differ from exponential law.

One more distinctive feature is the complex shells vibrations, which differ in form from each other, after the removal of excitation impulse almost on the entire interpulse time interval. In acoustics this phenomenon is called reverberation (derived from Latin *reverbero*—reflect). Note that in architectural acoustics this phenomenon is conditioned by multiple reflection of sound energy from the walls in a room, in hydroacoustics—reflection from the air bubbles in water, the sea surface, and the seafloor.

In this case reverberation is caused by the exchange of sound energy between the shells through the surrounding medium. Figures 2.45(e) and 2.45(f) illustrates the particle velocities of the second and third shells when $q = 5$. It shows how long the reverberation process can last after the removal of electric excitation impulse. We can see that reverberation process lasts long enough (for almost $3\overline{\tau}$), exponentially decaying.

It is noteworthy that the end shells vibration amplitude is almost by 1.5 times more than the amplitude in the case of the continuous operation of the grating. That is why if we ignore this phenomenon and the "surges" of the amplitude of the shells particle velocity, which are indicated above, it can lead to the shell damage while in operation.

The case when each shell is fed from a separate electric generator was considered above. But in many practical cases, all the shells or a part of the shells can be excited by one common generator. The effect of interaction of the shells mediated by electric circuits occurs in this situation. The influence of this effect on the energetic and directional properties of the grating is significant and comparable with the influence of acoustic interaction of the shells mediated by the field. Quantitative data about interactions of the shells mediated by electric circuits can be found in works [74,75].

2.5 SOUND RADIATION BY A GRATING COMPOSED OF FREE-FLOODING PIEZOCERAMIC SHELLS

The study of the sound field created by free-flooding cylindrical shells* of finite length has a rich history. Giving the credit to the papers of Rayleigh [1], we can point out many articles [59–64], where the problem of sound radiation by rigid cylindrical shell was investigated. Certain particle velocity is given at the surface of the shell. Different analytic methods were developed and interesting quantitative data were obtained in these works. It was determined on the cusp of 50's–60's that free-flooding cylindrical shells

*We have in mind shells with the inner volume open to the surrounding space through open butt ends that provide an opportunity to study the sound emitted by the inner walls of a shell. In foreign literature, such open shells are most frequently called free-flooding cylindrical shells. In contrast to open shells, closed shells (these shells were considered in the previous paragraph, see Fig. 2.33) have rigid mufflers at the butt ends. These mufflers keep the surrounding medium out of the shells inner volume and therefore exclude sound emission by the inner walls.

made of piezoelectric material can serve as efficient radiators of sound energy in a liquid medium [65–68]. In particular, Mc Mahon, G.W. [68] showed experimentally that sound emission efficiency can be increased sufficiently and the radiation frequency band can be broadened by using several such shells.

As it would seem, these facts had to stimulate an increase of interest in the solution of problems on sound emission by a system of free-flooding cylindrical shells. However, there was no increase in the number of publications, which is apparently connected, with the difficulties arising in the case of solution of such problems. Therefore, there is at the present time only a small number of papers, of which Refs. 69 and 70 deserve our attention first of all. In Ref. 69 quantitative data on the radiation impedances of two identical rigid pipes are given and in Ref. 70 two pipes made of piezoelectric material are considered. Data connected mainly with the value of the current appearing in the electrical circuit of one of the pipes in the process of sound emission by the other are given here. In concluding a short review of the published papers we can infer that the problem of sound emission of free-flooding shells is still very topical.

One of the possible approaches to the solution of the "through" problems on sound emission of free-flooding cylindrical piezoceramic shells is considered below. Electrical voltages at the electrodes are given instead of particle velocities at the shell walls. Let us use the same problem-solving procedure as in the previous paragraph, when the grating formed of closed piezoceramic shells was considered. Thus, all the notations are the same as above.

Let us consider a linear grating consisting of T identical free-flooding thin cylindrical piezoceramic shells ($\tau = 1, 2, \ldots, T$—shell number). The grating is immersed in ideal compressible liquid with wave resistance ρc (see Fig. 2.46 and compare with Fig. 2.33).

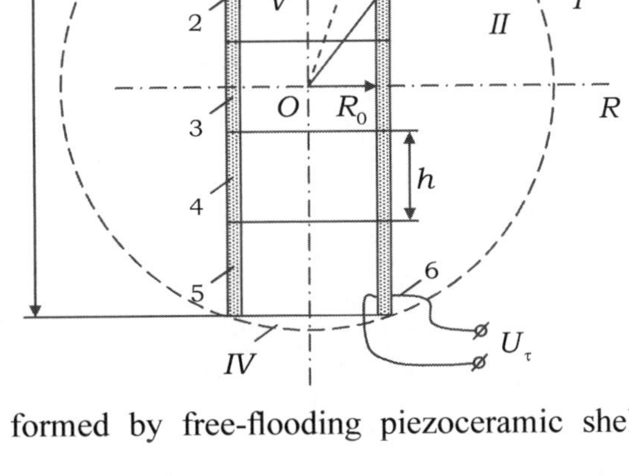

FIG. 2.46: Grating formed by free-flooding piezoceramic shells; 1–5—shells, 6—electrical circuits

We shell restrict ourselves to the particular but important case of the grating step being equal to the shell height. We shell assume that the shells a radially polarized and perform only radial oscillations under the action of the electrical voltages U_τ applied to the electrodes. These oscillations are described by Eq. (2.89), to which linear algebraic Eq. (2.90) corresponds. These equations were used in the previous paragraph 2.4. All the constants and variables that are in Eqs. (2.89) and (2.90), remain the same.

To solve the problem we shell use the modified method of partial areas [53]. According to this method we introduce the spherical (r, θ) and cylindrical (R, z) coordinate systems with the common center in the point O. We divide the whole area of the field into the following subareas (Fig. 2.46): subarea I ($r \geq r_0$, $0 \leq \theta \leq \pi$); subarea II ($r \leq r_0, R \geq R_0, \theta_0 \leq \theta \leq \pi - \theta_0, |z| \leq H$); subarea III ($r \leq r_0, z \geq H, 0 \leq \theta \leq \theta_0$, $R \leq R_0$); subarea IV ($r \leq r_0$, $z \geq -H$, $\pi - \theta_0 \leq \theta \leq \pi$, $R \leq R_0$); subarea V ($|z| \leq H$, $R \leq R_0$). We assume the velocity potentials in these subareas to be in the form of the following sets of particular solutions of the Helmholtz equation:

$$\Phi_1 = \sum_{n=0}^{\infty} A_n h_n^{(1)}(kr) P_n(\cos\theta), \tag{2.95}$$

$$\Phi_2 = \sum_{m=0}^{\infty} B_m M_0^{(1)}(k_m R) \cos\frac{m\pi}{2H}(z+H) + \sum_{\nu=0}^{\infty} C_\nu i_{q_\nu}(kr) \Delta_{q_\nu}(\cos\theta), \tag{2.96}$$

$$\Phi_3 = \sum_{\beta}^{\infty} F_\beta J_0(\delta_\beta R) e^{ik_\beta z} + \sum_{\tau}^{\infty} E_\tau i_{m_\tau}(kr) P_{m_\tau}(\cos\theta), \tag{2.97}$$

$$\Phi_4 = \sum_{\zeta}^{\infty} S_\zeta J_0(\delta_\zeta R) e^{-ik_\zeta z} + \sum_{\gamma}^{\infty} G_\gamma i_{m_\gamma}(kr) P_{m_\gamma}(\cos\theta), \tag{2.98}$$

$$\Phi_5 = \sum_{l}^{\infty} D_l M_0^{(1)}(k_l R) \cos\alpha_l z + \sum_{g}^{\infty} \left[W_g \sin k_g z + Y_g \cos k_g z\right] J_0(\delta_g R). \tag{2.99}$$

All symbols in the expressions (2.95) and (2.96) are identical to the symbols in (2.87) and (2.88) respectively; δ_β—roots of the equation $J_0'(\delta_\beta R) = 0$ when $R = R_0$; δ_ζ—roots of the equation $J_0'(\delta_\zeta R) = 0$ when $R = R_0$; m_τ—roots of the equation $P'_{m_\tau}(\cos\theta) = 0$ when $\theta = \theta_0$; m_γ—roots of the equation $P'_{m_\gamma}(\cos\theta) = 0$ when $\theta = \pi - \theta_0$; $k_\beta = (k^2 - \delta_\beta^2)^{1/2}$; $k_\zeta = (k^2 - \delta_\zeta^2)^{1/2}$; $J_0'(\delta_g R) = 0$ when $R = R_0$; $k_g = (k^2 - \delta_g^2)^{1/2}$; $\alpha_l = l\pi/2H$; $k_l = (k^2 - \alpha_l^2)^{1/2}$;

$$M_0^{(1)}(k_l R) = \begin{cases} H_0^{(1)}(k_l R), & k \geq \alpha_l, \\ K_0(k_l R), & k < \alpha_l. \end{cases}$$

It is evident that the first sums in (2.97) and (2.98) allow us to meet the matching conditions at the surfaces $0 \leq R \leq R_0$, $z = \pm H$, and the second sums—at the surfaces $r = r_0$, $0 \leq \theta \leq \theta_0$, $\pi - \theta_0 \leq \theta \leq \pi$. The first sum in (2.99) allows us to meet boundary condition at the inner surfaces of the sells $R = R_0$, $|z| \le H$, and the second sum—matching conditions at the surfaces $0 \le R \le R_0$, $z = \pm H$. The sums in formulas (2.95) and (2.96) allow us to meet matching conditions at the boundary of subareas I, II and boundary conditions at the outer surfaces of the shells. It was discussed in the previous paragraph. Thus, the use of the parts of coordinate surfaces of spherical and cylindrical coordinate systems, allows us to meet all boundary conditions and matching conditions at the boundaries of subareas.

We shall now formulate the system of functional equations combining continuity conditions of the field components at the boundaries of the subareas and the boundary conditions at the surfaces of the shells. Including above mentioned, we have:

$$\frac{\partial \Phi_1}{\partial r} = \begin{cases} \partial \Phi_3/\partial r, & 0 \le \theta \le \theta_0, \quad r = r_0, \\ \partial \Phi_2/\partial r, & \theta_0 \le \theta \le \pi - \theta_0, \quad r = r_0, \\ \partial \Phi_4/\partial r, & \pi - \theta \le \theta \le \pi, \quad r = r_0, \end{cases}$$

$$\begin{gathered} \Phi_3 = \Phi_1, \quad 0 \le \theta \le \theta_0, \quad r = r_0, \\ \Phi_2 = \Phi_1, \quad \theta_0 \le \theta \le \pi - \theta_0, \quad r = r_0, \\ \Phi_4 = \Phi_1, \quad \pi - \theta_0 \le \theta \le \pi, \quad r = r_0, \\ \Phi_3 = \Phi_5, \quad z = H, \quad 0 \le R \le R_0, \\ \Phi_4 = \Phi_5, \quad z = -H, \quad 0 \le R \le R_0, \\ \partial \Phi_3/\partial z = \partial \Phi_5/\partial z, \quad z = H, \quad 0 \le R \le R_0, \\ \partial \Phi_4/\partial z = \partial \Phi_5/\partial z, \quad z = -H, \quad 0 \le R \le R_0, \end{gathered} \tag{2.100}$$

$$-\frac{\partial \Phi_2}{\partial R} = \begin{cases} \upsilon_1, & H - h \le z \le H, \quad R = R_0, \\ \upsilon_2, & H - 2h \le z \le H - h, \quad R = R_0, \\ \dots\dots\dots\dots & \\ \upsilon_T, & -H \le z \le H, \quad R = R_0, \end{cases}$$

$$-\frac{\partial \Phi_5}{\partial R} = \begin{cases} \upsilon_1, & H - h \le z \le H, \quad R = R_0, \\ \upsilon_2, & H - 2h \le z \le H - h, \quad R = R_0, \\ \dots\dots\dots\dots & \\ \upsilon_T, & -H \le z \le h - H, \quad R = R_0. \end{cases}$$

After the system (2.100) is made algebraic [with account of motion equation of the shells (2.90)], we obtain the infinite system of linear algebraic equations with respect to the unknown complex coefficients, that enter into the expressions (2.95)—(2.99).

In order to conduct the analysis of the radiation field created by the free-flooding shells, the calculations were made for the following parameters: $h/R_0 = 1/3$, $N_\tau = 8$ H/V, $Q = 15$, $M_\tau = 1.5$ kg, $\omega_{0\tau} = \pi \times 1.4 \times 10^4$ s^{-1}, $C_{0\tau} = 0.085$ μF, $\rho c = 1.5 \times 10^6$ kg/m^{2}s, $U_\tau =$ const. The total number of unknowns in algebraic system was 88. In this case, the discrepancy of the field components at the borders of the subareas (except the points in the vicinity of the ribs $r = r_0$, $\theta = \theta_0$, $\theta = \pi - \theta_0$) was not larger than 1–3%. The discrepancy in power in the near and far zones was less than 2–4%.

Let us look at Fig. 2.47, which shows the frequency dependencies of the normalized (with respect to the electrical voltage) acoustic power, radiated by a grating made of three shells.

Let us compare curves 1 and 4, which characterize the total radiated powers of the grating made of free-flooding shells and the grating made of the same but closed shells. The grating of closed shells has only one maximum of radiated power in the region of the shells resonance. The grating of free-flooding shells has two sharp radiation maximums. The high-frequency maximum is caused by the resonance of the shells, and low-frequency maximum is due to the resonance of the liquid column bounded by the inner volume of the grating. Strictly speaking, the law frequency maximum is caused by the resonance of the system "shells—liquid column inside the grating—surrounding medium." However, understanding this we shell use the words "the resonance of the liquid column."

The fact of appearance of the liquid column resonance was well studied [1,42,65–68,72,73]. The situation is much more important as the level of radiated power in the area of this resonance (using the adopted method of shells excitation) is very high. The grating of free-flooding shells (all other factors being equal) has a wider band and is energetically more efficient than the one of closed shells. This important conclusion agrees with the experimental data given in Ref. [68]. Let us give our attention to the curves 2 and 3 in Fig. 2.47, which characterizes the energy contribution of the outer and inner walls of the shells. Correlating the curves 2 and 3, we can see that there is "a division of labor" of a sort between these walls. In the area of the high frequency maximum of energy radiation the outer walls of the shells give the main contribution (to 80%) and the contribution of the inner walls is small, and decreases rapidly with frequency increase. It is different in the area of the law frequency maximum of energy radiation. Here the main part of the energy is radiated by the inner walls and the contribution of the outer walls is small

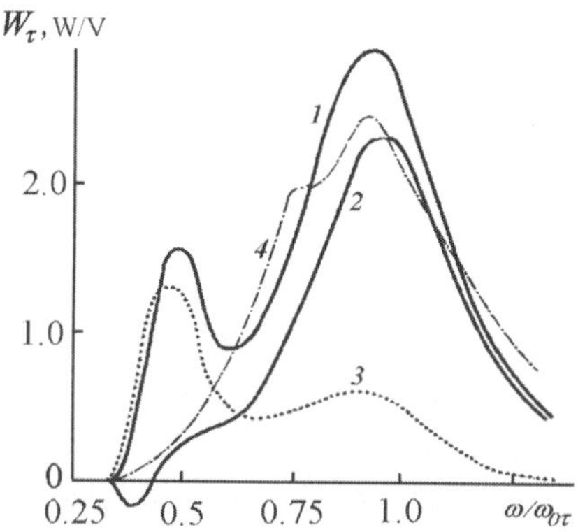

FIG. 2.47: Frequency dependencies of radiated acoustic power of a grating made of three free-flooding shells; 1—total power, 2, 3—powers radiated by the outer and inner walls of the shells respectively, 4—total power, radiated by a grating made of three closed shells

and decreases rapidly with frequency decrease, reaching zero level at $\omega/\omega_{0\tau} \approx 0.43$. The special feature is that at $\omega/\omega_{0\tau} < 0.43$ only the inner walls emit sound and the outer walls absorb the field energy. The total radiated power decreases rapidly with the frequency decrease on account of this effect, and the efficiency of the grating under consideration, as a sound radiator, becomes insignificant.

The dependencies established above also remain true for longer gratings with large number of shells. For example, the analysis of calculation data obtained for the grating of five similar shells ($T = 5, 2H/h = 5$) allowed us to ascertain, that the high-frequency maximum of radiated power resulting from the resonance of the shell stayed at practically the same frequency as for the grating of three shells. However the low-frequency maximum shifted down and was found to be at a frequency which is about 1.6 times lower than the corresponding frequency in the three-component grating. This is a natural phenomenon, because the length of the five-component grating (hence, the length of column of liquid inside of the grating) is about 1.67 times larger than the three-component one.

Figure 2.48 illustrates the amplitudes and phases of the particle velocities of the shells in the three-component grating (in view of the symmetry of the problem relative to the plane $z = 0$, Figs. 2.48 and 2.49 show the data only for the central and the outmost shells). Amplitudes and phases of particle velocities of the free-flooding shells depend comparatively weakly on their position in the grating over the entire frequency range under consideration. Unlike the closed shells, the free-flooding shells do not absorb the energy from the field. Thus the grating of free-flooding shells has an advantage in energetic efficiency comparatively to the grating of closed shells. This important conclusion is strengthen by the data in Fig. 2.49, where frequency dependencies of electroacoustic efficiency of the free flooding and closed shells are demonstrated. The efficiency of free-flooding shells in low-frequency region, where the resonance of the fluid column takes place, is higher than the efficiency of closed shells. In the region of shell region the efficiency of free-flooding shells do not have the "gaps," characteristic for the closed shells. These gaps are determined by the effect of acoustic interaction which results in the fact

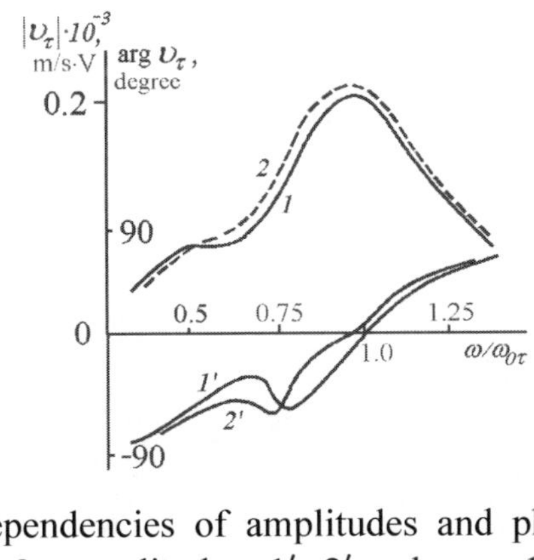

FIG. 2.48: Frequency dependencies of amplitudes and phases of particle velocities of free-flooding shells; 1, 2—amplitudes, 1′, 2′—phases; 1, 1′—central shell, 2, 2′—outermost shell

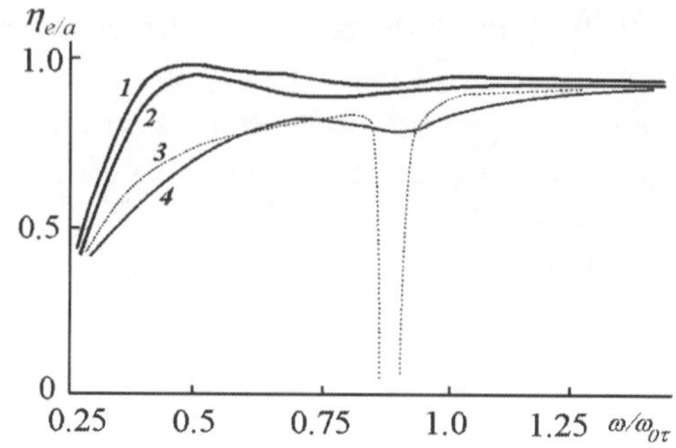

FIG. 2.49: Frequency dependencies of electroacoustic efficiencies of the shells; 1, 2—for free-flooding shells, 3, 4—for open shells; 1, 3—central shells, 2, 4—outermost shells

that the central shell changes the regime of sound energy emission to its absorption from the field.

We now consider some of the properties of the far field of the grating. The sound pressure maximum is in the plane $\theta = \pi/2$, and the minimum is in the direction z-axis. The most important feature is very weak dependence of the shape of directivity pattern $R(\theta)$ and the concentration coefficient K (see curve 1 in Fig. 2.50) on frequency. For example, in band wider than two octaves ($0.28 \leq \omega/\omega_{0\tau} \leq 1.2$) the width of the main side lobe of the directivity pattern (at the pressure level 0.7 of the maximum) changes in a very small ranges $76 \pm 8°$. The concentration coefficient also changes in the small range 3–3.7. For these reasons, the main part of energy is radiated into the equatorial sector ($0.3\pi \leq \theta \leq 0.7\pi$), and the part of energy radiated into polar sectors is small. The relative pressure level $\sigma = R(\theta = 0)/R(\theta = \pi/2)$ in the direction of the lattice symmetry axis is close to zero (curve 3 in Fig. 2.50). That agrees with results of works [72,73], where it is shown that for free-flooding shells of finite length the far sound field

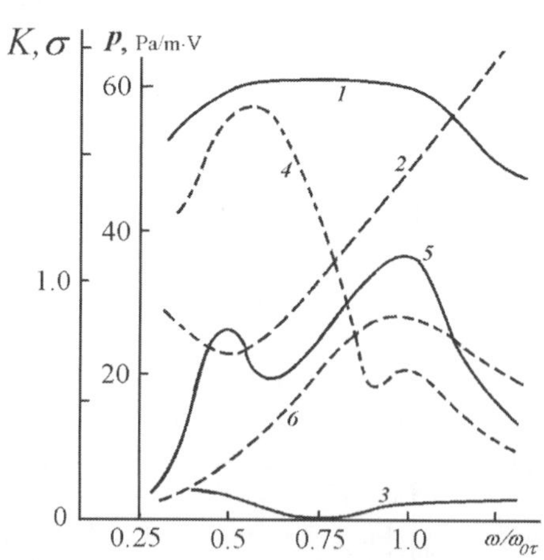

FIG. 2.50: Frequency dependencies: 1, 2—of concentration coefficient, 3, 4—the relative level of pressure in the direction of z axis, 5, 6—the pressure in the plane $\theta = \pi/2$; 1, 3, 5—for the grating of free-flooding shells; 2, 4, 6—for the grating of closed shells

in the direction of their symmetry axis is equal to zero. The frequency dependencies of the values K and σ for the grating of closed shells are shown in Fig. 2.50 (curves 2 and 4) in order to demonstrate more descriptively the sharp difference of the far field of a free flooding shells grating from the far field of a closed shells grating. A conclusion can be drawn from the behavior analysis on the sufficient dependence of the far field parameters of a closed shell grating on the frequency. At low frequencies the value $K < 1$, and $\sigma > 1$, which suggests that the main part of the energy is radiated into the polar sectors. Significant alteration of the field distribution in the far field takes place with the increase in the frequency and, beginning from the frequency $\omega/\omega_{0\tau} \geq 0.7$, the grating radiates mainly into the equatorial sector. However, in this frequency region the pressure level in the direction of z-axis remains significant. In contrast, the acoustic parameters of the grating of free-flooding shells depend weekly on that frequency. Figure 2.50 (curve 5) shows the frequency dependency of the pressure in the plane $\theta = \pi/2$, normalized to one meter and to the value of electric voltage. The similar dependence for a grating of closed shells is given here also (curve 6) these curves repeat qualitatively the parts of the curves 1 and 4 in Fig. 2.47. But at the same time they allow us to determine the sensitivity from the pressure in radiation regime.

We shall note in conclusion that the attention was given above to only one but widely spread mode of shells excitation in a grating. The electrical voltage of equal amplitude and phase is conducted to each of them. Setting other values for electrical voltage one can obtain the new data for the particle velocities of the shells and sound field created by them. However, within the framework of the developed approach one can find an answer for another important question: what values of U_τ are necessary to provide given values of the velocities υ_τ. It is enough for this to accept the values υ_τ to be known (preset) in Eqs. (2.90) and (2.100). Then the values of U_τ are unknown. It was done before when studying the acoustical properties of the grating of closed shells.

2.6 SOUND RADIATION BY PIEZOCERAMIC FREE-FLOODING SHELL IN THE VICINITY OF THE SCREENS OF FINITE SIZE

In the previous paragraph we considered the acoustic properties of the grating of free-flooding piezoceramic shells on the premise that there are no bodies in the vicinity of the grating. In the real radiating systems, some constructional parts are often situated near the shells (for example, when they are in acoustic buoys). Consequently, we will consider below the influence of the bodies of finite sizes on the characteristics of the sound field created by the single open piezoceramic shell.

We consider axisymmetric problem of sound emission by free-flooding piezoceramic shell with height h. The screens in the form of spherical segments, Fig. 2.51 are situated near the butt ends of the shell at the distance H. As usually, we shell consider that such radiating system is immersed in an ideally compressible liquid with density ρ and sound velocity c. The piezoceramic shell is radially polarized. The electrodes, at which electrical voltage $U(t)$ is applied, are situated on its cylindrical surface.

Such statement of the problem allows us to determine the important practical characteristics: acoustical pressure, power, the vibration velocity of the shell surface. We also

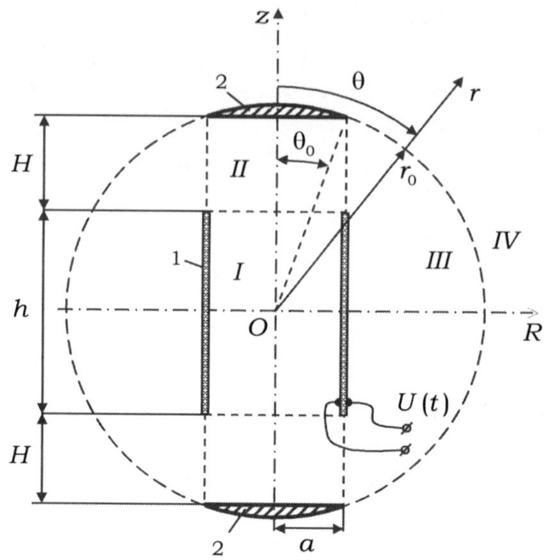

FIG. 2.51: Geometry of the problem: 1—shell, 2—screens in the form of spherical segments

can estimate the influence of the properties of the screens surfaces and the distance from the screens to the shell on these characteristics.

To solve this problem we introduce two coordinate systems with a common center O: spherical (r, θ) and cylindrical (R, z). Each screen has two boundary surfaces: outer $(r = r_0,\ 0 \le \theta \le \theta_0$, or $\pi - \theta_0 \le \theta \le \pi)$ and inner $(0 \le R \le a,\ z = \pm\ (h/2 + H))$. We assume that outer surfaces of the screens are acoustically rigid, and inner surfaces—acoustically rigid or acoustically soft. For brevity we will call these screens acoustically rigid or soft respectively.

The general solution of the problem is based on the partial domains method. We partition all the domain of the sound field existence into four partial domains (see Fig. 2.51). The first domain I is bounded by the inner side surface of the shell $R = a$, $|z| \le h/2$ and the end surfaces $z = \pm\ h/2$, $0 \le R \le a$. The second domain II consists of two cylindrical zones, situated between the ends of the shell $z = \pm\ h/2$, $0 \le R \le a$ and inner surfaces of the screens $z = \pm\ (h/2 + H)$, $0 \le R \le a$. The third domain III, limited by the cylindrical surface $R = a$, $|z| \le h/2 + H$ and the part of spherical surface $r = r_0$, $\theta_0 \le \theta \le \pi - \theta_0$. The fourth IV (outer) domain is determined by inequality $r \ge r_0$.

Let us consider the continuous mode of the shell under the action of harmonic electrical voltage. We represent the velocity potentials in partial domains in the form of the superposition of particular solutions of Helmholtz equations in the corresponding system of coordinates. We will write the expressions for sound fields, bearing in mind that the problem is symmetrical with respect to the plane $z = 0$. For domain I

$$\Phi_I = \sum_{n=0}^{\infty} A_n \cos\left(\frac{2n\pi z}{h}\right) J_0(\eta_n R) + \sum_{n=0}^{\infty} B_n J_0(\mu_n R) \cos(\xi_n z). \tag{2.101}$$

Here $\eta_n = \sqrt{k^2 - (2n\pi/h)^2}$, $\xi_n = \sqrt{k^2 - \mu_n^2}$, $k = \omega/c$. The values of μ_n are the roots of the equation $J_0(\mu_n a) = 0$. Solution (2.101), is the superposition of two Fourier series. The first of them, due to the completeness and orthogonality of the set of functions $\cos(2n\pi z/h)$, $n = 0, 1, 2, ...$ on the segment $|z| \le h/2$, provides the fulfillment of boundary conditions on the inner cylindrical surface of the shell. The second Fourier series, due to the analogous properties of the set of the functions $J_0(\mu_n R)$, $n = 0, 1, 2, ...$, $0 \le R \le a$, provides the sound fields matching between domains I and II on the surfaces $z = \pm h/2, 0 \le R \le a$.

We write the general solution for domain II as follows

$$\Phi_{II} = \sum_{n=0}^{\infty} M_n \cos\left[\frac{n\pi}{H}\left(|z| - \frac{h}{2}\right)\right] J_0(\beta_n R) \qquad (2.102)$$
$$+ \sum_{n=0}^{\infty} J_0(\mu_n R)\left[G_n \cos\left(\xi_n\left(|z| - \frac{h+H}{2}\right)\right) + F_n \sin\left(\xi_n\left(|z| - \frac{h+H}{2}\right)\right)\right],$$

where $\beta_n = \sqrt{k^2 - (n\pi/H)^2}$. The first Fourier series in expression (2.102), due to the set of functions $\cos[(n\pi/H)(|z| - h/2)]$, $n = 0, 1, 2, ...$, $h/2 \le |z| \le h/2 + H$, provides the matching conditions for the fields on the cylindrical boundary between domains II and III. The second series provides matching conditions for the fields on the surfaces $z = \pm h/2, 0 \le R \le a$ and boundary condition on the inner surfaces of the screens $z = \pm\ (h/2 + H), 0 \le R \le a$.

We represent the general solution in domain III as follows

$$\Phi_{III} = \sum_{n=0}^{\infty} D_n \cos\left(\frac{n\pi z}{h/2 + H}\right) H_0^{(1)}(\gamma_n R) + \sum_{n=0}^{\infty} E_n T_{\nu_n}(\cos\theta)\, j_{\nu_n}(kr). \quad (2.103)$$

Here $j_{\nu_n}(kr)$—spherical Bessel functions of first kind, $H_0^{(1)}(\gamma_n R)$—Hankel functions of first kind, $\gamma_n = \sqrt{k^2 - (n\pi/\,(h/2 + H))^2}$. Notation $T_{\nu_n}(\cos\theta)$ corresponds to the expression

$$T_{\nu_n}(\cos\theta) = P_{\nu_n}(\cos\theta) + L_n Q_{\nu_n}(\cos\theta), \qquad (2.104)$$

where $P_{\nu_n}(\cos\theta)$ and $Q_{\nu_n}(\cos\theta)$—Legendre functions of the first and second kind. The values ν_n and L_n are defined from two equations: $dT_{\nu_n}(\cos\theta)/d\theta = 0$ where $\theta = \theta_0$ and $\theta = \pi - \theta_0$. Note, that formally they can be replaced with the equations $T_{\nu_n}(\cos\theta) = 0$ when $\theta = \theta_0$ and $\theta = \pi - \theta_0$. But the numerical experiment showed, that it almost does not influence on the efficiency of the algorithms, which are used for quantitative estimation of the sound fields radiated by the shell.

The first series in expression (2.103), due to the set of functions $\cos[n\pi z/(h/2 + H)]$, $n = 0, 1, 2, ...$, $|z| \le h/2 + H$, provides matching conditions at the cylindrical surface $R = a$, $|z| \le h/2 + H$. The second series, due to the set of functions $T_{\nu_n}(\cos\theta)$, $n = 0, 1, 2, ...$, $\theta_0 \le \theta \le \pi - \theta_0$, provides the boundary conditions at the spherical surface $r = r_0$, $\theta_0 \le \theta \le \pi - \theta_0$.

We represent the field in outer domain IV as follows:

$$\Phi_{IV} = \sum_{n=0}^{\infty} C_n P_n (\cos\theta)\, h_n^{(1)} (kr). \tag{2.105}$$

Here $h_n^{(1)} (kr)$—spherical Hankel functions of first kind. Due to the symmetry of the problem with respect to the plane $z = 0$, we should use only the even functions $P_n (x)$ and $T_n (x)$ in the expressions given above where $x = \cos\theta$ changes at segment $[-1, 1]$ when the angle θ changes within limits $[0, \pi]$.

As in the previous paragraphs, when describing the radial vibrations of piezoceramic shells, we suppose that:

1. the thickness of the shell wall is less than the radius of its middle surface. It allows us to examine the radial vibrations of the middle surface of the shell (the mechanical and electrical values change in radial direction can be neglected);

2. the height of the shell is less than its radius. Due to this fact, the energy of bending deformations can be neglected and we can consider that the displacement of its middle surface does not depend on the z coordinate.

Consequently, for harmonic electrical voltage $U (t) = U_0 \exp(-i\omega t)$, we can transform the differential equation of the vibrations of thin shell into the equation of the form [see Eqs. (2.89) and (2.90)]

$$\Delta F + \upsilon Z = NU_0, \tag{2.106}$$

where υ—the amplitude of radial particle velocity of the shell, ΔF—the difference between the medium reaction force at inner surface and medium reaction force at outer surface of the shell. It is determined by the difference between the pressures on the corresponding surfaces

$$\Delta F = -i\omega\rho \left[\int\limits_{(S+)} \Phi_{III} dS - \int\limits_{(S-)} \Phi_I dS \right]. \tag{2.107}$$

It was performed the outer surface $(S+)$ integration and the inner surface $(S-)$ integration. In Eq. (2.106), we take

$$Z = -i\omega m \left(1 - \frac{\omega_0^2}{\omega^2} + i\frac{\omega_0}{\omega Q}\right), \quad N = \frac{2\pi h d_{31}}{s_{11}^E},$$

where d_{31}, s_{11}^E—the piezoelectric modulus and compliance modulus of piezoelectric material of the shell; m, N, Z, Q—the mass, electromechanical coupling factor, intrinsic mechanical impedance and mechanical Q-factor of the shell respectively; $\omega_0 = 2\pi f_0$—the circular frequency of the shell in a vacuum.

Let us write the system of functional equations, which contains the equation of the shell vibrations, field continuity conditions on the boundaries of partial domains and boundary conditions on the surfaces of the screens and shell:

$$-i\omega\rho\left[\int\limits_{(S+)}\Phi_{III}dS-\int\limits_{(S-)}\Phi_{I}dS\right]+\upsilon Z=NU_0,$$

$$-\left.\frac{\partial\Phi_I}{\partial R}\right|_{S-}=\upsilon,\quad R=a,\quad |z|\le h/2,$$

$$-\frac{\partial\Phi_I}{\partial z}=-\frac{\partial\Phi_{II}}{\partial z},\quad z=\pm\,h/2,\quad 0\le R\le a,$$

$$\Phi_I=\Phi_{II},\quad z=\pm\,h/2,\quad 0\le R\le a,$$

$$-\frac{\partial\Phi_{II}}{\partial z}=0\ \text{ or }\ \Phi_{II}=0,\ z=\pm\,h/2,\ 0\le R\le a,$$

$$\Phi_{II}=\Phi_{III},\quad R=a,\quad h/2\le|z|\le h/2+H, \tag{2.108}$$

$$-\frac{\partial\Phi_{III}}{\partial R}=\begin{cases}-\dfrac{\partial\Phi_{II}}{\partial R}, & R=a,\quad h/2\le|z|\le h/2+H,\\ \upsilon, & R=a,\quad |z|\le h/2,\end{cases}$$

$$\Phi_{IV}=\Phi_{III},\quad r=r_0,\quad \theta_0\le\theta\le\pi-\theta_0,$$

$$-\frac{\partial\Phi_{IV}}{\partial r}=\begin{cases}-\dfrac{\partial\Phi_{III}}{\partial r}, & r=r_0,\quad \theta_0\le\theta\le\pi-\theta_0,\\ 0, & r=r_0,\quad 0\le\theta\le\theta_0,\\ 0, & r=r_0,\quad \pi-\theta_0\le\theta\le\pi.\end{cases}$$

The conversion of functional Eqs. (2.108) to an algebraic set of equations can be performed using the orthogonality of the corresponding sets of functions. We obtain the infinite system of linear algebraic equations of second kind in unknowns A_n, B_n, C_n, D_n, E_n, F_n, G_n, M_n, and υ. This system is bulky and we will not write its explicit form. A similar system of equations is considered in work [78]. The problem of sound radiation by the grating of piezoceramic shells with the end shields in the form of truncated spherical sector was considered in this work. The solution algorithm was described and the obtained results accuracy was estimated. Consequently, we will not consider this question. Let us proceed to the analysis of quantitative data.

Let us take the following parameters of the shell were: $a = 0.075$ m, $h = 0.05$ m, $m = 1.5$ kg, $h/2a = 1/3$, $Q = 25$, $f_0 \approx 7440$ Hz. The shell is made of piezoceramic CTBS-3 with the following characteristics: $d_{31} = 1.6\times10^{-10}$ C/N, $s_{11}^E = 11.3\times10^{-12}$ m^2/N and it is immersed in a medium with density $\rho = 10^3$ kg/m^3 and sound velocity $c = 1500$ m/s.

Figure 2.52 shows the frequency dependencies of normalized radiated acoustic power of the shell. The shell is situated near the rigid screens with different values of ratio H/h. Here $W^* = (NU_0)^2/(\rho cS)$, S—the area of shell surface. The dashed curves with small dashes show the power radiated by the outer side surface of the shell, and the dashed curves with large dashes show the power radiated by its inner side surface. The continuous lines show full power radiated by the shell. Inasmuch as it is the

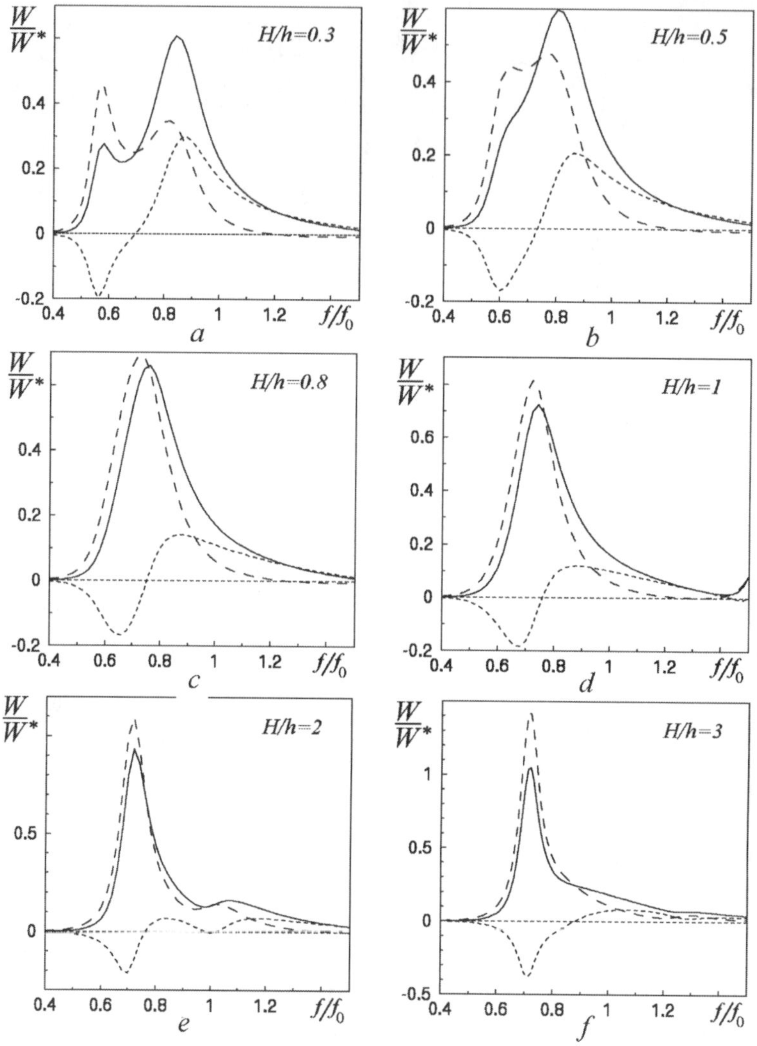

FIG. 2.52: Frequency dependencies of acoustic power, radiated by the shell with acoustically rigid shields: dashed curves with small dashes—radiation power of the outer surface, dashed curves with large dashes—radiation power of the inner surface, continuous line—full power

free-flooding shell, there is an acoustic interaction between the inner and outer side surfaces of the shell mediated by surrounding medium.

Figure 2.52 shows the dependency analysis, which allows us to list the following characteristics of the considered system acoustic properties.

1. The maximum level of radiating power of the shell depends weakly on the distance between the shell and the screens. It may be noted that there is a little increase in power with an increase of H/h.

2. The form of the curves, which determine the radiation by the outer surface of the shell remains the same with increase of H/h. There is some critical frequency (it lies in the range $f/f_0 = 0.7...0.8$). When the ratio H/h changes, outer side surface absorbs energy from the field at frequencies bellow critical frequency, and outer side surface emits energy at frequencies above this frequency.

3. In all the frequency range, the inner surface of the shell always radiates energy in acoustic medium.

4. The frequency dependencies of radiation power of the inner surface and the shell as a whole have complex (with two maximums) character, when the screens are relatively close to the shell ($H/h \leq 0.5$). When moving the screens away from the shell, these dependencies take the typical form of the resonance curves (with one maximum).

5. When $H/h \geq 1$ the frequency dependencies of acoustic power vary only slightly. It is the direct consequence of the fact that the influence of the screens on the shell reduces with increase in distance between them.

Now let us consider the curves in Fig. 2.53, which correspond to the case when there are acoustically soft screens. Here the following features can be observed.

1. The influence of the soft screens on the acoustic properties of the shell is more significant than the influence of the rigid screens. The reason of it is pressure phase inversion, when the wave is reflected from the soft surface.

2. The radiation power level of the shell depends on the distance between the shell and the screens. The zone $0.7 \leq H/h \leq 0.9$ is particularly unfavorable, because the radiation level decreases sharply.

3. Under certain conditions, the inner surface of the shell can absorb energy from a field or emit energy in a medium in contrast to the rigid screens. As it follows from Fig. 2.53, when $H/h < 0.8$ outer surface of the shell emits energy in a medium and inner surface—absorbs it. The form of the power frequency dependencies is sharp resonance. The resonance maximums decrease when $H/h \to 0.8$. As an example of the energy flow distribution near the shell when $H/h < 0.8$, Fig. 2.54(a) shows the intensity vector field when $H/h = 0.5$ at frequency $f_W/f_0 = 0.72$ (f_W—frequency of maximum radiation power of the shell). The length and direction of the arrows on the graphs define the relative level and direction of intensity vector in the points, which correspond to the beginning of the arrows (only the quarter of considered acoustic problem is shown due to the symmetry of the problem).

4. When $H/h > 0.8$ the situation changes. The inner surface of the shell emits energy in a medium, and the outer surface absorbs it from the field. The radiation power of the shell increases with an increase of value H/h. A typical example of the flows of energy in this case is represented in Fig. 2.54(b) for $H/h = 1$, $f_W/f_0 = 0.72$.

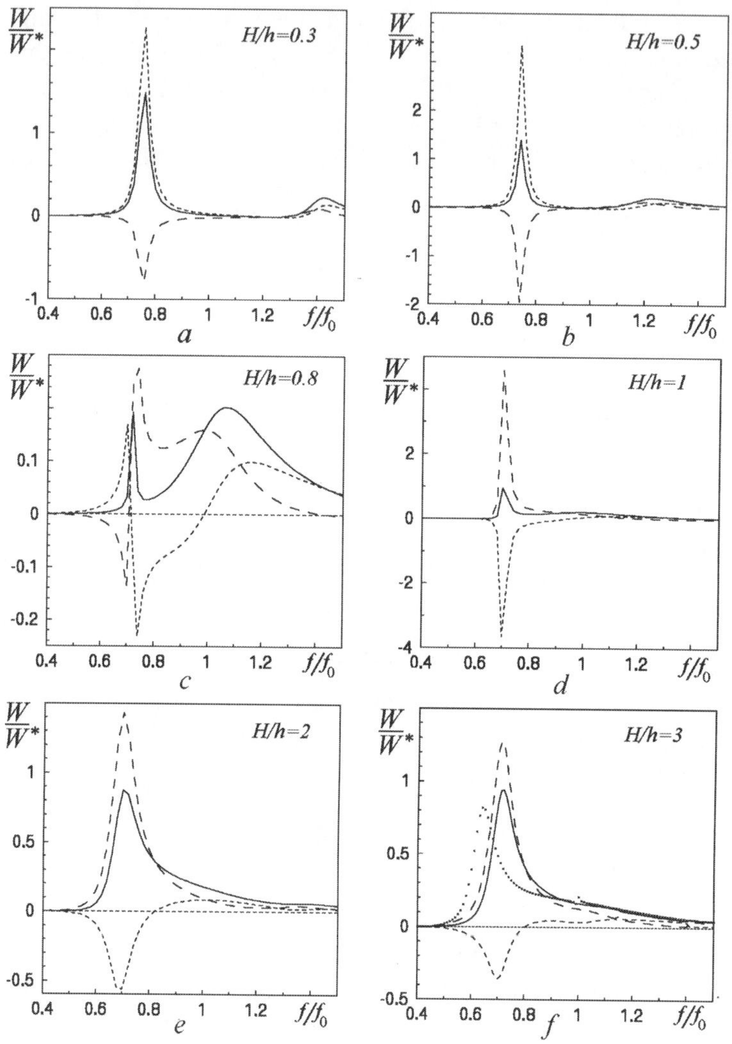

FIG. 2.53: Frequency dependencies of acoustic power, radiated by the shell with acoustically soft shields: dashed curves with small dashes—radiation power of the outer surface, dashed curves with large dashes—radiation power of the inner surface, continuous line—full power, dotted line (on bottom right graph)—full power in a free space

5. When $H/h > 2$ the width of the resonance curves increases and the curves become more close to the corresponding curves for the case of the rigid screens. It means that when $H/h > 2$ the influence of the screens on the shell becomes weak [compare with the dependency for the case when there are no screens, Fig. 2.53(f)].

Figures 2.55 and 2.56 illustrate the dependencies of maximum radiation power (at frequency f_W) and normalized amplitude of the particle velocity of the shell on the value of ratio H/h, where $\upsilon^* = NU_0/(\rho c S)$.

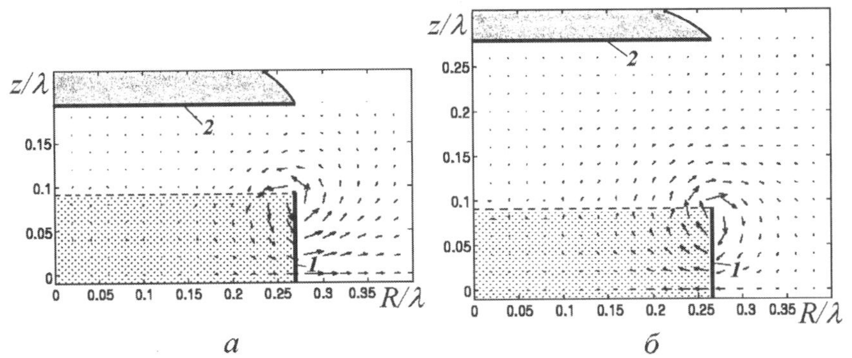

FIG. 2.54: The near field of intensity vector when $f_W/f_0 = 0.72$ (the screens are acoustically soft): (a)—$H/h = 0.5$, (b)—$H/h = 1$

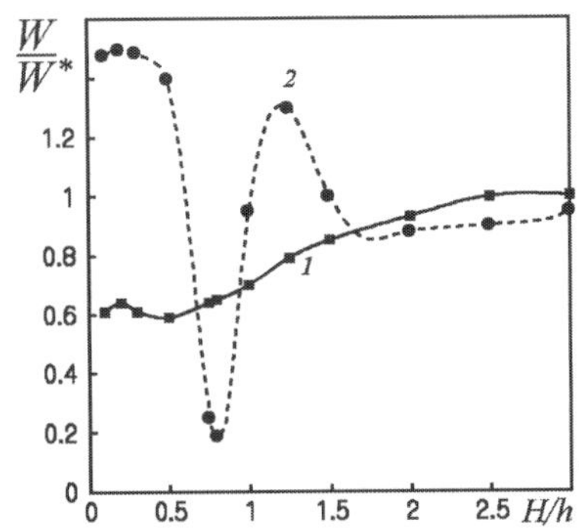

FIG. 2.55: Dependency of maximum radiation power of the shell on the value of H/h: 1—acoustically rigid screens, 2—acoustically soft screens

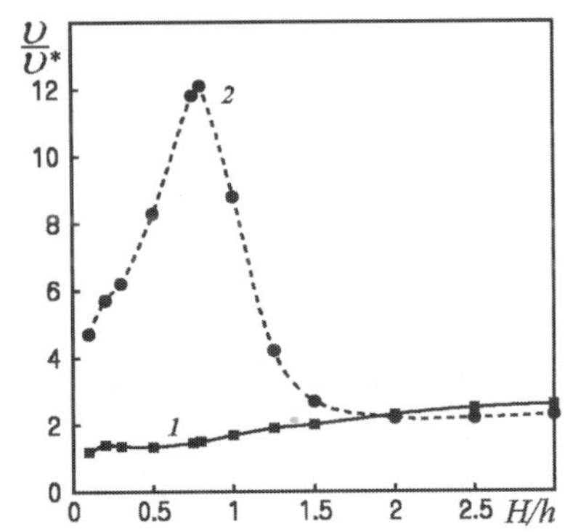

FIG. 2.56: Dependency of the corresponding amplitude of the shell particle velocity on the value of H/h: 1—acoustically rigid screens, 2—acoustically soft screens

We can see an increase of the dependencies of the acoustic power and velocity amplitude of the shell on the value of H/h in the case of the rigid screens (curves 1). The indicated dependencies become nonuniform (see curves 2) because of the presence of the acoustically soft screens, specially in the vicinity of $H/h = 0.8$. Here the amplitude of the shell velocity is maximum, and radiation power is minimum. It points to the acoustic shirt circuit between the inner and outer surfaces of the shell. In practice, it may lead to the mechanical break-down of the shell due to the mechanical stresses, arising in the shell, which exceed allowable values.

2.7 SOUND RADIATOR IN THE FORM OF A DISC

The problem of sound radiation by a circular disc holds a specific place in acoustics. It is impossible now to say when it was first formulated. But we can say that as early as in the times of John Rayleigh it aroused interest, and even John Rayleigh paid a lot of attention to this problem [1]. A lot of acousticians then worked on this problem, the first were McLachlan, H. Stenzel, L.Ya. Gutin, E.T. Hanson, S.N. Rzhevkin, E.L. Shenderov [3,72,80–83].

The question is why this problem has been relevant for more than 150 years. From our point of view, it is caused by two reasons. Firstly, because of wide practical application of this problem. In many practical situations the field, created by vibrating elements of acoustic devices, can be represented as the field of vibrating circular disk. Such elements are, for example, the diffusers of electrodynamics and electrostatic loudspeakers [6,80,84], core and laminated piezoceramic transducers of hydro-acoustic arrays [38]. Besides, the data on the disk radiation field are often used for the estimation of the fields, created by vibrating surfaces of different building structures, vehicles and machinery.

The second reason is that the problem of sound radiation by circular disc is of physical and mathematical interest. Study of the far and near fields of disc radiation is instructive and allows us to demonstrate the interesting peculiarities of such complex wave process as sound diffraction. Besides, the circular disc has simple geometrical form. Thus, the problem of sound radiation by the disc is often used as a check problem to compare the effectiveness of the different methods of solution.

Let us mention briefly the existing methods of the solution of the problem on sound radiation by circular disc. The disc is in an infinite rigid plane screen and there is no space between the disc and the screen (i.e., particle velocity at the surface of the screen is considered to be equal to zero). It is often assumed that the disc is a plane and infinitely thin object. The disc does not have a mass and bending stiffness. The screen with the disc is situated in an ideal acoustic medium with density ρ and sound velocity c. The disc surface executes harmonic vibrations with circular frequency ω and amplitude, which is defined as

$$\upsilon_{\mathbf{n}}(S) = \upsilon_0 g(S), \tag{2.109}$$

here index $\mathbf{n}$ indicates the direction of the outward normal to the surface of the disc, υ_0—vibration amplitude, having dimension of velocity, $g(S)$—dimensionless function determining the velocity distribution over the disc surface.

The simplest and well-studied situation is when the value of the velocity amplitude over the entire disc surface is constant ($g(S) = 1$). In this case, the vibrating disc is often called "piston diaphragm" or just "piston" [3]. The sound field, created by arbitrary shaped piston, can be defined basing on Huygens principle. The rigorous analytical representation of the pressure field based on Huygens principle is given as the Kirchhoff integral [3,4,11,18]:

$$p(M) = \frac{1}{4\pi} \int_S \left(\frac{\partial p}{\partial n} \frac{\exp(ikr)}{r} - p \frac{\partial}{\partial n} \left(\frac{\exp(ikr)}{r} \right) \right) dS, \qquad (2.110)$$

where $p(M)$—pressure in some point of observation M, r—the distance from the point of observation M to the element dS of the piston surface S, $k = \omega/c$—wave number. Unfortunately, it is difficult to use formula (2.110) for calculation of sound fields, because it is necessary to set pressure and particle velocity on the piston surface. In such a case we can say that a problem is "over-determined" [11]. But Rayleigh showed that when a piston is in an infinite rigid screen, expression (2.110) can be reduced to an integral (Rayleigh integral):

$$p(M) = -\frac{i\omega\rho}{2\pi} \upsilon_0 \int_S g(S) \frac{\exp(ikr)}{r} dS. \qquad (2.111)$$

Integral (2.111) can be evaluated only numerically for arbitrary form of the piston and arbitrary velocity distribution over its surface. When the piston is circular and the velocity distribution over its surface is uniform, integral (2.111) over the piston surface can be represented as a single integral over the piston boundary [3,4]. Basing on these facts, Rayleigh obtained the simple analytical expression for the radiation impedance of the circular piston [1]

$$Z_r^{(P)} = \rho c S \left(1 - \frac{2J_1(2ka)}{2ka} - i \frac{2\mathbf{S}_1(2ka)}{2ka} \right); \qquad (2.112)$$

which is known in modern literature as Rayleigh formula. Here $J_1(2ka)$—Bessel function, $\mathbf{S}_1(2ka)$—Struve function, $S = \pi a^2$, a—piston radius. Note that the relatively simple formulas for the estimation of the square piston field radiation were obtained later [85].

It may be noted that the problem on radiation by the piston, situated in rigid infinite screen, is mathematically equivalent to the problem on radiation by pulsing piston without a screen. It follows from the symmetry condition of the problem. By pulsing piston is meant infinitely thin disc. The particle velocities, which are set on the both surfaces of this disc, have the same dimensions but are directed in opposite sides.

It is important to note the works of McLachlan [80] and specially Stenzel, H. [81]. By the end of the 1930s they performed a large amount of work on the computation of near field, created by circular piston. These data facilitated progress in the creating of hydroacoustic arrays for the detecting devices during the Second World War and just after the war.

L.Ya. Gutin and E.T. Hanson [82,83] took a completely different approach for the estimation of pulsing field and oscillating piston (without screens). By oscillating piston is meant infinitely thin disc. The particle velocities, which are set on the both surfaces of this disc, have the same dimensions and direction. Due to the antisymmetry condition, the problem of oscillating piston radiation is mathematically equivalent to the problem of radiation by the piston, situated in an infinite acoustically soft screen (i.e. the screen, at the surface of which pressure is equal to zero). Gutin and Hanson used Helmholtz equation in spheroidal system of coordinates for degenerated case, when one of the spheroid axes tended to zero (it is so called oblate spheroid system of coordinates [22]). These authors obtained the expressions for the radiation impedance and sound field, created by pulsing and oscillating pistons, represented by series in the spheroidal functions. In particular, Gutin determined radiation impedance, with satisfactory accuracy for practical purposes, up to the wave dimension of piston diameter approximately equal to 1.3. But this approach was not developed because of the difficulties in work with spheroid functions.

Relatively recently (in 1989), Shenderov E.L. took one more approach to the solution of the problem on sound radiation by oscillating piston [72]. He used Hankel integral transformation. The initial problem reduces to the solution of pair integral equations in the unknown sound pressure function. Then the approximate numerical solution of initial problem was obtained. An important benefit of this approach is the possibility to consider the cases of nonuniform particle velocity distribution over the piston surface. For example, Shenderov estimated the radiation field of a circular membrane, fixed at the edge, which was vibrating at its first mode [72].

Having carried out the analysis of the sources of literature, we can say that each of the considered above approaches to the solution of problem on sound radiation by circular piston is very interesting and has its own advantages and disadvantages.

The purpose of this paragraph is to consider an effective approach [4], allowing us to find the solutions for the large class of problems, where the main component is an oscillating piston or disc. It is based on the partial domains method. Particle velocity distribution over the disc surface is arbitrary.

2.7.1 Disc in an Infinite Flat Screen

We assume that a disc with radius a is situated in an infinite acoustically rigid, or acoustically soft screen (Fig. 2.57). The disc executes harmonic vibrations with frequency ω. The amplitude distribution over its surface is defined by Eq. (2.109).

Let us locate the origin of the spherical (r, θ, ψ) and cylindrical (R, ψ, z) coordinate systems in the center of the disc. Following the partial domains method, we partition all the sound field existence domain into two domains: domain I—$r \leq a$, $\theta = [0, \pi/2]$, $\psi = [0, 2\pi]$; domain II—$r \geq a$, $\theta = [0, \pi/2]$, $\psi = [0, 2\pi]$. In each of these domains the pressure field is represented as superposition of Helmholtz equation particular solutions. Assuming that the considered problem is symmetric with respect to the ROz plane, the field in domain II can be represented as follows

$$p_{II} = \sum_{m=0}^{\infty} \sum_{n=0}^{\infty} B_{mn} h_{\nu_n}^{(1)} (kr)\, P_{\nu_n}^{(m)} (\cos\theta) \cos(m\psi), \tag{2.113}$$

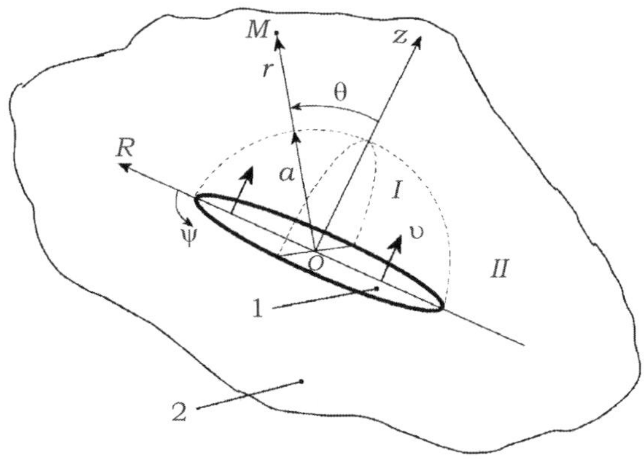

FIG. 2.57: Disc in an infinite plane screen (general case); 1—disc, 2—screen

where $h_{\nu_n}^{(1)}(kr)$—Hankel spherical functions of first kind, $P_{\nu_n}^{(m)}(\cos\theta)$—associated Legendre functions of first kind [3,22]. The choice of index value ν_n, namely $\nu_n = 2n$, $n = 0, 1, 2, \ldots$ or $\nu_n = 2n+1$, $n = 0, 1, 2, \ldots$ is determined by the boundary condition at the surface of rigid or soft screen respectively and by the index value $m = 0, 1, 2, \ldots$. Solution (2.113) is general for corresponding domain II, because of the completeness and orthogonality of the selected set of angle functions at the surface $\theta = [0, \pi/2]$, $\psi = [0, 2\pi]$, $r \geq a$. The matching conditions of the sound fields at the boundaries of domains I and II can be fulfilled due to the coefficients $B_{m\nu_n}$ selection.

The field in domain I can be represented as the sum of two summands:

$$p_I = p_I^{(1)} + p_I^{(2)}. \tag{2.114}$$

By analogy with expression (2.113) the constituent $p_I^{(1)}$ can be written as follows

$$p_I^{(1)} = \sum_{m=0}^{\infty}\sum_{n=0}^{\infty} A_{mn} j_{\nu_n}(kr) P_{\nu_n}^{(m)}(\cos\theta)\cos(m\psi), \tag{2.115}$$

where $j_{\nu_n}(kr)$—spherical Bessel functions of first kind. The choice of the index values ν_n is determined by the boundary conditions at the disc surface and index value m. Spherical Bessel functions of the first kind are used as radial functions in Eq. (2.115). These functions describe standing waves in bounded domain I and provide finite value of pressure at the point $r = 0$.

The constituent $p_I^{(2)}$ should be represented as the superposition of propagating waves and nonuniform waves, moving from the disc surface along the Oz axis:

$$p_I^{(2)} = \sum_{m=0}^{\infty}\sum_{n=1}^{\infty} C_{mn} J_m\left(\mu_n^{(m)} R\right)\exp\left(i\sqrt{k^2 - \left(\mu_n^{(m)}\right)^2} \times z\right)\cos(m\psi), \tag{2.116}$$

here the values $\mu_m^{(n)}$ can be defined from the equation solution $J_n'(x) = dJ_n(x)/dx = 0$ when $x = \mu_n^{(m)} a$, $n = 0, 1, 2, \ldots$, $m = 1, 2, \ldots$. The expression for $p_I^{(2)}$ and $p_I^{(1)}$ allows us to fulfill the sound field matching conditions at the boundary of domains I and II and fulfill the boundary condition (2.109) at the disc surface

$$\frac{1}{i\omega\rho}\frac{\partial p_I^{(1)}}{\partial z}\bigg|_{z=0} + \frac{1}{i\omega\rho}\frac{\partial p_I^{(2)}}{\partial z}\bigg|_{z=0} = \upsilon_z(S). \tag{2.117}$$

Unknown coefficients A_{mn}, B_{mn}, and C_{mn} in (2.113), (2.115), and (2.116) can be determined from the system of functional equations. It combines boundary condition (2.117) and the following matching conditions of the sound fields at the boundary of partial domains I and II:

$$p_I = p_{II}, \quad r = a, \quad \theta = [0, \pi/2], \quad \psi = [0, 2\pi], \tag{2.118}$$

$$\frac{\partial p_I}{\partial r} = \frac{\partial p_{II}}{\partial r}, \quad r = a, \quad \theta = [0, \pi/2], \quad \psi = [0, 2\pi]. \tag{2.119}$$

We substitute expressions (2.113), (2.115) and (2.116) in (2.117)–(2.119) and we use the completeness and orthogonality properties of the corresponding angle functions. We obtain infinite system of linear algebraic equations with respect to the unknown complex coefficients $A_{m\nu_n}$, $B_{m\nu_n}$, and C_{mn}. Infinite system of linear algebraic equations can be solved by reduction method. The convergence of obtained solution can be determined in the course of numerical experiment while investigating the matching conditions of the sound fields at the boundary of domains I and II.

Let us now consider some particular cases that are of practical importance. We assume that particle velocity is the same over the whole surface of the disc (piston) $\upsilon_z(S) = \upsilon_0$. In this case the problem becomes axially symmetric and it is enough to consider it in the ROz plane (Fig. 2.58). The expressions for the sound fields in partial domains I and II become simpler:

$$p_{II} = \sum_{n=0}^{\infty} B_n h_{\nu_n}^{(1)}(kr)\, P_{\nu_n}(\cos\theta), \tag{2.120}$$

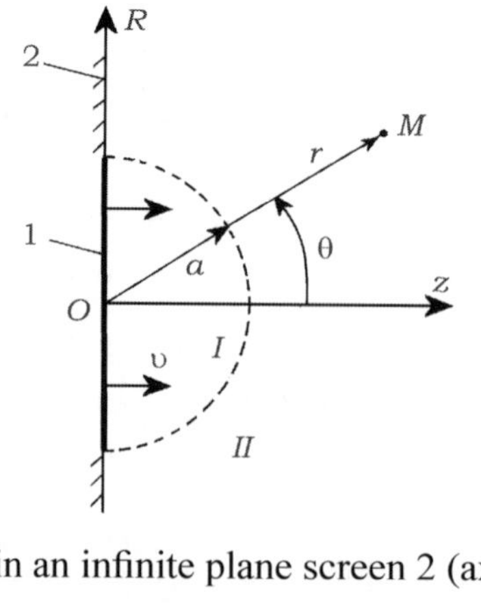

FIG. 2.58: Disc 1 in an infinite plane screen 2 (axisymmetric case)

$$p_I = \sum_{n=0}^{\infty} A_n j_{2n}(kr) P_{2n}(\cos\theta) + \rho c \upsilon_0 \exp(ikr\cos\theta), \qquad (2.121)$$

where $\nu_n = 2n$, $n = 0, 1, 2, \ldots$ for the case of acoustically rigid screen and $\nu_n = 2n+1$, $n = 0, 1, 2, \ldots$ for the case of acoustically soft screen.

Let us first consider the convergence of the obtained solution. Let us determine the discrepancy of matching conditions of the sound fields at the boundary of domains I and II. Average discrepancy values in pressure $\delta_p = (p_I - p_{II}) / |p_{II}|$ and particle velocity $\delta_\upsilon = (\upsilon_{rI} - \upsilon_{rII}) / |\upsilon_{rII}|$ at the boundary of domains I and II are given in Table 2.2 ($ka = 7$). When performing the reduction of infinite system of algebraic equations the summation index varies from 0 to N. It should be noted that the discrepancy increases slightly near the edge of the disc ($r = a$, $\theta = 90°$) (especially for the case of acoustically soft screen). It is a predictable result within the framework of the considered mathematical model. The mentioned calculations give an idea about the point-by-point convergence of the solution in the near field of the disc.

We turn now to an integral estimation of the convergence process. We compare the acoustic power W_1 at the disc surface (domain I) and acoustic power W_2 in the far field of the disc (domain II). The discrepancy computations in power $\delta_W = (W_1 - W_2) / W_2$ are represented in Table 2.3. We can see that there is satisfactory convergence of the solution.

We proceed now to the analysis of the numerical data characterizing the disc sound field. Figures 2.59(a) and 2.59(b) illustrates dependencies of the normalized pressure amplitude $\tilde{p} = |p| / (\rho c \upsilon_0)$ along the z axis for small ($ka = 1$) and large enough ($ka = 7$) wave radius of the disc. Curves 1 correspond to the case of acoustically rigid screen, curves 2 correspond to the case of acoustically soft screen. The points correspond to the calculation by Eq. (2.122) for the pressure amplitude on the disc axis in acoustically rigid screen. Equation (2.122) is obtained by Rayleigh [1,3]:

TABLE 2.2: Estimates of the residuals for pressure and velocity

	Acoustically rigid screen		**Acoustically soft screen**	
	$N = 10$	$N = 20$	$N = 10$	$N = 20$
$\lvert \mathrm{Re}\delta_p \rvert$	10^{-9}	5×10^{-15}	0.016	0.009
$\lvert \mathrm{Im}\delta_p \rvert$	0.012	0.003	0.024	0.01
$\lvert \mathrm{Re}\delta_\upsilon \rvert$	2×10^{-9}	10^{-14}	0.022	0.02
$\lvert \mathrm{Im}\delta_\upsilon \rvert$	0.003	0.0007	0.032	0.022

TABLE 2.3: Estimate of the residual for power

	Acoustically rigid screen		**Acoustically soft screen**	
	$N = 10$	$N = 20$	$N = 10$	$N = 20$
$\lvert \delta_p \rvert$	10^{-7}	5×10^{-14}	10^{-3}	10^{-4}

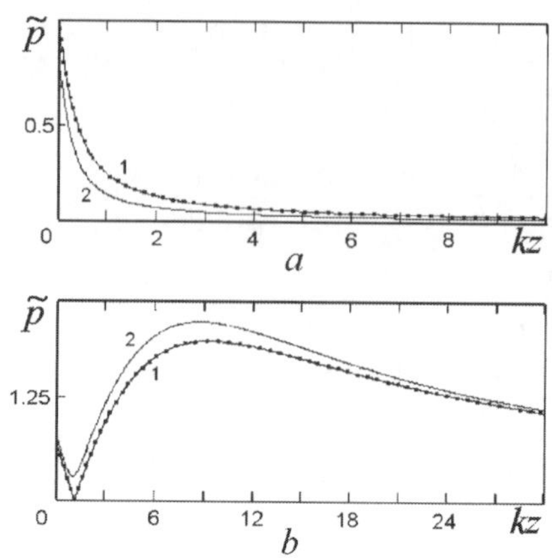

FIG. 2.59: Sound pressure distribution along the disc axis (points—calculation by Eq. (2.122)); (a)—$ka = 1$, (b)—$ka = 7$; 1—rigid screen, 2—soft screen

$$\frac{|p|}{\rho c v_0} = \left| 2 \sin\left(\frac{\sqrt{(ka)^2 + (kz)^2} - kz}{2} \right) \right| . \tag{2.122}$$

As we can see, two variants of the pressure amplitude calculation on the disc axis in the acoustically rigid screen match together within graphical accuracy. Figure 2.60 shows the analogical results for the pressure amplitude along the surface of the disc and screen.

Figure 2.61 shows the directivity patterns in pressure for two values of ka. For the small wave dimensions of the disc [$ka = 1$, Fig. 2.61(a)] the pattern shapes depend essentially on the properties of the screen. When the screen is rigid, the disc is almost non-directional (i.e., it radiates like monopole). When the screen is soft, the disc is directional (it radiates like dipole). When ka increases [Fig. 2.61(b)], the directivity patterns match together with graphical accuracy for both variants of the screen. These facts are

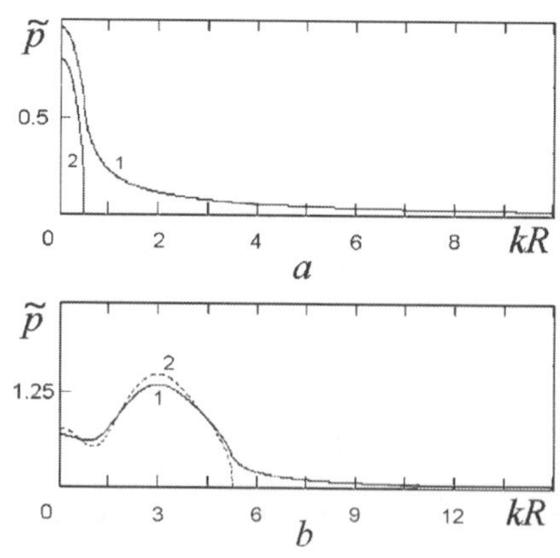

FIG. 2.60: Sound pressure distribution along the surface of the disc and screen; (a)—$ka = 1$, (b)—$ka = 7$; 1—rigid screen, 2—soft screen

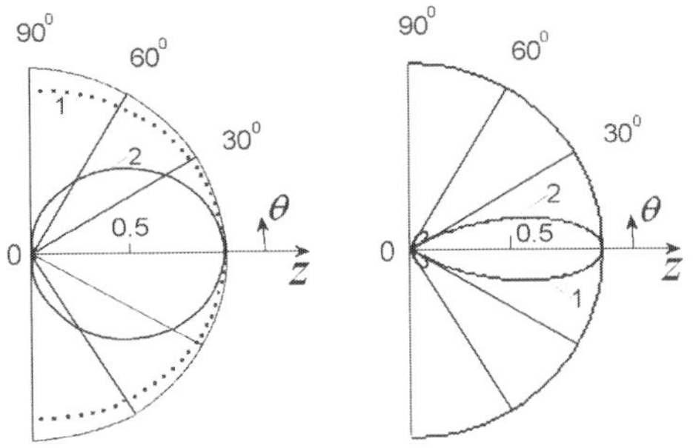

FIG. 2.61: Directivity patterns of the disc in the screen; (a)—$ka = 1$, (b)—$ka = 7$; 1—rigid screen, 2—soft screen

well known [3,72]. They result from the fact that when the disc wave radius is sufficiently large, the sound energy in the far field concentrates along the disc axis. Consequently, the influence of the properties of the screen becomes unessential.

Now we turn attention to such important integral characteristic of sound radiators as radiation impedance. As it is known [3,4,72], it is determined by the expression

$$Z_r = \frac{1}{|\tilde{\upsilon}|^2} \int\limits_S p\upsilon^* dS. \tag{2.123}$$

Here $\tilde{\upsilon}$—the particle velocity of the disc surface in reference point, and υ^*—the complex conjugate value of particle velocity υ. In this case, radiation impedance can be written as follows:

$$Z_r = \rho c S \left[1 + \frac{2}{(ka)^2} \sum_{n=0}^{\infty} A_n P_{2n}(0) \int\limits_0^{ka} j_{2n}(\xi)\, \xi d\xi \right]. \tag{2.124}$$

Figure 2.62(a) shows real $R' = \mathrm{Re}\,(Z_r/\rho c S)$ (curves 1, 2) and imaginary $X' = \mathrm{Im}\,(Z_r/\rho c S)$ (curves $1', 2'$) parts of dimensionless radiation impedance of the disc in acoustically rigid and acoustically soft screens, as the functions of the disc wave radius ka. Figure 2.62(b) illustrates the analogical dependencies for the situation, when the particle velocity distribution over the disc surface corresponds to the first mode of the vibrations of circular membrane fixed along the counter line, i.e., $\upsilon\,(r, \theta = \pi/2) = \upsilon_0 J_0\,(\eta r/a)$. The η parameter is determined as the first root of the equation $J_0\,(\eta) = 0$ and its value is $\eta = 0.7655\pi$. When we compare the graphs in Figs. 2.62(a) and 2.62(b), we can see that the components level of radiation impedance in the case when the disc is fixed along the counter line is lower than when disc moves like piston.

It is interesting to compare the calculation of disc radiation impedance in an infinite rigid screen by Eq. (2.124) and by Rayleigh formula (2.112). For this purpose the discrepancy calculation of imaginary and real parts of impedance Z_r were carried out as the function of the ka:

$$\delta_1 = \left| \mathrm{Re}\left(Z_r - Z_r^{(R)}\right) / \mathrm{Re} Z_r^{(R)} \right|,$$

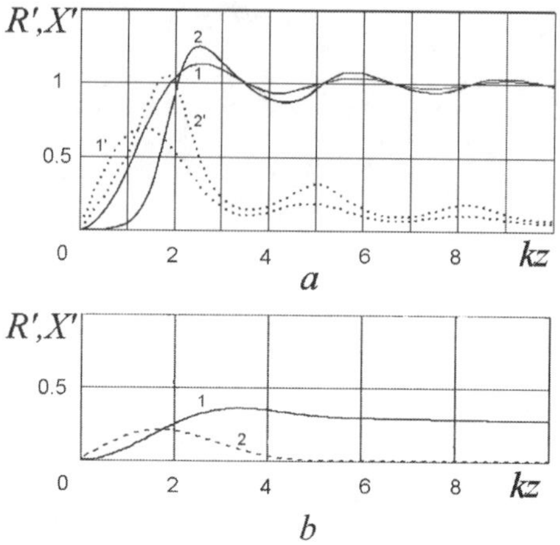

FIG. 2.62: Dimensionless radiation impedance of the disc in an infinite screen; (a)—for $\upsilon(S) = \upsilon_0$; (b)—for $\upsilon(S) = \upsilon_0 J_0(\eta(r/a))$, $\eta = 0.7655\pi$; 1, 2—real parts, $1', 2'$—imaginary parts, $1, 1'$—rigid screen, $2, 2'$—soft screen

$$\delta_2 = \left|\mathrm{Im}\left(Z_r - Z_r^{(R)}\right) / \mathrm{Im} Z_r^{(R)}\right|$$

(here superscript (R) means, that impedance is calculated by Rayleigh formula). It turned out that (see Fig. 2.63) when $ka \leq 10$, discrepancy is $\delta_1 < 3 \times 10^{-14}$, and discrepancy is practically determined by imaginary parts Z_r. The value of discrepancy decreases rapidly when the number N increases. When $ka = 10$ the value $\delta_2 = 0.04$ for $N = 20$, and for $N = 25$ we have $\delta_2 = 0.027$. With a decrease of wave dimension ka discrepancy decreases substantially.

2.7.2 Disc in a Hemispherical Screen

Let us consider the situation when the disc is situated in a screen of finite sizes. For example, the disc is situated in a hemispherical screen, with the same radius a. The screen is acoustically rigid or acoustically soft (see Fig. 2.64). We assume that velocity on the disc surface is uniform, i.e., $\upsilon_z(S) = \upsilon_0$. For the case when the disc is in an infinite screen (Fig. 2.58), the disc radiates only in semi-space. In the situation, illustrated in

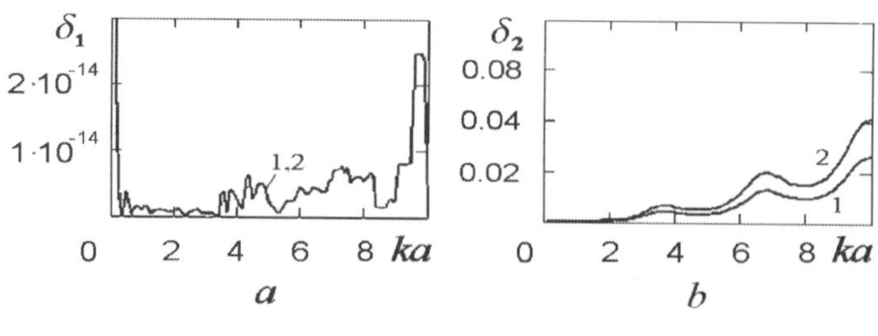

FIG. 2.63: The values of discrepancy between our data and calculation by Rayleigh formula (2.112); 1—when $N = 25$, 2—when $N = 20$

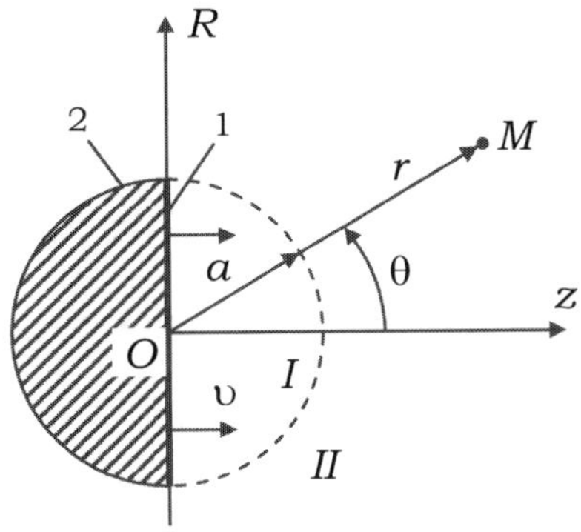

FIG. 2.64: Disc 1 in hemispherical screen 2

Fig. 2.64, the disc radiates sound in the whole space. Consequently, domain II is now the space surrounding the sphere with radius a. The field in domain I is determined by expression (2.121). Bearing it in mind, we write the following expression for the field in domain II:

$$p_{II} = \sum_{n=0}^{\infty} B_n h_n^{(1)}(kr) P_n(\cos\theta). \tag{2.125}$$

The system of functional equations, which determines the matching conditions of the sound fields at the boundary of partial domains I and II for the case of rigid hemisphere, should be written as follows:

$$p_I = p_{II}, \quad r = a, \quad \theta = [0, \pi/2], \quad \psi = [0, 2\pi] \tag{2.126}$$

$$\frac{\partial p_{II}}{\partial r} = \begin{cases} \dfrac{\partial p_I}{\partial r}, & r = a, \quad \theta = [0, \pi/2], \quad \psi = [0, 2\pi], \\ 0, & r = a, \quad \theta = [\pi/2, \pi], \quad \psi = [0, 2\pi], \end{cases} \tag{2.127}$$

and for the case of soft hemisphere it should be written as follows:

$$p_{II} = \begin{cases} p_I, & r = a, \quad \theta = [0, \pi/2], \quad \psi = [0, 2\pi], \\ 0, & r = a, \quad \theta = [\pi/2, \pi], \quad \psi = [0, 2\pi]. \end{cases} \tag{2.128}$$

$$\frac{\partial p_I}{\partial r} = \frac{\partial p_{II}}{\partial r}, \quad r = a, \quad \theta = [0, \pi/2], \quad \psi = [0, 2\pi]. \tag{2.129}$$

The calculations analogical to those, which were done for the disc in an infinite screen, are represented bellow. Figure 2.65 illustrates the pressure amplitude distribution along the disc axis. Figure 2.66 shows amplitude distribution over the disc surface. We compare these data with the data in Figs. 2.59 and 2.60. We can see that the change of the plane screen properties (rigid or soft), has a greater impact on the near field of the disc than the corresponding change of the properties of the hemisphere surface.

Figure 2.67 shows directivity patterns of the disc in a hemispherical screen.

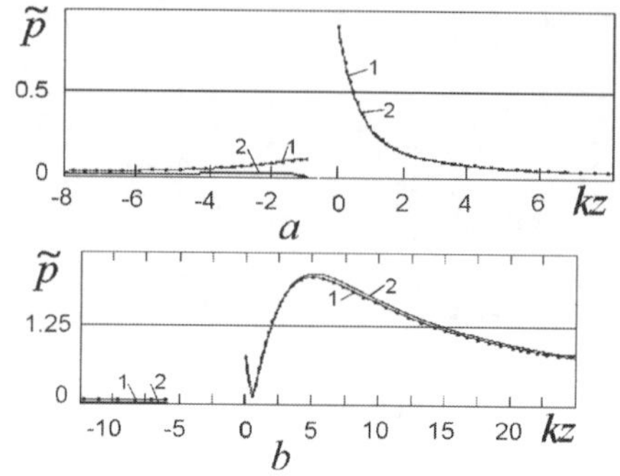

FIG. 2.65: Normalized sound pressure distribution along the disc axis; (a)—$ka = 1$, (b)—$ka = 7$; 1—rigid hemisphere, 2—soft hemisphere

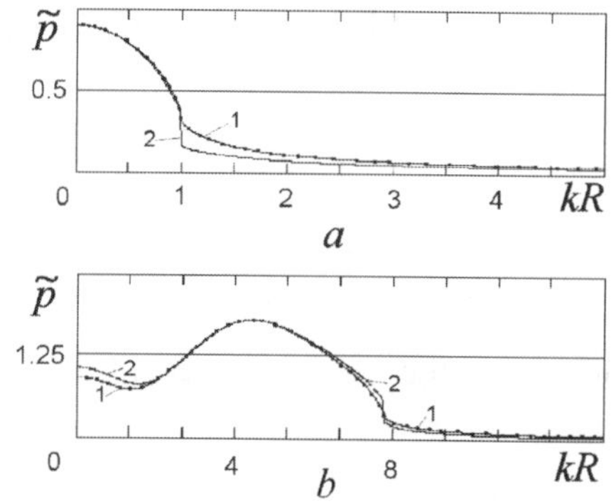

FIG. 2.66: Normalized sound pressure distribution along the disc surface; (a)—$ka = 1$, (b)—$ka = 7$; 1—rigid hemisphere, 2—soft hemisphere

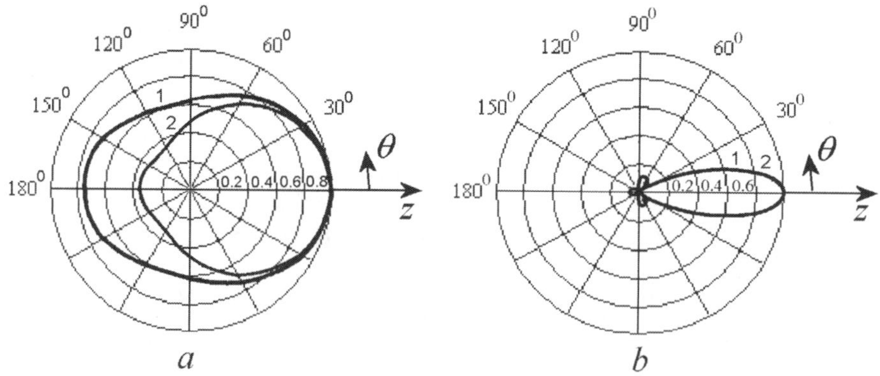

FIG. 2.67: Directivity pattern of the disc in a hemispherical screen; (a)—$ka = 1$, (b)—$ka = 7$; 1—rigid hemisphere, 2—soft hemisphere

As we can see, when the wave dimension of the disc is small, [Fig. 2.67(a)] the influence of the hemisphere properties is significant. The directional properties of the acoustically soft hemisphere are stronger than in the case of acoustically rigid hemisphere. In particular, the back lobe, in the case of the soft hemisphere, is almost half as large as in the case of the rigid hemisphere. The graphs for the disc on hemisphere and in an infinite screen are different [compare Figs. 2.67(a) and 2.61(a)]. When the wave radius is comparatively large [$ka = 7$, Fig. 2.67(b)], the hemisphere properties are of no importance. Here the directivity patterns almost coincide with the directivity patterns of the disc in an infinite plane screen [compare Figs. 2.67(b) and 2.61(b)].

The radiation impedance of the disc on hemisphere as the function of wave radius ka is shown in Fig. 2.68. Let us compare these graphs with the analogous graphs for the disc in an infinite screen [Fig. 2.62(a)]. As we can see, the shapes of the curves are similar, but they differ from each other in the domain of the small wave values ($ka \leq 2$). The real parts Z_r are much different from each other when the disc is in an infinite plane rigid or soft screen. For the disc on hemisphere this difference is negligible.

2.7.3 Disc in a Finite Circular Screen

As we mentioned before (and it is well known from literature [6,72,80,84]), an oscillating disc, as a sound radiator, is less effective in the domain $ka \leq 2$, than a pulsating disc. It is related to the fact that the pressure on its surfaces is of different signs. As the result of this, the wave effect, known as acoustical shirt circuit, arises [3,6,84]. This effect leads to a considerable decrease of radiation impedance. Consequently, such devices as the electrodynamics sound radiators are usually placed in the screens. It leads to the weakening of acoustic shirt circuit effect and to the increasing of radiation impedance of electrodynamics radiator at low frequencies. Bellow we will consider the degree of impact that a ring acoustically rigid screen has on its radiation impedance using the example of oscillating disc.

We assume that the oscillating disc, performing reciprocating motion, is situated in a ring screen (Fig. 2.69). The outer radius of the screen is equal (b), and the inner radius is equal to the disc radius a. The surface $\theta = \pi/2$, $\psi = [0, 2\pi]$, $r > b$ (dashed lines in Fig. 2.69) is acoustically soft for the case of oscillating disc. Then we have three partial domains: domain I—$r \leq a$, $\theta = [0, \pi/2]$; domain II—$a \leq r \leq b$, $\theta = [0, \pi/2]$; domain III—$r \geq b$, $\theta = [0, \pi/2]$.

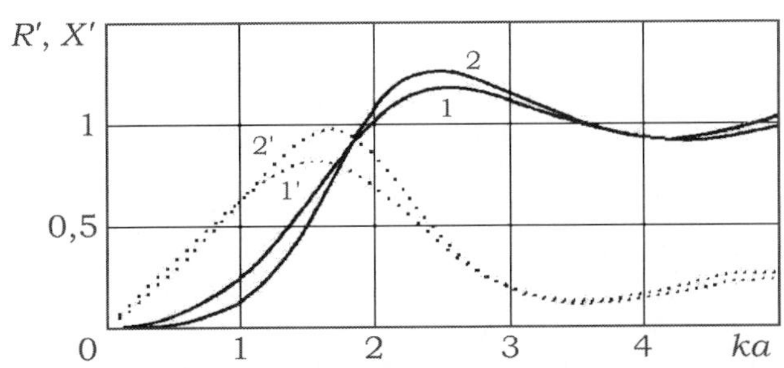

FIG. 2.68: Normalized real (curves 1, 2) and imaginary (curves $1'$, $2'$) parts of the disc radiation impedance in hemispherical screen; 1, $1'$—rigid hemisphere, 2, $2'$—soft hemisphere

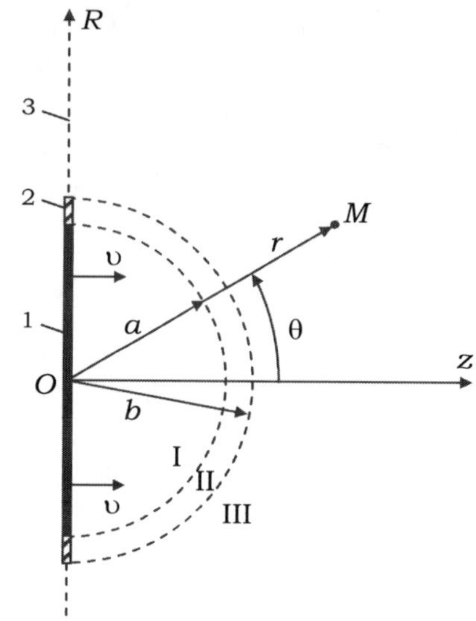

FIG. 2.69: Oscillating disc in a ring screen of finite sizes; 1—disc, 2—ring rigid screen, 3—infinite plane acoustically soft screen

The pressure field in domain III is determined by formula (2.120), and it is determined by (2.121) in domain I. The field in domain II should be written as two sets of particular solutions of Helmholtz equation, because this domain is bounded by two surfaces: $\theta = [0, \pi/2]$, $r = a$ and $\theta = [0, \pi/2]$, $r = b$. Consequently,

$$p_{II} = \sum_{n-0}^{\infty} C_n j_{2n}(kr) P_{2n}(\cos\theta) + \sum_{n-0}^{\infty} D_n y_{2n}(kr) P_{2n}(\cos\theta), \tag{2.130}$$

where $y_{2n}(kr)$—spherical Bessel functions of the second kind (they are often called as Neumann functions).

System of functional equations, which determines the sound fields matching at the boundaries of the partial domains, is of the following form

$$p_I = p_{II}, \quad r = a, \quad \theta = [0, \pi/2], \quad \psi = [0, 2\pi], \tag{2.131}$$

$$\frac{\partial p_I}{\partial r} = \frac{\partial p_{II}}{\partial r}, \quad r = a, \quad \theta = [0, \pi/2], \quad \psi = [0, 2\pi]. \tag{2.132}$$

$$p_{II} = p_{III}, \quad r = b, \quad \theta = [0, \pi/2], \quad \psi = [0, 2\pi], \tag{2.133}$$

$$\frac{\partial p_{II}}{\partial r} = \frac{\partial p_{III}}{\partial r}, \quad r = b, \quad \theta = [0, \pi/2], \quad \psi = [0, 2\pi]. \tag{2.134}$$

It is interesting to note, that the solution of the posed problem can be obtained in another way. It is necessary to combine domains I and II in one partial domain. We suppose that the particle velocity at the disc surface and at the rigid ring screen surface is set in the form of one function

$$\upsilon_n = \left\{ \begin{array}{ll} \upsilon_0, & 0 \le R \le a \\ 0, & a \le R \le b \end{array} \right\}. \tag{2.135}$$

As it is seen from Fig. 2.70, when the screen is absent, when $k(b-a) = 0$, real part of radiation impedance is rather small and is about 0.01. Consequently, the energetic efficiency of the oscillating disc, as a sound radiator, is also rather low. But, the real part of impedance increases rapidly with an increase of the value of $k(b-a)$. When the value is $k(b-a) \approx 1.2$ it reaches the level 0.25. It corresponds to the level of a pulsating disc when the value of $ka \approx 0.5$ [compare with curve 1 in Fig. 2.62(a)]. For $k(b-a) \ge 1.2$ the real part of radiation impedance vibrates, approaching the value of 0.25 asymptotically.

The influence of the screen on the directivity pattern depends heavily on the wave dimension of the disc ka. For the small values of ka, the screen influences significantly on the directivity pattern. When the value of ka is large, the influence of the screen is imperceptible. This question was examined in details in monograph [72], but the solution of this task was found by the other method.

2.7.4 Disc in a Free Space

Let us consider the disc vibration in a free space, without any screens (Fig. 2.71). The amplitudes and phases of the particle velocities on the both surfaces of the disc can differ from each other. The particle velocity distribution on the disc surface can be different. Let us consider the particular case, when particle velocity is distributed uniformly on the both surfaces of the disc.

We suppose that the amplitude of the particle velocity of the right hand side of the disc $\upsilon_1 = \upsilon_0$, and the amplitude of the left hand side $\upsilon_2 = \upsilon_0 \xi \exp(-i\alpha)$, see Fig. 2.71. Here the coefficient ξ determines the ratio of velocity amplitudes $\xi = |\upsilon_2|/|\upsilon_1|$ on the both sides of the disc, and the parameter $\alpha \ge 0$ points to the fact that the vibrations of the disc left hand side delay with respect to the right hand side. It is evident that if $\xi = 1$ and $\alpha = 0$, we have the problem on the sound radiation by pulsating disc, if $\xi = 1$ and $\alpha = \pi$—the problem of sound radiation by oscillating disc.

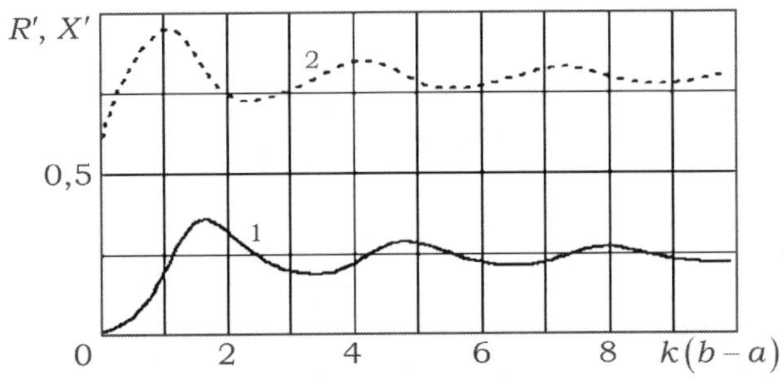

FIG. 2.70: Real (curve 1) and imaginary (curve 2) parts of the dimensionless radiation impedance of oscillating disc in the rigid ring screen of finite sizes, when the value $ka = 0.5$ is fixed and the value of the width of the screen is variable $k(b-a)$

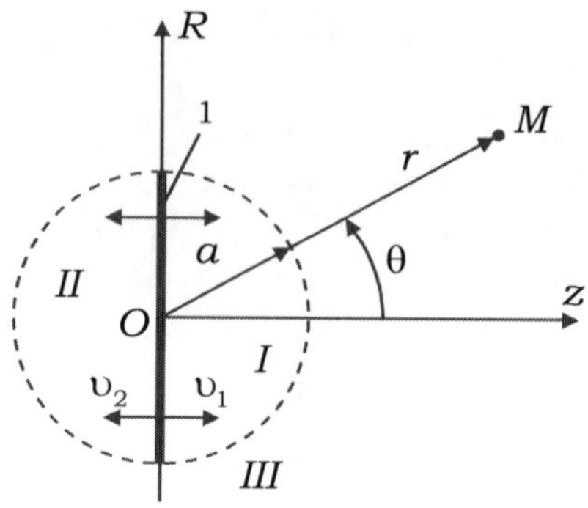

FIG. 2.71: Disc 1 in a free space near the screen

Let us partition the whole domain of the sound field existence into three partial domains: domain I—$r \leq a$, $\theta = [0, \pi/2]$; domain II—$r \leq a$, $\theta = [\pi/2, \pi]$; domain III—$r \geq a$, $\theta = [0, \pi]$. The following boundary conditions must be fulfilled on the surfaces of the disc:

$$\frac{1}{i\omega\rho}\frac{\partial p_I}{\partial \mathbf{n}} = \frac{1}{i\omega\rho}\frac{\partial p_I}{\partial z} = \upsilon_0, \tag{2.136}$$

$$\frac{1}{i\omega\rho}\frac{\partial p_{II}}{\partial \mathbf{n}} = -\frac{1}{i\omega\rho}\frac{\partial p_{II}}{\partial z} = \upsilon_0 \xi \exp(-i\alpha), \tag{2.137}$$

where $\mathbf{n}$—the outer normal to the disc surface. Bearing in mind boundary conditions (2.136) and (2.137), the expressions for pressure in each of the partial domains can be written as follows:

$$p_I = \sum_{n=0}^{\infty} A_n j_{2n}(kr) P_{2n}(\cos\theta) + \rho c \upsilon_0 \exp(ikr\cos\theta), \tag{2.138}$$

$$p_{II} = \sum_{n=0}^{\infty} B_n j_{2n}(kr) P_{2n}(\cos\theta) + \rho c \upsilon_0 \xi \exp(-i\alpha) \exp(-ikr\cos\theta), \tag{2.139}$$

$$p_{III} = \sum_{n-0}^{\infty} C_n h_n^{(1)}(kr) P_n(\cos\theta). \tag{2.140}$$

It is to be noted that the plane waves in expressions (2.138) and (2.139) are the waves, propagating from the surface of the disc.

Two variants of the sound fields matching conditions at the boundaries of partial domains are possible for the considered problem. In the first variant we match the pressure at the whole surface of the spheres with radius a (Fig. 2.71), and in the second variant we match particle velocity. Let us select the second variant, i.e., we have the following conditions at the boundary of partial domains:

$$p_{III} = p_I, \quad r = a, \quad \theta = [0, \pi/2], \quad \psi = [0, 2\pi], \tag{2.141}$$

$$p_{III} = p_{II}, \quad r = a, \quad \theta = [\pi/2, \pi], \quad \psi = [0, 2\pi], \tag{2.142}$$

$$\frac{\partial p_{III}}{\partial r} = \begin{cases} \dfrac{\partial p_I}{\partial r}, & r = a, \quad \theta = [0, \pi/2], \quad \psi = [0, 2\pi], \\ \dfrac{\partial p_{II}}{\partial r}, & r = a, \quad \theta = [\pi/2, \pi], \quad \psi = [0, 2\pi]. \end{cases} \tag{2.143}$$

The conversion of functional Eqs. (2.141) and (2.142) to an algebraic set of equations is performed on the base of completeness and orthogonality of the function $P_{2n}(\cos\theta)$, $n = 0, 1, 2, \ldots$ on the intervals $\theta = [0, \pi/2]$ and $\theta = [\pi/2, \pi]$. The conversion of Eq. (2.143) to an algebraic equation is performed on the base of completeness and orthogonality of the functions $P_n(\cos\theta)$, $n = 0, 1, 2, \ldots$ on the interval $\theta = [0, \pi]$.

We paid a lot of attention to such important characteristics of pulsating and oscillating discs as their radiation impedances, pressure distribution on their radiating surfaces, and directivity patterns. Now we dwell on the character of the distribution of the particle velocities vectors $\mathbf{V} = (V_R, V_z)$ and the intensities $\mathbf{I} = (I_R, I_z)$ of the sound field near the disc surface. The projections of the indicated vectors on the axis of coordinates were estimated as

$$V_R = \frac{1}{i\omega\rho}\frac{\partial p}{\partial R}, \quad V_z = \frac{1}{i\omega\rho}\frac{\partial p}{\partial z}, \quad I_R = \frac{1}{2}\mathrm{Re}\,(pV_R^*), \quad I_z = \frac{1}{2}\mathrm{Re}\,(pV_z^*). \tag{2.144}$$

Figure 2.72 shows the fields of particle velocity and intensity near the disc for the wave radius $ka = 2$ and the parameters values $\xi = 1$, $\alpha = 0$. It corresponds to the pulsating disc.

The analogous fields for the oscillating disc are represented for comparison in Fig. 2.73 (here $ka = 2$, $\xi = 1$, $\alpha = \pi$). Note that the beginning of the arrow corresponds to the point, at which the vector value is determined, and the length of the arrow is proportional to the modulus of the vector. Only the hemisphere $R \geq 0$ (half-disc) is shown in the figures because of the sound field symmetry with respect to the z axis. It is clear that for the pulsating disc (Fig. 2.72), the velocity and intensity vectors are totally symmetrical with regard to the $z = 0$ plane. This means that there is no interaction between the sound fields of the left and right semi-spaces. The problem of the pulsating disc in a free space

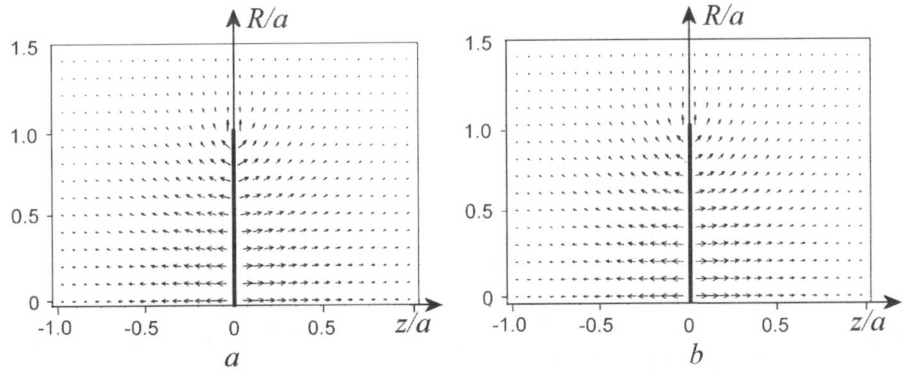

FIG. 2.72: Particle velocity amplitude vectors field (a) and intensity vectors (b) in the vicinity of the pulsating disc ($ka = 2$, $\xi = 1$, $\alpha = 0$)

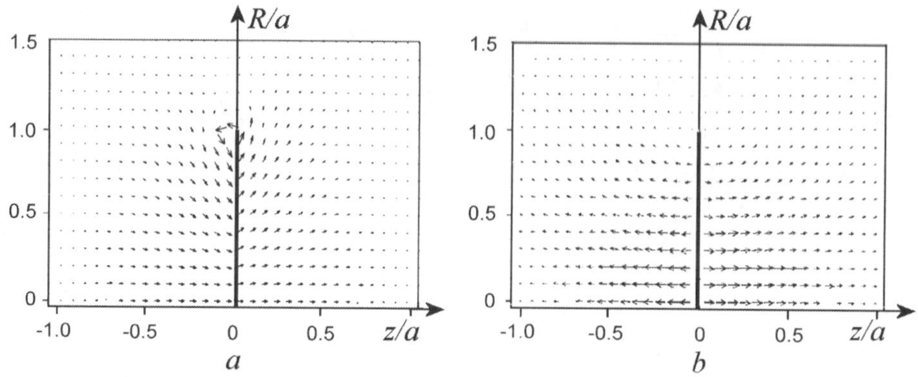

FIG. 2.73: Fields of the particle velocity amplitude vectors (a) and intensity vectors (b) in the vicinity of oscillating disc ($ka = 2$, $\xi = 1$, $\alpha = \pi$)

is equivalent to the problem of the disc vibration in an infinite plane rigid screen. A characteristic feature of the fields in Fig. 2.72 is that the vectors in the center of the disc are directed along the z axis. Vectors turn into the R axis direction when approaching the edge of the disc. In such a case the velocities amplitude and intensities levels change slightly.

The situation is different for an oscillating disc Fig. 2.73. Here, as expected, the particle velocity field becomes antisymmetrical with respect to the $z = 0$ plane. When approaching the edge of the disc, the particle velocity vectors amplitude increases and the vectors turn to the R axis [see Fig. 2.73(a)]. The most critical situation is near the edge of the disc. Surrounding medium is "pumped" from one disc surface to the other. It is a direct consequence of the oscillating character of the disc motion, which creates the opposite in sign pressure on its surfaces. It is this phenomenon that is the essence of the acoustical shirt circuit effect. The intensity field of the oscillating disc [see Fig. 2.73(b)] is symmetrical with respect to the $z = 0$ plane as in the case of pulsating disc. The intensity level falls smoothly almost to zero, when moving from the center of the disc to its edge. This points to the fact that all the energy of the disc edge vibrations is discharged on "pumping" the medium from one of its surfaces to the other. Consequently, the acoustic energy is not radiated in a surrounding medium. This is the reason why the oscillating disc is less effective than the pulsating disc, when all other parameters are common. With an increase of the wave size of oscillating disc, the negative impact of acoustic shirt circuit weakens. When $ka \geq 2$ the efficiency of the oscillating disc becomes comparable with the efficiency of pulsating disc [see the curves of radiation impedance in Fig. 2.62(a)].

Let us consider another interesting variant of the disc motion, when one of its sides is vibrating, and the other is blocked ($\xi = 0$). Such disc is called a one-sided disc and it was considered in different works [72,82], where its radiation impedance was examined. This disc is interesting from the practical viewpoint because unidirectional radiator is used in different acoustical devices. We are interested in the fields of particle velocity and intensity in the vicinity of such disc. As it is seen from Fig. 2.74, the amplitude of

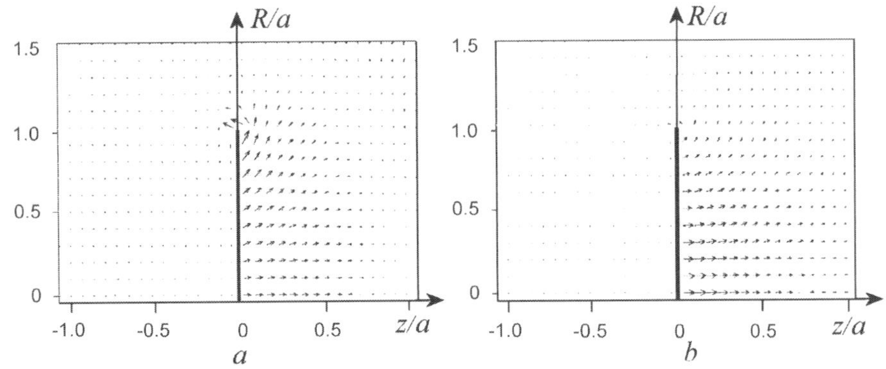

FIG. 2.74: Field of vectors of particle velocity amplitude (a) and intensity vectors (b) in the vicinity of a one-sided disc ($ka = 2$ and $\xi = 0$)

the particle velocity vectors increases when moving from the center of the disc to its edge. The vectors turn into the direction of the R axis, and they form a kind of "fan" at the edge of the disc.

Particle velocity is equal to zero at the backside of the disc. Intensity decreases near the edge of the disc, and some part of the energy "flows" over the edge of the disc into the back space. It can be assumed that, despite the absence of the vibrations of the back side of the disc, these two factors—the "fan" of particle velocities and the "flow" of sound energy over the disc edge yield that the sound is radiated by the disc in the back direction. We turn now to the directivity pattern of the one-sided disc [see Fig. 2.75(a)]. As we can see, relative back radiation level is rather noticeable and is almost 0.45. Thus, in the domain of small and moderate values of ka, the one-sided disc is not a unidirectional radiator. With further increase of ka the level of back radiation decreases and it can be considered almost unidirectional.

The question arises—whether it is possible to make the disc unidirectional by means of the proper choice of parameters ξ and α when ka is moderate? A series of numerical experiments was conducted, and we found that, at least theoretically, it can be realized.

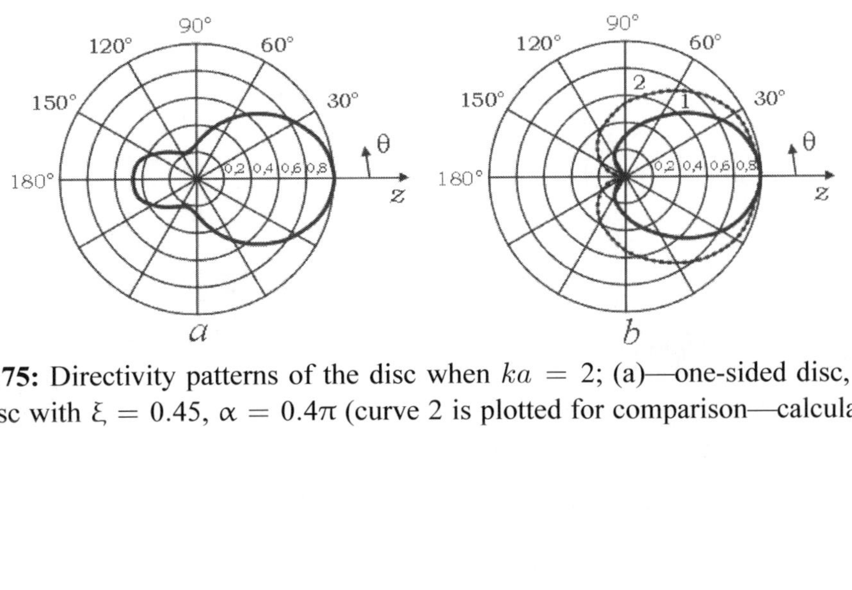

FIG. 2.75: Directivity patterns of the disc when $ka = 2$; (a)—one-sided disc, $\xi = 0$; (b)—disc with $\xi = 0.45$, $\alpha = 0.4\pi$ (curve 2 is plotted for comparison—calculated cardioid)

As an example, the directivity pattern for the case when $\xi = 0.45$, and $\alpha = 0.4\pi$, is shown in Fig. 2.75(b). A cardioidal pattern is also given here for comparison (curve 2). As we can see, a disc with such an amplitude-phase particle velocity distribution becomes a unidirectional sound radiator. Its directivity pattern is similar to cardioid, but it is narrower.

Let us, in conclusion, consider the frequency dependencies of the full normalized power radiated by disc $\widetilde{W} = \upsilon_0^2 R'/\rho c S$ when $\upsilon_0 = 1$ [Fig. 2.76(a)] and its concentration coefficient K [Fig. 2.76(b)]. Note that the curves of acoustic power must coincide with the curves for the real part of radiation impedance, because the power is determined as $W = \upsilon_0^2 \mathrm{Re} Z_r$.

As it follows from the represented graphs, for $ka \geq 2$, the power of pulsating disc 1 and oscillating disc 2 is twice as much as one-sided disc 3. It is due to the fact that the pulsating and oscillating discs radiate sound from both their surfaces, and the one-sided disc radiates only from one surface. Disc 4 radiates higher power than the one-sided disc ($\xi = 0.45$, $\alpha = 0.4\pi$). Both its surfaces vibrate, but with different efficiency. The situation is different in the domain $ka \leq 2$. The power radiated by oscillating disc decreases rapidly, and when $ka \leq 1.5$ its level becomes the lowest due to the acoustic shirt circuit. As for the pulsating and one-sided discs, the ratio of their powers remains the same as for the case of large values of ka.

We turn now to the analysis of the frequency dependency of concentration coefficient K. As one would expect (see, for example, [42]), when $ka \to 0$ the concentration coefficient of the oscillating disc tends to three, and the concentration coefficient of the pulsating and one-sided discs tends to one. For the large values of ka the concentration coefficients of pulsating disc 1 and oscillating disc 2 increase and become almost equal. The concentration coefficients of one-sided disc 3 and disc 4 also become equal, but they are twice as much as the concentration coefficients of disc 1 and disc 2.

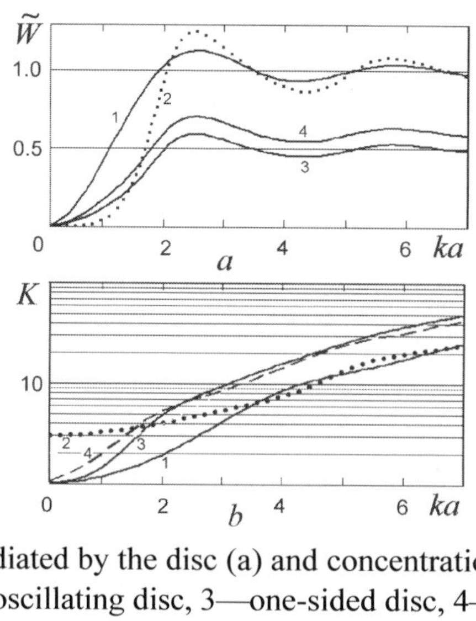

FIG. 2.76: Full power radiated by the disc (a) and concentration coefficient (b) for $\upsilon_0 = 1$; 1—pulsating disc, 2—oscillating disc, 3—one-sided disc, 4—disc with the parameters $\xi = 0.45$, $\alpha = 0.4\pi$

2.8 SOUND RADIATOR IN THE FORM OF THE ROD PIEZOCERAMIC TRANSDUCER IN HEMISPHERICAL SCREEN. PROBLEM SOLUTION INCLUDING THE PLATE FLEXURAL DEFORMATIONS

The radiator in the form of a disc, situated in a hemispherical screen, was considered above, in Section 2.7.2. Let us consider now the case, which is more interesting on the practical side. A radiator in the form of a rod piezoceramic transducer is "installed" in a hemispherical screen. As it is known, the rod piezoceramic transducers (Tonpliz—type transducer) are widely used in hydro-acoustics for making hydroacoustic arrays of different application [5,7,9,10,16,17,29,32–35,38,57,66,67]. As a rule, they consist of a piezoceramic rod and a metal plate, which is rigidly fixed on it. This plate is necessary for the better coordination of the mechanical impedance of piezoceramic rod with radiation impedance. It allows us to augment the radiation power and to improve the efficiency of such electro-acoustic transducer. But when the plate is rather large, the flexural deformations occur. L.Ya. Gutin showed in monograph [86], that the flexural deformations of the plate cause the noticeable variation of the rod transducer resonance frequency. The other authors (see, for example, [87]) confirmed this fact experimentally. The influence of the plate flexural deformations on the rod transducer is considered in work [88]. An equivalent scheme of such transducer is proposed in this work. But it is not possible to explain the influence of the plate flexural deformations on the mechanical and acoustical parameters of the transducer and to estimate this phenomenon strictly, using the results obtained in the works of reference.

A strict solution of the problem of the rod transducer radiation, including the plate flexural deformations, is given bellow. The influence of the plate deformations on the acoustic parameters of the transducer is also considered.

Let us consider a rod transducer, Fig. 2.77. There is a round plate at one of the rod's ends. The other end is fixed on a rigid base (it is not able to move along the Oz axis). The transducer is situated in a hemispherical rigid screen and is immersed in an ideal compressible liquid with wave impedance ρc. The plate can execute vibrations along the Oz axis as a single whole. The peripheral (thin) part of the plate $R_1 \leq R \leq R_0$ can execute the flexural vibrations.

When describing the movement of the peripheral part of the plate, we use classical model of thin plates. The elastic ring plate is fixed rigidly, at the boundary $R = R_1$. Its outer edge is free $R = R_0$.

The particle velocity at the radiating surface of the plate $z = 0, 0 \leq R \leq R_0$ can be represented as follows:

$$\upsilon = \begin{cases} \upsilon_0; & 0 \leq R \leq R_1, \quad z = 0, \\ \upsilon_0 + \upsilon(R); & R_1 \leq R \leq R_0, \quad z = 0, \end{cases} \tag{2.145}$$

where υ_0 and $\upsilon(R)$—piston and bending parts of the plate particle velocity. The function $\upsilon(R)$ should be the solution of the differential equation of the bending vibrations of thin plates [89] for the following conditions:

$$\upsilon(R) = 0; \ R = R_1, \tag{2.146}$$

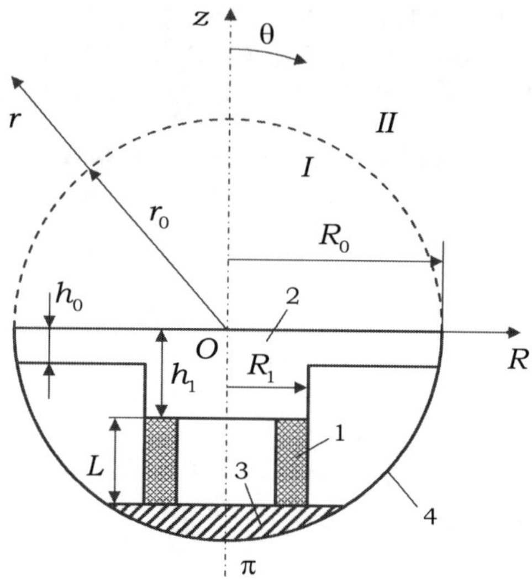

FIG. 2.77: Rod transducer in hemispherical screen; 1—piezoceramic rod, 2—round plate (head mass), 3—rigid base (tail mass), 4—hemispherical screen (housing)

$$d\upsilon(R)/dR = 0; \; R = R_1, \tag{2.147}$$

$$-D\left[\frac{\partial}{\partial R}\nabla^2\upsilon(R) - (1-\sigma)\frac{1}{R}\frac{\partial\upsilon(R)}{\partial R}\right] = 0; \; R = R_0, \tag{2.148}$$

$$-D\frac{\partial}{\partial R}\nabla^2\upsilon(R) = 0; \; R = R_0. \tag{2.149}$$

Here $D = Eh_0^3/12(1-\sigma^2)$, $\nabla^2 = \partial^2/\partial R^2 + (1/R)\partial/\partial R$, E, and σ—Young modulus and Poisson ratio of the plate material.

The particle velocity of the plate at the interval $R_1 \leq R \leq R_0$ can be represented by the series in the complete system of functions of the natural modes of the plate vibrations. Then

$$\upsilon(R) = \sum_{l=1}^{\infty} \upsilon_l f(\lambda_l R). \tag{2.150}$$

For boundary conditions (2.146)–(2.149) the expression for the natural vibrations has the following form

$$f(\lambda_l R) = g_{1l}J_0(\lambda_l R) + g_{2l}N_0(\lambda_l R) + g_{3l}I_0(\lambda_l R) + g_{4l}K_0(\lambda_l R), \tag{2.151}$$

where the constants $g_{1l}, g_{2l}, g_{3l}, g_{4l}$ and the natural numbers λ_l are determined by known method [89].

Now we turn our attention to the notation format of the transducer outer sound field. Let us partition all the domain of the rod transducer radiation field existence into two subdomains: domain I $(0 \leq r \leq r_0, 0 \leq \theta \leq \pi/2)$ and domain II $(r \leq r_0, 0 \leq \theta \leq \pi)$.

In each of these domains the velocities potentials can be represented as follows:

$$\Phi_1 = \sum_{\nu=1}^{\infty} B_\nu j_{m_\nu}(kr) P_{m_\nu}(\cos\theta) + \sum_{\tau=1}^{\infty} C_\tau J_0(\alpha_\tau R)\exp(-ik_\tau z), \tag{2.152}$$

$$\Phi_2 = \sum_{n=0}^{\infty} A_n h_n^{(1)}(kr) P_n(\cos\theta). \tag{2.153}$$

Expression (2.153) is the general solution of the problem of sound radiation by the sphere. The summands in expression (2.152) are the solutions in spherical and cylindrical coordinate systems with the common coordinate origin in the point O (see Fig. 2.77). Such representation allows us to fulfill matching conditions at the boundary $r = r_0, 0 \leq \theta \leq \pi/2$ due to the choice of the coefficients B_ν. The choice of the coefficients C_τ allows us to fulfill boundary condition at the surface $z = 0, 0 \leq R \leq R_0$.

The unknown complex coefficients $A_n, B_\nu, C_\tau, \upsilon_l$ are determined from the following system of functional equations:

$$\Phi_1 = \Phi_2;\ r = r_0,\ 0 \leq \theta \leq \pi/2, \tag{2.154}$$

$$\frac{\partial \Phi_2}{\partial r} = \begin{cases} \dfrac{\partial \Phi_1}{\partial r}; & 0 \leq \theta \leq \pi/2, \quad r = r_0, \\ 0; & \pi/2 \leq \theta \leq \pi, \quad r = r_0, \end{cases} \tag{2.155}$$

$$-\frac{\partial \Phi_1}{\partial z} = \upsilon;\ z = 0,\ 0 \leq R \leq R_0, \tag{2.156}$$

$$\nabla^2\nabla^2\upsilon - \lambda_l^4 \upsilon = \frac{\omega^2 \rho_1}{D}\Phi_1;\ z = 0, R_1 \leq R \leq R_0, \tag{2.157}$$

where ρ_1—the density of the plate material. The system of functional Eqs. (2.154)–(2.157) can be transformed into an infinite system of linear algebraic equations of second kind:

$$B_\nu j_{m_\nu}(kr_0) N_{\nu\nu} - \sum_{n}^{\infty} A_n h_n^{(1)}(kr_0) N_{\nu n} + \sum_{\tau}^{\infty} C_\tau N_{\nu\tau} = 0,$$

$$A_n k h_n'^{(1)}(kr_0) N_{nn} - \sum_{\nu}^{\infty} B_\nu j'_{m_\nu}(kr_0) N_{\nu n} - \sum_{\tau}^{\infty} C_\tau(\alpha_\tau N_{n\tau} + ik_\tau \bar{N}_{n\tau}) = 0,$$

$$iC_\tau k_\tau N_{\tau\tau} + \sum_{l}^{\infty} \upsilon_l N_{l\tau} = -\upsilon_0 N_\tau, \tag{2.158}$$

$$\frac{\rho}{\rho_1 h_0}\left[\sum_{\nu}^{\infty} B_\nu P_{m\nu}\left(\cos\frac{\pi}{2}\right) N_{\nu l} + \sum_{\tau}^{\infty} C_\tau N_{l\tau}\right] - \upsilon_l\left[1 - \left(\frac{\omega_l}{\omega}\right)^2\right] N_{ll} = \upsilon_0 N_l,$$

here

$$N_{\tau\tau} = \int_0^{R_0} J_0^2(\alpha_\tau R) R dR,\ N_{l\tau} = \int_{R_1}^{R_0} J_0(\alpha_\tau R) f(\lambda_l R) R dR,$$

$$N_\tau = \int_0^{R_0} J_0(\alpha_\tau R) R dR, \quad N_{\nu n} = \int_0^1 P_{m_\nu}(\cos\theta) P_n(\cos\theta) d(\cos\theta),$$

$$N_{\nu\tau} = \int_0^1 P_{m_\nu}(\cos\theta) J_0(\alpha_\tau r_0 \sin\theta) \exp(-ik_\tau r_0 \cos\theta) d(\cos\theta),$$

$$N_{\nu\nu} = \int_0^1 [P_{m_\nu}(\cos\theta)]^2 d(\cos\theta), \quad N_{nn} = \int_{-1}^1 [P_n(\cos\theta)]^2 d(\cos\theta),$$

$$N_{n\tau} = \int_0^1 P_n(\cos\theta) \sin\theta \, [J_0'(\alpha_\tau r_0 \sin\theta)] \exp(-ik_\tau r_0 \cos\theta) d(\cos\theta),$$

$$\bar{N}_{n\tau} = \int_0^1 P_n(\cos\theta) \cos\theta \, [J_0(\alpha_\tau r_0 \sin\theta)] \exp(-ik_\tau r_0 \cos\theta) d(\cos\theta),$$

$$N_{\nu l} = \int_{R_1}^{R_0} j_{m_\nu}(kR) f(\lambda_l R) R dR, \quad N_{l\tau} = \int_{R_1}^{R_0} J_0(\alpha_\tau R) f(\lambda_l R) R dR,$$

$$N_{ll} = \int_{R_1}^{R_0} [f(\lambda_l R)]^2 \, R dR, \quad N_l = \int_{R_1}^{R_0} f(\lambda_l R) R dR,$$

here $k_\tau = \sqrt{k^2 - \alpha_\tau^2}$, m_ν—the roots of the equation $P'_{m_\nu}(\cos\theta) = 0$ when $\theta = \pi/2$, ω_l—natural frequencies of the bending vibrations of the plate in a vacuum, α_τ—the roots of equation $J_0'(\alpha_\tau R) = 0$ for $R = R_0$; the prime on the Bessel and Hankel functions means that it is necessary to take a derivative with respect to full argument.

The reduction method was used in solving such system (2.158). All the calculation data given bellow are determined for $n = 0, 1, \ldots, 9$, $l = 1$, $\nu = 1, 2, \ldots, 5$, $\tau = 1, 2, \ldots, 5$. Boundary conditions and matching conditions (2.154)–(2.157) are fulfilled with accuracy not less than 5% of maximum value of determined characteristics.

We use the classical computational scheme for a rod transducer [86], which is showed in Fig. 2.78. To determine the velocity of the plate vibrations υ_0, it is necessary to know the radiation impedance Z. The impedance is determined on the base of the solution of corresponding diffraction problem, when $\upsilon_0 = \text{const}$. Let's proceed similarly: we set $\upsilon_0 = 1$, solve problem (2.158), and we determine the radiation impedance by the known formula [8]:

$$Z = \frac{1}{|\upsilon_0|^2} \int_0^{R_0} p\upsilon^* R dR, \tag{2.159}$$

where p—sound pressure, υ^*—complex-conjugate value of υ.

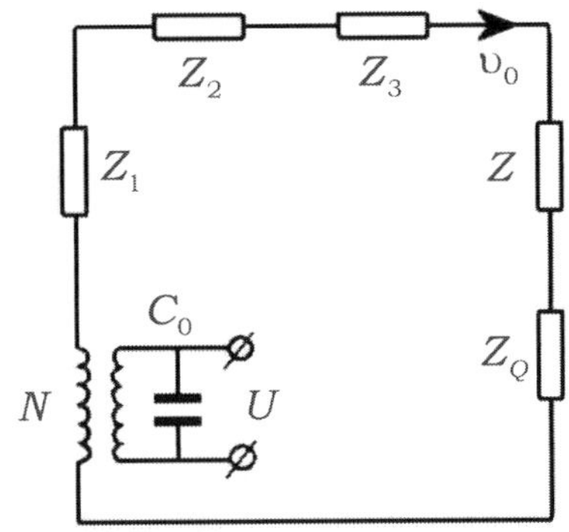

FIG. 2.78: Equivalent electromechanical scheme of the transducer; $Z_1 = -i\rho_2 c_2 S_2/\sin kL$, $Z_2 = i\rho_2 c_2 tg(kL/2)$, $Z_3 = i\omega m$, N—transformation coefficient, C_0—electrical capacity of clamped piezoelectric rod, U—electrical voltage, Z—radiation impedance, Z_Q—plate bending impedance, m—plate mass, $\rho_2 c_2$—wave impedance of piezoelectric material, S_2—cross section of piezoelectric rod

Before proceeding to the analysis of the results of the calculation of the value of Z, let us discuss the new element of the equivalent scheme-impedance Z_Q (see Fig. 2.78). It appears when there is bending deformation of the plate. When there is no bending deformation of the plate (thickness compressibility of the plate can be neglected) its impedance has pure inertial character. It is defined as the ratio of inertial force F to the piston velocity of the plate:

$$Z_H = \frac{F}{\upsilon_0} = \frac{m}{\upsilon_0}\frac{d\upsilon_0}{dt} = i\omega m. \tag{2.160}$$

Shearing force appears in our case on the contour $R = R_1$ due to the presence of velocity $\upsilon(R)$. Using the known expression for F_Q [89], we define impedance

$$Z_Q = \frac{1}{\upsilon_0}F_Q(R = R_1) = -i\frac{\upsilon_l 2\pi R_1}{\omega\upsilon_0}D\lambda_l^3\frac{\partial f(\lambda_l R)}{\partial R}, \quad R = R_1, \; l = 1. \tag{2.161}$$

It is interesting to study the peculiarities of the value of Z_Q without acoustic load, i.e., when $Z = 0$. System (2.158) is simplified maximally in this case, because there is no need in the first three equations. And there is no need in the member, which is within the first square brackets in the fourth equation. There is a possibility to relate υ_0 to υ_1:

$$\upsilon_1 = -\upsilon_0 N_1\left\{\left[1 - \left(\frac{\omega_1}{\omega}\right)^2\right]N_{11}\right\}^{-1}. \tag{2.162}$$

As it follows from the last expression, when approaching the plate flexural frequency ω_1, velocity υ_1 increases sharply. When $\omega \to \omega_1$ the value of $\upsilon_1 \to \infty$, because the dissipation losses in the material of the plate were not taken into account. Bearing in mind (2.162), Eq. (2.161) takes the form

$$Z_Q = -i\frac{S_E}{\omega\left[1-(\omega_1/\omega)^2\right]}, \qquad S_E = 2\pi R_1 D\lambda_1^3 \frac{N_1}{N_{11}}\frac{\partial f(\lambda_1 R)}{\partial R}|_{R=R_1}. \tag{2.163}$$

As we can see, when $\omega < \omega_1$ the impedance Z_Q has the nature of mass, and when $\omega > \omega_1$, it has the nature of elasticity. We interpret the value of S_E as the equivalent elasticity of the peripheral part of the plate (the equivalent mass is $m_E = S_E/\omega_1^2$). The expression for Z_Q than describes the impedance of parallel circuit, consisting of the mass m_E and elasticity S_E.

Basing on the carried out analysis, we can make a series of important and useful conclusions. Firstly, the rod transducer with a plate, executing bending vibrations, is a complex system of the coupled mechanical circuits. The significant changes in its natural frequencies spectrum are the consequence of it. As an example, consider how the lowest natural frequencies change. Figure 2.79 illustrates the typical motion of the reactive part of the complex mechanical impedance of the transducer without taking into account the bending of the plate. If there are no bending deformations of the plate, the transducer spectrum is characterized by the interchange of resonances and anti-resonances of longitudinal vibrations. The lowest resonance has frequency ω_0, the anti-resonance ω_{01} follows then and so on. If there is bending deformation of the plate, when the inequality $\omega_0 < \omega_1 < \omega_{01}$ is true, the frequency of the lowest resonance of the transducers decreases and takes the value $\bar{\omega}_0$. The reason of it is the occurrence of the impedance Z_Q, having the nature of mass when $\omega < \omega_1$. Then the anti-resonance of the transducer follows at frequency $\omega = \omega_1$. This anti-resonance is caused by the resonance of the plate bending. The resonance ω_2 results from the equality of the elastic resistance, which is typical for Z_Q when $\omega > \omega_1$, and the inertial resistance, which determines the mechanical impedance of the transducer vibrating longitudinally in the range $\omega_0 < \omega < \omega_{01}$.

Thus, the plate banding leads to the reduction of transducer resonance lowest frequency and to the occurrence of another two resonance frequencies.

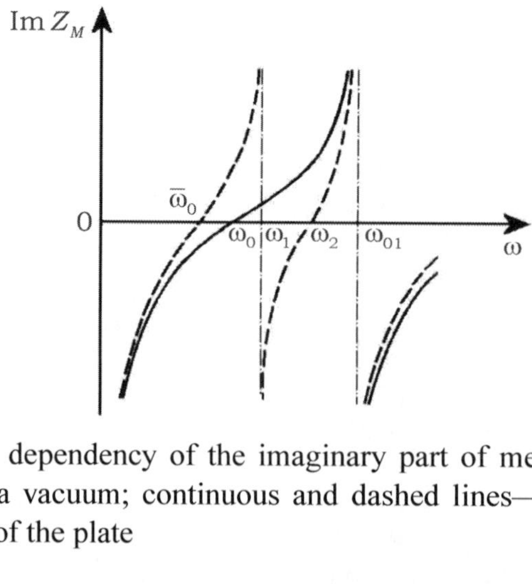

FIG. 2.79: Frequency dependency of the imaginary part of mechanical impedance of the rod transducer in a vacuum; continuous and dashed lines—including and without including the bending of the plate

The dependency of the change of the transducer lowest resonance frequency on the geometrical parameters of the peripheral part of the plate is of interest in practice. Figure 2.80 shows such dependencies, calculated for the following parameters:

$$\rho_2 c_2 S_2 = 9.4 \times 10^4 \text{ kg/s}, \; L = 0.11 \text{ m}, \; R_0 = 0.104 \text{ m}, \; R_1 = 0.05 \text{ m},$$

$$\rho_1 c_1 = 2.26 \times 10^7 \text{ kg/(m}^2\text{s)}, \; h_1 = 0.06 \text{ m}, \; \rho c = 1.5 \times 10^6 \text{ kg/(m}^2\text{s)}, \; N = 10 \text{ N/V}.$$

As we can see, if the bending of the plate is not included, the lowest resonance frequency of the transducer increases in a linear fashion with a decrease in thickness. It is natural, because with a decrease of h_0 the weight of the plate decreases. The situation is different when the bending of the plate is included. When the thickness of the plate is relatively thick, the resonance frequencies in both considered cases are close. With a decrease in the thickness of the plate the value of $\bar{\omega}_0$ reaches a maximum (for $h_0 \approx 0.0175$ m), then it decreases and asymptotically approaches curve 3. Such behavior of the lowest resonance frequency of the transducer is in good agreement with experimental data [87].

Let us consider now the impedance Z_Q behavior, when a transducer works in a medium. We perform the transformations, using the fourth equation in system (2.158) and expression (2.161), and we obtain

$$Z_{Q,\rho c \neq 0} = -i \frac{S_E}{\omega \left[1 - (\omega_1/\omega)^2\right]} \left\{ 1 - \frac{\rho}{\rho_1 h_0 N_1} \int_{R_1}^{R_0} \Phi_1 f(\lambda_1 R) R dR \right\}. \qquad (2.164)$$

In the last expression, the second member in curved brackets determines the contribution of the outer field to the value of impedance Z_Q. The impedance Z_Q is of reactive nature in a vacuum. If there is a surrounding medium, the impedance is complex. For illustrative purpose, Fig. 2.81 (curves 1) shows the frequency dependencies of the value of Z_Q for the indicated above parameters of the transducer for $h_0 = 0.02$ m. We can see that the

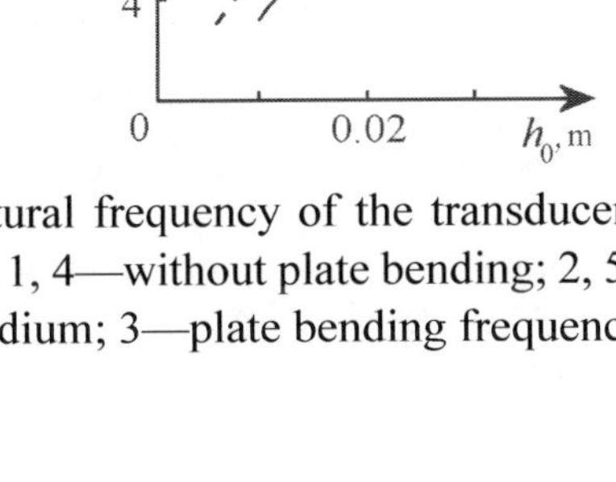

FIG. 2.80: The lowest natural frequency of the transducer versus the thickness of the peripheral part of the plate; 1, 4—without plate bending; 2, 5—with plate bending; 1–3—in a vacuum, 4, 5—in a medium; 3—plate bending frequency, $\bar{f}_0 = \bar{\omega}_0 / 2\pi$

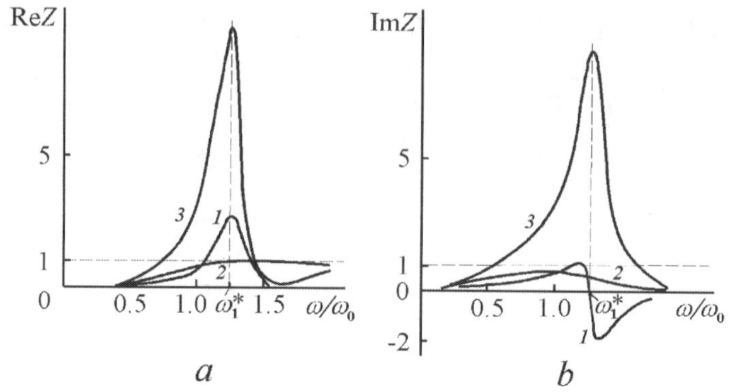

FIG. 2.81: Frequency dependencies of the impedancesnormalized to $\rho c S$; (a)—active part, (b)—reactive component; 1—impedance Z_Q, 2, 3—radiation impedance without the plate bending and with the plate bending, S—plate area, ω_0—resonance frequency of the transducer in a vacuum without including the plate bending

active part Z_Q has a clearly pronounced maximum at the resonance frequency of the plate flexural vibrations in the medium ω_1^*. The reactive part Z_Q is equal to zero and it changes sign. This figure represents also the frequency dependencies of the radiation impedance Z (with the plate bending and without it). The behavior of radiation impedance without the plate bending (curves 2) is well known [86]. The behavior of radiation impedance with the plate bending (curve 3) is more interesting for consideration.

The main feature here is a sharp increase of the active and reactive components of Z at the frequency ω_1^*. Such behavior of Z can be explained as follows. As it was pointed above, the impedance Z_Q can be represented by parallel contour. It is the reason of the sharp decrease in the velocity υ_0 at the frequency ω_1^* and relatively large increase in the velocity υ_1. This fact prevents sound pressure decrease on the plate of transducer. Consequently, the numerator of expression (2.159) changes weekly, and the denominator decreases sharply. It leads to an increase of the value of Z at the frequency ω_1^*.

The quantitative data on the changing of the lowest natural frequency of the transducer when passing from vacuum to acoustic medium can be of practical interest. We can answer this question if we compare the curves 1–5 in Fig. 2.80. As it follows from these data, when the transducer works in the medium, its lowest natural frequency drops considerably. This fact is caused by the relatively large value of the reactive component of the impedance Z, which has the nature of mass.

Let us make the comparative analysis of the far field of the rod transducer with the plate bending and without the plate bending. Figure 2.82 illustrates these data. This analysis allows us to draw the important conclusions. The plate bending leads to

- the decrease of the sound pressure in the domain of the lowest natural frequency of the transducer;

- the decrease of the sound pressure in the domain of the resonance frequency of the flexural vibrations of the plate;

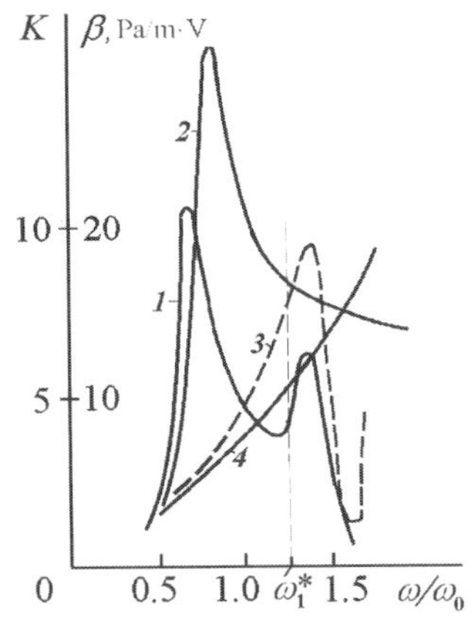

FIG. 2.82: Frequency dependencies of the transducer sensitivity (curves 1, 2) and concentration coefficient (curves 3, 4); 1, 3—with the plate bending, 2, 4—without the plate bending

- the appearance of the second maximum of the sound pressure at the frequencies above ω_1^*.

We turn our attention now to the peculiarities of the directivity pattern and the concentration coefficient of the transducer, Fig. 2.83. At the frequencies $\omega < \omega_1^*$ the plate bending influences slightly the directivity pattern and concentration coefficient of the transducer. But the change in these characteristics becomes substantial, beginning with the frequency ω_1^* and above. Here the directivity pattern maximum can be turned through 180°, and the value of the concentration coefficient can decrease sharply. As the detailed analysis shows, the reason for these facts is an increase (with frequency increase) in the phase differential between the velocities υ_0 and $\upsilon(R)$. In the domain $\omega \geq \omega_1^*$ the phase differential may attain 90° and more.

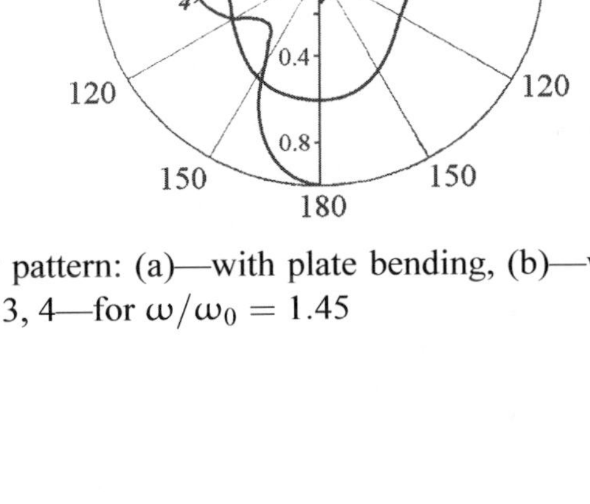

FIG. 2.83: Directivity pattern: (a)—with plate bending, (b)—without plate bending; 1, 2—for $\omega/\omega_0 = 0.77$, 3, 4—for $\omega/\omega_0 = 1.45$

Summarizing the performed analysis, we can draw a conclusion that the plate bending influences all the parameters of the rod transducer. If this fact is not taken into account, the grave errors can appear in the estimation of the directional properties and energetic properties of the transducer.

In conclusion, we note that the materials given in the second chapter, are based on the results of works, performed by authors at different times [4,28,42–48,50,53,56,71,74,75,77–79,90,91].

REFERENCES

1. Rayleigh, J., *The Theory of Sound*, New York: Dover Publications, vols. 1–2, 1945. [in Russian]
2. Skudrzyk, E., *Fundamentals of Acoustics,* New York: Springer, 1971.
3. Rzhevkin, S.N., *A Course of Lectures on the Theory of Sound*, Moscow: Moscow State University Press, 1960. [in Russian]
4. Grinchenko, V.T., Vovk, I.V., and Matsypura, V.T., *Basic Principles of Acoustics,* Kyiv: Naukova Dumka, 2007. [in Russian]
5. Hueter, T.F., Twenty years in underwater acoustics: Generation and reception, *J. Acoust. Soc. Am.*, vol. **51**, no. 3, pp. 1025–1040, 1972.
6. Furduev, V.V., *Electroacoustics,* Moscow: Gostehizdat, 1948. [in Russian]
7. Sverdlin, G.M., *Applied Hydroacoustics*, Leningrad: Sudostroenie, 1976. [in Russian]
8. Smaryshev, M.D., *Directivity of the Hydroacoustic Antennas*, Leningrad: Sudostroenie, 1973. [in Russian]
9. Horton, J.W., *Fundamentals of Sonar,* Annapolis, Meryland: Unated States Naval Institute, 1959.
10. Camp, L., *Underwater Acoustics,* New York: Wiley, 1970.
11. Shenderov, E.L., *Wave Problems of Hydroacoustics*, Leningrad: Sudostroenie, 1972. [in Russian]
12. Alekceev, B.N., Increasing the bandwidth of a cylindrical piezoceramic transducer by means of two transition layers, *Sov. Phys. Acoust.*, vol. **22**, no. 2, pp. 101–104, 1976.
13. Dianov, D.B. and Kuzmenko, A.G., Investigation of possibilities for broadening transmission band of cylindrical piezoelectric ceramic transducers, *Sov. Phys. Acoust.*, vol. **16**, no. 2, pp. 197–200, 1970.
14. Glazanov, V.E., *Hydroacoustic Arrays Shielding*, Leningrad: Sudostroenie, 1986. [in Russian]
15. Skudrzyk, E., *Simple and Complex Vibratory Systems,* Pensilvania: The Pensilvania State University Press, 1968.
16. Bogorodskiy, V.V., (Ed.) *Underwater Electroacoustic Transducers (Engineering): Guide*, Leningrad: Sudostroenie, 1983. [in Russian]
17. Leyko, A.G., Shamarin, Y.E., and Tkachenko, V.P., *Underwater Acoustic Arrays*, Kiev: Avanpostprim, 2000. [in Russian]
18. Hönl, H., Maue, A., and Westpfahl, K., *Theorie der Beugung*, Berlin: Springer, 1961.

19. Kantorovich, L.V. and Krylov, V.I., *Approximate Methods of Higher Analysis*, New York: Intersience Publisher, 1958.
20. Bateman, H., *Higher Transcendental Functions*, vol. 2, New York: McGraw-Hill, 1953.
21. Gradshteyn, I.S. and Ryzhik, I.M., *Table of Integrals, Series, and Products*, New York: Academic Press, 2014.
22. Abramowitz, M. and Stegun, I.A., *Handbook of Mathematical Functions with Formulas, Graphs, and Mathematical Tables*, vol. 9., New York: Dover, 1972.
23. Polya, G. and Szegö, G.P., *Problems and Theorems in Analysis*, vol. 1, New York: Springer, 1972.
24. Hobson, E.W., *The Theory of Spherical and Ellipsoidal Harmonics*, Cambridge: Cambridge University Press, 1931.
25. Bateman, H., *Higher Transcendental Functions*, vol. I, New York: McGraw-Hill, 1953.
26. Vaynshteyn, L.A. and Belkina, M.G., Double reduction method and infinite systems of linear equations for expansion coefficients of the required function with singularities, *Report. AS USSR*, vol. **194**, no. 4, pp. 794–797, 1970. [in Russian]
27. Grinchenko, V.T. and Luneva, S.A., Sound field of a baffled circular cylinder, *Sov. Phys. Acoust.*, vol. **26**, no. 3, pp. 462–466, 1980.
28. Vovk, I.V., Exitation of a sound field by a transducer loaded on a liquid wedge, *Sov. Phys. Acoust.*, vol. **17**, no. 2, pp. 311–314, 1971.
29. Meson, W.P. and Thurston, R.N., (Eds.) *Physical Acoustics*, vol. 1, part A, New York: Academic Press, 1964.
30. Morse, P.M., *Vibrations and Sound*, New York: Mc-Graw-Hill, 1948.
31. Mandelshtam, L.I., *The Lectures on Oscillation Theory*, Moscow: Nauka, 1972.
32. Sverdlin, G.M., *Hydroacoustic Transducers and Arrays*, Leningrad: Sudostroenie, 1980. [in Russian]
33. Woollett, R., Ultrasonic Transducers, underwater sound transducers, *Ultrason.*, vol. **8**, no. 4, pp. 243–253, 1970.
34. Woollett, R., Tends and problem in sonar design, *IEEE Trans. Sonics Ultrason.*, vol. **10**, no. 3, pp. 116–124, 1963.
35. Bjørnø, L., European Commission, *Proc. of the Second European Conf. on Underwater Acoustics*, Denmark: Lyngby, vols. 1–2, 4–9 July, 1994.
36. Grinchenko, V.T. and Senchenko, I.V., Sound radiation by partially shielded piezoceramic shells, *Appl. Mech.*, vol. **18**, no. 2, pp. 15–21, 1982.
37. Weaver, W., Timoshenko, S.P., and Young, D.H., *Vibration Problems in Engineering*, New York: Wiley, 1990.
38. Evtyutov, A.P., Kolesnikov, A.E., Lyalikov, A.P., Mitko, V.B., Ponomarenko, V.I., Prostakov, A.L., Sverdlin, G.M., Smatyshev, V.D., and Tarasyuk, Yu.F., *Handbook of Hydroacoustics*, Leningrad: Sudostroenie, 1988. [in Russian]
39. Love, A.E.H., *A Treatise on the Mathematical Theory of Elasticity*, Cambridge: Cambridge University Press, 2013.
40. Goldenveizer, A.L., *Theory of Elastic Thin Walled Shells*, Moskow: Nauka, 1979. [in Russian]

41. Überall, H., Sound scattering by elastic cylindrical shells, *J. Acoust. Soc. Am*, vol. **39**, no. 2, pp. 272–275, 1965.

42. Grinchenko, V.T. and Vovk, I.V., *The Wave Problems of Sound Scattering by Elastic Shells*, Kiev: Naukova Dumka, 1986. [in Russian]

43. Vovk, I.V., Grinchenko, V.T., and Maiatskii, V.I., Sound field of an infinite circular cylindrical transducer partially coated with an acoustically compliant layer, *Sov. Phys. Acoust.*, vol. **18**, pp. 308–311, 1973.

44. Vovk, I.V., Sound radiation from a cylinder partially enclosed in an open annular layer, *Sov. Phys. Acoust.*, vol. **26**, no. 3, pp. 198–201, 1980.

45. Vovk, I.V., Nontraditional method of resonant frequency control of low-frequency underwater acoustic transducers, *Proc. of the European Conf. on Underwater Acoustics*, pp. 631–634, Luxembourg, 14–18 Sept., 1992.

46. Vovk, I.V. and Oleynik, V.N., Sound radiation by piezoceramic shell filled with liquid with non-symmetrical insertion, *Acoust. J.*, vol. **40**, no. 2, pp. 220–224, 1994. [in Russian]

47. Vovk, I.V., Grinchenko, V.T., and Oleynik, V.N., On one method of construction of a directed low-frequency hydroacoustic radiator, *Proc. of the European Conf. on Underwater Acoustics*, pp. 573–578, Luxembourg, 14–18 Sept., 1992.

48. Vovk, I.V. and Oleynik, V.N., Sound radiation by cylindrical piezoelastic shell with asymmetric insertion, *J. Acoust. Soc. Am.*, vol. **99**, no. 1, pp. 133–138, 1996.

49. Alekseev, V.K. and Lependin, L.F., Sound field of a pulsating ring on a cylinder, *Sov. Phys. Acoust.*, vol. **13**, no. 1, p. 98, 1967.

50. Vovk, I.V., Grinchenko, V.T., and Kotsyuba, V.S., Method of estimating acoustical properties of a cylindrical radiator of finite height, *Sov. Phys. Acoust.*, vol. **21**, no. 5, pp. 434–436, 1975.

51. Kozyrev, V.A. and Shenderov, E.L., Radiation impedance of a cylinder of finite height, *Sov. Phys. Acoust.*, vol. **26**, no. 3, pp. 230–236, 1980.

52. Sandman, B.E., Fluid-loading influence coefficients for a finite cylindrical shell, *J. Acoust. Soc. Am.*, vol. **60**, no. 6, pp. 1256–1264, 1976.

53. Vovk, I.V. and Grinchenko, V.T., On expansion of partial region method possibilities in application to problems of radiation and wave scattering, *Sov. Phys. Acoust.*, vol. **35**, no. 1, pp. 29–36, 1989.

54. Vovk, I.V. and Grinchenko, V.T., Sound radiation by a number of coaxial piezoceramic shells, *Sov. Phys. Acoust.*, vol. **37**, no. 2, pp. 259–269, 1991.

55. Boriseyko, V.A., Martynenko, V.S., and Ulitko, A.F., Toward the theory of piezoceramic shells, *Math. Phys.*, **21**, pp. 71–76, 1977. [in Russian]

56. Boriseiko, V.A., Grinchenko, V.T., and Ulitko, A.F., Relations of electroelasticity for piezoceramic shells of revolution, *Int. Appl. Mech.*, vol. **12**, no. 2, pp. 126–131, 1976.

57. Kikuchi, E., (Ed.), *Ultrasonic Transducers,* Tokyo: Corona Publishing Company, 1969.

58. Timoshenko, S.P. and Woinowsky-Krieger, S., *Theory of Plates and Shell,* New York: McGraw-Hill, 1959.

59. Williams, W.E., Diffraction by a cylinder of finite length, *Proc. Cambridge Philosofical Soc.*, vol. **52**, pp. 332–335, 1956.

60. Sherman, C.H. and Parke, N.G., Acoustic radiation from a thin torus with application to the free-flooding ring transducer, *J. Acoust. Soc. Am.*, vol. **38**, no. 5, pp. 715–722, 1965.

61. Gots, A.A. and Dianov, D.V., Acoustic field of spherical belt, *Report VIII National Acoustical Conf.*, Moscow: Nauka, pp. 100–103, 1973. [in Russian]

62. Shenderov, E.L., Sound radiation associated with axisymmetric oscillations of a finite open tube, *Sov. Phys. Acoust.*, vol. **35**, no. 1, pp. 82–87, 1989.

63. Basovskii, V.G., Gomilko, A.M., and Meleshko, V.V., Sound radiation by a rigid cylindrical shell of finite length, *Sov. Phys. Acoust.*, vol. **35**, no. 6, pp. 573–576, 1989.

64. Glazanov, V.E., Radiation impedance of a water-filled cylinder, *Sov. Phys. Acoust.*, vol. **35**, no. 5, pp. 467–469, 1989.

65. Kendig, P.M., Advanced transducer developments, in: *Underwater Acoustics,* Albers, V.M., Ed., Chapter 2, New York: Plenum Press, 1967.

66. Rijnla, H.A.J., Modern transducers, theory and practice, *Underwater Acoustic and Signal Processing,* pp. 225–242, 1986.

67. Zelesak, J.F. and Rogers, P.H., Low-frequency radiation characteristics of free-flooded ring transducer with application to a low-frequency directional hydrophone, *J. Acoust. Soc. Am.,* vol. **56**, pp. 1052–1058, 1974.

68. McMahon, G.W., Performance of open ferroelectric ceramic cylinders, *J. Acoust. Soc. Am.*, vol. **36**, no. 3, pp. 528–533, 1964.

69. Tomilina, T.M., Radiation impedance of a hollow cylindrical radiator, *Sov. Phys. Acoust.*, vol. **35**, no. 1, pp. 73–75, 1989.

70. Rogers, F.H., Matematical model for a free-flooding piezoelectric cylinder transducer, *J. Acoust. Soc. Am.*, vol. **80**, no. 1, pp. 13–18, 1986.

71. Vovk, I.V., Sound radiation by coaxial system of interacting piezoceramic shells, *All-Union Symp. on the Interaction between the Acoustic Waves and Elastic Bodies,* Tallinn: Nauka, pp. 51–54, 1989. [in Russian]

72. Shenderov, E.L., *Sound Radiation and Scattering*, Leningrad: Sudostroenie, 1989. [in Russian]

73. Basovskii, V.G. and Vovk, I.V., Sound radiation by a finite grating consisting of open rigid cylindrical shells, *Acoust. Phys.*, vol. **44**, pp. 497–506, 1998.

74. Vovk, I.V. and Grinchenko, V.T., Interaction of electroacoustic transducers in an array, *Acoust. Phys.,* vol. **42**, pp. 159–165, 1996.

75. Vovk, I.V. and Grinchenko, V.T., Electroacoustic transducers interaction for different ways of electric excitation, *Report at 11th All–Union Conf., Section "A"*, Moscow: Nauka, pp. 143–146, 1990. [in Russian]

76. Vovk, I.V., Grinchenko, V.T., and Savin, E.O., Sound radiation by system of coaxial piezoceramic cylindrical shells, *Report Academy of Science of Ukrainian SSR*, Kiev: Naukova Dumka, pp. 72–76, 1991. [in Russian]

77. Vovk, I.V., Sound radiation by a finite grating by free-flooding cylindrical piezo-ceramic shells, *Sov. Phys. Acoust.*, vol. **38**, no. 3, pp. 234–237, 1992.

78. Vovk, I.V. and Matsypura, V.T., Sound radiation by the grating, formed by coaxial cylindrical piezoceramic shells with end-shields. Part 1. Theory, *Acoust. Visnyk*, vol. **4**, no. 2, pp. 11–17, 2001. [in Russian]

79. Vovk, I.V. and Matsypura, V.T., Sound radiation by an open piezoceramic ring near the screens of finite sizes, *Acoust. Visnyk*, vol. **7**, no. 2, pp. 25–35, 2004. [in Russian]

80. McLachlan, N.W., *Loud Speakers: Theory, Performance, Testing, and Design,* Oxford: Clarendon Press, 1934.

81. Stenzel, H., *Leitfaden zur Berechnung der Schallvorgänge*, Berlin: Springer, 1939.

82. Gutin, L.Ya., On the sound field of piston radiators, *Tech. Phys. J.*, vol. **7**, no. 10, pp. 1096–1106, 1937. [in Russian]

83. Hanson, E.T., Ellipsoidal function and their application, *Phil. Trans. Roy. Soc. Ser. A*, vol. **232**, pp. 223–228, 1933.

84. Iofe, V.K., *Electroacoustics*, Moscow: Svyazizdat, 1954. [in Russian]

85. Chetaev, D.A., The impedance of a rectangular piston, vibrating in an infinite plane screen, *Appl. Math. Mech.*, vol. **15**, pp. 439–444, 1951. [in Russian]

86. Gutin, L.Ya., *The Theory of Piezoelectric Vibrators Used in Hydroacoustics. Selected Works,* Leningrad: Sudostroenie, 1977. [in Russian]

87. Rijnja, H.A., The design of sensitive hydrophones, *Acoustica*, vol. **33**, no. 1, pp. 1–9, 1976.

88. Butler, J.L., Cipolla, W.R., and Brown, W.D., Radiating head flexure and it's effect on transducer performance, *Acoust. Soc. Am.*, vol. **70**, no. 2, pp. 500–503, 1981.

89. Khronin, D.V., *Theory and Calculation of Vibrations in Aircraft Engines*, Moscow: Mashinostroenie, 1970. [in Russian]

90. Vovk, I.V., Grinchenko, V.T., and Matsypura, V.T., A new approach to the estimation of acoustical properties of a sound radiator in the form of a disc, *Acoust. Visnyk*, vol. **11**, no. 3, pp. 13–26, 2008. [in Russian]

91. Vovk, I.V. and Grinchenko, V.T., Influence of flexural deformations of the radiating head on the sound field of a rod transducer, *Sov. Phys. Acoust.,* vol. **34**, no. 5, pp. 471–475, 1988.

CHAPTER 3

Diffraction of Sound by the Infinite Gratings Composed of Elastic Shells

"What do you say, Watson?"
I shrugged my shoulders. "I must confess that I am out of my depths," said I.

Arthur Conan Doyle, "The Adventure of the Stockbroker's Clerk"

3.1 STATEMENT OF PROBLEM. TYPES OF THE CONSIDERED SHELLS

The acoustic screens* are widely used for protection from in-plant noise, and for acoustic correction of the concert halls and the auditoriums. In the manufacture of the acoustic arrays and the electro-acoustic transducers, the screens are used for controlling their sound field. For example, they can be used to make the arrays with a required pattern function, or to weaken interaction between the transducers [1–5].

Only acoustically hard screens are used in aeroacoustics. These screens are made of the materials, the wave impedance of which is higher than the wave impedance of air [1,3–5]. It is impossible to make a material, the wave impedance of which is smaller than the wave impedance of air. In hydroacoustics, by contrast, the acoustically soft screens are often used. The wave impedance of these screens materials is smaller than the wave impedance of water. The materials, which are usually used, are the rubber layers with closed air-bubble voids or cylindrical conduits [6]. The compliance (volume compressibility) of such screens is high in water and, as a result, they have the broad bandwidth of effective sound reflection. But high compressibility is also the main reason of the sound reflection properties rapid decrease with an increase of hydrostatic pressure. That is why they can be applied at depth down to 100 m, when the hydrostatic pressure is to 10 atm. Thus, it was necessary to find new materials and designs for making the screens capable of acceptable sound reflection at great depth of the ocean.

In 1957, it was found that one of the ways of making the acoustically soft screens in liquid is to use the elastic shells (made of metal or firm plastic) with natural frequencies close to the frequencies of incident waves [7]. And later this idea was developed in works [8–18]. It was found that due to this property of the shells it is possible to make the acoustically soft screens for the acoustic arrays and the transducers in a limited frequency range. The screens of elastic shells can work more effectively at much greater depth than the screens of porous rubber layers.

*We consider only the sound-reflecting screens. Sound-absorbing screens are not considered.

ISBN: 978–1–56700–459–5

The statement of the problems, which will be considered in this chapter, is used in a large number of writings on sound wave diffraction. Refer to [19–24], and to the bibliography of these writings.

Harmonic sound wave is incident on the grating of elastic shells that, is immersed in ideal compressible liquid with wave impedance ρc. The unknown quantities are the characteristics of scattered sound field, the dependence of these characteristics on frequency, and the vibration mode of the elastic shells walls.

We will consider one class of elastic shells in the form of hollow bars. The walls of the bars are made of thin plates attached at the edges through non deformable supports. Figure 3.1 illustrates the examples of such constructions.

The question arises as to whether it is possible to use the shells of classical form (by the shells of classical form we mean the smooth, cylindrical or elliptic shells and the other similar shells). To answer this question, we will carry out a comparative analysis of the vibrating systems of the classical shells and of the shells represented in Fig. 3.1(a). Figure 3.2 shows a cross section of three shells. The shells are infinite along the axis normal to the plane of the figure. Two lower order natural modes of these shells are showed in Fig. 3.2. We will give the frequency formulas for these modes [25–29]:

1. for a cylindrical shell

$$\omega_{n=0} = \frac{1}{r}\left(\frac{E_1}{\rho_1}\right)^{1/2}; \quad \omega_{n=2} = \frac{7{,}2h}{r^2}\left\{\frac{E_1}{12\rho_1(1-\sigma^2)}\right\}^{1/2},$$

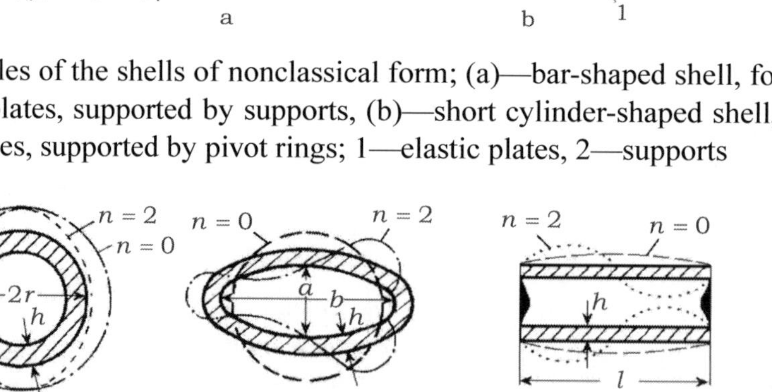

FIG. 3.1: Examples of the shells of nonclassical form; (a)—bar-shaped shell, formed by two rectangular plates, supported by supports, (b)—short cylinder-shaped shell, formed by two round plates, supported by pivot rings; 1—elastic plates, 2—supports

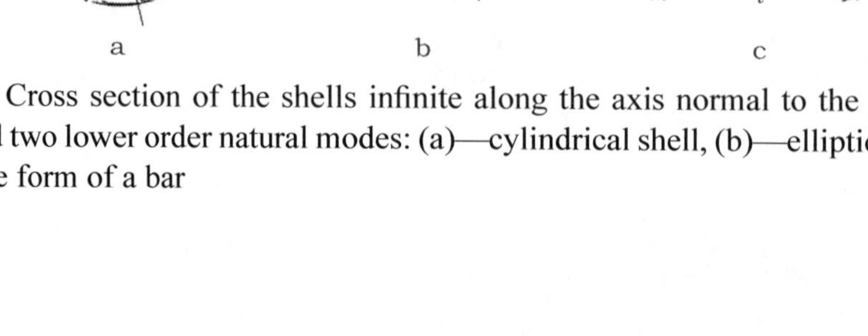

FIG. 3.2: Cross section of the shells infinite along the axis normal to the plane of the figure, and two lower order natural modes: (a)—cylindrical shell, (b)—elliptic shell, (c)—shell in the form of a bar

2. for an elliptic shell

$$\omega_{n=0} = \frac{(1{,}64)^2 h}{b^2} \left\{ \frac{E_1}{12\rho_1(1-\sigma^2)} \right\}^{1/2},$$

$$\omega_{n=2} = \frac{(2{,}75)^2 h}{b^2} \left\{ \frac{E_1}{12\rho_1(1-\sigma^2)} \right\}^{1/2},$$

3. for a bar

$$\omega_n = \frac{n^2 h}{l^2} \left\{ \frac{E_1}{12\rho_1(1-\sigma^2)} \right\}^{1/2},$$

E_1—elasticity modulus; ρ_1—density; σ—Poisson ratio of the material of a shell.

The sound reflecting properties of the grating of shells are obtained due to high volume compressibility of the shells. As the Fig. 3.2 shows, it is preferable to use the first vibrating mode of a hollow bar and a cylindrical tube. There are no shells surface areas vibrating in antiphase in this case. An elliptic tube is also applicable (with $n = 0$), because its surface areas vibrating in antiphase are of unequal size (as it is shown in [29], the inequality is be bigger, the bigger is ratio a/b). When vibrating modes are complex (for example, $n = 2$), the shells surfaces, vibrating in antiphase, are equal (or close) in size. As a result, the volume compressibility of the shells is close to zero and the desired sound-proofing is absent.

Let us compare the expressions for the natural frequencies of the lover vibration modes of the shells. As we can see, the first natural frequency of a cylindrical tube is inversely proportional to its radius. The natural frequencies of a hollow bar and an elliptical tube are proportional to the thickness of the walls and inversely proportional to the squared linear dimension of a cross section. These are the fundamental distinctions. Taking into account the physics of water, metals and plastics, when $n = 0$, it is impossible to get small wave dimensions of a cylindrical tube diameter in water $2r/\lambda_1$ (here $\lambda_1 = c/f_{n=o}$, c—sound velocity in water). For example, a steel tube has $2r = 0.5$ m, and wave diameter in water $2r/\lambda_1 \approx 1$ when $f_{n=o} = 3$ kHz. It follows that the increment of the grating of cylindrical tubes cannot be less than wave-length in water. Consequently, the homogeneity of the grating surface properties cannot be ensured. The screening of nonfunctional surfaces of the transducers with such grating is impossible because their dimensions do not exceed 0.5–1 of the wave-length in water.

The natural frequencies of a hollow bar and a tube of elliptical cross-section depend on the ratio of their two geometrical sizes. It ensures small full sizes and small wave sizes of their cross-section, and therefore, small increment of the grating. All other conditions being equal, the hollow bars are more preferable than the tubes with elliptical cross-section, because the later (as it was mentioned above) have small surface areas vibrating in antiphase. As a result, the volume compressibility of such tubes decreases.

On the whole, the results of qualitative analysis lead to the conclusion that the hollow bars are more applicable for use in hydroacoustic equipment.

The study of sound diffraction on elastic shells is based on solving the acoustical and mechanical problems simultaneously. We will consider the specific character of such conjugate problems in the context of solving the problem of sound diffraction on elastic shells of classical form. If the sound field affecting the elastic shell is known, the solution of the problem can be divided into stages [30,31]:

1. the choice of the coordinate system in which the variables are separated in Helmholtz equation and one of the coordinate surfaces coincides with the surface of the elastic shell;

2. presentation of the sound field near the shell as superposition of two fields—incident field and scattered field; at that, it is necessary to meet two conditions: scattered field must satisfy Sommerfeld condition [32], and both fields must be represented by their natural wave functions, typical for external field existence domain;

3. the choice of a mechanical model, describing the motion of a shell and presentation of its walls particle velocity in the form of the natural vibrating modes superposition;

4. formulation of functional equations representing conjugation conditions on the shell surface: conjugation of liquid particle velocity components normal to the surface and the shell, and conjugation of sound pressure in liquid and mechanical stress in the shell, normal to its surface;

5. the algebraization of functional equations basing on the completeness and orthogonality of the wave functions; and thus we obtain finite or infinite system of linear algebraic equations in the unknown coefficients of the series for wave functions;

6. the investigation of the wave properties of acoustic-mechanical systems on the basis of quantitative data analysis.

If we deal with a group of the shells rather then with one shell, we can obtain complete solution of the problem for many shells of indicated classical forms, using the addition theorem for the wave functions [33–35]. Such solution cannot be performed if the form of the shells is not classical, but as shown in Fig. 3.1. There is no such a coordinate system in which the surface of the shell could be described by one coordinate surface.

Hereafter we will consider the problems of sound waves diffraction by the grating composed of the closed elastic shells (as illustrated in Fig. 3.1). We will gradually increase the complexity of considered tasks.

3.2 GRATING COMPOSED OF ELASTIC BARS WITH ONE ELASTIC PLATE

3.2.1 Problem Solution

We consider the periodic grating composed of hollow bars of infinite length. The bars have no spacing between them. Each bar is formed with thin elastic plate (plate 1), attached at the edges through nondeformable supports 2. Figure 3.3 illustrates a cross section of the grating in a plane normal to its surface.

Let a plane sound wave, with velocity potential Φ_0, be incident at the grating at an angle θ. The wave excites flexural vibrations of the plates and vibrations of each bar as a single whole along the Ox axis.

Such grating is simple and not very interesting on the practical side. But the problem solving procedure and the mechanism of sound transmission through the grating can easily be analyzed on this example. The obtained quantitative results will be used for the analysis of more complex gratings.

Let us introduce the following assumptions and simplifications to avoid the problem solution overloading. The interior of the bars is assumed to be vacuum (further on we will also consider the case when there is a medium inside the bars), and the plates are supposed to be made of isotropic, ideal-elastic material. The last assumption is an idealization of real materials, but it is justified and widely used in studies of interaction between the sound waves, the elastic plates, and the shells in liquid. The reason of it is that on the one hand metals are characterized by low dissipative losses [30,36], and on the other they are characterized by relatively high values of the radiation impedance active components of the elastic plates and shells performing bending vibrations in liquid [30]. The losses of acoustic energy of the plates (shells) due to reemission into surrounding medium are therefore higher than the acoustic energy losses in plate's material because of dissipation. Thus the latter can be ignored. But it is to be noted that when we consider the vibrations of elastic units in gas, this assumption is false (see, for example [37]) because the wave impedance of gas is much lower than the wave impedance of liquid. Acoustic energy losses due to reemission into surrounding medium may be the same as dissipative losses in the material of elastic units.

Taking the grating configuration, illustrated in Fig. 3.3, into account, we choose Cartesian coordinate system to describe sound scattering process. Due to the absence of spacing between the elements of the grating, its frontal and back surfaces can be brought into coincidence with the coordinate surfaces $x = 0$ and $x = l$, respectively. Thus, we partition the entire domain of sound field existence into two partial domains: domain $x \leq 0$ and domain $x \geq l$. There is an interchange of energy between these partial domains due to the movement of the grating elements.

Let us consider now the sound fields in specified partial domains. As it is known, there are different forms for representing a sound field. For example, we can use Kirchhoff

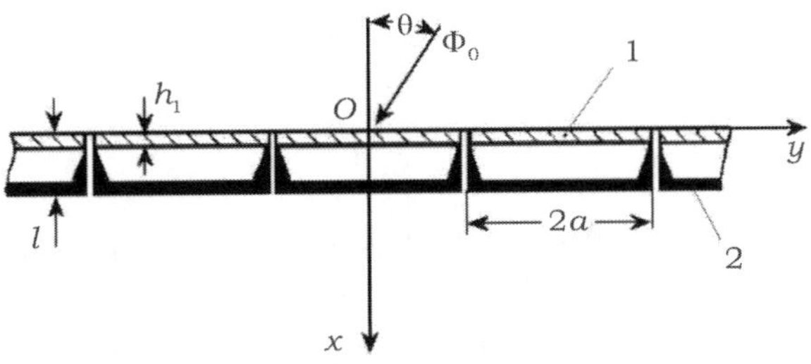

FIG. 3.3: Grating of bars, formed with one plate and a support: 1—plate, 2—support

integral, Green function and so on [32,38]. We use the possibility of representing the field in different forms to simplify matching boundary conditions on the grating surface. The grating periodicity leads to the periodicity of sound field. Bearing it in mind, we represent the sound fields as the superposition of the plane modes propagating at different angles to axis Ox. We then represent the velocity potential in domain ($x \leq 0$) as the sum of the incident plane wave and the superposition of reflected plane waves, and it can be represented as superposition of outgoing plane waves behind the grating, in domain $x \geq l$. Taking into consideration that the fields in both domains are periodical relative to y coordinate, we represent the velocity potentials in the indicated domains in the following form:

$$\begin{aligned}\Phi_1 &= \Phi_0 + e^{iky\sin\theta} \sum_{n=-\infty}^{\infty} A_n e^{-ik_n x} e^{i(\pi/a)ny}; \quad x \leq 0,\\ \Phi_2 &= e^{iky\sin\theta} \sum_{n=-\infty}^{\infty} B_n e^{ik_n(x-l)} e^{i(\pi/a)ny}; \quad x \geq l,\end{aligned} \tag{3.1}$$

here

$$\Phi_0 = e^{ik(x\cos\theta + y\sin\theta)}; \quad k_n = \left[k^2 - \left(k\sin\theta + \frac{\pi}{a}n\right)^2\right]^{1/2}; \quad n = 0;\ \pm 1;\ \pm 2; \ldots,$$

A_n, B_n—unknown complex coefficients, characterizing amplitude and phase of nth plane wave. Time dependence, as before, is represented as $e^{-i\omega t}$. Functions Φ_1 and Φ_2 are continuous in the corresponding domains, and they are the solution of Helmholtz equation. They represent the superposition of running and inhomogeneous waves.

Now we represent the vibration of the plates, forming the walls of the bars. At first we specify the properties of the plates. The plates are assumed to be thin. The thickness vibrations therefore can be neglected in the frequency band concerned, and we can take into account only the bending vibrations of the plates [30,39]. In this case, the compliance of the bars will be high (in comparison with water) and the grating will have good sound-attenuating properties.

These considerations enable us to write the differential equation of the thin plate bending vibrations:

$$\frac{d^4 \upsilon^{(q)}(y)}{dy^4} - k_{iz1}\upsilon^{(q)}(y) = i\frac{\omega}{D_1}p^{(q)}; \quad q = 0;\ \pm 1;\ \pm 2; \ldots, \tag{3.2}$$

where $\upsilon^{(q)}(y)$—plate particle velocity of the qth bar;

$$k_{iz1} = \left[\frac{\omega\rho_1 h_1}{D_1}\right]^{1/4}; \quad D_1 = \frac{E_1 h_1^3}{12(1-\sigma^2)},$$

$p^{(q)}$—sound pressure difference on the outer and interior sides of the plates of the qth bar.

We represent the particle velocity of the plates by series of natural modes of the plates in vacuum. The vibration modes of the plates must satisfy the homogeneous equation for

(3.2), and the boundary conditions at the edges of the plates. To specify the problem, the plates are supposed to be simply supported at the edges. In this case, there are the following conditions at the edges of the plate

$$\upsilon^{(q)}(y) = 0; \quad \frac{\partial^2 \upsilon^{(q)}(y)}{\partial y^2} = 0; \quad y = \pm a + 2qa. \tag{3.3}$$

The following functions satisfy conditions (3.3) and Eq. (3.2) with right hand side equal to zero.

$$\psi_t(y) = \sin \frac{\pi t}{2a}(y + a); \quad t = 1;\ 2;\ 3; \ldots, \tag{3.4}$$

These functions are the natural functions of simply supported vibrating plates. Taking into consideration the form of function (3.4), and also the ability of the bars to vibrate as a single whole along the Ox axis, we represent the particle velocity of the plate of qth bar in the following form:

$$\upsilon^{(q)}(y) = e^{i2aq\sin\theta}\left[\upsilon_o^{(q)} + \sum_{t=1}^{\infty} \upsilon_t^{(q)} \sin\frac{\pi t}{2a}(y + a - 2qa)\right], \tag{3.5}$$

where $\upsilon_o^{(q)}$—particle velocity of qth bar as a single whole along the Ox axis.

Now we direct attention to the phase factor before the square brackets in Eq. (3.5). It determines the delay of an oblique-incident plane wave arrival to a qth bar. The possibility to determine this delay in such a simple way follows from the periodicity of the grating and Floquet theorem [33].

The unknown complex coefficients, that are in series (3.1), (3.5), are determined from the following system of functional equations. The system contains the conjugation conditions of the particle velocities of the plates and surrounding medium, and also the conjugation conditions of the mechanical stress in the plates and the acoustic pressure in medium:

$$\begin{gathered}
\upsilon^{(q)}(y) = \partial\Phi_1/\partial y; \quad x = 0; \quad -a + 2qa \le y \le a + 2qa, \\
\upsilon_0^{(q)} e^{i2kaq\sin\theta} = \partial\Phi_2/\partial y; \quad x = l; \quad -a + 2qa \le y \le a + 2qa, \\
\frac{d^4\upsilon^{(q)}(y)}{dy^4} - k_{iz}\upsilon^{(q)}(y) = i\frac{\omega^2}{D_1}\rho_1\Phi_1; \quad x = 0; \quad -a + 2qa \le y \le a + 2qa, \\
M\upsilon_0^{(q)} e^{i2kaq\sin\theta} = \rho \int_{-a+2qa}^{a+2qa} \left[\Phi_1(x = 0) - \Phi_2(x = l)\right] dy,
\end{gathered} \tag{3.6}$$

M—the total mass of the bar per unit of its length, $M = M_{pl} + M_{\text{sup}}$, M_{pl}, M_{sup} plate mass and support mass respectively.

The system (3.6) is algebraized in common way basing on the completeness and orthogonality of the functions $e^{i(\pi/a)ny}$ and $\sin(\pi t/2a)(y + a - 2qa)$ on the interval $-a + 2qa \le y \le a + 2qa$.

Substituting expressions (3.1) and (3.5) in system (3.6) and using the orthogonality of the functions, we obtain the following infinite system, where $q = 0$:

$$
\begin{gathered}
A_n k_n N_n - i\upsilon_0^{(0)} N_n^* - i\sum_{t=1}^{\infty} \upsilon_t^{(0)} N_{nt} = -\delta k \cos\theta, \\
\upsilon_0^{(0)} \frac{M}{\rho} - \sum_{n=-\infty}^{\infty} A_n N_n^* + \sum_{n=-\infty}^{\infty} B_n N_n^* = \delta_n, \\
B_n k_n N_n = i\upsilon_0^{(0)} N_n^*, \\
\upsilon_t^{(0)} a Z_t - \upsilon_0^{(0)} \delta_t (\theta = 0) + \beta \sum_{n=-\infty}^{\infty} A_n N_{nt} = -\beta\delta_t.
\end{gathered}
\tag{3.7}
$$

Here:

$$
N_n = \begin{cases} 2a; \ n = m; \\ 0; \ n \neq m; \end{cases} \quad \delta_n = \begin{cases} 2a; \ n = 0; \\ 0; \ n \neq 0; \end{cases} \quad \beta = \frac{\rho}{\rho_1 h_1},
$$

$$
Z_t = \left(\frac{\omega_t}{\omega}\right)^2 - 1; \ \omega_t^2 = \frac{D_1}{\rho_1 h_1}\left(\frac{\pi t}{2a}\right)^4,
$$

$$
\delta_t = \frac{\pi t}{2a} \frac{e^{-ika\sin\theta} - (-1)^t e^{ika\sin\theta}}{(\pi t/2a)^2 - (k\sin\theta)^2}; \quad \delta = \frac{2\sin(ka\sin\theta)}{k\sin\theta}.
$$

We determine the value $\upsilon_0^{(0)}$ from the third equation of system (3.7), and then we substitute it in the first and the last equations, and thus obtain the following system of equations

$$
\begin{gathered}
A_n k_n N_n - B_n k_n N_n - i\sum_{t=1}^{\infty} \upsilon_t^{(0)} N_{nt} = \delta_n k \cos\theta, \\
B_n \frac{ik_n M N_n}{\rho N_n^*} + \sum_{m=-\infty}^{\infty} N_m^* (A_m - B_m) = -\delta, \\
\upsilon_t^{(0)} a Z_t + B_n i k_n \delta_t (\theta = 0) \frac{N_n}{N_n^*} + \beta \sum_{m=-\infty}^{\infty} A_m N_{mt} = -\beta\delta_t.
\end{gathered}
\tag{3.8}
$$

When we solve system (3.8) we obtain the quantitative characteristics of scattered sound field and the particle velocities of the plates and bars. In the same time this system has the properties typical for more complex gratings.

In view of the mentioned circumstances, it is important to investigate system (3.8) to find out whether it is possible to solve the problem by reduction method or by successive approximation. We can use for that a well-known approach, proposed by Koch [40,41]. This approach is based on checking the convergence of the series composed of the matrix coefficients of infinite system.

As we can see, system (3.8) is rather complex, and the analysis of the general case is a complicated problem. Thus we confine ourselves to special case analysis, which is important on the practical side, when $\theta = 0$. In this case, the initial physical problem becomes symmetrical with respect to plane $y = 0$ (Fig. 3.3), so system (3.8) can be simplified. Omitting intermediate computations, we obtain:

$$A_n + A_0 \frac{1}{ikM/2a\rho - 1} - \frac{i}{2a} \sum_{t=1}^{\infty} \upsilon_t^{(0)} \frac{\overline{N_{nt}}}{k_n} - \frac{\delta^*}{k} \left[1 - \frac{1}{ikM/2a\rho - 1} \right], \quad (3.9)$$

$$\upsilon_t^{(0)} - iA_0 \frac{k\delta_t}{[ikM/2a\rho - 1]\, Z_t} + \frac{2\beta}{a} \sum_{m=-\infty}^{\infty} A_n \frac{\overline{N_{nt}}}{Z_t} = \frac{\delta_t}{Z_t} \left[\beta - \frac{k}{a\,[ikM/2a\rho - 1]} \right],$$

where

$$\delta^* = \begin{cases} 1; & n = 0; \\ 0; & n \neq 0; \end{cases} \quad \overline{N_{nt}} = (-1)^n \frac{\pi t}{a} \left[\left(\frac{\pi t}{2a} \right)^2 - \left(\frac{n\pi}{a} \right)^2 \right]; \quad t = 1;\ 3;\ 5; \ldots.$$

System (3.9) consists of two infinite systems with two series of unknowns A_0, A_1, A_2, ... and $\upsilon_1^{(0)}, \upsilon_3^{(0)}, \upsilon_5^{(0)}, \ldots$. The results obtained from the systems with a single series of unknowns [40,41] are true for the systems with two series of unknowns. The systems with a single series of unknowns are of the following form

$$x_n + \sum_m^{\infty} C_{nm} x_m = b_n,$$

And we will estimate the convergence of the double series $\sum_{t,n} \left| \overline{N_{nt}}/k_n \right|^2$ and $\sum_{t,n} \left| \overline{N_{nt}}/Z_t \right|^2$, without turning attention to the frequency rates, at which one of the values of Z_t becomes zero. The system can be solved by reduction method when the value of Z_t becomes zero (for some values of index $t = T$ and frequency $f = f^*$). When the values of frequency and index are finite, the systems are quasi-regular. To prove their resolvability, it is necessary to make the same computations and reasoning as bellow for system 3.9. In view of the expressions for k_n, $\overline{N_{nt}}$ and by taking $q = t/2$, we have

$$\sum_{q,\, n} \left| \frac{\overline{N_{nq}}}{k_n} \right|^2_{n,\, q \gg 1} = \sum_{q,\, n} \frac{q^2}{n^2 \left(n^2 - q^2\right)^2} = \sum_q q^2 \sum_n \frac{1}{n^2 \left(n^2 - q^2\right)^2}. \quad (3.10)$$

Let us write the expression under the summation sign, with index n, as three summands

$$\frac{1}{n^2 \left(n^2 - q^2\right)^2} = \frac{1}{q^4 n^2} - \frac{1}{q^4 \left(n^2 - q^2\right)} + \frac{1}{q^2 \left(n^2 - q^2\right)^2}. \quad (3.11)$$

Taking into account expression (3.11), we obtain

$$\sum_q q^2 \sum_n \frac{1}{n^2 \left(n^2 - q^2\right)^2}$$
$$= \sum_q \frac{1}{q^2} \sum_n \frac{1}{n^2} - \sum_q \frac{1}{q^2} \sum_n \frac{1}{n^2 - q^2} + \sum_q \sum_n \frac{1}{\left(n^2 - q^2\right)^2}. \quad (3.12)$$

The first series in the right hand side of expression (3.12) converges. We can show that the second and the third series in the right hand side of expression (3.12) also converge. Using the expression given in work [42], it is not difficult to determine the sums of the following series:

$$\sum_{n=1}^{\infty} \frac{1}{n^2 - q^2} = \frac{1}{\pi q^2} - \frac{1}{2q} \operatorname{ctg} \pi q,$$
$$\sum_{n=1}^{\infty} \frac{1}{(n^2 - q^2)^2} = \frac{\pi}{4q^2 \sin^2 \pi q} + \frac{1}{4q^3} \operatorname{ctg} \pi q - \frac{1}{\pi q^4}. \tag{3.13}$$

In view of the last expressions, and the fact that $q = t/2 = 0.5, 1.5, 2.5, \ldots$ we obtain

$$\sum_q \frac{1}{q^2} \sum_n \frac{1}{n^2 - q^2} = \frac{1}{\pi} \sum_q \frac{1}{q^4},$$
$$\sum_q \sum_n \frac{1}{(n^2 - q^2)^2} = \frac{\pi}{4} \sum_q \frac{1}{q} - \frac{1}{\pi} \sum_q \frac{1}{q^4}. \tag{3.14}$$

The convergence of the double series $\sum\limits_{q,n} \left| \overline{N_{nt}/Z_q} \right|^2$ is much faster than the convergence of the series $\sum\limits_{q,n} \left| \overline{N_{nt}/k_n} \right|^2$, and it follows from the estimation

$$\sum_{q,n} \left| \frac{\overline{N_{nt}}}{Z_q} \right|^2_{n,\, q \gg 1} = \text{const} \sum_q \frac{1}{q^6} \sum_n \frac{1}{(n^2 - q^2)^2}. \tag{3.15}$$

We can make the following conclusions, basing on the analysis of double series convergence and on the convergence of the series $\sum\limits_t |\delta_t/Z_t|^2$:

1. the solution of the boundary problem, which was set, exists;

2. we can use reduction method or the method of successive approximations to obtain the values of the desired number of the first unknowns in expressions (3.9).

We should follow the certain problem's criteria when choosing finite system order. Such criteria depend on the character of the problem to solve.

3.2.2 Analysis of Scattered Field Structure and of the Motion of Grating Elements

We proceed with the quantitative analysis of the acoustic properties of a grating. First of all we work out a long-wave (approximate) solution of system (3.9), by taking $2a \ll \lambda_1$, where $\lambda_1 = c/f_1$ and f_1—the first natural frequency of the plates in vacuum. There is only one propagating plane wave in the semi-spaces $x \leq 0$ and $x \geq l$, when $\lambda \geq \lambda_1$, because $\text{Im}k_n = 0$ only when $n = 0$. There are only decaying (inhomogeneous) waves when $n \neq 0$, because $\text{Re}k_n = 0$.

That is why we can take only the unknowns with numbers $n = m = 0$ and $t = 1$. Then the solution of system (3.9) becomes trivial and can be represented in explicit form. Omitting simple computations, we write the expression for sound transmission k_{ins} and reflection k_{ref} coefficients when $\theta < \pi/2$:

$$k_{\text{ins}} = B_0 = \left\{1 + \frac{1}{2Z_1}\left[1 - \frac{M\alpha^2}{2a\rho_1 h_1}\right] - \frac{i}{2}\left[\frac{\beta}{kZ_1\cos\theta} + \frac{kM\alpha^2\cos\theta}{2a\rho}\right]\right\}^{-1}, \quad (3.16)$$

$$k_{\text{ref}} = A_0 - 1 + \left(1 - \frac{iMk\alpha^2\cos\theta}{2a\rho}\right)$$
$$/\left\{1 + \frac{1}{2Z_1}\left[1 - \frac{M\alpha^2}{2a\rho_1 h_1}\right] - \frac{i}{2}\left[\frac{\beta}{kZ_1\cos\theta} + \frac{kM\alpha^2\cos\theta}{2a\rho}\right]\right\},$$

where $\alpha = ka\sin\theta/\sin(ka\sin\theta)$.

It follows from expression (3.16): when $\omega \to 0$, $Z_1 \to (\omega_1/\omega)^2 \to \infty$, $k_{\text{ins}} \to 1$, $k_{\text{ref}} \to 0$, and when $\omega \to \omega_1$ value of $Z_1 \to 0$; and

$$k_{\text{ins}} \to 2Z_1/\left(1 - \frac{M\alpha^2}{2a\rho_1 h_1} - i\frac{\beta}{k\cos\theta}\right) \to 0, \quad (3.17)$$

$$k_{\text{ref}} \to \left[-1 + 2Z_1\left(1 - \frac{Mk\alpha^2\cos\theta}{2a\rho}\right)\right]/\left(1 - \frac{M\alpha^2}{2a\rho_1 h_1} - i\frac{\beta}{k\cos\theta}\right) \to -1.$$

Due to the estimations in system (3.17), we can find the interesting and important properties of the grating of hollow elastic bars:

1. the grating is acoustically transparent at frequencies that are lower than the first natural frequency;
2. in frequency band close to the first natural frequency of the plates, transmissivity of the grating is low;
3. the phase of the reflection coefficient in the region of the first natural frequency of the plates is equal π; in other words, the grating behaves like an acoustically soft surface at these frequencies.

Equations (3.16) and (3.17) are approximate, and the question arises as to whether the calculation is accurate. We will answer this question bellow, basing on the comparison of the results calculated by (3.16) and the numerical solution of system (3.8) using the unknowns of higher order. We can estimate the accuracy of the calculations performed by Eqs. (3.16) and (3.17), checking whether the law of conservation of energy is fulfilled:

$$|k_{\text{ref}}|^2 + |k_{\text{ins}}|^2 = 1. \quad (3.18)$$

Note that we can prove that system (3.8) can be solved by reduction method by checking equality (3.18).

Substituting expressions (3.16) in equality (3.18) we obtain

$$|k_{\text{ref}}|^2 + |k_{\text{ins}}|^2 = 1 - \chi,$$

where

$$\chi = \frac{2}{Z_1}\left(1 + \frac{M\alpha^2}{2a\rho_1 h_1}\right) \Big/ \left\{\left[2 + \frac{1}{Z_1}\left(1 + \frac{M\alpha^2}{2a\rho_1 h_1}\right)\right]^2 + \left[\frac{\beta}{kZ_1 \cos\theta} + \frac{kM\alpha^2 \cos\theta}{2a\rho}\right]^2\right\}.$$

We give an estimate to the value of χ for the following two cases:

$$\begin{aligned} &\omega \to 0; \ \chi \to \left(\frac{\omega}{\omega_1}\right)^2 (1/2 + \alpha^2), \\ &\omega \to \omega_1; \ Z_1 \to 0; \ \chi \to Z_1 \frac{2(1 + 2\alpha^2)}{(1 - 2\alpha^2)^2 + (\beta/k\cos\theta)^2}. \end{aligned} \tag{3.19}$$

As it follows from expressions (3.19), the law of conservation of energy is fulfilled adequately. For example, for simplicity, we set $\theta = 0$, at frequency $\omega/\omega_1 = 0.1$ we have $\chi = 0.015$ and $|k_{\text{ref}}|^2 + |k_{\text{ins}}|^2 = 0.985$, and at frequency $\omega/\omega_1 = 1$ we have $\chi = 0$.

Analysis revealed that the general quantitative properties of the sound fields in regions $x < 0$ and $x > l$ are determined only by several first terms of the corresponding series.

Let us analyze the quantitative data, calculated by a computer on the basis of numerical solution of system (3.8). The system (3.8) reduction order was determined by the numerical experiment and the calculated data comparison. It was found that if the system is reduced at the level $|n| = N \geq (4-5)\,ka$ and $t = T > 4ka$ it almost does not change the first four significant figures of the coefficients A_N, B_N, $\upsilon_T^{(0)}$, $\upsilon_0^{(0)}$ and provides equality (3.18) performance with accuracy not less than 10^{-3}–10^{-4}.

The computations were performed for the situation when the grating is immersed in water with wave impedance $\rho c = 1.5 \times 10^6$ kg/m^{2}s, and the plates are made of titanium alloy [43] with density $\rho_1 = 4.5 \times 10^3$ kg/m^3, Young modulus $E = 1.16 \times 10^{11}$ Pa and Poison ratio $v = 0.3$. Here $M_{\text{sup}} = 0.1 M_{pl}$.

Figure 3.4 illustrates the curves, calculated on the basis of the numerical solution of system (3.8), characterizing the frequency dependence of sound transmission coefficient through the grating when $2a/\lambda_1 = 0.1$ and $2a/\lambda_1 = 0.06$ at different angles of wave incidence. As we can see, the transmissivity of the grating decreases with an increase of angle θ, and this is reflected in an increase in the frequency bandwidth of low transmissivity. The data in Fig. 3.4 confirm the conclusions made above, that the transmissivity of the grating is high at frequencies that are lower than the natural frequency of the plates. With an increase of frequency, the transmissivity of the grating decreases gradually and reaches its minimum near the natural frequency of the plates, and then we can observe an increase in the sound transmissivity. The frequency of the minimum transmissivity of

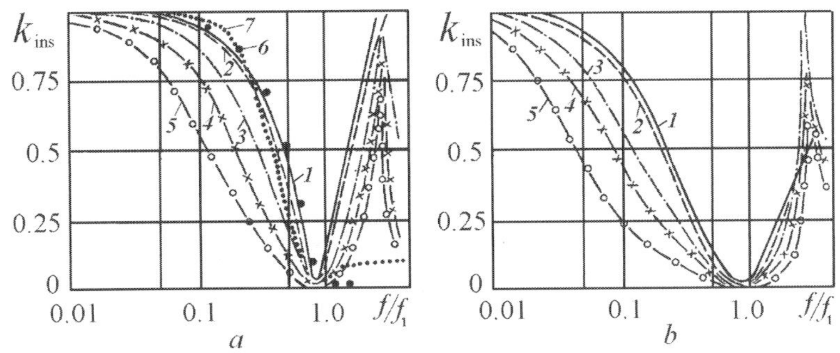

FIG. 3.4: Frequency dependence of sound transmission coefficient $M_{\text{sup}} = 0.1M_{pl}$; (a)—for $2a/\lambda_1 = 0.1$, (b)—for $2a/\lambda_1 = 0.06$; curves 1, 2, 3, 4, 5—correspond to θ equal 0, 30, 60, 70, and 80 degrees; curve 6 (marked with large dotes)—approximate calculation by (3.16); curve 7—for $M_{\text{sup}} = 10M_{pl}$

the grating and the natural frequency f_1 are mismatched slightly because of the presence of liquid mass added to the plate.

Comparing the curves in Figs. 3.4(a) and 3.4(b), we can make certain, that the frequency band of law transmissivity of the grating essentially depends on wave dimension of the plate $2a/\lambda_1$. With increasing $2a/\lambda_1$ (if $f_1 = \text{const}$), the bandwidth of insulation of the grating decreases. The reason of it is that the screens of the plate increases with an increase of the plate size and, as a consequence, the Q-factor of mechanical vibration system becomes higher.

For detailed consideration of some physics of sound transmission through the grating of elastic bars, we analyze the frequency dependence of the normalized (relative to ρc) input impedance of the grating Z_{in} at normal incidence of a plane sound wave. Analyzing Fig. 3.5, we can make certain, that the input impedance reactive component $\text{Im}Z_{in} = X$ has mass character in the domain $f/f_1 < 1$.

This is because of the following circumstances. The compliance of the walls of the bars is small at low frequencies, and a bar screens plays the major part in the process

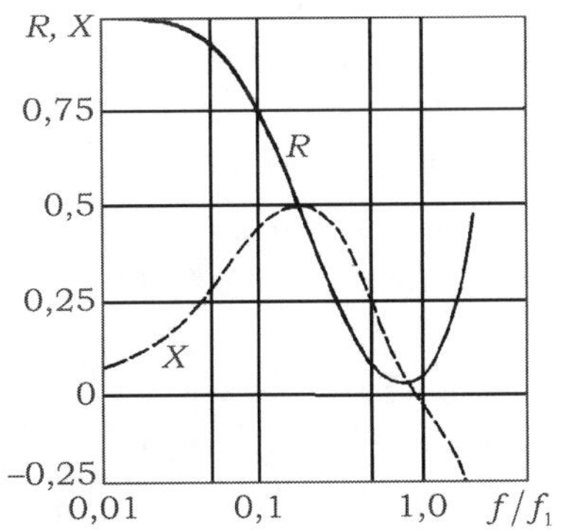

FIG. 3.5: Frequency dependencies of input impedance components of the grating, when $\theta = 0$ and $2a/\lambda_1 = 0.1$

of sound transmission. That is why the determinant motion is the vibration of the bar as a single whole along the Ox axis. The value X therefore increases in direct ratio to frequency when $f/f_1 \ll 1$. From this point on, the elasticity of the plates plays important role and the rate of increase of the value of X decreases. The value of X reaches an extremum in the domain $f/f_1 \approx 0.2$, than it falls down and crosses abscesses near $f/f_1 \approx 1$. It goes to prove that there is a resonance of mechanical vibrating system: hollow bars and surrounding medium. After resonance, the value X has elastic character.

Frequency dependence of input impedance active component $\mathrm{Re}Z_{in} = R$ is similar to frequency dependence of k_{ins}. Thus, in the region of the resonance of the system bars–water, the input impedance of the grating is of less importance than the wave impedance of water. That is why the grating has high sound-attenuating properties and its surface becomes similar in acoustic properties to acoustically soft surface.

Acoustic properties of the grating of hollow bars were considered above. The investigation of kinematics of the elastic system elements also gives us an insight into the physical processes that take place in such systems. We therefore analyze the amplitudes and phases of the particle velocities of the plates and bars for the most typical frequency domains. Namely: in the domain of low frequencies $f/f_1 \ll 1$, of resonance frequency $f/f_1 \approx 1$ of the frequencies higher than the resonance $f/f_1 > 1$, when the grating becomes acoustically transparent. The data for the titanium plate, with wave dimension $2a/\lambda_1 = 0.1$, at normal incidence of the plane wave on the grating, is shown in Table 3.1.

As the table shows, the amplitude of bar vibrations as a single whole at low frequencies is almost equal to the particle velocity amplitude in the incident wave. At the same time, the plates are bending in antiphase to bar motion. But, as $|\upsilon_0^{(0)}| > |\upsilon_1^{(0)}|$, absolute plate motion coincide in direction with bar motion, and its amplitude is equal $|\upsilon_0^{(0)}| - |\upsilon_1^{(0)}|$. Thus the bar motion as a single whole prevails. This is the reason why the transmissivity of the grating at these frequencies is high.

The situation is different in the region of resonance of the plates. The small amplitude of the bar vibration as a single whole is about 5% of the particle velocity amplitude of the medium in the incident wave. The amplitude of the bending plates, on the other hand, is large. In spite of the fact that the incident wave excites the sizable flexural vibrations of the plate, it cannot induce the vibrations of the support, although there is a mechanical linkage between the plate and the support. This paradoxical fact is the cause of the law transmissivity of the grating. At frequencies $f/f_1 > 1$, when the value k_{ins} is close to one, we can observe sizable flexural vibrations, and they are imparted to the support

TABLE 3.1: Amplitudes and phases of the vibrational speed of the plate

f/f_1	**Amplitude of the particle velocity in the incident wave, m/s**	$\left\|\upsilon_1^{(0)}\right\|$, **m/s**	$\arg_1^{(0)}$, **degree**	$\left\|\upsilon_0^{(0)}\right\|$, **m/s**	$\arg_0^{(0)}$, **degree**
0.1	0.0126	0.00915	224	0.0122	13.5
1.0	0.126	0.4	270	0.0056	264
2.66	0.335	0.935	348	0.303	308

through the ball joint. The result is that the vibration amplitude of the support becomes close to the vibration amplitude of the medium in the incident wave.

3.3 GRATING COMPOSED OF SEPARATED BARS WITH ONE ELASTIC PLATE

Let us consider more complicated case, when the bars are spaced relative to each other. The gaps between the bars are field with surrounding medium. These gaps complicate the processes of interaction between the elastic elements of the grating and the exchange of energy between front and back semi-spaces. In particular, the grating sound-insulation becomes worse, because sound energy is transmitted from front to back semi-spaces not only due to the bars vibration as a single whole, but also because of sound transmission through the gaps between them.

Figure 3.6 illustrates the cross-section of the considered grating (compare with Fig. 3.3). It is impossible to superpose the surface of the grating with one of the coordinate surfaces of Cartesian reference system because of spacing between the bars. This circumstance complicates the solution of diffraction problem.

To overcome these difficulties, we have to partition the sound field existence domain into subdomains (partial domains), the boundaries of which are the sections of the coordinate surfaces of Cartesian reference system. The sound field in each domain can be represented in terms of the natural wave functions of these domains. It is necessary to conjugate the sound pressures and the particle velocities on the boundary of subdomains and to algebraize the corresponding functional equations.

The solution scheme of the problem of diffraction by grating of elastic shells, which was described, is used in this monograph for considering the wide variety of acoustic problems. This general solution method of boundary problems for Helmholtz equation is effective for the fields scattered by elastic elements. It is necessary that specific solutions contain the elements enabling us to meet the matching conditions on the surface of deformable plates.

We turn again to Fig. 3.6. Let us partition the whole sound field existence domain into the following partial domains: semi-spaces $x \leq 0$ and $x \geq 0$, and an infinite number of subdomains, which are the gaps between the bars. We confine ourselves to the case

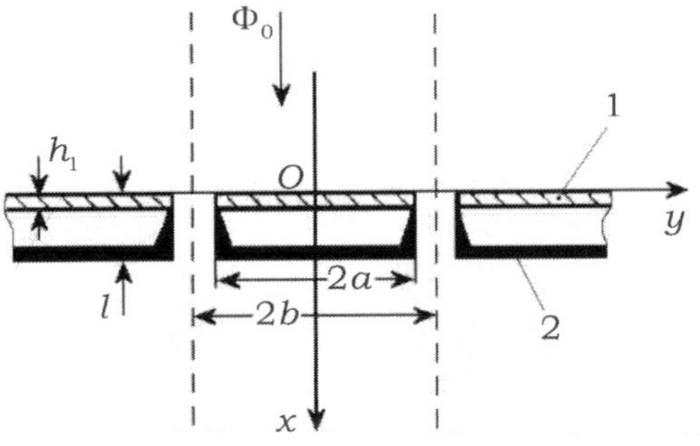

FIG. 3.6: Grating of separated bars; 1—plates, 2—supports

of normal incidence of the sound wave on the grating $\theta = 0$. Taking this restriction and the symmetry properties of the grating into account, we can replace the original problem with an equivalent problem of sound propagation in a flat waveguide with acoustically rigid walls.

The walls of such a waveguide are plotted in Fig. 3.6 with dashed lines. The partial domains can be defined as $x \le 0$ and $|y| \le b$; domain $0 \le x \le l$, $a \le x \le b$; domain $0 \le x \le l$, $-a \ge x \ge -b$; domain $x \ge l$, $|y| \le b$ (note, that it is not necessary to consider the third region due to the symmetry of the waveguide and the incident wave with respect to Ox axis).

Assuming that the supports are nondeformable, and the plates are hinge supported at the edges, we formulate the boundary conditions:

$$\begin{gathered} \partial\Phi/\partial y = 0;\ \ |y| = b;\ \ |x| < \infty, \\ \partial\Phi/\partial y = 0;\ \ |y| = a;\ \ 0 \le x \le l, \\ \upsilon^{(1)}(y) = d^2\upsilon^{(1)}(y)/dy^2 = 0;\ \ x = 0;\ \ |y| = a, \end{gathered} \tag{3.20}$$

where $\upsilon^{(1)}(y)$—particle velocity of the plate.

The sound fields can be represented by natural wave functions of plane waveguide in the partial domains that were selected [3]. Bearing in mind the boundary conditions (3.20), we obtain

$$\begin{gathered} \Phi_1 = e^{ikx} + \sum_{n=0}^{\infty} A_n e^{ik_n x} \cos\alpha_n y, \\ \Phi_2 = \sum_{p=0}^{\infty} \left[B_p \cos k_p x + C_p \sin k_p x\right] \cos\alpha_p (y - a), \\ \Phi_4 = \sum_{m=0}^{\infty} D_m e^{ik_m (x-l)} \cos\alpha_m y, \end{gathered} \tag{3.21}$$

where $k_n = \left(k^2 - \alpha_n^2\right)^{1/2}$; $k_m = \left(k^2 - \alpha_m^2\right)^{1/2}$; $k_p = \left(k^2 - \alpha_p^2\right)^{1/2}$; $\alpha_n = \pi n/b$; $\alpha_m = \pi m/b$; $\alpha_p = \pi p/(b - a)$.

Taking into account conditions (3.20), we represent the particle velocity of the plates by series expansions in their mode configurations in vacuum:

$$\upsilon^{(1)}(y) = \upsilon_0 + \sum_{s=0}^{\infty} \upsilon_s^{(1)} \cos\alpha_s y;\ \ \alpha_s = (2s + 1)\,\pi/a, \tag{3.22}$$

where υ_0—the amplitude of the bar motion particle velocity as a single whole along the axis Ox.

The unknown complex coefficients in series (3.21) and (3.22) are determined from the following system of functional equations:

$$\Phi_1 = \Phi_2; \quad x = 0,$$
$$a \le y \le b,$$
$$\Phi_2 = \Phi_4; \quad x - l,$$
$$-\partial\Phi_1/\partial x = \begin{cases} -\partial\Phi_2/\partial x; & a \le y \le b; \quad x = 0, \\ \upsilon^{(1)}(y); & 0 \le y \le a; \quad x = 0, \end{cases}$$
$$-\partial\Phi_4/\partial x = \begin{cases} -\partial\Phi_2/\partial x; & a \le y \le b; \quad x = l, \\ \upsilon_0; & 0 \le y \le a; \quad x = l, \end{cases} \tag{3.23}$$
$$M\upsilon_0 = \rho \int\limits_0^a [\Phi_1(x=0) - \Phi_4(x=l)]\, dy; \quad M = M_{pl} + M_{\text{sup}},$$
$$\frac{d^4\upsilon^{(1)}(y)}{dy^4} - k_{iz1}\upsilon^{(1)}(y) = -\frac{\omega^2\rho}{D_1}\Phi_1; \quad x = 0; \quad 0 \le y \le a.$$

The notation in system (3.23) is the same as in Eqs. (3.2) and (3.6).

System (3.23) is algebraized in the same way as system (3.8). As a result, we obtain the infinite system of linear equations of the second kind in the unknowns $A_n, B_p, C_p, \upsilon_s^{(1)}, \upsilon_0$. We will not write down the system of equations. Let us consider the problem solution at a long wave approximation, setting $kb \ll 1$, $l < b$, $n = m = p = s = 0$. After uncomplicated but bulky calculations, we have

$$k_{\text{ins}} \approx \frac{2(\varepsilon + \bar{\gamma})}{2(\varepsilon + \bar{\gamma}) - ikl(1 + \varepsilon\bar{\gamma} + \varepsilon^2) - i\left[\bar{\beta}(1-\varepsilon)/kZ_1^{(1)}\right][2(\varepsilon + \bar{\gamma}) + 2kl(\bar{\alpha} - 1)]},$$
$$k_{\text{ref}} \approx k_{np}\left[1 - kl\frac{\varepsilon kl + i}{\varepsilon + \bar{\gamma}}\right] - 1, \tag{3.24}$$

where $\varepsilon = (1 - a/b)$—perforation coefficient, $\bar{\gamma} = \rho la/M$—the ratio of the mass of ambient liquid, displaced by the bar, to the bar mass (per unit length of the bar),

$$\bar{\beta} = \frac{4\rho}{\pi^2\rho_1 h_1}; \quad \bar{\alpha} = \frac{a\rho_1 h_1}{M}; \quad Z_1^{(1)} = \left(\frac{\omega_1^{(1)}}{\omega}\right) - 1.$$

The calculation of k_{ins} by (3.24) gives the following values for acoustic characteristics of the grating:

$$\omega \to 0; \quad kl \to 0; \quad kZ_1^{(1)} \to \frac{\left(\omega_1^{(1)}\right)^2}{\omega} \to \infty; \quad k_{\text{ins}} \to 1; \quad k_{\text{ref}} \to 0, \tag{3.25}$$
$$\omega \to \omega_1^{(1)}; \quad Z_1^{(1)} \to 0; \quad k_{\text{ins}} \to \frac{2(\varepsilon + \bar{\gamma})kZ_1^{(1)}}{-i\bar{\beta}[2(\varepsilon + \bar{\gamma}) + i2kl(\bar{\alpha} - 1)]} \to 0; \quad k_{\text{ref}} \to -1.$$

It follows from the last two expressions that, in spite of the spacing between the bars, the grating retains its sound attenuating properties in the domain of the first natural frequency of the plates. At the same time the value of k_{ins} essentially depends on the size of the spacing between the bars. The values of k_{ins} and ε are in complicated interdependence, and that is why it is worthwhile to make a quantitative analysis, basing on the results of calculation by (3.24). All the calculations were performed for the following values: $2af_1/c = 0.1$; $l/2a = 0.2$; $\rho_1 = 4.5 \times 10^3$ kg/m^3; $E_1 = 1.16 \times 10^{11}$ Pa; $\sigma_1 = 0.3$; $\rho c = 1.5 \times 10^6$ kg/m s; $\bar{\gamma} = 1/3$; $\bar{\alpha} = 1/2$; $\beta = 0.27$ and were presented in the Table 3.2.

The data of Table 3.2 for $\varepsilon = 0$ were obtained as a result of calculations by formula (3.16) that holds for the case, when there is no spacing between the bars. As the table shows, the increasing of the spacing, all other parameters being equal, leads to the increasing of the transparency of the grating at the frequencies bellow the first natural frequency of the plate in vacuum. But in the vicinity of the first natural frequency of the plates, the sound insulation of the grating is rather effective even with high degree of perforation ($\varepsilon = 0.5$).

The gratings of elastic elements can be used in practice if they have high reflectivity in some frequency band.

The frequency band of effective sound reflection is the band, where the condition $|k_{\text{ins}}| \le k_0$ is met for some given value of k_0. We often take $k_0 = 0.2$ in acoustics. In this case only 4% of incident wave energy travel into the second semi-space on the boundary of the frequency band of effective sound insulation.

We can determine the band of effective sound insulation of considered grating using the data shown in Fig. 3.7, where the curves 1, 2, and 3 correspond to the values

TABLE 3.2: Estimations of the transmission coefficient of sound

Parameter	$f/f_1^{(1)}$											
	0.25				0.5				0.71			
ε	0	0.1	0.25	0.5	0	0.1	0.25	0.5	0	0.1	0.25	0.5
$\|k_{\text{ins}}\|$	0.82	0.87	0.92	0.95	0.46	0.58	0.63	0.78	0.26	0.32	0.39	0.5

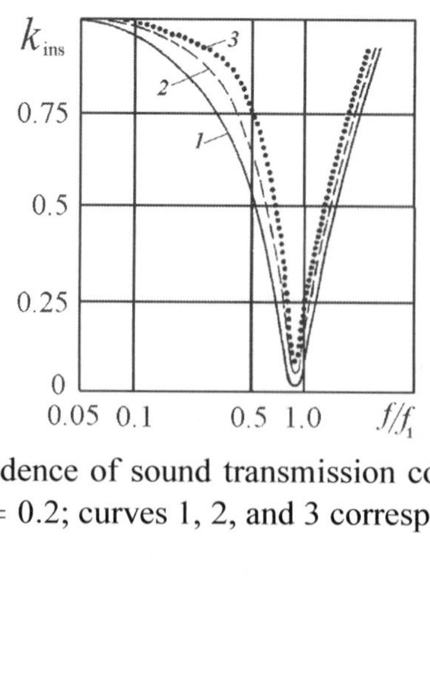

FIG. 3.7: Frequency dependence of sound transmission coefficient through the grating for $2a/\lambda_1 = 0.1$ and $l/2a = 0.2$; curves 1, 2, and 3 correspond to perforation coefficient ε equal 0, 0.25, and 0.5

of perforation coefficient $\varepsilon = 0, 0.25$, and 0.5. The sound transmission coefficient frequency dependencies are obtained on the basis of numerical solution of equation system 3.23 for different values of the perforation coefficient. We can observe that with increase of perforation coefficient, the bandwidth of effective sound insulation narrows, and the sound transmission coefficient increases at the resonance frequency of the elastic plates in the grating. This frequency is below the natural frequency of the plates in vacuum. As the data in Fig. 3.7 shows, the degree of perforation affects scarcely the frequency dependence of k_{ins} in the frequency domain $f > f_1^{(1)}$. We can change the properties of the grating by changing its geometric parameters only at low frequencies. It is also possible to control the properties of the grating by adding the elastic elements, changing their position relative to the direction of the incident sound wave and so on. The quantitative analysis of the influence of these factors is given in the following paragraphs of the chapter.

3.4 GRATING COMPOSED OF BARS WITH TWO ELASTIC PLATES

It would appear reasonable that an expansion in the number of the elastic walls in a bar leads to an increase of its elasticity, and improvement of sound-attenuating properties of the grating. We will consider the grating formed by bars with two (generally unequal) plates, normal to the plane of the grating (Fig. 3.8).

The solution of the problem of the plane sound wave diffraction by such grating is similar to the solution of the problem discussed above. Thus we will omit the details, and will point out the main distinctions.

First of all, boundary conditions have to be supplemented with one more condition concerning the fixation of the second (back) plate of the bar. Assuming that it is hinging, we have

$$\upsilon^{(2)}(y) = d^2\upsilon^{(2)}(y)\big/dy^2 = 0; \quad x = l; \quad |y| = a, \tag{3.26}$$

where $\upsilon^{(2)}(y)$—particle velocity of the second plate.

Secondly, we represent the particle velocity of the second plate by series of natural vibration modes of the plate in vacuum. Taking into account the accepted condition (3.26), we have

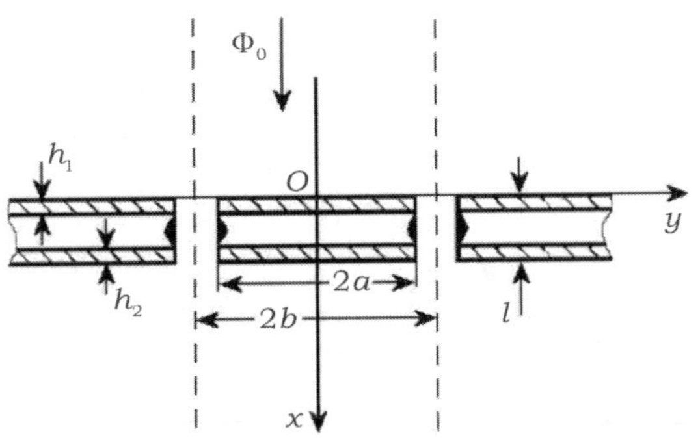

FIG. 3.8: Grating formed by bars of two plates

$$\upsilon^{(2)}(y) = \upsilon_0 + \sum_{t=0}^{\infty} \upsilon_t^{(2)} \cos \alpha_t y; \quad \alpha_t = (2t+1)\pi/a. \tag{3.27}$$

System (3.23) must be supplemented with an equitation showing the relation between the particle velocity of the second plate and the sound pressure on its surface:

$$\frac{d^4\upsilon^{(2)}(y)}{dy^4} - k_{iz2}^4 \upsilon^{(2)}(y) = \frac{\omega^2 \rho}{D_2} \Phi_4; \quad x = l; \quad 0 \le y \le a, \tag{3.28}$$

where

$$k_{iz2} = \left(\frac{\omega^4 \rho_2 h_2}{D_2}\right)^{1/4}; \quad D_2 = \frac{E_2 h_2^3}{12\left(1 - \sigma_2^2\right)}.$$

The fourth equation of the system (3.23) must be substituted by the following:

$$-\frac{\partial \Phi_4}{\partial x} = \begin{cases} -\dfrac{\partial \Phi_2}{\partial x}; & a \le y \le b; \quad x = l, \\ \upsilon^{(2)}(y); & 0 \le y \le a; \quad x = l. \end{cases} \tag{3.29}$$

The procedure for deriving the algebraic equations from the system of functional equations, that includes all changes and additions to the system (3.23), is the same as in the previous paragraphs. We use the algebraic equations to find the values of infinite series coefficient. The equations forming the infinite system will be changed slightly. The procedure for deriving equations is simple although bulky problem, which can be solved by expending the simple functions in a Fourier series. We will analyze the results, which are obtained on the basis of the equations solution, without giving the explicit equations.

Important information about the acoustic properties of the grating concerned can be obtained when considering the law-frequency approximate solution.

In this case, the infinite system is substituted by the finite system with only one coefficient from each group. Using the derived finite system, we estimate an approximate value of the coefficient of the sound transmission through the grating.

By taking $kb \ll 1, l < b$, we find

$$k_{\text{ins}} \approx \frac{Q}{F + G}, \tag{3.30}$$

where

$$Q = 2\left[\varepsilon + \bar{\gamma} + \bar{\gamma}(1 - \varepsilon)\frac{16\bar{\gamma}}{\pi^2 Z_1^{(2)}}\right],$$

$$F = 2(\varepsilon + \bar{\gamma}) - ikl\left(1 + \varepsilon\bar{\gamma} + \varepsilon^2\right) + \frac{2}{\pi Z_1^{(2)}}\left[\bar{\gamma}\frac{4}{\pi^2} + l\beta_2(1 - \varepsilon)\right] - i(\varepsilon + \bar{\gamma})\frac{2\beta_2(1 - \varepsilon)}{\pi k Z_1^{(2)}},$$

$$G = -i(1 + \varepsilon)\frac{\bar{\beta}_1}{\pi k Z_1^{(1)}}\left[2(\varepsilon + \bar{\gamma}) + \frac{16\bar{\gamma}(1 - \varepsilon)}{\pi^2 Z_1^{(2)}} + 2kl(\bar{\alpha} - 1)\left[1 + i\frac{2\beta_2(1 - \varepsilon)}{\pi k Z_1^{(2)}}\right]\right],$$

$$\bar{\beta}_1 = \frac{4\rho}{\pi^2 \rho_1 h_1}; \quad \beta_2 = \frac{4\rho}{\pi^2 \rho_2 h_2}; \quad Z_1^{(2)} = \left(\frac{\omega_1^{(2)}}{\omega}\right) - 1,$$

The rest notation is the same as in the Eq. (3.24).

Data comparison of the approximate long-wavelength and the exact solutions, which was made in the previous paragraph, shows that an approximate solution gives the qualitatively true acoustic properties of the grating. Hence, the qualitative analysis of the expression for k_{ins} has to be done in some typical special cases.

For $\omega \to 0$, $|k_{\text{ins}}| \to 1$. The second elastic wall that appears in the bars can't affect the acoustic properties of the grating. Only the mass properties of its units are of importance in the low-frequency region. If the frequency of the incident wave is lower than the front elastic plate natural frequency $\omega < \omega_1^{(1)}$ and, in the same time, the second (back) plate natural frequency is high enough $\omega_1^{(2)} \gg \omega_1^{(1)}$, than the acoustic properties of the grating concerned become identical to the properties of the grating with one elastic wall. The expression for the transmission coefficient, which follows from the Eq. (3.30), is completely congruent with the expression (3.24) in this case. We perform the computation on the presumption that the natural frequency of the back plate is lower than the front plate's natural frequency. We can conclude on the basis of the k_{ins} values analysis, that the acoustic properties of a grating are almost independent of the location of an elastic unit in the walls of the bar.

The case when the front and back plates are identical is of interest.

In this case, the approximation equation for calculating the transmission coefficient shows the decrease of the grating transmissivity near the natural frequency of the plate. The more complete quantitative and qualitative description of the situation can be obtained on the basis of the numerical solution of the corresponding infinite systems. In the next paragraphs we will pay great attention to the analysis of the peculiarities of the elastic plate's oscillations in the walls of the bars and relationship between the oscillation mode and the acoustic properties of the grating. Some typical data, received in the course of solving the infinite system equations, is given below.

Figure 3.9(a) illustrates frequency dependence of k_{ins} when the plates are identical ($2a/\lambda_1 = 0.1$, $l/2a = 0.2$). The plates are made of titanium alloy, with the following mechanical properties [43]: $\rho_1 = \rho_2 = 4.5 \times 10^3$ kg/m^3; $E_1 = E_2 = 1.16 \times 10^{11}$ Pa; $\sigma_1 = \sigma_2 = 0.3$. Perforation ratios ε equal 0.15, 0.35, and 0.5 correspond to the curves 1, 2, and 3.

As one would expect, with an increase of perforation ratio, the frequency domain, where the acoustic transmissivity of the grating is low, decreases. But it is significant, that despite the spacing between the bars, sound-attenuating properties of the grating do not get worse, comparing with the grating where the bars do not have the spacing between them. It is the direct consequence of increasing the number of the elastic plates in the bars. Such bars have higher compliance comparing with the bars with only one elastic plate. It is instructive to compare this quantitative data with the results of the investigation of the sound waves diffraction by a grating of acoustically rigid bars [38], and the diffraction of the electromagnetic waves by a grating of the bars with the ideal acoustic transmissivity for the case of E—polarization [45], which is equivalent in acoustics

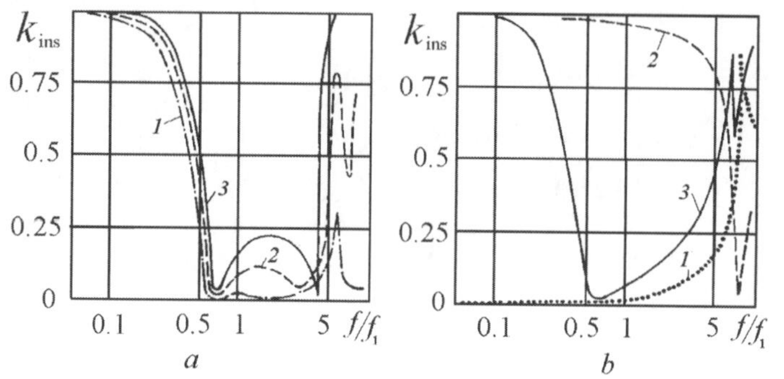

FIG. 3.9: Frequency dependence of the transmission coefficient of sound passing through the grating formed by bars; (a)—bars, formed by two identical plates, $2a/\lambda_1 = 0.1$ and $l/a = 0.1$, perforation ratios ε equal 0.15, 0.35, and 0.5 correspond to the curves 1, 2, and 3; (b)—curves 1 and 2 correspond to the ideal acoustically compliant and acoustically rigid bars, curve 3 corresponds to the bars with two elastic plates, the results are calculated with $2af_1/c = 0.06$ and $\varepsilon = 0.5$

to the case of the compliant bars [38]. Curves 1 and 2 in Fig. 3.9(b) correspond to the cases of acoustically rigid and compliant bars (the data was taken from the investigations [38,45]), and curve 3 corresponds to the hollow elastic bars. The results were calculated for $\varepsilon = 0.5$ and $2af_1/s = 0.06$.

The analysis of these curves indicates that in the domain $f/f_1^{(1)} \ll 1$ the grating formed by the hollow bars is similar, in acoustic characteristics, to a grating of rigid bars.

But from $f/f_1^{(1)} \approx 0.4$ to $f/f_1^{(1)} \approx 3$, the grating becomes similar in its properties to a grating of acoustically compliant bars. There is no such correspondence at higher frequencies. The elastic bars interaction and the influence of the surrounding medium on the bars become more complicated owing to the fact that the high vibration modes of the plates begin to play major part.

3.5 GRATING COMPOSED OF THE BARS WITH ELASTIC PLATES NORMAL TO THE PLANE OF THE GRATING

The cases when the elastic walls of the bars are parallel to the plane of the grating were examined in the previous chapters. Increasing perforation coefficient of the grating leads to decrease in the bandwidth of effective insulation. The distance between the elements of the grating can be increased because of practical needs. For example, there is a need in a grating transparent for water and nontransparent for sound. The elastic elements of the gratings considered above are the plates executing flexural vibrations, and that is why it is possible to make the grating composed of small elements in comparison with sound wave length in the medium.

It is therefore volume compressibility of the elements that determines the acoustical properties of the grating. If that is so, than we can change the orientation of elastic elements of the grating relative to the incident sound field without change of acoustic properties of the grating.

Let us consider the case when the elastic walls of the bar are normal to the plane of the grating, Fig. 3.10. It is evident that an interaction effect between the plates will play the major part in such a grating. In this case, we can expect an increase in bandwidth of effective sound insulation of the grating and a greater influence of the medium between the elements of the grating on the natural frequencies of the plates.

It is possible to control the resonance and acoustic properties of the grating by changing the grating spacing. We will examine the quantitative data bellow and will show the possibility of such control in practice.

We proceed now to the problem solution. The entire field existence domain will be partitioned into the following partial domains: domain $x \leq 0$ and domain $x \geq l$, and an infinite number of domains, each of which is limited by the volume of spacing between the neighboring bars. Due to the symmetry of the considered grating, the velocity potentials in the specified partial domains can be represented in the following form:

$$\begin{gathered}\Phi_1 = e^{ik(x\cos\theta + y\sin\theta)} + e^{iky\sin\theta}\sum_{n=-\infty}^{\infty} A_n e^{-ik_n x} e^{i(n\pi/b)y}, \\ \Phi_2 = e^{iky\sin\theta}\sum_{n=-\infty}^{\infty} D_n e^{-ik_n(x-l)} e^{i(n\pi/b)y},\end{gathered} \tag{3.31}$$

$$\Phi_q = 2e^{2ikbq\sin\theta}\Bigg\{\sum_{p=0}^{\infty}\left[B_p\cos + C_p\sin k_p x\right]\cos\frac{\pi p}{2a}(y - 2bq + a)$$

$$+\sum_{v=1}^{\infty}\sin\frac{v\pi}{l}x\left[E_v\cos\left[k_v(y-2bq+a)\right] + F_v\sin\left[k_v(y-2bq+a)\right]\right]\Bigg\},$$

where

$$k_n = \left\{k^2 - v\left(\frac{\pi n}{l} - k\sin\theta\right)^2\right\}^{1/2};\ k_p = \left\{k^2 - \left(\frac{\pi p}{2a}\right)^2\right\}^{1/2};\ k_v = \left\{k^2 - \left(\frac{v\pi}{l}\right)^2\right\}^{1/2},$$

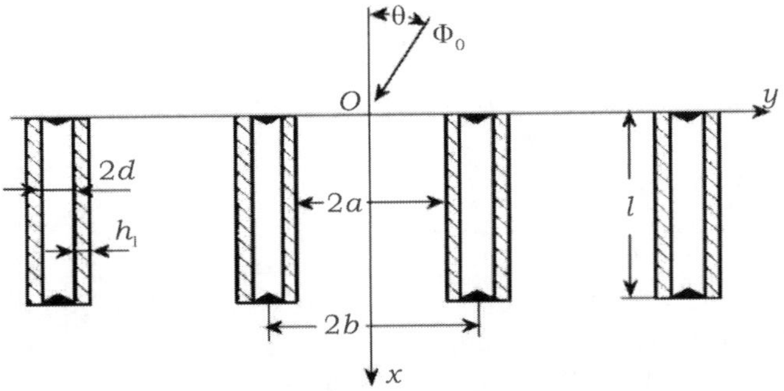

FIG. 3.10: Grating composed of the bars with elastic walls normal to the plane of the grating

and $q = 0, \pm 1, \pm 2, \ldots$—enumerates the gaps between the bars.

Note that the sound field in the gaps is represented as superposition of two infinite sums over the indices p and v, while in expression (3.21) was only one infinite sum over the index p. It is due to the properties of the gap walls surface. The walls of the gaps of the grating shown in Fig. 3.8, were formed with nondeformable (acoustically rigid) supports. That is why it is enough to represent the sound field as (3.21) to meet the boundary conditions on the walls of the gaps and the matching conditions of the sound field components on the inlet and the outlet of the gap. In the present case, the walls are not acoustically rigid but elastic. Consequently, it is necessary to introduce an additional sum over the index v in the expression for Φ_q to meet the matching conditions on the surfaces of the elastic plates.

We represent the particle velocities of the plates, as in the cases considered above, by series of natural modes of the plates in vacuum. We obtain an expression for the hinged supported plates

$$\upsilon_{q\tau}(x) = e^{2ikbq\sin\theta} \sum_{v=1}^{\infty} \upsilon_{q\tau v} \sin\frac{v\pi}{l}x, \tag{3.32}$$

where τ—the number of the plate in the gap ($\tau = 1$ for the upper plate and $\tau = 2$ for the lower plate).

The unknown complex coefficients in expressions (3.31) and (3.32) are to be determined from the following system of functional equations:

$$\begin{gathered}
-\partial\Phi_q/\partial y = \upsilon_{q\tau}(x); \quad 0 \le x \le l; \quad |y - 2bq| = a, \\
\Phi_1 = \Phi_q; \quad x = 0; \quad |y - 2bq| \le a, \\
\Phi_2 = \Phi_q; \quad x = l; \quad |y - 2bq| \le a, \\
-\frac{\partial\Phi_1}{\partial x} = \begin{cases} -\dfrac{\partial\Phi_q}{\partial x}; & |y - 2bq| \le a; \quad x = 0, \\ 0; & a \le |y - 2bq| \le 2(b - a); \quad x = 0, \end{cases} \\
-\frac{\partial\Phi_2}{\partial x} = \begin{cases} -\dfrac{\partial\Phi_q}{\partial x}; & |y - 2bq| \le a; \quad x = l, \\ 0; & a \le |y - 2bq| \le 2(b - a); \quad x = l, \end{cases} \\
\frac{d^4\upsilon_{q\tau}(x)}{dx^4} - k_{iz}^4\upsilon_{q\tau}(x) = -\frac{\omega^2\rho}{D}\Phi_q; \ |y - 2bq| = a, \\
k_{iz} = \left(\frac{\omega^4\rho_1 h_1}{D_1}\right)^{1/4}; \quad D_1 = \frac{E_1 h_1^3}{12(1 - \sigma^2)}.
\end{gathered} \tag{3.33}$$

System (3.33) is replaced with equivalent infinite system of algebraic equations of the second kind. This operation is omitted. We proceed with analysis of the acoustic properties of the grating in low-frequency approximation when the inequalities $l \ll \lambda_1$, $2a/l \le 1$ and $(b - a)/b \ll 1$ are satisfied. In this case we can confine ourselves to the first terms in series (3.31), (3.32) and the coefficients $A_0, D_0, E_1, B_0, C_0, \upsilon_{0r1}$ can be determined analytically. Omitting the intermediate calculations, we write the expression for sound transmission coefficient k_{ins} through the grating and its normalized input impedance Z_{ins} when $\theta = 0$:

$$k_{\text{ins}} \approx \frac{1}{\xi kl + i}\left(\xi kl + i(\xi - 1/2)(kl)^2\right),$$

$$\arg k_{\text{ins}} = \frac{(\xi kl)^2 - \xi}{kl(\xi^2 - \xi - 1/2)}; \quad Z_{in} = R + iX,$$

$$R = \frac{(\xi^2 - 1)(kl)^2}{[kl(\xi - 1)]^2 + 4}; \quad X = -\frac{2(\xi + 1)kl}{[kl(\xi - 1)]^2 + 4}, \tag{3.34}$$

$$\xi = \frac{\bar{Z}_1 - 2}{4}; \quad \bar{Z}_1 = \alpha[Z_1\beta + \text{ctg}\alpha]; \quad Z_1 = 1 - (f_1/f)^2,$$

$$\alpha = \pi a/l; \quad \beta = \frac{\pi\rho_1 h_1}{\rho l}.$$

In accordance to (3.34), when $f \to 0$ we have $k_{\text{ins}} \to 1$, $R \to 1$, $X \to 0$. Thus the grating is almost transparent at low frequencies. It behaves like the gratings considered in the previous paragraphs. Now we estimate the parameters of the elastic plates–surrounding medium system at resonance frequency f_{rez}.

Such estimation was made in Sections 3.3 and 3.4 as frequency f tended to f_1. It was justified, because the natural frequency of the plates–surrounding medium system differed slightly from the natural frequency of the plates in vacuum (refer to Figs. 3.4 and 3.7, for example). It is impossible to do the same in the case in question. The resonance frequency of the plates is influenced significantly by the medium due to of the bars orientation in the grating. Consequently, we determine the lowest resonance frequency f_{rez1} of the plates–surrounding medium system. The expression for X from (3.34) can be used for this purpose. Setting $X = 0$, we have $\xi = -1$, where

$$f_{\text{rez1}} = f_1\left\{\frac{\alpha\beta}{\alpha\beta + \alpha\text{cth}\alpha + 2}\right\}^{1/2}. \tag{3.35}$$

It follows from this expression that, when $2a/l \to 0$ the value of $f_{\text{rez1}} \to 0$. Further data about the nature of change of f_{rez1} against $2a/l$ is shown in Fig. 3.11 for the case $l/\lambda = 0.1$. As one can see, when $2a/l \approx 0.5$, resonance frequency of the plates–liquid system makes up about 40% of the resonance frequency of the plates in vacuum. A further decrease of value of f_{rez1} is observed when $2a/l < 0.5$. This important property of the grating is the consequence of enhancement of interaction effect between the elastic walls of neighboring bars approaching each other.

This interaction manifests itself in the following way. Even weak vibrations of the plates give rise to sizable vibrations of liquid in the gap along the axis Ox, when the values of $2a/l$ are small. The smaller is ratio $2a/l$, the larger is ratio of the plate's vibrations amplitude to the vibrations of liquid in the gap between the bars. Thus, the gap filled with liquid together with the plates form a kind of transformer of particle velocities with

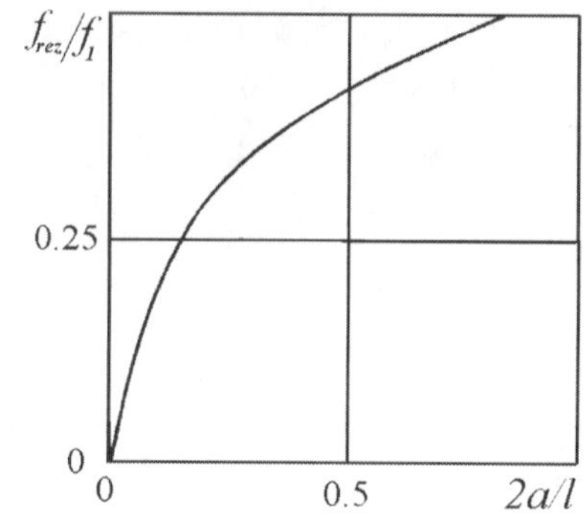

FIG. 3.11: Plates–surrounding medium system resonance frequency versus spacing between the bars

a transformation ratio approximately proportional to l/a. The transformation ratio can be estimated more precisely, considering that the plate does not vibrate like a piston but it vibrates according to the law $\sin \pi x/l$. The mass of the liquid in the gap, which executes vibrations along the Ox axis, is added to the plates and co-oscillates together with them. That is why the amplitude of the liquid vibrations along the Ox axis is multiplied approximately by $(l/a)^2$. The mass of liquid is reduced in proportion to a/l with a decrease in distance between the neighbouring bars. Yet still the mass of the liquid aded to the plates encreases. This results in the reduction of resonance frequency of the plates—liquid system. It is not difficult to give the following estimations for the transmission ratio and the impedance active part when $f \to f_{\text{rez}1}$, using the value of $\xi = -1$. In this case $R \to 0$ and $k_{\text{ins}} \to kl$. This goes to show, that the grating is an effective sound insulator in the region of the resonance frequency.

We proceed to the detailed analysis of the acoustic properties of the grating, basing on the computer calculated data. Figure 3.12 shows typical frequency dependencies of

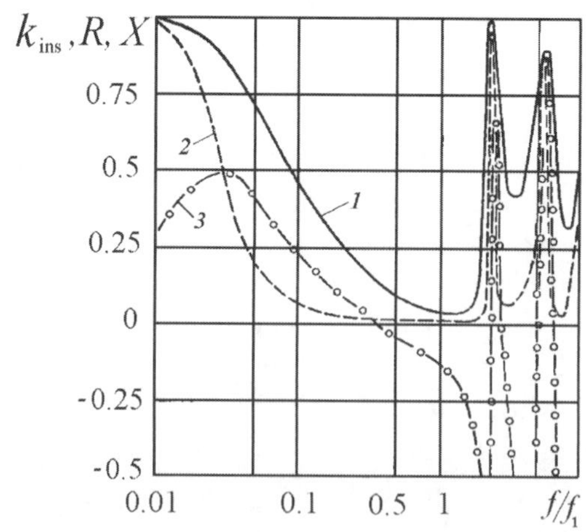

FIG. 3.12: Frequency dependence of sound transmission coefficient (curve 1), active (curve 2), and reactive (curve 3) components of input impedance of the grating

the sound transmission coefficient passing through the grating and the components of its input impedance.

The grating is made of titanium alloy $\rho_1 = 4.7 \times 10^3$ kg/m^3, $E_1 = 1.16 \times 10^{11}$ Pa [43] and is immersed in water. We take the following geometric dimensions of the grating $l/\lambda_1 = 0.05$, $2a/l = 1$, and $l/a = 1.05$. The curve, which characterizes frequency dependence of value of X change, periodically crosses x-axis, changing its sign. It means that there are the resonances in the plates–liquid system. The values of k_{ins} and R take on the minimum and maximum interleaved one after another in the frequency domain corresponding to these resonances. The values of k_{ins} and R are small when the value of X crosses the x-axis smoothly, and they are large when crossing is abrupt. This circumstance points to the resonance interleaving (resonance is followed by antiresonance and so on). It follows from Foster's theorem [44].

The general acoustic and mechanical properties of the grating can be found out by means of the analysis of expressions (3.34) and the dependencies showed in Figs. 3.11 and 3.12. It is necessary to examine the field in the vicinity of the grating and the plates surface particle velocity to get a deeper insight into physics at resonance and antiresonance frequencies of the plates-liquid system. For this purpose, the distribution of the modulus of sound pressure and particle velocity of liquid in the vicinity of one period of the grating when $\theta = 0$ is represented in Figs. 3.13 and 3.14. The values of pressure phase and particle velocities are also indicated here, near the curves, in the interval $0 \leq x \leq l$. As the data in Fig. 3.13 shows, a standing wave arises in the region of resonance ($f/f_1 \approx 0.4$) in front of the grating because of an incident and a grating-reflected waves interference. The amplitude of sound pressure in a standing wave is close to $2p_0$ (where p_0—sound pressure in an incident wave) because the value of k_{ins} at these frequencies is small (k_{ref} is large respectively). The sound pressure decreases abruptly in the beginning of the gap ($x \approx 0$) due to the elasticity of the plates but at the end of the gap ($x \approx l$) it increases a little and remains constant in amplitude behind the grating. The value of the phase of pressure is almost constant along the whole length of the gap.

The distribution of longitudinal particle velocity in front of the grating [Fig. 3.13(b)] is also characterized by the presence of standing wave. But in the region $x \approx 0$ (in contrast to sound pressure) the value of $|\upsilon_x| \approx 2|\upsilon_0|$ (where υ_0—particle velocity of liquid in the incident wave). The above-mentioned characteristics of sound field prove the conclusion, that the grating is similar in its properties to the ideally soft acoustic surfaces on which there are the following equalities $p = 0$ and $\upsilon = 2\upsilon_0$.

Particle velocity decreases rapidly at the end of the gap and remains constant behind the grating. The value of the phase of longitudinal particle velocity and of sound pressure, vary slightly along the whole length of the gap. Transverse particle velocity component υ_y of liquid in front of the grating and behind the grating is equal to zero [Fig. 3.13(b)]. The value of υ_y distribution in the gap, along the axis Ox almost repeats the plate vibration mode at first natural frequency viz $\sin \pi x/l$. The maximum amplitude υ_y is on the surface of the plates ($y = \pm\, a$) and it does not exceed the value of υ_0.

According to the calculation, the particle velocity of the plates in the frequency domain $f/f_1 \approx 0.4$ is determined by the first term of series (3.32), and the amplitude υ_y on the surface of the plate is close to the vibration amplitude in the incident wave.

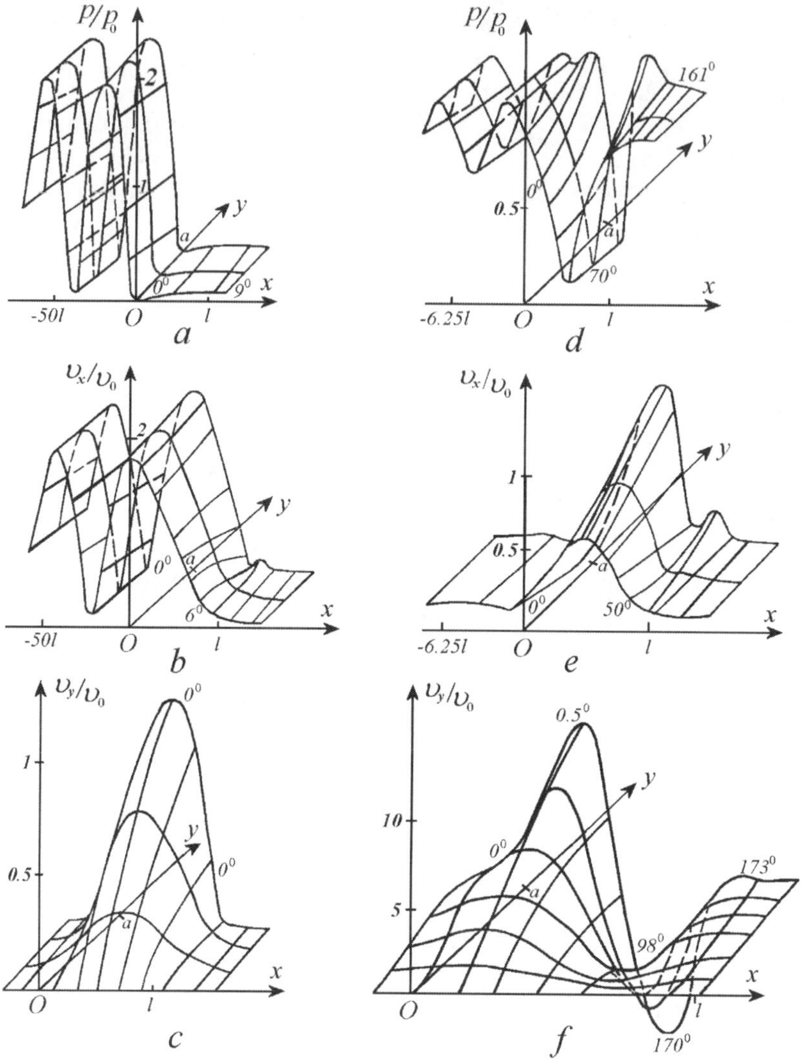

FIG. 3.13: Pressure distribution (a, d) and the particle velocity components (b, c, e, f) in the vicinity of one period of the grating; (a, b, c)—for $f/f_1 \approx 0.4$; (d, e, f)—for $f/f_1 \approx 2.5$

It is possible to diagram the flow path of liquid in the gap and vibration mode of the plate in the region of the first resonance of elastic plates-liquid system [Fig. 3.14(a)] due to the analysis of distribution of the sound field components in the vicinity of the grating. There is rather different distribution of the sound field components at the second natural frequency of plates-liquid system when $f/f_1 \approx 2.5$ (Fig. 3.13). The level of sound pressure in front of the grating and behind the grating is little different from p_0. This is because of high transmissivity of the grating. The pressure in the gap falls dramatically and takes on minimum value when $x \approx l/2$, and then it increases dramatically but its phase is opposite in sign. Thus, each half of the plate is under pressure of an opposite

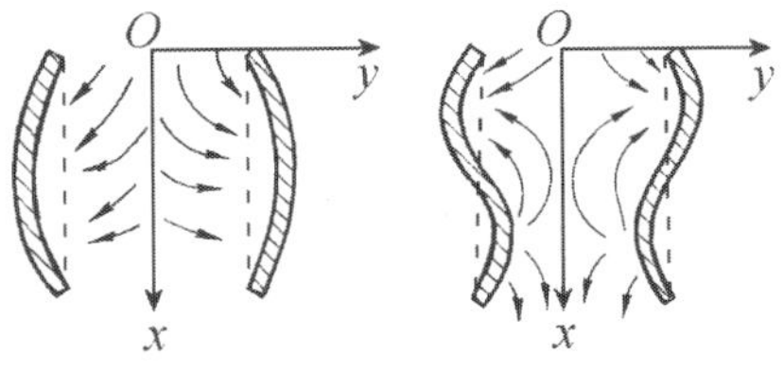

FIG. 3.14: Vibration mode of the plates and medium flow path between the bars: (a)—for $f/f_1 \approx 0.4$, (b)—for $f/f_1 \approx 2.5$

sign. As a result, the distribution of the value of υ_y along the gap repeats the vibration mode along the gap, at the second natural frequency viz $\sin 2\pi x/l$.

The x-dependence of the change of the value of υ_x along the gap is different from the x-dependence of pressure p [compare Figs. 3.13(d) and 3.13(e)]. It is significant that the amplitudes of the values of υ_y and υ_x can be an order of magnitude greater than the amplitude υ_0 at the second natural frequency of the plates-liquid system. It means that the quality-factor of mechanical vibration plates-liquid system is large at this frequency, that is caused by an acoustic short circuit of the plates surface areas vibrating in antiphase (and equal in area). The liquid in the gap executes oscillating vibrations mainly along the axis Ox, refer to Fig. 3.14(b).

The results of the analysis of the next following resonances of elastic plates-liquid system lead to the conclusion that the transmissivity of the grating reaches its minimum when the vibration modes of the plates with odd index n dominate and the transmissivity reaches its maximum when the vibration modes with even index n dominate. To be noticed is that the grating has the best sound attenuating properties in the region of the first resonance of the plates-liquid system. The sound attenuating properties become worse at the next following resonances (when the vibration modes with index $n = 3, 5, 7, ...$ dominate in turn). This is due to the presence of the plate's surface areas vibrating in antiphase.

We proceed with analysis of the sound attenuating properties of the grating dependency on its physical, geometric and wave parameters. The value of $\Delta = \log_2(f_{\text{top}}/f_{\text{bot}})$, is the assessment criterion of sound attenuating properties of the grating where f_{top} and f_{bot} are upper and lower frequency range limits respectively (in the vicinity of f_{rez1}), where the condition $k_{\text{ins}} \le 0.2$ is met. Although the value Δ is non-dimensional, it is easy to verify that it is numerically equal to the frequency band represented in octaves. Hereafter the value of Δ will be named as the bandwidth of effective sound insulation of the grating. Figure 3.15 shows computational data characterizing the grating properties depending on the grating spacing, the length of the elastic plates and their material. In all cases $a/b = 0.95$. Block curves correspond to $2b/l = 2$, dashed and dash-and-dot curves correspond to $2b/l = 1$ and $2b/l = 0.67$, respectively; the wave impedance of plates material is $\rho_1 c_1 = 1.37 \times 10^7$ kg/m^{2}s [for the data in Fig. 3.15(a)], $\rho_1 c_1 = 2.34 \times 10^7$ kg/m^{2}s [for the data in Fig. 3.15(b)] and $\rho_1 c_1 = 3.96 \times 10^7$ kg/m^{2}s [for the data in Fig. 3.15(c)].

The data analysis allows formulating the following regularities:

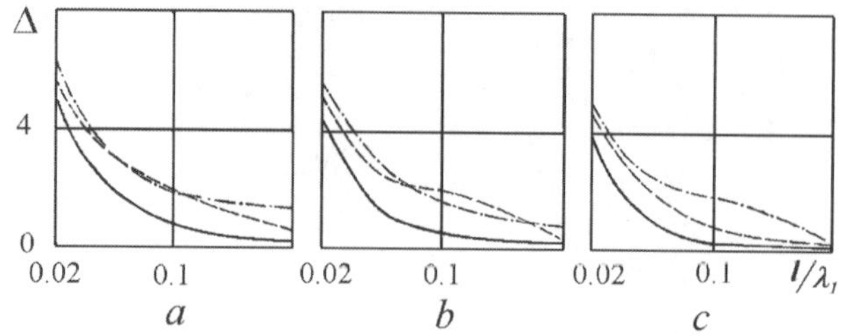

FIG. 3.15: Bandwidth of effective sound insulation *vs* grating spacing, elastic plate's length and the material of the plates when $a/b = 0.95$; block curves—for $2b/l = 2$, dashed for $2b/l = 1$, dashed-and-dot for $2b/l = 0.67$; (a)—for $\rho_1 c_1 = 1.37 \times 10^7$ kg/m^{2}s, (b)—for $\rho_1 c_1 = 2.34 \times 10^7$ kg/m^{2}s, (c)—for $\rho_1 c_1 = 3.96 \times 10^7$ kg/m^{2}s

1. the insulation band increases with decreasing wave dimension of the plates forming the bar; $\Delta \approx \text{const}\,(\lambda_1/l)^{\mu}$, where $1 \leq \mu \leq 2$;

2. the insulation band increases with decreasing wave impedance $\rho_1 c_1$ of the plate's material;

3. the insulation band increases with decreasing distance between the walls of the bars, if $2b/l \geq 1$; it is true for the steel plates with the smaller values of $2b/l$, and it is true only for some values of l/λ_1 for the lighter materials.

The first two indicated above regularities are conditioned by the following physical property—the compliance of the plate's increases with decrease of the wave impedance of the plate's material and decrease of their wave-length (f_1 is fixed). As a consequence, the sound-insulating attributes of the grating improve. The third regularity is more complex. One might assume that decreasing of $2b/l$ leads to increase in number of elastic elements on a unit of the grating surface when the values of $2b/l$ are relatively large, and the interaction between the neighboring bars is not great. It results in an increase in compliance of the grating surface, and as a consequence, in an increase in sound insulation of the grating. When the value of $2b/l$ is small, the interaction between the bars plays the major role. As a result, the further increase of the value of Δ becomes considerably slower.

3.6 INFLUENCE OF THE AIR INSIDE THE SHELLS ON THE ACOUSTIC PROPERTIES OF THE GRATING

In the foregoing studies of the gratings, the interior of the shells was considered to be vacuum. This assumption is an idealization. On the practical side, it is important to know how the presence of a medium alters the acoustic properties of a grating. It is particularly interesting in the cases when compressed air is injected in the interior volume of the bars, compensating the pressure in the medium surrounding the grating when using the grating deeply in the ocean. To answer this question, one of the gratings, which were studied above, will be considered. For example, the grating showed in Fig. 3.10. We suppose that the bars are filled with gas with wave impedance $\rho_0 c_0$. This task will be rather simple, if we confine the discussion to the case of normal incidence on the grating of a sound wave. Due to the symmetry of the grating, it is enough to consider only one period of the grating $|y| \leq b$, and to partition the domain of the field existence in to four partial domains: the first domain $x \leq 0$, $|y| \leq b$ the second domain $x \geq l$, $|y| \leq b$ the third domain $0 \leq x \leq l$, $|y| \leq a$ the fourth domain $0 \leq x \leq l$, $a + h \leq |y| \leq b$. Supposing that plates are simply supported at the edges by nondeformable supports, let us right the boundary conditions in the hollow bars:

$$\begin{gathered} \partial\Phi/\partial x = 0; \quad x = 0, \quad x = l, \quad a + h \leq |y| \leq b, \\ \partial\Phi/\partial y = 0; \quad 0 \leq x \leq l, \quad |y| = b, \end{gathered} \tag{3.36}$$

and out of the bars

$$\begin{gathered} \partial\Phi/\partial x = 0; \quad x = 0, \quad x = l, \quad a + h \leq |y| \leq b, \\ \partial\Phi/\partial y = 0; \quad x \leq 0, \quad x \geq l, \quad |y| = b. \end{gathered} \tag{3.37}$$

Using these boundary conditions for the case of normal incidence of the plane wave on the grating, we represent the sound fields in the partial domains and the particle velocity of the plates as the following series;

$$\Phi_1 = e^{ikx} + \sum_{n=0}^{\infty} A_n e^{-ik_n x} \cos\frac{\pi n}{b} y,$$

$$\Phi_2 = \sum_{m=0}^{\infty} D_m e^{ik_m (x-l)} \cos\frac{\pi m}{b} y,$$

$$\Phi_3 = \sum_{p=0}^{\infty} [B_p \cos k_p x + C_p \sin k_p x] \cos\frac{\pi p}{a} y + \sum_{\nu=1}^{\infty} E_\nu \sin\frac{\pi \nu}{l} x \cos k_\nu y,$$

$$\Phi_4 = \sum_{q=0}^{\infty} F_q \cos\frac{\pi q}{l} x \left[\cos k_q (y - a - h) + \mathrm{tg} k_q (b - a - h) \sin k_q (y - a - h)\right],$$

$$\upsilon(x) = \sum_{s=1}^{\infty} \upsilon_s \sin\frac{\pi s}{l} x, \tag{3.38}$$

where

$$k_n = \left[k^2 - \left(\frac{\pi n}{b}\right)^2\right]^{1/2}; \quad k_m = \left[k^2 - \left(\frac{\pi m}{b}\right)^2\right]^{1/2}; \quad k_\nu = \left[k^2 - \left(\frac{\pi \nu}{l}\right)^2\right]^{1/2},$$

$$k_q = \left[k_0^2 - \left(\frac{\pi q}{l}\right)^2\right]^{1/2}; \quad k_p = \left[k^2 - \left(\frac{\pi p}{a}\right)^2\right]^{1/2}; \quad k_0 = 2\pi/c_0.$$

Unknown complex coefficients in the series (3.38) are to be determined from the the following system of functional equations:

$$-\partial\Phi_3/\partial y = \upsilon(x); \quad 0 \leq x \leq l, \quad |y| = a,$$

$$-\partial\Phi_4/\partial y = \upsilon(x);\quad 0 \le x \le l,\quad |y| = a + h,$$

$$-\frac{\partial\Phi_1}{\partial x} = \begin{cases} -\partial\Phi_3/\partial x; & |y| \le a,\quad x = 0; \\ 0; & a \le |y| \le b,\quad x = 0, \end{cases}$$

$$-\frac{\partial\Phi_2}{\partial x} = \begin{cases} -\partial\Phi_3/\partial x; & |y| \le a,\quad x = l, \\ 0; & a \le |y| \le b,\quad x = l, \end{cases} \tag{3.39}$$

$$\Phi_1 = \Phi_3;\quad x = 0,\quad |y| \le a,$$

$$\Phi_2 = \Phi_3;\quad x = l,\quad |y| \le a,$$

$$\frac{d^4\upsilon(x)}{dx^4} - k_{iz}^4\upsilon(x) = -\frac{\omega^2}{D_1}\left[\rho\Phi_3(|y| = a) - \rho_0\Phi_4(|y| = a + h)\right];\quad 0 \le x \le l.$$

The structure of the expression (3.38) enables us to derive the infinite system of the linear algebraic equations from the system of the functional Eqs. (3.39). Here we use the completeness and orthogonality properties of the trigonometric functions that are in the expression (3.38). This procedure is similar to the procedure for the hollow shells, which was described above. The reduction method is used to make choice of the finite system order. This choice is based on comparing the results, obtained when the number of equations is systematically increased.

Omitting the details of the computations, we will analyze the calculation results.

Frequency dependencies, characterizing the acoustic transmissivity of the grating, versus the properties of the medium filling the bars, are plotted in Fig. 3.16. Here $l/\lambda_1 = 0.04, 2a/l = 0.4, d/l = 0.1$, and $\rho_1 c_1/\rho c = 15$; curves 1, 2, 3, 4, and 5 correspond to $\rho_0 c_0/\rho c$, equal 0.15, 0.067, 0.033, 0.012, and 0.00028 (note, that $\rho_0 c_0/\rho c$ approximately correspond to air pressure inside the shells—500, 300, 150, 40, and 1 kg/cm^2). It is obvious, that sound-attenuating properties of the grating deteriorate when the value of

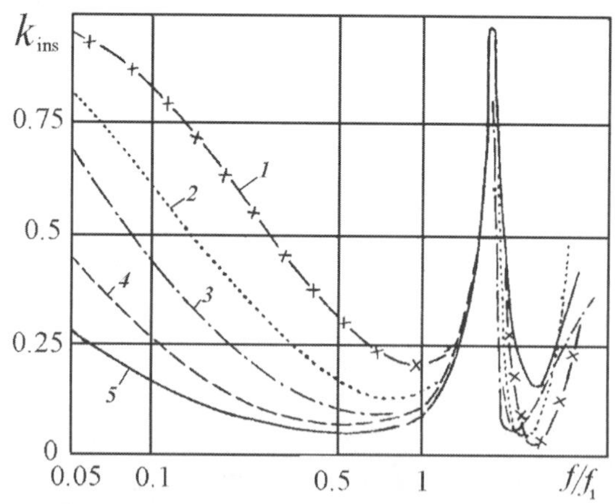

FIG. 3.16: Acoustic transmissivity of the grating vs properties of the medium filling the bars; $l/\lambda_1 = 0.04$, $2a/l = 0.4$, $d/l = 0.1$, $\rho_1 c_1/\rho c = 15$; curves 1, 2, 3, 4, and 5 correspond to $\rho_0 c_0/\rho c$, equal 0.15, 0.067, 0.033, 0.012, and 0.00028

the ratio $\rho_0 c_0/\rho c$ increases. This fact admits a simple physical interpretation true for the interior of the bar of short wavelength. The gas inside the bar can be characterized by a certain elasticity, which is added to the elasticity of the plates. This fact decreases admittance of the bar as a whole, and, as a result, degrades the sound-attenuating properties of the grating. It is important to note, that when $\rho_0 c_0/\rho c = 0.00028$, that, corresponds to the ratio of air wave impedance, in case of normal atmospheric pressure, to water wave impedance, sound-attenuating properties of the grating are the same as in the case when the interior of the bars is vacuum ($\rho_0 c_0/\rho c = 0$). This fact is of great practical significance, because it enables to estimate sound-attenuating properties of the grating without considering the air inside the bars (at least when it is less than $\rho_0 c_0/\rho c \approx 0.001$). With increasing the value $\rho_0 c_0/\rho c$, the frequency bandwidth of the efficient acoustic-insulation decreases, and the minimum transmittance increases. And also we can observe the rise in the resonance frequency of the plate–water system. Figure 3.17 illustrates the influence of the medium filling the bars on the natural frequency of the system "plates—surrounding medium." The ratio of the first natural frequency of the bars with the medium filling them to the first natural frequency without the medium is shown on Y-axis. $l/\lambda_1 = 0.04$, $2a/l = 0.4$, $d/l = 0.1$, and $\rho_1 c_1/\rho c = 15$. Since the interior of the bar has short wavelength, the most significant is the elasticity of the medium inside the bars added to the elasticity of the system.

As it follows from the previous consideration, the frequency at which transmission ratio increases sharply, is the second natural frequency of the system "bars—liquid". The natural oscillations have such forms that the volume of spacing between the elastic bars does not change and the influence of the stiffness of the medium filling the bars is unessential. It is shown in Fig. 3.16 that the peak of the transmission ratio frequency dependence doesn't depend on the stiffness of a medium filling the bars. According to the calculation results, the transparency of the grating is maximum when $f/f_1 \approx 2.3$. The data in Fig. 3.16 can be explained otherwise. Basing on these data, we can say that acoustic-insulation bandwidth of the grating formed by elastic bars is to a certain extent proportionate to the frequency interval between the first and the second natural frequencies. Consequently, one of the ways of improving the bandwidth attributes of the grating is to use the constructions and the fixing of elastic bars that increase the interval between natural frequencies. Sound-attenuating properties of the grating deteriorate when an interior volume of the bar increases (value $\rho_0 c_0/\rho c$ is fixed). The elasticity of a medium inside the bar increases in this case. Figure 3.18 illustrates how the frequency band of the effective acoustic insulation of the grating changes versus the wave impedance of a medium filling the bar and an interior volume of the bar. The results obtained here

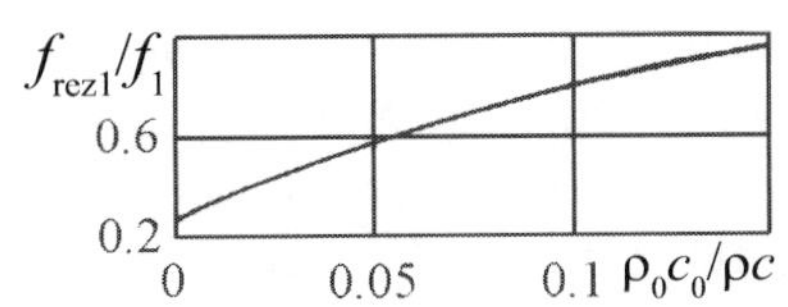

FIG. 3.17: Influence of the medium filling the bars, on the natural frequency of the system bars–surrounding medium for $l/\lambda_1 = 0.04$, $2a/l = 0.4$, $d/l = 0.1$, and $\rho_1 c_1/\rho c = 15$

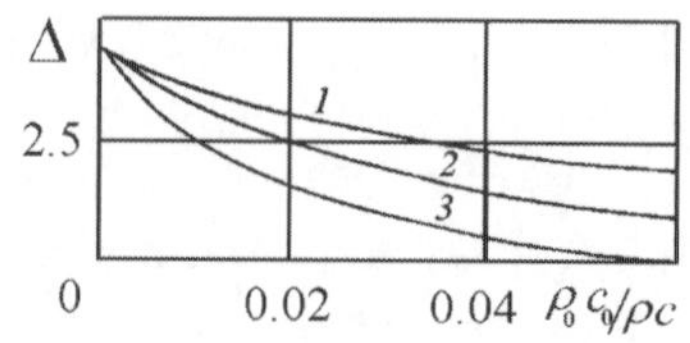

FIG. 3.18: Frequency band of effective acoustic insulation of the grating *vs* wave impedance of the medium filling the bar and interior volume of the bar; the values of d/l are 0.1, 0.05, and 0.025 and they correspond to the curves 1, 2, and 3; the rest characteristics are the same as in Fig. 3.17

should prove useful for constructing the gratings situated in a medium at high hydrostatic pressure. To protect the bars against fracture, an excess pressure is created in their interior volume by injection of compressed gas, compensating pressure in a medium surrounding the grating (water).

3.7 METHODS OF EFFICIENCY-ENHANCEMENT OF THE GRATINGS

3.7.1 Passive Method

Acoustic properties of the one-layer gratings of hollow elastic bars were investigated in the previous paragraphs. Performance potential of such sound-proofing system, which is characterized by the frequency dependence of sound transmission coefficient, was investigated on the basis of the analysis of an enormous volume of calculation data.

Efficiency of the grating made of elastic elements depends on many factors. The thorough investigation of these factors and the quantitative estimates of acoustic characteristics of the gratings let us conclude that these gratings can be successfully used for sound—proofing. It is often necessary to expend the frequency band of effective insulation when such gratings are used for sound—proofing. A natural solution of this problem is to use two or more gratings, in which the elastic bars generally have different natural frequencies. In such multi-layer grating, interaction effects between the layers play a significant part, influencing the degree of broadening of the insulation band and all the acoustic properties of the grating.

Taking all considerations into account, we will investigate the acoustic properties of the multi-layer gratings considering, as an example, the double-layer gratings. In this paragraph, we will consider the gratings made of bars with elastic walls normal to the plane of the gratings (Fig. 3.19).

Let a wave be incident on the double-layer grating at arbitrary angle θ. The partial domains can be defined as follows: the first domain $x \leq 0$; the second domain $0 \leq x \leq l$; the third domain $l \leq x \leq a$; the fourth domain $a \leq x \leq r$; the fifth domain $x \geq r$.

Summarizing the expressions derived in the previous paragraphs, we can write the velocity potentials in these domains and the particle velocities of the plates as follows:

$$\Phi_1 = e^{ik(x\cos\theta - y\sin\theta)} + e^{iky\sin\theta} \sum_{n=-\infty}^{\infty} A_n e^{-i[k_n x - (\pi n/d)y]},$$

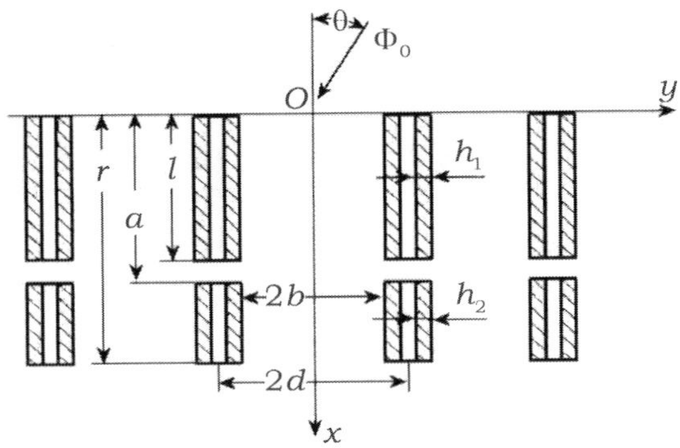

FIG. 3.19: Double-layer grating

$$\Phi_2 = e^{2ikdq\sin\theta}\left\{\sum_{p=1}^{\infty} B_p \sin\frac{\pi p}{l}x\cos\left[k_p(y-2dq+b)\right]+\right.$$

$$\left.+\sum_{s=0}^{\infty} C_s\cos(k_s x-\beta_s)\cos\frac{\pi s}{2b}(y-2dq+b)\right\},$$

$$\Phi_3 = e^{-iky\sin\theta}\sum_{n=-\infty}^{\infty}\left[D_n e^{-ik_n(x-l)}+E_n e^{ik_n(x-l)}\right]e^{i(\pi n/d)y}, \qquad (3.40)$$

$$\Phi_4 = e^{2ikdq\sin\theta}\left\{\sum_{t=1}^{\infty} F_t\sin\frac{\pi l(x-a)}{r-a}\cos[k_t(y-2dq+b)-\alpha_t]\right.$$

$$\left.+\sum_{m=0}^{\infty} G_m\cos[k_m(x-a)-\alpha_m]\cos\left[\frac{\pi m}{2b}(y-2dq+b)\right]\right\},$$

$$\Phi_5 = e^{-iky\sin\theta}\sum_{n=-\infty}^{\infty} T_n e^{i[k_n(x-r)+(\pi n/d)y]},$$

where

$$k_n^2 = k^2-\left(\frac{\pi n}{d}-k\sin\theta\right)^2;\; k_s^2 = k^2-\left(\frac{\pi s}{2b}\right)^2;\quad k_p^2 = k^2-\left(\frac{\pi p}{l}\right)^2,$$

$$k_m^2 = k^2-\left(\frac{\pi m}{2b}\right)^2;\; k_t^2 = k^2-\left(\frac{\pi t}{r-a}\right)^2.$$

We suppose that the elastic plates in the grating are hinge supported at edges. Their particle velocities are represented by a series in the proper vibration mode in vacuum

$$\upsilon_{jq}^{(\tau)} = e^{2ikdq\sin\theta}\sum_{p=1}^{\infty}\upsilon_{jqp}^{(\tau)}\sin\frac{\pi p}{l}x, \qquad (3.41)$$

j enumerates the layers of the grating, q enumerates the gaps between the bars, τ is the number of the plate in the bar, $\tau = 1$—for the upper plate, and $\tau = 2$—for the lower plate in the bar.

The unknown complex coefficients in the expressions (3.40) and (3.41) are determined from the following system of functional equations, which is the generalization of the system (3.33) for a double-layer grating:

$$
\begin{gathered}
\Phi_1 = \Phi_2; \ \ |y| \le b, \ \ x = 0, \\
\frac{\partial \Phi_1}{\partial x} = \begin{cases} \dfrac{\partial \Phi_2}{\partial x}; & |y| \le b, \ \ x = 0, \\ 0; & b \le |y| \le d, \ \ x = 0, \end{cases} \\
\Phi_2 = \Phi_3; \ \ |y| \le b, \ \ x = l, \\
\frac{\partial \Phi_3}{\partial x} = \begin{cases} \dfrac{\partial \Phi_2}{\partial x}; & |y| \le b, \ \ x = l, \\ 0; & b \le |y| \le d; \ \ x = l, \end{cases} \\
\Phi_3 = \Phi_4; \ \ |y| \le b, \ \ x = a, \\
\frac{\partial \Phi_3}{\partial x} = \begin{cases} \dfrac{\partial \Phi_4}{\partial x}; & |y| \le b, \ \ x = a, \\ 0; & b \le |y| \le d; \ \ x = a, \end{cases} \\
\Phi_5 = \Phi_4; \ \ |y| \le b, \ \ x = r, \\
\frac{\partial \Phi_5}{\partial x} = \begin{cases} \dfrac{\partial \Phi_4}{\partial x}; & |y| \le b, \ \ x = r, \\ 0; & b \le |y| \le d; \ \ x = r, \end{cases} \\
-\frac{\partial \Phi_2}{\partial y} = \upsilon_{1q}^{(1)}; \ \ y = b, \ \ 0 \le x \le l, \\
-\frac{\partial \Phi_2}{\partial y} = \upsilon_{1q}^{(2)}; \ \ y = -b, \ \ 0 \le x \le l, \\
-\frac{\partial \Phi_4}{\partial y} = \upsilon_{2q}^{(1)}; \ \ y = b, \ \ a \le x \le r, \\
-\frac{\partial \Phi_4}{\partial y} = \upsilon_{2q}^{(2)}; \ \ y = -b, \ \ a \le x \le r, \\
\frac{d^4 \upsilon_{1q}^{(1)}}{dx^4} - k_1^4 \upsilon_{1q}^{(1)} = -\frac{\omega^2 \rho}{D_1} \Phi_2; \ \ y = b, \ \ 0 \le x \le l, \\
\frac{d^4 \upsilon_{1q}^{(2)}}{dx^4} - k_1^4 \upsilon_{1q}^{(2)} = -\frac{\omega^2 \rho}{D_1} \Phi_2; \ \ y = -b, \ \ 0 \le x \le l, \\
\frac{d^4 \upsilon_{2q}^{(1)}}{dx^4} - k_2^4 \upsilon_{2q}^{(1)} = -\frac{\omega^2 \rho}{D_2} \Phi_4; \ \ y = b, \ \ a \le x \le r,
\end{gathered}
\qquad (3.42)
$$

$$\frac{d^4 \upsilon_{2q}^{(2)}}{dx^4} - k_2^4 \upsilon_{2q}^{(2)} = -\frac{\omega^2 \rho}{D_2} \Phi_4; \quad y = -b, \quad a \leq x \leq r.$$

The infinite system of algebraic equations that follows from (3.42) is rather bulky because of the large number of partial domains. This system is similar, in structure and in coefficients properties, to the one-layer gratings considered above, and it is not necessary to write it in an explicit form. We solve the system of equations using reduction method, and we proceed to the quantitative analysis of a double layer grating. The analysis will be based on calculated data $\rho_1 c_1/\rho c = 15$, $a = 1.01l, b = 0.5l, d = 0.57l, l = r - a$, and $\theta = 0$. The sound-transmission characteristics of the double layer grating can be represented by some typical frequency dependences of the transmission coefficient which are given in Fig. 3.20. Curve 2 in this figure corresponds to the case, in which both layers of the grating are identical $h_1 = h_2, l = r - a = 0.1(c/\bar{f}_1), \bar{f}_1 = (f_1^{(1)} + f_1^{(2)})/2; f_1^{(1)}$ and $f_1^{(2)}$—are the first natural frequencies of the plates in vacuum, in the first and the second layers respectively. The frequency dependence of k_{ins} (curve 1) for the single-layer grating, which was considered in the previous paragraph, is also shown in this figure for their comparison.

The case of two identical layers in the grating is primarily of interest in connection with the fact that the interaction between layers is most conspicuous in this case. This situation is similar to the situation in the oscillatory system with two degrees of freedom when there is strong coupling between the partial systems [3]. All couplings between the partial systems are strong, because these systems are identical. Accordingly, a comparison of curves 1 and 2 gives a clear picture of the quantitative and qualitative influence of interaction between layers on the sound insulation of the grating. In the interval of relatively law frequencies the presence of the second layer decreases the value of the sound transmission coefficient. But at a frequency $f = 0.5\bar{f}_1$, an abrupt increase in the sound transmissivity is observed as a consequence of the strong interaction between layers. Naturally, if the resonance frequencies of the plates in the layers of the grating are

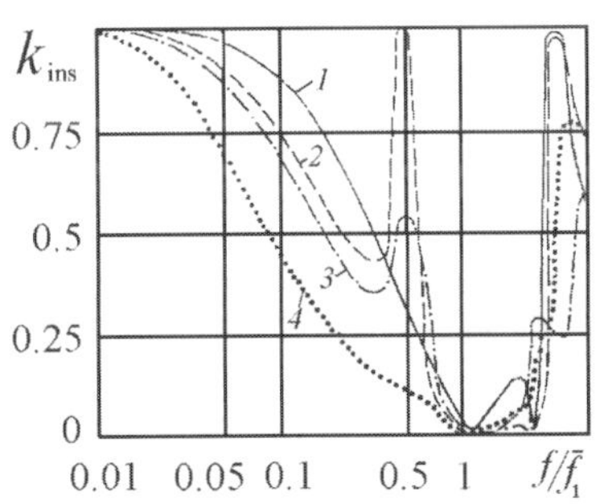

FIG. 3.20: Sound transmission coefficient vs frequency. $\rho_1 c_1/\rho c = 15, a = 1.01l, b = 0.5l, d = 0.57l, l = r - a$, and $\theta = 0$; curve 1 corresponds to single-layer grating, curve 2 corresponds to double-layer grating with two identical layers, curve 3 refers to the case in which the natural frequency of the plates in the first layer is $0.833\bar{f}_1$, and for the second layer it is—$1.2\bar{f}_1$, curve 4 refers to the case in which the natural frequency of the first layer plates is $0.625\bar{f}_1$ and for the second layer it is—$1.6\bar{f}_1$.

separated, the coupling between layers is lessened, and the peak transmissivity in frequency range $f = 0.5\bar{f}_1$ is lower. This tendency is clear perceived in the analytical data. Curve 3 in Fig. 3.20 refers to the case in which the natural frequency of the plates in the first layer is $0.833\bar{f}_1$, and for the second layer it is $1.2\bar{f}_1$. A more important characteristic of a frequency-separated grating, is the fact that, a decrease in the value of acoustic transmissivity of the grating at low frequencies is observed along with a reduction of the transmissivity peak. Precisely this fact can be used to achieve a substantial increase in the bandwidth of effective insulation by a suitable separation of the natural frequencies of the plates associated with different layers of the grating. A quantitative estimate of the degree of broadening of the insulation band is given by curve 4 for the case in which the natural frequency of the plates in the first layer is $0.625\bar{f}_1$ and for the second layer it is $1.6\bar{f}_1$.

Deeper insight into certain aspects of the process of sound transmission through a double-layer grating can be gained on the basis of an analysis of the frequency dependence of its input impedance Z. The real part R and the imaginary part X for the three cases represented in Fig. 3.20 (curves 1, 2, 3) are given in Fig. 3.21, respectively. The general behavior of the frequency dependence of R emulates the frequency dependence of the sound transmission coefficient. The frequency dependence of the reactance component X of the input impedance largely complements and explains the behavior of the curves in Fig. 3.20.

Curve 1 characterizes the variation of X for a single-layer grating. It has a single point with $X = 0$, (in the investigated frequency range), corresponding to resonance of the system plates-liquid. The case of a grating with two identical layers (curve 2) is characterized by a complex frequency dependence. Here we observe three points at which $X = 0$: the firs resonance frequency $f_{r1} \approx 0.3\bar{f}_1$, the first antiresonance frequency $f_{a1} \approx 0.5\bar{f}_1$, and the second resonance frequency $f_{r2} \approx 0.63\bar{f}_1$. If we now refer to curve 2 in Fig. 3.20, we can readily verify that the frequencies f_{r1} and f_{r2} correspond to relative minimum of the transmissivity of the grating, and f_{a1}—to the maximum of this quantity. Separation of the natural frequencies of the plates in layer significantly changes the frequency dependence of the imaginary part of the input impedance. Even for

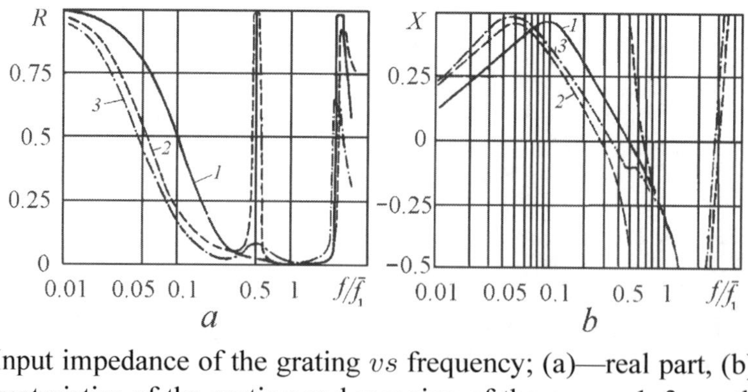

FIG. 3.21: Input impedance of the grating *vs* frequency; (a)—real part, (b)—imaginary part; all characteristics of the grating and meaning of the curves 1–3 are the same as in Fig. 3.20

a relatively small separation, corresponding to curve 3 in Fig. 3.21, only one resonance of the system plates-liquid remains at frequency $f_{r1} \approx 0.35\bar{f}_1$.

In the result of interaction between the plates in the layers, two resonance frequencies appear and one antiresonance frequency appear between them, instead of one resonance frequency of the plates-liquid system. The vibration mode of the plates in the region of these three frequencies corresponds to their first natural vibration mode. In this case, the plates vibrate in phase at the resonance frequencies and they vibrate strictly in antiphase at the antiresonance frequency. At resonance frequencies the fluid executes vibrations in the space between the bars of the grating with a volume change (that is the key condition for law acoustic transmissivity of the grating). There is no volume change at antiresonance frequency and in consequence of this the grating becomes almost completely acoustically transparent. Comparing the Figs. 3.22 and 3.14 we can make certain that the flow paths of the liquid in the space between the bars are to a certain extent similar to each other. Nevertheless they are of different appearance origin (at least at antiresonance frequency). In a single-layer grating, these flow paths have place when the second natural vibration mode of the plates dominate. The same flow paths appear in a double layer grating at first natural vibration mode of the plates as a result of interaction between the plates from different layers of the grating.

We have considered in details the situation when the first natural mode of the plates dominates. As the calculations show, the similar situations arise at higher vibration modes of the plates. The analysis of these cases is not very interesting. Only the frequency range, where the first natural mode of the plates dominates, is of importance for sound proofing.

In conclusion we note the following circumstance. The frequency characteristics of the double-layer grating were improved through use of different resonance frequency in each layer of the grating. It was implemented because the elastic walls of the shells in each layer differ in geometric or physical parameters. The identical shells can also be used in both cases but the increment size for each layer should be different. Then, as the calculation data show, the frequency characteristics of a double-layer grating will be better than the single-layer grating frequency characteristics.

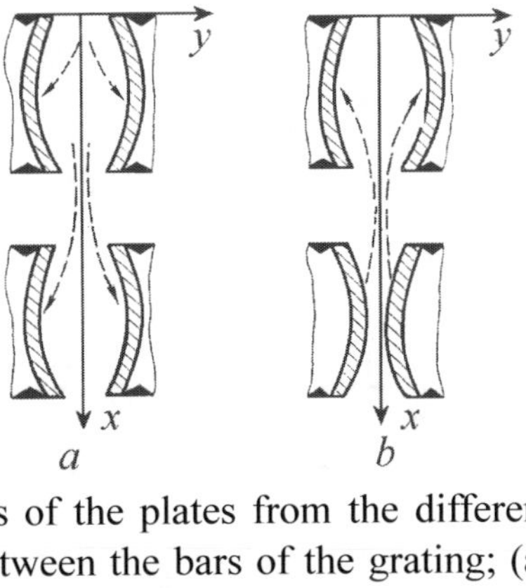

FIG. 3.22: Vibration modes of the plates from the different layers of the grating and flow path of the medium between the bars of the grating; (a)—at frequency of the first resonance, (b)—at frequency of the first antiresonance

3.7.2 Active Method

One of the active methods, which is of interest in practice, is using the elastic walls of the shells completely or partly made of piezoelectric material. If the electric field of the frequency of an incident wave is applied to such walls, the walls will bend. It is an equivalent to a certain additional external pressure applied to the walls [11].

As an example, we consider one-layer grating, pictured in Fig. 3.10, supposing that each elastic plate has three layers: middle layer made of passive isotropic material, and outer layers made of piezoelectric material, for example of piezoceramics, Fig. 3.23. We call attention to the plus and minus signs which indicate the direction of polarization of the piezoceramic layers. The plate can be made to bend due to such polarization, and the indicated way of voltage application to the electrodes.

The local system of coordinates zO_1x is introduced to depict the motion of the plates. We suppose that a three-layer plate, forming the elastic wall of the grating, Fig. 3.23, is thin enough and can be depicted with the help of classical plate-bending theory [46]. According to the construction of the grating, plane deformation takes place in the O_1y—direction that is $\varepsilon_y = 0$. In accordance with thin plate theory, the plate equations can be written in such form [47,48]:

$$\begin{aligned} \varepsilon_x &= S_{11}^E \sigma_x + S_{12}^E \sigma_y + d_{31} E_z, \\ 0 &= S_{12}^E \sigma_x + S_{11}^E \sigma_y + d_{31} E_z, \end{aligned} \tag{3.43}$$

where ε_x—strain tensor components, S_{ij}^E—elastic compliance tensor components, determined at constant electric field, d_{31}—piezoelectric modulus, σ_x, σ_y—stress tensor components, E_z—electric field strength.

The coefficients in Eqs. (3.42) are the piecewise constant function of coordinate z. For middle passive layer $d_{31} = 0$.

We assume that electric field in piezoceramics is given in accordance with electric load and electrodes position:

$$E_z = \begin{cases} E_0 e^{-i\omega t}; \ h_1 \le z \le h_1 + h_2, \\ 0; \ |z| \le h_1, \\ -E_0 e^{-i\omega t}; \ -h_1 - h_2 \le z \le -h_1, \end{cases} \tag{3.44}$$

where E_0—complex amplitude of electric field strength.

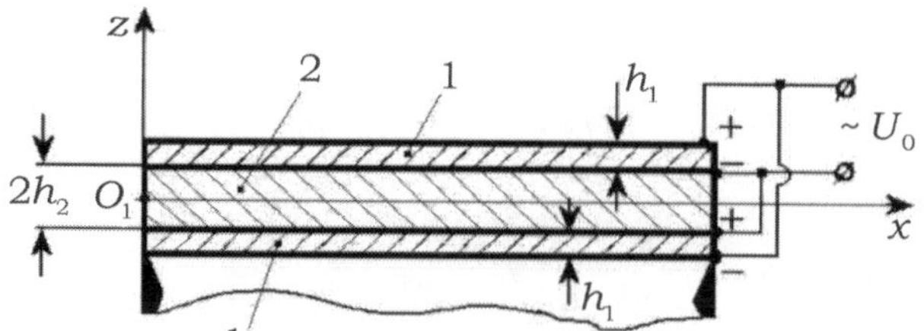

FIG. 3.23: Three layer plate; 1—piezoelectric layers, 2—the layer of passive material

Additional electric field strength, arising due to the relation between electric and mechanical values in piezoelectric ceramics, is negligible for almost all the frequencies in the range under consideration. We determine voltage σ_y from the second equation of system (3.43) and substitute it into the first equation. Thus we obtain relation between voltages, deformations, and electric field:

$$\sigma_x = \frac{S_{11}^E}{(S_{11}^E)^2 - (S_{12}^E)}\varepsilon_x - \frac{d_{31}}{S_{11}^E - S_{12}^E}E_z. \tag{3.45}$$

Assuming that mechanical properties of the piezoceramic layers of the plate and of the passive plate differ slightly, we can describe the plate deformations in the whole block with a help of the Kirchhoff hypothesis. Thus we can right the expression for strain distribution ε_x through the thickness of the plate:

$$\varepsilon_x = -z\frac{\partial^2 w(x,t)}{\partial x^2}, \tag{3.46}$$

$w(x,t)$—flexural deformation of the middle layer of the plate $z = 0$.

After performing the elementary transformations, we obtain the relation between the bending moment of the plate and the curvature of the middle layer of the plate:

$$M_x = -D\frac{\partial^2 w(x,t)}{\partial x^2} + SE_z. \tag{3.47}$$

Where

$$D = \frac{S_{11}^E}{(S_{11}^E)^2 - (S_{12}^E)^2}\frac{2}{3}\left[(h_1+h_2)^3 - h_1^3\right] + \frac{2Eh_1^3}{3(1-\nu^2)},$$

$$S = \frac{d_{31}}{S_{11}^E + S_{12}^E}\left[h_1^2 - (h_1+h_2)^2\right],$$

E and ν—modulus of elasticity and the Poisson ratio of the middle passive layer of the plate.

Taking into consideration that electric field does not depend on x-coordinate, and bearing in mind formula (3.47), we can write the equation for the plate element

$$D\frac{\partial^4 w(x,t)}{dx^4} + M\frac{\partial^2 w(x,t)}{\partial t^2} = i\omega\rho\tilde{\Phi}(x,t). \tag{3.48}$$

Here $M = 2\rho_1 h_1 + 2\rho_2 h_2$, and ρ_1, ρ_2, ρ—are the densities of the middle layer, outer piezoceramic layers, and surrounding medium respectively. It was taken into account that the plate vibrates in acoustic medium. Acoustic load is written in the right-hand side of the Eq. (3.48), where $\tilde{\Phi}(x,t)$ is the velocity potential of the particles of the medium.

Oscillating speed amplitude $\upsilon(x)$ for harmonic motion is determined from the equation

$$\frac{d^4\upsilon(x)}{dx^4} - k_1^4\upsilon(x) = -\frac{\omega^2\rho}{D}\Phi(x), \tag{3.49}$$

where the wave number is $k_1^4 = M\omega^2/D$.

This equation is the same as the motion equation of the element of the plate without piezoeffect. Piezoeffect influences the boundary conditions. In view of the expression (3.47), the boundary conditions of the hinged edges include formula

$$\frac{d^2\upsilon(x)}{dx^2} + i\omega S E_z = 0, \;\; x = 0, \;\; x = l. \tag{3.50}$$

We represent the general solution for $\upsilon(x)$ by series in the vibrational eigenfunctions of the plate without piezoeffect:

$$\upsilon(x) = Ax(x - l) + \sum_{s=1}^{\infty} a_s \sin\frac{\pi s}{l}x, \tag{3.51}$$

A and a_s—unknown complex coefficients.

Due to the invariable A selection, expression (3.51) allows to meet the boundary conditions for the bending moment at the plate edges. From Eq. (3.50) we determine

$$A = i\frac{\omega S E_0}{2D}. \tag{3.52}$$

Zero condition for $\upsilon(x)$ at the edges of the plate is satisfied. The summand $Ax(x - l)$ have to be represented by a series in the function system $\sin(\pi s x/l)$, when we substitute expression (3.51) into the motion Eq. (3.49). The expansion of (3.52) together with the expansion of $\Phi(x)$ let us to algebraize Eq. (3.49) together with conjugation conditions and boundary conditions. These transformations are similar to the ones given above.

We are now ready to analyze the quantitative data on acoustic properties of the gratings made of active elements. The calculations are made for the case of the normal incidence of the flat wave on the grating. The geometric and physical parameters have the following values: $l/\lambda_1 = 0.1$, $2b/l = 1$, $b/a = 1.1$, $h_1/l = 0.027$, $h_2/l = 0.008$, $\rho_1 = 4.5\times10^3$ kg/m^3, $E_1 = 1.16\times10^{11}$ Pa, $\rho c = 1.5\times10^6$ kg/m^{2}s, $\nu = 0.3$. The incident wave pressure is 0.1 Pa. The geometric parameters of the grating are showed in Figs. 3.10 and 3.23. The piezoceramic layers of the plates are considered to be made of ceramic PZT-4 [43] with the following parameters $\rho_2 = 7.5 \times 10^3$ kg/m^3, $S_{11}^E = 1.2 \times 10^{-11}$ m^2/N, $S_{12}^E = -0.405 \times 10^{-11}$ m^2/N, $d_{31} = -1.54 \times 10^{-10}$ C/N.

The goal of the calculations is to estimate the capability to control the transmissivity of a grating by means of the voltage applied to the piezoceramic layers. Generally speaking, it is beyond question that we can control the transmissivity of the grating. Now we have to find out the attainable level of the widening of the bandwidth of effective sound insulation at moderate electric field in the active elements of the plate. It is also necessary to indicate the easily realized methods of control of the phase characteristics of the applied field. We call attention to the fact that in expressions (3.44) E_0 is a complex value. That is why sound transmission coefficient is determined in calculations for each value of the frequency of the incident wave in some variation interval of $E_0 h_2$. The increment of going through the values is 0.28 in the modulus, and 20° in the phase. From the resulting array of values we choose the minimum values and the corresponding values of the modulus and phase. The frequency dependence of the coefficient k_{ins} is plotted in Fig. 3.24. The frequency dependences of the modulus and argument E_0h_2 that ensure k_{ins} (curves 1 and 2, respectively) are plotted in Fig. 3.25. Notice that the curve oscillations illustrated in Fig. 3.24 are elicited by the selection of a comparatively large increment in changing the values of the amplitude and phase in the calculations. Analyzing these data and comparing them with the case when $E_0h_2 = 0$ (dashed line in Fig. 3.24), we can prove that acoustic transmissivity of the grating can be controlled by means of electric field applied to the piezoceramic plates.

We call attention to another important consideration. The additional load, created on the plates by means of the given technique, effectively improves the sound insulation of the grating in the frequency interval $f \leq f_1$. At higher frequencies $f > f_1$ an improvement of sound insulation is not observed. This fact is attributable to the following circumstances. For $f \leq f_1$ the dominant flexural mode of the plates is $\sin(\pi x/l)$. By applying the voltage to the plates, we promote an increase in this mode of their flexural deformation since the function $x(x-l)$ is close to the function $\sin(\pi x/l)$. The situation is altogether different in the frequency interval $f > f_1$. Here the higher plate modes begin to play a major role. The effective electromechanical coupling factor decreases essentially because of inconsistency between the vibration mode and the uniform distribution of the external voltage. Vibrations of the piezoceramic plates at high frequencies with uniform electrode coating are weak.

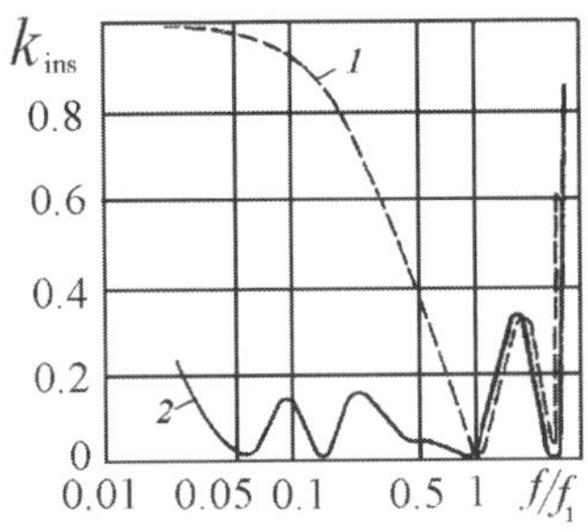

FIG. 3.24: Sound transmission coefficient versus frequency; 1—for $E_0h_2 = 0$, 2—for $E_0h_2 \neq 0$

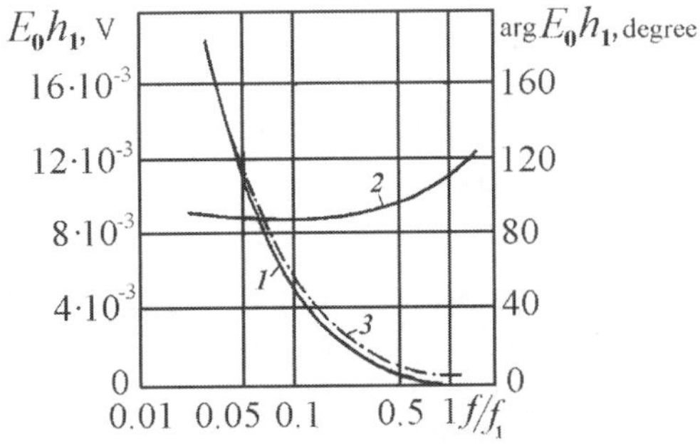

FIG. 3.25: Frequency dependencies of the modulus (curve 1) and phase (curve 2) of electric field on piezoactive plates, ensuring values k_{ins} versus frequency plotted in Fig. 3.24 (curve 2); curve 3—approximate formulas were used in calculation

As it was mentioned above, the higher vibration modes do not change the volume and thus the compliance of the grating. The elements of the grating are small in comparison with the wave-length. And the grating becomes a bad acoustic reflector. The sound attenuating properties of the grating rest the same as the properties of the passive elements in the frequency range $f > f_1$. It is proved by the almost entire matching of the dashed and block curves in Fig. 3.24 for $f/f_1 > 1$. It should be noted that the effective electromechanical coupling coefficient, characterizing acoustic properties of the grating, can be controlled. We can correlate the field distribution with the characteristics of the induced vibration modes, dividing the electrodes into parts and changing their polarity. Thus the effective electromechanical coupling coefficient increases greatly. But if the corresponding vibration mode is not related with the shell volume change, the reflectivity of the grating will not change significantly.

We can obtain rather simple formula showing the relationship between the sound transmission coefficient and the electric field in piezoceramic gratings, for the frequency range $f < f_1$. This formula will be obtained if we use only the first terms of the series representing the acoustic field and the particle velocity of the plates. Thus we find

$$k_{\text{ins}} = -\frac{1}{\xi kl + i}\left\{\xi kl + \frac{kl}{2}\tau U_0 + i\left[(kl)^2(\xi - 1/2) - \tau U_0\right]\right\}, \tag{3.53}$$

where $\tau = (2\pi)^2(S/D)f_1^2\,(l/\pi)^2\,(M/300h_2)$, U_0—complex amplitude of potential difference, applied to the electrodes of the piezoceramic plates.

Using ratio (3.53), we obtain an approximate solution of the inverse problem. Using the given the value of sound transmission coefficient we determine the potential difference on the electrodes. For example, by taking $k_{\text{ins}} = 0$, we find

$$\begin{aligned} |U_0| &= \frac{kl}{\tau}\left\{\left[\frac{kl}{2}(\xi - 1)\right]^2 + \xi^2\right\}^{1/2}, \\ \arg U_0 &= \operatorname{arctg}\frac{-2\xi}{kl(\xi - 1)}. \end{aligned} \tag{3.54}$$

The values of the potential difference modulus, determined from the approximation formula are given in Fig. 3.25 (curve 3). This curve almost coincides with the curve obtained by rigorous calculations. In conclusion, it is essential to note that the active method, discussed above, enables us to change the acoustic properties of the grating from its full acoustic transmissivity to the soundproofness in the given frequency band.

3.8 GRATING COMPOSED OF SHELLS OF FINITE DIMENSIONS

The gratings of shells of infinite length were considered above. Now we consider the grating of elastic shells of finite dimensions, formed with two thin round plates. The shells are attached at the edges through nondeformable pivot ring. The shells of this type have a number of advantages with regard to manufacturing process, comparing with the other types of shells, and thus they are of practical interest. We investigate the acoustic properties of an infinite plane grating composed of the indicated shells located in the nods of hexagonal mesh (Fig. 3.26). In this case, the shells in the grating are the most tightly packed, and so the maximum sound insulation is provided. It does not complicate the solution of the considered problem if the plates have different geometric and physical characteristics. We can change the properties of the grating by changing these values.

We restrict ourselves to the case of normal incidence of a plane sound wave on the grating. As the analysis performed in the previous paragraphs shows, the sound attenuating properties of the grating for the case of oblique incidence of the plane wave are better than for normal incidence. By improvement of sound attenuating properties of the grating we mean that the transmissivity of the grating decreases in the vicinity of principal resonance frequency, and the frequency bandwidth, where the transmission coefficient is less than some given value, increases. As it was mentioned above, the value of the pressure transmission coefficient is usually $k_{\text{ins}} = 0.2$.

Due to the symmetry of the grating, we can consider the problem of sound diffraction by a shell that is inside of an infinite, hexagonal wave guide with acoustically rigid boundaries [49], instead of the problem of sound diffraction by the grating. Such a wave guide is indicated with dashed lines in Fig. 3.26. To simplify the problem we have to replace the hexagonal waveguide with a round waveguide with radius $r_1 = (S/\pi)^{1/2}$, where S—cross section area of the hexagonal waveguide. It is justified when the maximum transverse size of the hexagonal waveguide does not exceed the half wave length. Figure 3.27 shows the longitudinal cross section of the round waveguide with elastic shell in it.

We partition all the field existence domain in the waveguide into the following partial domains: the first domain $r \le r_1$, $x \le 0$; the second (ring shaped) domain $r_0 \le r \le r_1$, $0 \le x \le l$; the third domain $r \le r_1$, $x \ge l$. We represent the sound field in these domains by Fourier series in the normal modes of round and ring waveguides with acoustically rigid walls:

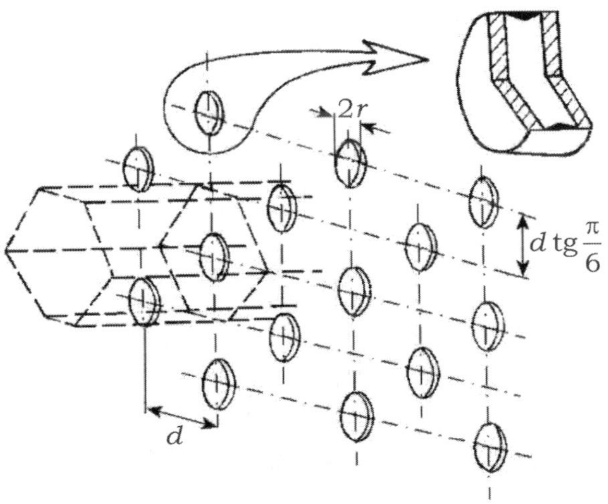

FIG. 3.26: The grating composed of shells in the shape of short cylinders with round plates (see also Fig. 3.1)

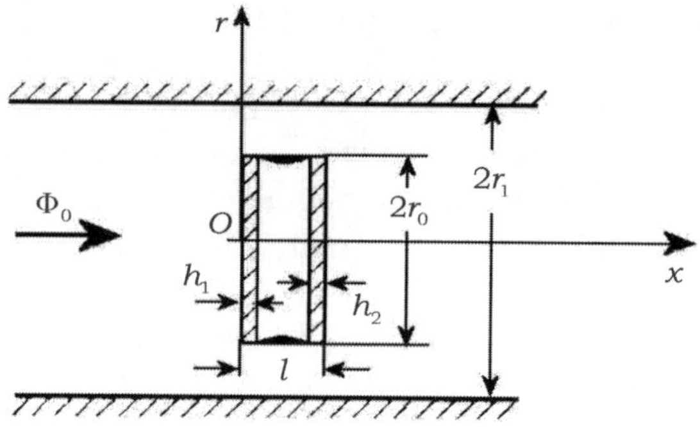

FIG. 3.27: Longitudinal cross section of the round waveguide with a shell in the shape of a short cylinder with the round plates

$$\Phi_1 = e^{ikx} + \sum_{\nu=1}^{\infty} A_\nu J_0(\alpha_\nu r) e^{-ik_\nu x},$$

$$\Phi_2 = \sum_{p=1}^{\infty} \Delta_p(r)\,[E_p \sin k_p x + F_p \cos k_p x], \tag{3.55}$$

$$\Phi_3 = \sum_{\mu=1}^{\infty} C_\mu J_0(\alpha_\mu r) e^{ik_\mu (x-l)},$$

where $\Delta_p(r) = J_0(\alpha_p r) - B_p N_0(\alpha_p r)$; $k_\nu = (k^2 - \alpha_\nu^2)^{1/2}$; $k_p = (k^2 - \alpha_p^2)^{1/2}$; $k_\mu = (k^2 - \alpha_\mu^2)^{1/2}$; $J_0()$, and $N_0()$—Bessel and Neumann functions.

We represent the particle velocities of the plates as the superposition of two components. The first component corresponds to the shell motion along the Ox, axis as a single whole, and the second component describes the bending vibrations of the plates and is represented by series expansions in the mode configurations of the plates in vacuum [39,46]:

$$\upsilon_\tau(r) = \upsilon_0 + \sum_{s=1}^{\infty} \upsilon_s^{(\tau)} \Delta_s^{(\tau)}(r); \ \tau = 1; 2, \tag{3.56}$$

where $\Delta_s^{(\tau)}(r) = J_0(\alpha_s r) - [J_0(\alpha_s)/I_0(\alpha_s)] I_0(\alpha_s)$: $I_0()$—modified Bessel function of the first kind.

The values of the coefficients α_ν, α_p, α_μ, α_s, and B_p in (3.55) and (3.56) are determined from the following boundary conditions:

$$\begin{aligned} &\partial\Phi_2/\partial(kr) = 0; \ 0 \le x \le l, \ r = r_1, \\ &\partial\Phi_1/\partial(kr) = 0; \ x \le 0, \ r = r_1, \\ &\partial\Phi_3/\partial(kr) = 0; \ x \ge l, \ r = r_1, \\ &\partial\Phi_2/\partial(kr) = 0; \ 0 \le x \le l, \ r = r_0, \\ &\upsilon_\tau(r) = \frac{d^2\upsilon_\tau(r)}{dr^2} + \frac{\sigma_\tau}{r}\frac{d\upsilon_\tau(r)}{dr} = 0; \ r = r_0; \ \begin{cases} x = 0, \\ \tau = 1, \end{cases} \ \begin{cases} x = l, \\ \tau = 2. \end{cases} \end{aligned} \tag{3.57}$$

The last condition in (3.57), describes the hinge-supported plates [39,46]. The other types of plate's supports can be regarded within the limits of the method that is used. The plates that are rigidly fixed at the edges will be considered below.

Taking into account conditions (3.57), the unknown complex coefficients in series (3.55) and (3.56) are determined from the following system of functional equations:

$$\begin{aligned}
&\Phi_1 = \Phi_2; \quad x = 0, \quad r_0 \le r \le r_1, \\
&\Phi_3 = \Phi_2; \quad x = l, \quad r_0 \le r \le r_1, \\
&-\frac{\partial \Phi_1}{\partial x} = \begin{cases} -\dfrac{\partial \Phi_2}{\partial x}; & r_0 \le r \le r_1, \quad x = 0, \\ \upsilon_1(r); & 0 \le r \le r_0, \quad x = 0, \end{cases} \\
&-\frac{\partial \Phi_3}{\partial x} = \begin{cases} -\dfrac{\partial \Phi_2}{\partial x}; & r_0 \le r \le r_1, \quad x = l, \\ \upsilon_2(r); & 0 \le r \le r_0, \quad x = l, \end{cases} \\
&\frac{M}{\rho}\upsilon_0 = \int_0^{r_0} \left[\Phi_1(x=0) - \Phi_3(x=l)\right] r\,dr, \\
&\nabla^2\nabla^2\upsilon_1(r) - k_1^4\upsilon_1(r) = -\frac{\omega^2\rho}{D_1}\Phi_1; \quad 0 \le r \le r_0, \quad x = 0, \\
&\nabla^2\nabla^2\upsilon_2(r) - k_2^4\upsilon_2(r) = -\frac{\omega^2\rho}{D_2}\Phi_3; \quad 0 \le r \le r_0, \quad x = l,
\end{aligned} \tag{3.58}$$

where $\nabla^2 = (\partial^2/\partial r^2) + (1/r)(\partial/\partial r)$; $k_\tau^4 = \omega^2 M_\tau / D_\tau$; $M_\tau = \rho_\tau h_\tau$; $M = M_1 + M_2 + \bar{M}$—total mass of the shell, $\bar{M}$—mass of the support.

The system of functional Eqs. (3.58) is replaced with a corresponding infinite system of linear algebraic equations of the second kind. Omitting this procedure, we proceed with analysis of acoustic properties of the considered grating. Figure 3.28 shows the

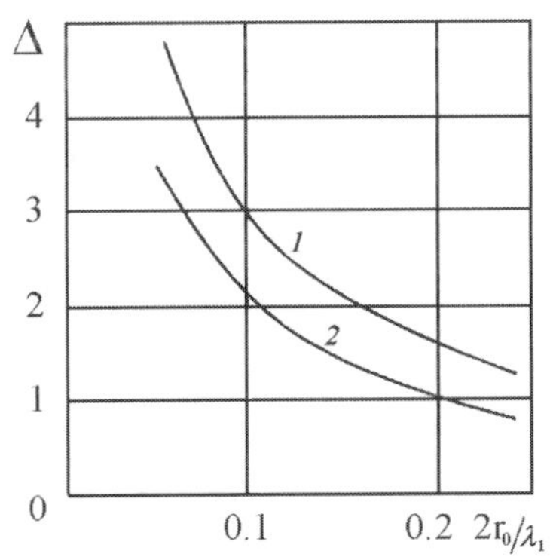

FIG. 3.28: Bandwidth of effective sound insulation of the grating *vs* wave diameter of the plates and type of the plates support when $\rho_1 c_1/\rho c = 15$, $h_1 = h_2$, $r_1/r_0 = 0.95$, $\bar{M} = 0$; 1—rigid fixing, 2—hinge support

calculated dependency of the bandwidth of effective sound insulation of the grating on the wave diameter of the plates and on the type of their support. The plates are assumed to be made of titanium alloy with $\rho_1 c_1/\rho c = 15$, and to be of equal thickness $h_1 = h_2$. The overlapping of the waveguide by the shell is calculated by formula $r_1/r_0 = 0.95$. In the calculations we neglect the mass of the supports in comparison with the mass of the plates. We set $\bar{M} = 0$.

The represented dependencies are similar to the dependencies of the shells considered above. The band of effective sound insulation of the grating Δ increases with a decrease of wave dimensions of the plate. It is to be noted that the type of the plates support has a great impact on the acoustic properties of the grating. If the edge of the plates is fixed rigidly (curve 1), the value of Δ can be increased noticeably in comparison to the case when the edge is hinge supported (curve 2). When performing the calculations for such type of support, the corresponding conditions in system (3.57) are replaced with the following:

$$\upsilon_\tau(r) = d\upsilon_\tau(r)/dr = 0; \quad r = r_0.$$

This fact can be explained as follows. At specified diameter and resonance frequency, the thickness of the round plate depends on the way of its edge fixing. If the plate is fixed rigidly, its thickness will be two times less than the thickness of the hinge supported plate when the natural frequencies of these plates are equal. Thus the rigidly fixed plate has larger dynamic compliance. And that is why bandwidth of effective insulation of the grating is large. The indicated increase in compliance of the plates plays the most important role at relatively low frequencies. Figure 3.29 illustrates the frequency dependencies of sound transmission coefficient through the grating of shells with the different types of the plates supports for $2r_0/\lambda_1 = 0.1$. Curve 1 corresponds to the case of simply supported plate, and curve 2 corresponds to rigidly fixed plate. Both plates have the same first natural frequencies. All the mentioned above distinctive features of the grating are readily seen. As much as the rigidly fixed plate is thinner than the simply supported plate, the mass of liquid added to the plate plays more important role and that is why the natural frequency of the plate in liquid is lower than natural frequency of the plate

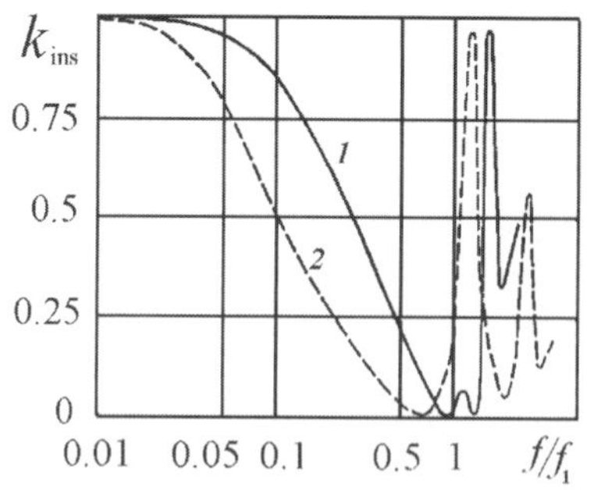

FIG. 3.29: Dependencies of sound transmission coefficient passing through the grating of shells with different plates supports for $2r_0/\lambda_1 = 0.1$ (the other parameters are the same as in Fig. 3.28); curve 1 corresponds to the case of hinged supported plate and curve 2—rigidly fixed plate

in vacuum. The acoustic properties of the gratings with the different plate's supports are almost identical in the region of relatively high frequencies.

We turn our attention to the analysis of the acoustic properties of the grating with allowances made for the kinematics of plates and bars. In the fifth paragraph this question was considered for the grating with the plates normal to the plane of the grating. It was shown that alternation of the maximum and the minimum transmissivities of the grating is connected with alternation of the resonance and the antiresonances frequencies of the plates-liquid system. The minimum transmissivity is in the region of the resonances of the pates-liquid system. The natural vibration modes of the plates with odd indices dominate here. The maximum transmissivity is in the region of the antiresonances of the pates-liquid system where the natural vibration modes of the plates with even indices dominate. When the plates are vibrating, the bar volume in the grating changes less in the latter case than in the first case. It is to be noted that only one natural vibration mode of the plates corresponds to each resonance or antiresonance.

We will study out the physical causes of occurrence of minimum and maximum transmissivity of the grating with plates normal to its surface. The grating considered in this paragraph will be analyzed.

Figure 3.30 shows the curves characterizing input impedance of the grating. If we compare these curves with the curves in Fig. 3.12, we can make certain that the frequency dependencies of the values of R and X for the gratings with different orientation of the plates are of the same qualitative character. Nevertheless, we will consider the particle velocity of both round plates, which form the shell, at resonance and antiresonance frequencies of the plates-liquid system (Fig. 3.30). Table 3.3 shows the data for the amplitude and phase characteristics of the first two terms in the series, which represent the particle velocities of the plates. According to the analysis of these data, the vibration mode of the plates almost coincides with the first vibration mode in vacuum at the frequency of the first resonance of the plates-liquid system ($f_{p1} \approx 0.5 f_1$). The role of the second mode is negligible. The plates are vibrating in antiphase, and as a consequence, the vibration mode of the shell is as it is shown in diagram. Thus, the shell vibrations are followed with the volume change at the first resonance of the elastic plates-liquid system. That is why the transmissivity of the grating is law.

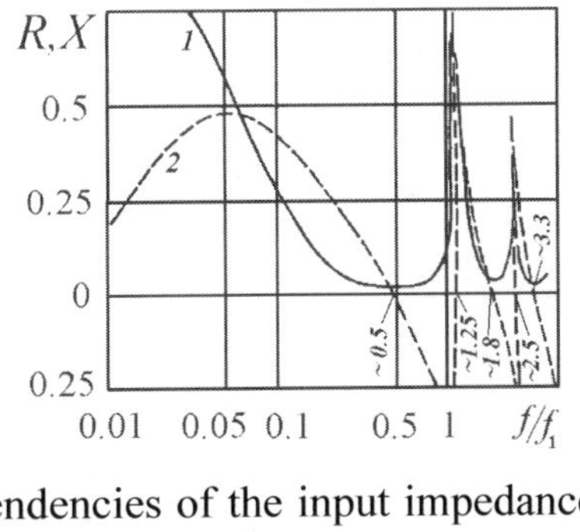

FIG. 3.30: Frequency dependencies of the input impedance when plates are hinge supported (the other parameters are the same as in Fig. 3.29); 1—real part, 2—imaginary part

TABLE 3.3: Amplitudes, phases of the vibrational velocity and the shape of oscillations of the plates

f	$\upsilon_1^{(1)}$		$\upsilon_2^{(1)}$		$\upsilon_1^{(2)}$		$\upsilon_2^{(2)}$		**Vibration mode of the plates**
	amplitude	**phase**	**amplitude**	**phase**	**amplitude**	**phase**	**amplitude**	**phase**	
$f_{p1} \approx 0.5f_1$	2.3	2.5°	0.08	—	0.98	185°	0.6	—	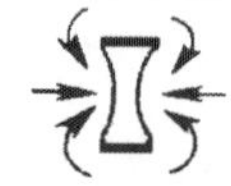
$f_{p2} \approx 1.25f_1$	39	193°	3.3	18°	39.7	197°	5.7	17°	
$f_{p3} \approx 1.8f_1$	2.1	180°	3.1	180.2°	4.1	180.7°	0.94	55°	
$f_{p4} \approx 2.5f_1$	10	183.7°	15	0°	15	182.7°	0.07	4°	

The vibration mode of the plates is determined by their first natural vibration mode in vacuum at the frequency of the first antiresonance ($f_{p2} \approx 1.25f_1$), as well as at the frequency of the first resonance. Nevertheless, both plates are vibrating in phase and there is almost no shell volume change. Liquid executes oscillating motion in the vicinity of the shell (the flow paths are indicated with arrows), and the grating becomes sound transparent.

Further analysis of the data in Table 3.3 shows that for the next following resonances and antiresonances of the plates-liquid system the situation is similar. But it should be noted that the vibration modes of the higher order have more complex structure.

It is an interesting special feature of the elastic shells vibrations in the grating, that the different vibration modes have different Q-factors. Q-factor is large at the antiresonance frequencies where the grating is sound transparent and it is relatively small at the resonance frequencies, corresponding to the minimum transmissivity. That is why the frequency domains where the grating is non-transparent are broader than the frequency domains where it is transparent.

In conclusion we note that the analysis of sound-reflecting properties of the grating is based on monograph [15] and original articles [8–14], published by the authors in different times.

REFERENCES

1. Furduev, V.V., *Electroacoustics*, Moscow: Gostehizdat, 1948. [in Russian]
2. Isakovich, M.A., *General Acoustics*, Moscow: Nauka, 1973. [in Russian]

3. Grinchenko, V.T., Vovk, I.V., and Matsypura, V.T., *Basic Principles of Acoustics*, Kiev: Naukova Dumka, 2007. [in Russian]

4. Osipov, G.L. and Yudin, E.Yu., (Eds.), *Noise Control in Buildings and Residential Areas*, Moskow: Stroyizdat, 1987. [in Russian]

5. Osipov, G.L. and Bobylev, V.N., (Eds.), *Sound Insulation and Sound Absorption,* Moskow: AST Astrel, 2004. [in Russian]

6. Glazanov, V.E., *Hydroacoustic Arrays Screening*, Leningrad: Sudostroenie, 1986. [in Russian]

7. Toulis, W.J., Acoustic refraction and scattering with compliant elements, *J. Acoust. Soc. Am.*, vol. **29**, pp. 1021–1033, 1957.

8. Vovk, I.V., Grinchenko, V.T., and Kononuchenko, L.A., Diffraction of a sound wave by a plane grating formed by hollow elastic bars, *Sov. Phys. Acoust.,* vol. **22**, no. 2, pp. 113–115, 1976.

9. Vovk, I.V. and Grinchenko, V.T., Diffraction of a plane wave by a double-layer grating of hollow elastic bars, *Sov. Phys. Acoust.,* vol. **24**, no. 6, pp. 480–482, 1978.

10. Vovk, I.V., Sound diffraction by a grating of gas-filled elastic bars, *Sov. Phys. Acoust.,* vol. **25**, no. 4, pp. 287–289, 1979.

11. Vovk, I.V., Calculation of the sound transmission coefficient through a grating with controlled-compliance elements, *Sov. Phys. Acoust.,* vol. **26**, no. 5, pp. 380–383, 1980.

12. Vovk, I.V., Diffraction of sound wave by a grating of rigid cylinders with elastic covers, *Pricladnaya Mechanika*, vol. **17**, no. 8, pp. 120–123, 1981.

13. Vovk, I.V. and Grinchenko, V.T., Relationship of the sound transmission of a grating of holow compliant bars to the cinematics of the bar walls, *Sov. Phys. Acoust.*, vol. **28**, no. 6, pp. 442–446, 1982.

14. Vovk, I.V., Kononuchenko, L.A., and Omelchenko, B.A., Sound diffraction by a finite grating of hollow elastic shells, vol. **30**, no. 4, pp. 270–273, 1984.

15. Grinchenko, V.T. and Vovk, I.V., *Wave Problems of the Sound Scattering on Elastic Shells,* Kiev: Naukova Dumka, 1986.

16. Radlinski, R.P. and Simon, M.M., Scattering by multiple grating of compliant tubes, *J. Acoust. Soc. Am.*, vol. **72**, pp. 607–614, 1982.

17. Radlinski, R.P., Scattering by multiple grating of compliant tubes in viscoelastic layer, *J. Acoust. Soc. Am.*, vol. **85**, pp. 2301–2310, 1989.

18. Radlinski, R.P. and Janus, R.S., Scattering from two and three gratings of densely packed compliant tubes, *J. Acoust. Soc. Am.*, vol. **80**, pp. 1803–1809, 1986.

19. Adonina, A.I. and Shestopalov, V.P., Diffraction of electromagnetic waves obliquely incident on a plane metallic grating with a dielectric layer (Electromagnetic wave diffraction of a dielectric layer with a plane metallic grating), *Sov. Phys. Tech. Phys.*, vol. **8**, pp. 479–486, 1963.

20. Kudasheva, O.A. and Sevryugova, N.V., Sound radiation by an infinite periodic slotted array, *Sov. Phys. Acoust.*, vol. **22**, no. 3, pp. 216–219, 1976.

21. Astapenko, V.M. and Maliuzhnets, G.D., Diffraction of a plane acoustic wave by a dense periodic grating (Plane acoustic wave diffraction by dense periodic grating, using crimped

surface scattering with Neumann and mixed boundary conditions), *Akusticheskii Zhurnal*, vol. **16**, pp. 354–363, 1970. [in Russian]

22. Shenderov, E.L., Sound transmission through a perforated plate of finite thickness (Sound passage through rigid screen of arbitrary wave thickness with apertures, using linear algebraic equations), *Sov. Phys. Acoust.*, vol. **16**, pp. 244–249, 1970.
23. Tversky, V., On the scattering of waves by on infinite grating, *Ann. Ant. Prop.*, vol. **10**, no. 6, pp. 217–223, 1962.
24. Brigham, G.A., Libuha, J.J., and Radlinsski, R.P., Analysis of scattering from large planar gratings of compliant cylindrical shells, *J. Acoust. Soc. Am.*, vol. **61**, no. 1, pp. 48–59, 1977.
25. Korenev, B.G. and Rabinovich, I.M., (Eds.), *Reference Book on the Dynamics of Structures,* Moskow: Stroyizdat, 1972. [in Russian]
26. Shimanskiy, Y.A., (Ed.), *Handbook on Structural Mechanics of Ship,* Leningrad: Sudpromgiz, 1958. [in Russian]
27. Weaver, W., Timoshenko, S.P., and Young, D.H., *Vibration Problems in Engineering*, 5th Edition, New York: Wiley, 1990.
28. Brigham, G.A., In-plane vibrations of tapered oval rings, *J. Acoust. Soc. Am.*, vol. **54**, no. 2, pp. 451–460, 1973.
29. Sato, K., Free flexural vibrations of an elliptical ring in its plane, *J. Acoust. Soc. Am.*, vol. **57**, no. 1, pp. 113–115, 1975.
30. Lyamshev, L.M., *Reflection of Sound by Thin Plates and Shells in Liquid*, Moskow: Academy of Sciences of the USSR, 1955. [in Russian]
31. Junger, M.C., Sound scattering by a thin elastic shells, *J. Acoust. Soc. Am.*, vol. **24**, no. 4, pp. 562–567, 1952.
32. Hönl, H., Maue, A., and Westpfahl, K., *Theorie der Beugung*, Berlin: Springer, 1961.
33. Abramowitz, M. and Stegun, I., *Handbook of Mathematical Functions with Formulas, Graphs, and Mathematical Tables,* New York: Dover, 1972.
34. Bateman, H., *Higher Transcendental Functions*, New York: McGraw-Hill, 1953.
35. Ivanov, E.A., *Diffraction of Electromagnetic Waves by Two Bodies*, Minsk: Nauka i Technika, 1968.
36. Golyamina, I.P., (Ed.), *Ultrasound: Small Encyclopedia*, Moskow: Sovetskaya Encyclopedia, 1979. [in Russian]
37. Lapin, A.D., Sound attenuation in waveguides/Review, *Akusticheskii Zhurnal,* vol. **21**, pp. 337–350, 1975.
38. Shenderov, E.L., *Wave Problems in Hydroacoustics*, Leningrad: Sudostroenie, 1972. [in Russian]
39. Gontkevich, V.S., *Natural Vibrations of Plates and Shells: Handbook*, Kiev: Naukova dumka, 1964. [in Russian]
40. Kantorovich, L.V. and Krylov, V.I., *Methods of Approximate Solution of Partial Differential Equations,* Moskow: Gostehizdat, 1936. [in Russian]
41. Kantorovich, L.V. and Krylov, V.I., *Approximate Methods of Higher Analysis,* New York: Intersience Publisher, 1958.

42. Gradshteyn, I.S. and Ryzhik, I.M., *Table of Integrals, Series, and Products*, New York: Academic Press, 2014.

43. Glazunov, S.G., Vazhenin, S.F., and Zyukov-Batyrov, G.D., *Application of Titanium in the National Economy,* Kiev: Technika, 1975.

44. Skudrzyk, E., *Simple and Complex Vibratory Systems,* Pennsylvania: The Pensilvania State University Press, 1968.

45. Masolov, S.A. and Taranov, I.E., Diffraction of electromagnetic waves by spatial periodic grating, composed of bars of rectangular cross-section, *Radiotechnika i Electronika*, vol. **9**, no. 1, pp. 53–60, 1964. [in Russian]

46. Timoshenko, S.P. and Woinowsky-Krieger, S., *Theory of Plates and Shells*, New York: McGraw-Hill, 1959.

47. Meson, W.P. and Thurston, R.N., (Eds.), *Physical Acoustics*, New York: Academic Press, 1964.

48. Boriseiko, V.A., Grinchenko, V.T., and Ulitko, A.F., Relation of electroelastisity for piezoceramic shells of revolution, *Sov. Appl. Mech.,* vol. **12**, no. 2, pp. 126–131, 1976.

49. Tyutekin, V.V., Propagation of elastic waves in medium with cylindrical conduits, *Acoust. J.,* vol. **2**, no. 3, pp. 214–219, 1956. [in Russian]

CHAPTER 4

The Distinctive Features of Sound Diffraction by Some Complex Bodies

"You know my methods. Apply them!"

Arthur Conan Doyle, "The Hound of the Baskervilles"

4.1 SOUND WAVES DIFFRACTION BY PARABOLIC REFLECTOR

The parabolic reflector has unique wave properties: reflector can transform a plane incident wave into a spherical wave, which concentrates in its focus. And vice versa—it transforms a spherical wave (if there is a source of spherical wave in its focus) into a plane wave. The parabolic reflector is widely used in the modern physics and engineering due to these properties. For example, the parabolic reflector is used in the receiving antennas of satellite television in radio engineering. There are a lot of location stations of different application in the military and civil spheres. The receiving antennas with parabolic reflectors are used in such stations. The parabolic reflectors are also used in medical acoustics. They are used, in particular, in the powerful sound radiators, which are applied to break the stones by noninvasive method (so called lithotripsy method). In hydro-acoustics the parabolic reflectors are often used as a part of the antennas of hydro-acoustic stations, used for navigation and submarines detection.

We are concerned with the reflectors used in acoustics. There are a lot of quantitative data in literature (see, for example, [1–6]) for the cases, when the reflector material is ideal (absolutely non-transparent). Reflector diameter D exceeds by far the wave length λ. There is not enough data in literature about the reflectors made of real (partially transparent) materials (see, for example, work [7]). This is particularly so with the reflectors, the dimensions of which are comparable with the wave length.

The purpose of this paragraph is to find the quantitative relations between the acoustical reflector parameters in the form of paraboloid of rotation and the degree of its material transparency. The range of the wave dimensions of reflector is rather wide.

Figure 4.1 shows reflector cross section with focus in the point O and all the geometrical parameters are indicated. Plane wave Φ_0 is propagated along the axis z. We assume that the reflector thickness h is rather small (the conditions $h \ll D$ and $h \ll \lambda$ are fulfilled). The reflector surface is characterized by local impedance

$$Z = -i\omega M f(\theta), \tag{4.1}$$

ISBN: 978–1–56700–459–5

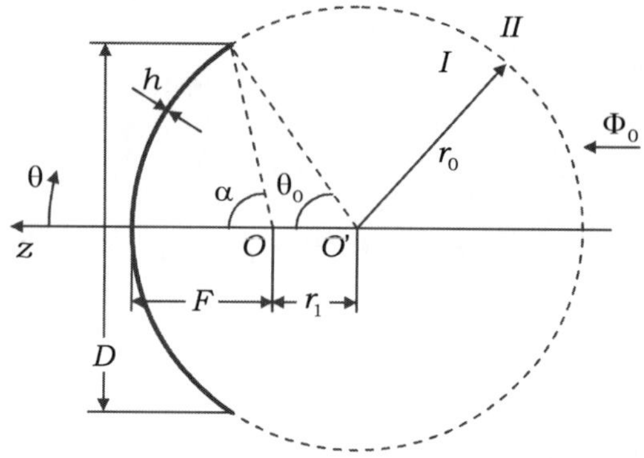

FIG. 4.1: Parabolic reflector

where $M = \rho_1 h$—area density, ρ_1—the density of reflector material, $f(\theta)$—the function of value M distribution on the reflector surface. Condition (4.1) characterizes the properties of a real reflector of small wave thickness, when the wave impedance of its material far exceeds the wave impedance of surrounding medium [8–10]. We confine ourselves to the consideration of so called shallow* reflector, when $\alpha < \pi/2$. The approximation of parabolic reflector by spherical reflector is possible then. And it is possible to use the spherical coordinate system with point O' as center (Fig. 4.1).

We will solve the problem by partial domains method. We partition the entire field existence domain into two domains. Domain I $(0 \le r \le r_0)$, and domain II $(r \ge r_0)$. We use the plane wave expansion by spherical wave functions [8,11]. We represent the velocity potentials as follows:

$$\Phi_I = \sum_{n=0}^{\infty} A_n j_n(kr) P_n(\cos\theta), \tag{4.2}$$

$$\Phi_{II} = \sum_{n=0}^{\infty} i^n (2n+1) j_n(kr) P_n(\cos\theta) + \sum_{n=0}^{\infty} B_n h_n^{(1)}(kr) P_n(\cos\theta). \tag{4.3}$$

It follows from condition (4.1) that each elementary reflector sector moves as a single unit under the influence of wave Φ_0. Consequently, the fields matching conditions on the surface $r = r_0$ will be as follows

$$\upsilon_{II} = \upsilon_I, \; 0 \le \theta \le \pi, \tag{4.4}$$

$$p_{II} = \begin{cases} p_I - Z\upsilon_I, & 0 \le \theta \le \theta_0, \\ p_I, & \theta_0 \le \theta \le \pi. \end{cases} \tag{4.5}$$

Here $\upsilon = -\partial\Phi/\partial r$, $p = -i\omega\rho\Phi$, ρ—surrounding medium density. Condition (4.4) allows us to relate the coefficients A_n and B_n:

*If $\alpha > \pi/2$, the reflector is called deep.

$$B_n = \frac{1}{h_n'^{(1)}(kr_0)}\left[A_n - i^n(2n+1)j_n'(kr_0)\right]. \tag{4.6}$$

The algebraization of functional Eq. (4.5) is performed using the properties of completeness and orthogonality of Legendre polynomial $P_n(\cos\theta)$, taking into account (4.6). We obtain the infinite system of algebraic equations of the second kind:

$$A_n + (kr_0)^2(2n+1)h_n'^{(1)}(kr_0)\frac{Z}{2\rho c}\sum_{m=0}^{\infty} A_m j'(kr_0)N_{nm} = (2n+1)i^n, \tag{4.7}$$

where $N_{nm} = \int\limits_{\cos\theta_0}^{1} f(\theta)P_n(\cos\theta)P_m(\cos\theta)d(\cos\theta)$, c—sound velocity in the medium. System (4.7) is used as initial to obtain the quantitative characteristics of the field, scattered by reflector.

Now we turn to the analysis of the parabolic reflector acoustic properties. All the numerical results correspond to the following parameters of the reflector and surrounding medium: $D/F = 2.8$, $r_0 = 2.14F$, $\alpha = 70°$, $f(\theta) = \text{const}$, $\rho = 10^3$ kg/m^3, $c = 1.5\times10^3$ m/s. The indicated value r_0 ensures (in the range of the considered frequencies) the absence of phase errors, which are more than $\lambda/12$ in the reflector opening. Such error is acceptable in practice [3] and it justifies the approximation of parabolic reflector by spherical reflector. The number of unknowns in system (4.7) is 36, when we solve this problem numerically.

Let us determine the influence of reflector transparency on its pressure gain coefficient K_p. Figure 4.2 shows the frequency dependencies of K_p for different values of M kg/m^2. Curve 4 [1] is given here for comparison, for an absolutely rigid (non-transparent) reflector. As we can see, there is a decrease of reflector gain coefficient with an increase of its transparency. For large M (curve 3), when the reflector transparency is close to

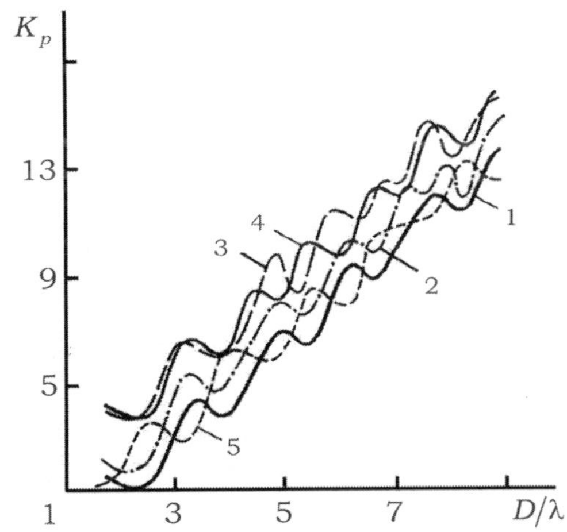

FIG. 4.2: Frequency dependency of reflector gain coefficient $D/\lambda = 2.6$; curves 1, 2, 3 correspond to $M = 106, 177, 1770$ kg/m^2; 4—absolutely rigid (non-transparent) reflector [1]; 5—$s = 2.1\times10^8$ kg/(ms^2)

zero, the coefficient K_p approaches the value of absolutely rigid (non-transparent) reflector (curve 4).

Now we turn to the analysis of the field in the area of reflector focus. Figure 4.3 shows the dependencies of normalized sound pressure (with respect to the pressure in focus) along the axis z. As we can see, for the large M, sound pressure distribution along the axis z is little different from the pressure calculated in work [1]. The situation is different when the reflector transparency becomes noticeable. The amplitude of sound pressure oscillation along the axis z increases with an increase of reflector transparency.

The influence of reflector transparency on the field in its focal region was considered above. The influence of reflector transparency on its directional properties is also of interest. We restate the initial problem to obtain the quantitative results. We assume that Q of point sources are in the area of reflector focus along the axis z. The representation of fields (4.2) and (4.3) in domains I and II should be changed. We exclude the series determining the field of plane wave in expression (4.3). Expression (4.2) should be supplemented with the series, which describe the fields of point sources [11]. Taking into account this changes, the fields in domains I and II can be represented as follows:

$$\Phi_1 = \sum_{n=0}^{\infty} A_n j_n(kr) P_n(\cos\theta) + ik \sum_{q=1}^{Q} \xi_q e^{i\psi_q} \sum_{n=0}^{\infty} (2n+1) j_n(kr^{(q)}) h_n^{(1)}(kr) P_n(\cos\theta), \quad (4.8)$$

$$\Phi_2 = \sum_{n=0}^{\infty} B_n h_n^{(1)} P_n(\cos\theta). \quad (4.9)$$

Here $r^{(q)}$—is radial coordinate of the qth point sound source (the sources are on the axis z, angular coordinate for all the sources $\theta^{(q)} \equiv 0$); ξ_q and ψ_q—amplitude and excitation phase of the qth point source.

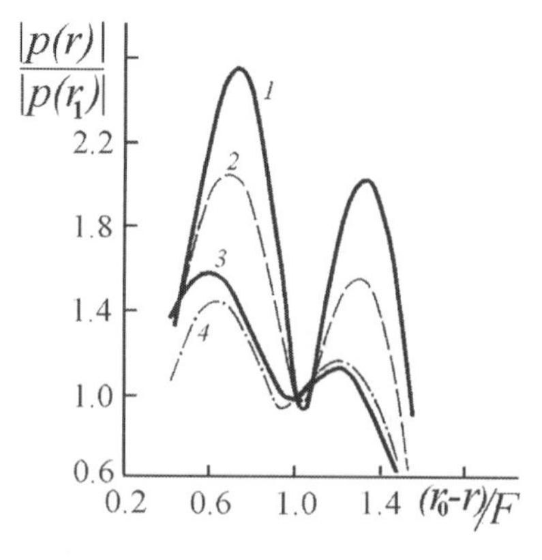

FIG. 4.3: Sound pressure distribution along the axis z

At first we consider the situation when there is only one point source in focus. Figure 4.4 shows the curves, giving the general idea of reflector transparency influence on its directivity pattern $R(\theta)$. The analysis of calculated data allows us to find some peculiarities of the value of $R(\theta)$ behavior when the frequency and reflector transparency change. We partition all the angle range $\pi - \theta$ into domains, to simplify the analysis. We obtain the following domains: main lobe domain, first side lobe domain, side radiation domain and geometrical shadow domain.

Let us consider the main lobe domain. The calculations show that starting from $D/\lambda \geq 2.5$, the width $\theta_{0.7}$ of the main lobe depends almost not at all on the reflector transparency (the value of $\theta_{0.7}$ is determined at level 0.7 by pressure). When $D/\lambda < 2.5$, an increase of reflector transparency leads to the main lobe broadening. For example, when $D/\lambda = 1.75$ the ratio of values $\theta_{0,7}$, calculated for $M = 106$ and $M = 1770$ kg/m^2, is 1.5.

The level of the first side lobe σ_1 increases with a decrease of M in the domain of the first side lobe, as it is seen from Fig. 4.5. Frequency dependency σ_1 has an oscillating character (it can be explained by the presence of interference effect between the

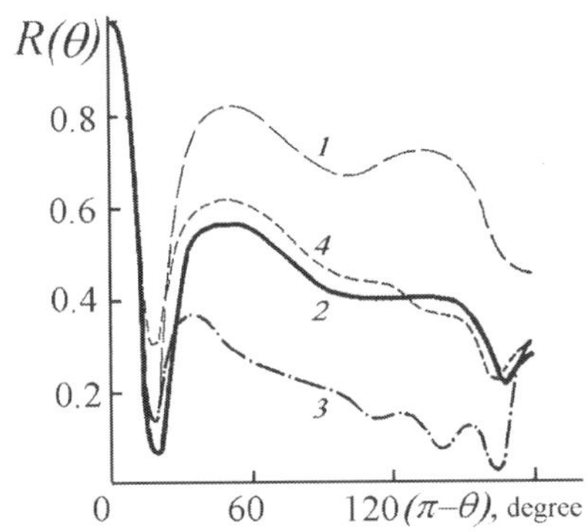

FIG. 4.4: Directivity pattern of reflector with one point source, which is situated in focus, $D/\lambda = 2.6$: curves 1, 2, 3—calculation for $M = 106, 177, 1770$ kg/m^2; 4—experiment for $M = 177$ kg/m^2

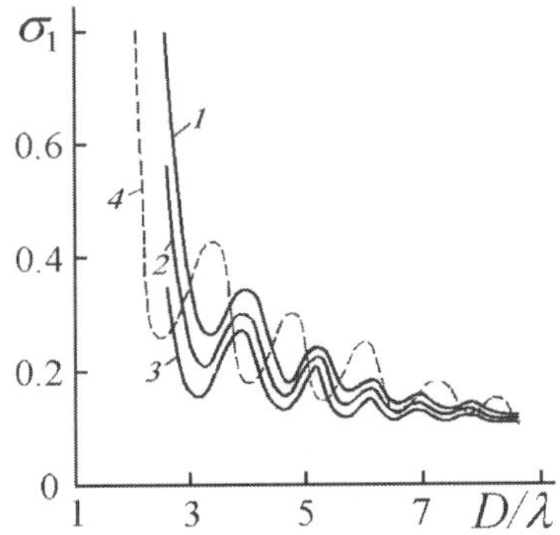

FIG. 4.5: Frequency dependency of the first side lobe level of directivity pattern; curves 1, 2, 3 correspond to $M = 106, 177, 1770$ kg/m^2; 4—$s = 2.1 \times 10^8$ kg/(ms^2)

incident and reflected waves). The oscillation amplitude decreases with a decrease of D/λ, because of the relative reduction of incident wave contribution. The middle level σ_1 approaches the value ~ 0.105, which coincides with the data in work [1].

The change of $R(\theta)$ is relatively small in the side radiation domain (angles $\pi - \theta$ about $90°$). That is why we confine ourselves to its value determination when $\pi - \theta = 90°$, which is denoted as σ_2. Figure 4.6 shows that value σ_2 reduces with an increase of frequency and a decrease of reflector transparency.

The change of $R(\theta)$ is of complex character in the domain of reflector geometrical shadow. It is caused by the interference effect between the wave passing through the reflector and the wave diffracted by its edge. The field estimation is done by mean value of $R(\theta)$ in the domain of shadow:

$$\sigma_3 = \frac{1}{1 - \cos\alpha} \int_0^{\alpha} R(\theta) \sin\theta d\theta.$$

The data in Fig. 4.7 determine the dependency of σ_3 on reflector frequency and transparency. Note that curve 3 characterizes the field level, caused by diffraction on the reflector edge.

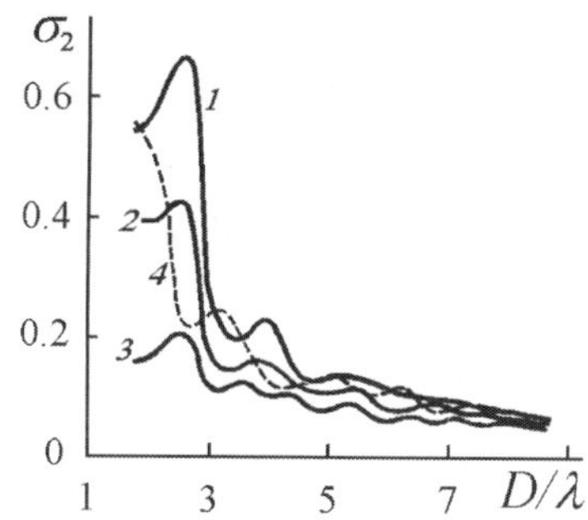

FIG. 4.6: Frequency dependency of the level of sidelight radiation: curves 1, 2, 3 correspond to $M = 106, 177, 1770$ kg/m^2; 4—$s = 2.1 \times 10^8$ kg/(ms^2)

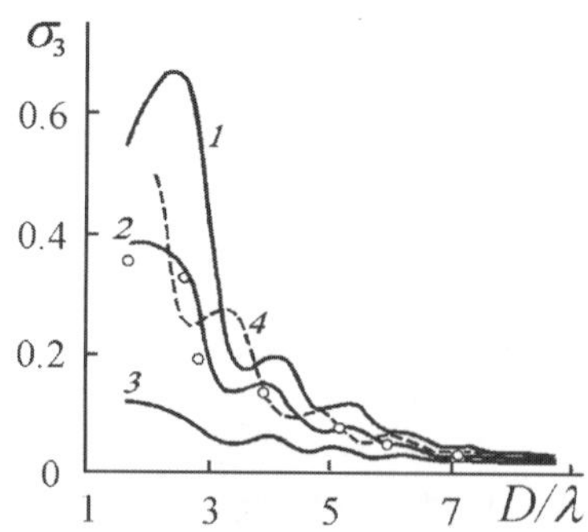

FIG. 4.7: Frequency dependency of radiation level in the domain of shadow: curves 1, 2, 3 correspond to $M = 106, 177, 1770$ kg/m^2; 4—$s = 2.1 \times 10^8$ kg/(ms^2); circles—experiment for $M = 177$ kg/m^2

The experimental tests were performed using the model of reflector in order to confirm the applicability of obtained results for the evaluation of acoustic properties of real reflectors. The obtained results were compared with the calculation data. The dummy of reflector was made of metal with density $7,1 \times 10^3$ kg/m^3 and it has the following parameters: $D/F = 2.8$, $\alpha = 70°$, $f(\theta) = \text{const}$, $M = 177$ kg/m^2. Piezoceramic sphere is the source of sound waves. It is situated in reflector focus. The sphere diameter does not exceed 0.07λ on the upper boundary of the considered frequency range. The reflector with the source of sound waves was in water. Figure 4.4 (curve 4) shows experimental directivity pattern, and Fig. 4.7—experimental values of σ_3 (indicated by circles). Comparing the curves 2 and 4 in Fig. 4.4, and the curve 2 with the curve marked with circles, in Fig. 4.7, we can see that the calculation data coincide with the experimental data.

We consider now the simplest grating, which is of practical interest. It is a linear grating, made of two point nondirective sound sources ($Q = 2$). We assume that the amplitudes and excitation phases of these sources are equal $\xi_1 = 1, \xi_2 = -1, \psi_1 = 0$, $\psi_2 = \beta kd$, respectively, where $d = (r^{(1)} - r^{(2)})$—the distance between the sources (assume that $r^{(1)} > r^{(2)}$), and β—real parameter, which varies within the limits from 0 to 1. When $\beta = 0$ we have dipole [12], which is characterized by directivity pattern in the form of "eight" when $d < \lambda/2$ (radiation maxima are directed along the axis z). If $\beta = 1$, the directivity pattern (when $d < \lambda/4$) takes the form of cardioids. The maximum of cardioids is in the positive z-axis direction and zero is in the opposite direction. The directivity pattern of two point sources has some intermediate form between "eight" and cardioids within the range $0 < \beta < 1$. Concentration factor is a maximum when $\beta = 0.33$. The directivity pattern can be considered as "optimal."

The calculations of the far field of reflector were performed for the indicated above values $\beta = 0, 0.33, 1$. Figures 4.8 and 4.9 show the results of this calculation, when the sources are on the axis z, in the points with coordinates $r^{(1)} = 1.157F$ and $r^{(2)} = 1.144F$.

We turn now to Fig. 4.8, which shows the directivity patterns of reflector with two sources in focal area for different parameter β values. Comparative analysis of the curves

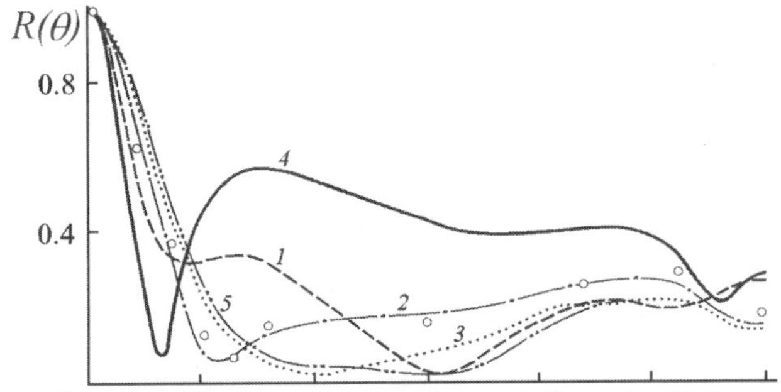

FIG. 4.8: Directivity patterns of reflector when $M = 177$ kg/m^2: curves 1, 2, 3 correspond to $\beta = 0, 1, 0.33$ when $Q = 2$; 4—one source in focus; 5—$Q = 6$; circles—experiment when $Q = 2$, $\beta = 1$

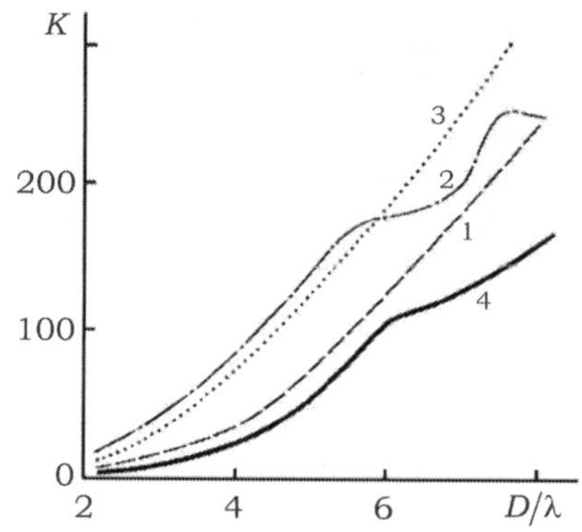

FIG. 4.9: Frequency dependencies of concentration coefficient: curves 1, 2, 3 for $Q = 2$, $\beta = 0,\ 1,\ 0.33$; 4—one sound source in focus

1, 2, 3, and 4 shows that when we use only two sources, which are correctly spaced and phased, it allows us to suppress sound radiation almost in the entire area of angles θ behind the main lobe of directivity pattern. We can also observe the other singularities of the directivity pattern of reflector for different relations of the phases of sources. When the sources are in anti-phase and they form the dipole ($\beta = 0$), the reflector directivity pattern has the low level in the area of angles θ, close to 90°. When $\beta = 1$ we can observe the low radiation level just behind the main lobe of directivity pattern ($25° \leq \theta \leq 45°$). When the value is "optimal" $\beta = 0.33$ the area of angles with the low radiation level, expands considerably and is within the limits $40° \leq \theta \leq 90°$.

So, we can draw a conclusion: when the two simple sources are on the axis of reflector in focal area and when their excitation is selected correctly, we can steer the directivity pattern of reflector.† The general level of radiation can be reduced in the area of angles θ behind the main lobe, and the radiation can be suppressed in a particular direction.

When analyzing the concentration coefficient K dependency on the reflector wave dimensions (Fig. 4.9), we can conclude that the nature of excitation of the sources also influences the concentration coefficient dependency on the reflector wave dimensions. The reflector with one sound source has less concentration coefficient than the reflector with two sound sources, especially when the parameter values are $\beta = 1$ and $\beta = 0.33$. Consequently, we can control the value of the concentration coefficient of reflector, when selecting the excitation character of the sources.

The simplest cases of the sources excitation are shown above. It is shown that we can influence the directional properties of reflector even if the number of sources is minimum ($Q = 2$). As the numerical experiments show, if we increase the number of sources and select the phases, we will be able to influence the directivity pattern of reflector more efficiently.

Figure 4.8 shows the directivity pattern (curve 5) for the case when there are six sources ($Q = 6$), situated on the axis of reflector in the points with coordinates $r^{(1)} =$

†Calculated and experimental data comparison in Fig. 4.8 validates this conclusion (the experiment was carried out for $\delta/\lambda = 0.04$ and $\delta/\lambda = 0.0154$, where δ—sources diameter).

$0.89F$, $r^{(2)} = 0.99F$, $r^{(3)} = 1.09F$, $r^{(4)} = 1.19F$, $r^{(5)} = 1.29F$, $r^{(6)} = 1.39F$. These points are characterized by the following amplitudes and excitation phases $\xi_1 = 1$, $\xi_2 = 1.17$, $\xi_3 = 1.31$, $\xi_4 = 1.35$, $\xi_5 = 1.04$, $\xi_6 = 0.83$, $\psi_1 = 0°$, $\psi_2 = 61°$, $\psi_3 = 126°$, $\psi_4 = 204°$, $\psi_5 = 255°$, $\psi_6 = 325°$. The comparison of the curves shows that, when there are a lot of sources, the area of angles θ, with the low level of radiation, can be expanded. It is to be noted that, when the surface density M decreases the level of the values of directivity pattern in the area of angles $30° \leq \theta \leq 180°$ increases. It is a natural phenomenon because the transparency of reflector increases.

Finally, it is interesting to consider the case, when the impedance is of elastic character $Z = isf(\theta)/\omega$. This is common for the thin layers of gas in liquid [8,9]. Here s [kg/(ms^2)]—surface elasticity. When investigating this case, conditions (4.4) and (4.5) on the surface $r = r_0$ should be replaced with the equalities

$$p_I = p_{II}, \quad 0 \leq \theta \leq \pi, \quad \upsilon_{II} = \begin{cases} \upsilon_I - \dfrac{p_I}{Z}, & 0 \leq \theta \leq \theta_0, \\ \upsilon_I, & \theta_0 \leq \theta \leq \pi. \end{cases}$$

The calculations of the values K_p and $R(\theta)$ are performed when $s = 2.1 \times 10^8$ kg/(ms^2) and $f(\theta) = \text{const}$ (this value of s provides the same sound transmission coefficient as in the case when $M = 106$ kg/m^2). Curves 1 and 5 in Fig. 4.2 correspond to the different properties of reflector surface. As we can see, the maximum oscillations of curve 1 correspond to the minimums of curve 5. The behavior of value σ_1, and the behavior of σ_2 to some extent, has the similar characteristic (see, Figs. 4.5 and 4.6). Comparing the data in Figs. 4.5–4.7, we can see that if the impedance has elastic character the general level of value σ_1 is a little higher. The values σ_3 are lower than in the case, when the impedance has the character of mass.

4.2 SOUND DIFFRACTION BY THE WEDGE OF FINITE DIMENSIONS

The problems of waves scattering by bodies of varying configuration are of significant theoretical and practical importance. The solution results may be the basic for the construction of location systems of the objects detection and their classification. Waves scattering by wedge bodies is of interest, because the form of such bodies is similar to the forms of some parts of the aviation or space objects. The problem of the waves scattering by wedge bodies of finite sizes is considered in this paragraph.

Let us consider the plane problem of scattering by wedge object (Fig. 4.10). We assume that the object has infinite extension along the axis, perpendicular to the figure plane. The object is a finite wedge with an angle $2\theta_1$. The sides of the wedge are closed by arc AB with radius a. The surfaces of the object are acoustically rigid. The wedge object is in an ideal medium with density ρ and sound velocity c. Let us introduce the polar coordinate system $rO\theta$ with the center in the angle point of the wedge.

We partition all the space of sound field existence into two domains: I—outside the circle of radius a, i.e., $r \geq a$, $0 \leq \theta \leq 2\pi$; II—sector of a circle of radius a, namely: $0 \leq r \leq a$, $\theta_1 \leq \theta \leq 2\pi - \theta_1$.

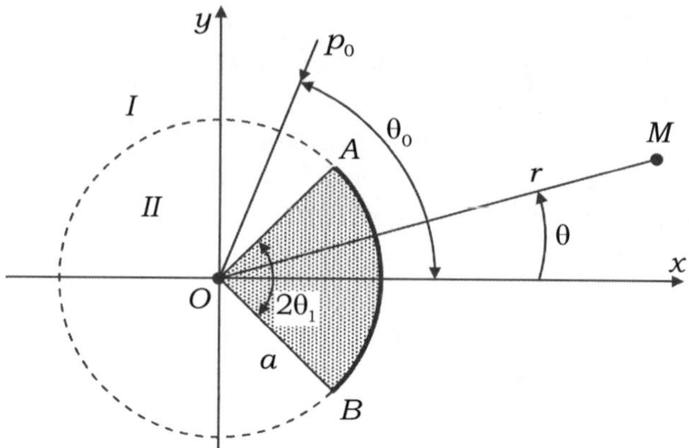

FIG. 4.10: Wedge object example

Let a plane harmonic wave with a unit pressure amplitude be incident on the wedge object in domain I

$$p_0 = \exp(-ikr\cos(\theta - \theta_0)). \tag{4.10}$$

The sign "minus" in the exponential index points out that the direction of the plane wave propagation is opposite to the direction of growth of radial coordinate r. Angles θ_0 and θ determine the angular direction of the wave incidence and the direction of the observation point.

The field of incident wave (4.10) can be written as follows [10]

$$p_0 = \sum_{n=0}^{\infty} (-i)^n \varepsilon_n J_n(kr)\cos(n(\theta - \theta_0)), \tag{4.11}$$

where $\varepsilon_0 = 1$, $\varepsilon_n = 2$, if $n > 0$, $J_n(kr)$—Bessel function of the first kind.

Scattered field appears when the plane wave interacts with the wedge object. We represent it in the form of superposition of cylindrical running waves $H_n^{(1)}(kr)\cos(n\theta)$, $n = 0, 1, 2, \ldots$ and $H_n^{(1)}(kr)\sin(n\theta)$, $n = 1, 2, 3, \ldots$. The angle functions $\cos(n\theta)$ and $\sin(n\theta)$ are necessary, because the angle of incidence θ_0 is arbitrary. Pressure field in domain I is as follows:

$$p_I = \sum_{n=0}^{\infty} (-i)^n \varepsilon_n J_n(kr)\cos(n(\theta - \theta_0)) + \sum_{n=0}^{\infty} A_n \frac{H_n^{(1)}(kr)}{H_n'^{(1)}(ka)}\cos(n\theta) + \sum_{n=1}^{\infty} B_n \frac{H_n^{(1)}(kr)}{H_n'^{(1)}(ka)}\sin(n\theta). \tag{4.12}$$

The field in domain II we represent as the superposition of standing waves:

$$p_{II} = \sum_{n=0}^{\infty} C_n \frac{J_{\alpha_n}(kr)}{J_{\alpha_n}(ka)}\cos(\alpha_n(\theta - \theta_1)), \tag{4.13}$$

when the values of α_n are determined from the boundary conditions on the rigid wedge borders $\partial p_{II}/\partial\theta = 0$ when $\theta = \theta_1$ and $\theta = 2\pi - \theta_1$. Consequently,

$$\alpha_n = \frac{\pi n}{2(\pi - \theta_1)}, \quad n = 0, 1, 2, ... \tag{4.14}$$

Let us write matching conditions on the interface of partial domains I and II (Fig. 4.10):

$$p_I = p_{II}, \ r = a, \ \theta = [\theta_1, \ 2\pi - \theta_1], \tag{4.15}$$

$$\frac{\partial p_I}{\partial r} = \begin{cases} \dfrac{\partial p_{II}}{\partial r}, & r = a, \quad \theta = [\theta_1, \ 2\pi - \theta_1], \\ 0, & r = a, \qquad |\theta| \le \theta_1. \end{cases} \tag{4.16}$$

We algebraize functional Eq. (4.15) due to the orthogonality of the system of functions $\cos(\alpha_n(\theta - \theta_1))$, $n = 0, 1, 2, ...$ on the interval $\theta = [\theta_1, 2\pi - \theta_1]$. We algebraize functional Eq. (4.16) two times: the first time—using the system of functions $\cos(n\theta)$, $n = 0, 1, 2, ...$, the second time—using the system $\sin(n\theta)$, $n = 1, 2, ...$ Both systems are orthogonal on the interval $\theta = [0, 2\pi]$. We obtain an infinite system of linear algebraic equations of the second kind in unknown coefficients A_n, B_n, C_n.

An infinite system of equations was solved by reduction method. To estimate the accuracy of the fulfillment of conditions (4.15) and (4.16). We find the discrepancy as the ratio of difference modulus of the field characteristics (pressure or particle velocity) to the left and to the right from the boundary between the partial domains to the plane wave amplitude (4.10). For example, pressure discrepancy $\delta_p = |p_I - p_{II}| / |p_0|$.

For example, Fig. 4.11 shows the pressure discrepancy graphs δ_p (solid line) and particle velocity discrepancy δ_υ (dashed line) along the circle with radius a: $ka = 15$, $\theta_1 = 45°$, $\theta_0 = 90°$. When calculating, 40 coefficients A_n, B_n, and C_n are kept. Angular range $\theta = [-\theta_1, \theta_1]$ corresponds to the rigid object surface (Fig. 4.10).

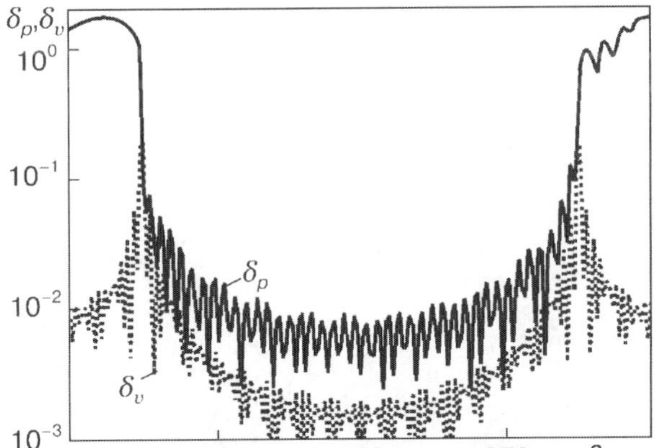

FIG. 4.11: Pressure discrepancy curves δ_p (solid curve) particle velocity discrepancy δ_υ (dashed) along the circle of radius a (Fig. 4.10): $ka = 15$, $\theta_1 = 45°$, $\theta_0 = 90°$

The discrepancy increases δ_υ in the vicinity of edges of wedge object $\theta = \pm\theta_1$, $r = a$. It is a logical result, because there is a singularity of particle velocity. When $ka = 15$, $\theta_1 = 45°$ the rigid boundaries of the object have the considerable wave dimensions. Consequently, pressure amplitude on this surface is approximately equal to the double amplitude of incident wave. If the discrepancy is $\delta < 0.1$, the fulfillment of boundary conditions on the interface between partial domains is satisfactory.

It should be noted that the curves in Fig. 4.11 are of "irregular" character and it means that the calculations are reliable. The limitation on amount of equations in the infinite system introduces an error when fulfilling the conditions for high harmonics.

When slowing the problem of sound scattering by an obstacle, the far sound field is usually explored. Scattered field here is a spherical expanding wave. It corresponds to the cylindrical expanding wave in the plane problem under consideration. According to (4.12), complex amplitude of scattered wave is as following

$$p_s(r,\theta,\theta_0) = A_0 \frac{H_0^{(1)}(kr)}{H_0^{\prime(1)}(ka)} + \sum_{n=1}^{\infty} \frac{H_n^{(1)}(kr)}{H_n^{\prime(1)}(ka)} [A_n \cos(n\theta) + B_n \sin(n\theta)]. \quad (4.17)$$

Taking into account the Hankel functions asymptotic behavior when $kr \to \infty$ formula (4.17) can be written as follows:

$$p_s(r,\theta,\theta_0) = \sqrt{\frac{2}{\pi kr}} \exp\left(ikr - i\frac{\pi}{4}\right) \quad (4.18)$$

$$\times \left[\frac{A_0}{H_0^{\prime(1)}(ka)} + \sum_{n=1}^{\infty} \frac{\exp(-in\pi/2)}{H_n^{\prime(1)}(ka)} (A_n \cos(n\theta) + B_n \sin(n\theta)) \right].$$

The intensity in the far field of scattered wave is determined by formula $I_{rs} = |p_s|^2/(2\rho c)$. Then, according to (4.18) we have

$$I_{rs}(r,\theta,\theta_0) = \frac{2I_0}{\pi kr} L(\theta,\theta_0) L^*(\theta,\theta_0), \quad (4.19)$$

where $I_0 = 1/(2\rho c)$—intensity of the plane incident wave of unit amplitude (4.10), asterisk—complex conjugation

$$L(\theta,\theta_0) = \frac{A_0}{H_0^{\prime(1)}(ka)} + \sum_{n=1}^{\infty} \frac{\exp(-in\pi/2)}{H_n^{\prime(1)}(ka)} (A_n \cos(n\theta) + B_n \sin(n\theta)). \quad (4.20)$$

The power of scattered wave, when integrating (4.20) over the circle of radius r is determined by relation

$$P_s(\theta_0) = \int_0^{2\pi} I_{rs}(r,\theta,\theta_0)\, r d\theta = \frac{2I_0}{\pi k} \int_0^{2\pi} L(\theta,\theta_0) L^*(\theta,\theta_0)\, d\theta. \quad (4.21)$$

The full cross-section of scattering is determined by the ratio of scattered wave power to the intensity of incident plane wave

$$\sigma_s(\theta_0) = \frac{P_s(\theta_0)}{I_0} = \frac{2}{\pi k} \int\limits_0^{2\pi} L(\theta, \theta_0)\, L^*(\theta, \theta_0)\, d\theta. \qquad (4.22)$$

Substituting (4.20) into (4.22), we obtain the following formula for the non-dimensional value of the full cross-section of dissipation:

$$\frac{\sigma_s(\theta_0)}{a} = \frac{2}{ka}\left[2\left|\frac{A_0}{H_0^{\prime(1)}(ka)}\right|^2 + \sum_{n=1}^{\infty} \frac{|A_n|^2 + |B_n|^2}{\left|H_n^{\prime(1)}(ka)\right|^2}\right]. \qquad (4.23)$$

It is of interest to find the positional cross-section of dissipation $\sigma(\theta, \theta_0)$. This value is equal to the ratio of power of dummy of non-directional cylindrical radiator with intensity equal to the intensity of scattered wave in the given direction θ, to the intensity of incident plane wave. According to the definition $\sigma(\theta, \theta_0)$, taking into account Eq. (4.19), we obtain the relation for the non-dimensional value of positional cross-section:

$$\frac{\sigma(\theta, \theta_0)}{a} = \frac{1}{a}\frac{2\pi r I_{rs}(r, \theta, \theta_0)}{I_0} = \frac{4}{ka} L(\theta, \theta_0)\, L^*(\theta, \theta_0). \qquad (4.24)$$

If we take angle $\theta = \theta_0$ in Eq. (4.24) we obtain the cross-section of back-scattering $\sigma_L(\theta_0) = \sigma(\theta = \theta_0, \theta_0)$.

We turn now to the analysis of numerical results. Figure 4.12(a) shows the values of the cross-section of dissipation σ_s, σ, σ_L as the functions of the angle of incidence of the plane wave θ_0 on the wedge object; $ka = 15$ (or $a \approx 2.4\lambda$), $\theta_1 = 45°$. The object has the considerable wave dimensions at such parameters. The arc AB and the wedge boundaries OA, OB are approximately equal in wave dimensions. As we can see, the full cross-section of dissipation σ_s (curve 1) almost does not depend on the angle of wave incidence θ_0. It is caused by the given dimensions of the object. The positional cross-section of scattering $\sigma(\theta = 0, \theta_0)$ (curve 2) characterizes the energy of scattered wave in the direction $\theta = 0$ as the function of the angle of incidence of plane wave θ_0. There is a "shade" lobe (the angle of incidence $\theta_0 = 180°$). The cross-section of back scattering σ_L (curve 3) is of specific interest.

When the angles of incidence are $\theta_0 \approx 0, ..., 70°$, curve 3 is a flat curve. It corresponds to the wave reflection from surface AB (Fig. 4.10) with large wave dimension (the arc length AB is about 3.75λ). The range of angles $\theta_0 \approx 70, ..., 180°$ may be defined as the reflection of the wave from the flat area OA of the wedge object surface. We can see here the considerable "irregularity" of curve 3. When the angle of incidence is $\theta_0 \approx 135°$ we have a so called "fleck." It means that there is a sharp increase of the reflected wave amplitude in the opposite direction. It is caused by the normal incidence of the plane wave on the boundary OA of large wave dimension.

We reduce the wave dimension of object, $ka = 0.7$; $\theta_1 = 45°$, Fig. 4.12(b). As we can see the values of the cross-section of scattering σ_s, σ, σ_L decrease and three curves become flat.

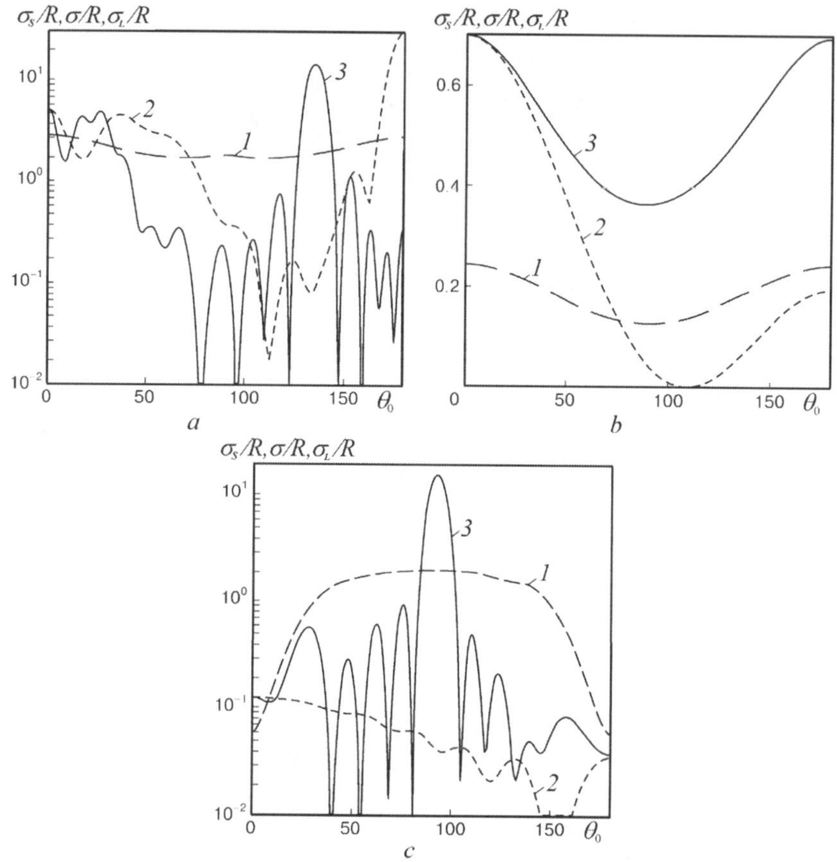

FIG. 4.12: Dependency of cross-section of scattering on the plane wave angle of incidence θ_0; 1—σ_s/a, 2—$\sigma(\theta = 0, \theta_0)/a$, 3—$\sigma_L/a$: (a)—$ka = 15$, $\theta_1 = 45°$, (b)—$ka = 0.7$, $\theta_1 = 45°$, (c)—$ka = 15$, $\theta_1 = 3°$

Figure 4.12(c) shows the graphs for the object with large value $ka = 15$, but with small angle $\theta_1 = 3°$. The geometry of object changes. The wave length of arc AB here is approximately ten times smaller than the value of ka and is approximately 0.24λ. As we can see, curve 1 for the full cross-section of scattering slopes steeply down in the vicinity of angles $\theta_0 = 0$ and $\theta_0 = 180°$. Positional cross section of scattering (curve 2) is less than in Fig. 4.12(a). The cross-section value of back scattering (curve 3) has pronounced peak in the vicinity of the angle of incidence $\theta_0 = 90°$.

Thus, the cross-section of scattering depends on the object geometry, wave dimensions of the surfaces, which form it, and the direction of wave incidence.

4.3 SOUND DIFFRACTION BY CORNER

The corner is widely used in physics and engineering. If the symmetry axis of corner coincides with the direction of the flat incident wave, it reflects this wave effectively in the opposite direction. This property is widely used in architectural acoustics for the sound field distribution correction in theater auditoriums and concert halls. The corner reflectors are widely used in radiotechnics and optics for precise measurements for topographical mapping, in marine navigation and so on. American astronauts mounted the mirror corner reflectors on the moon in 1969. If we radiate these reflectors by laser from the earth surface and receive the reflected ray we can find the distance from the earth to the moon. The corner reflectors are also used in corner antennas [3,4,7]. The source of cylindrical waves (or the series of sources) is placed inside the corner opening. The corner reflector transforms the cylindrical waves of the sources into the plane waves, which are radiated by corner antenna.

The indicated above corner properties are considered in this paragraph.

4.3.1 Problem Solution Construction

Let us consider the plane problem of the wave incidence on the corner of finite dimensions with the opening angle $2\theta_1$, Fig. 4.13. We assume that the surfaces of corner are acoustically rigid. The corner is immersed in the ideal medium with density ρ and sound velocity c. To construct the problem solution, we introduce the polar coordinate system $rO\theta$. According to the partial domains method all the space of sound field existence can be partitioned into three domains: I—exterior of the circle of radius a, i.e., $r \geq a$, $0 \leq \theta \leq 2\pi$; II—sector $0 \leq r \leq a$, $0 \leq \theta \leq 2\theta_1$; III—sector $0 \leq r \leq a$, $2\theta_1 \leq \theta \leq 2\pi$.

Let us consider two variants of the wave incidence on the corner: the first—the incidence of plane wave p_0 at an angle θ_0, the second—the incidence of cylindrical wave, created by the source S. The source is situates at an angle θ_1 at a distance R from the origin of coordinates O (Fig. 4.13).

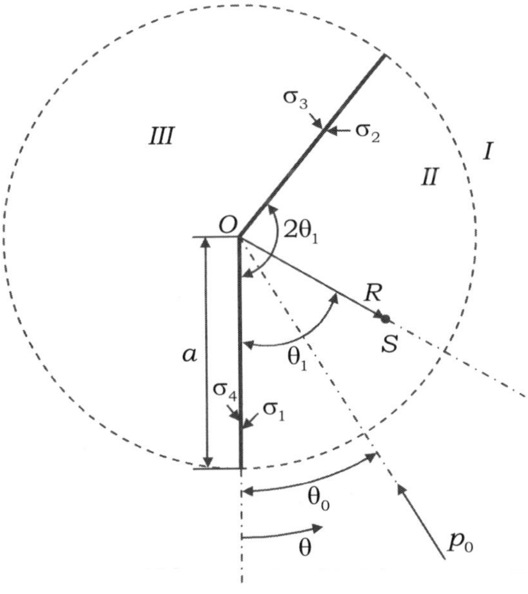

FIG. 4.13: Problem geometry

For the case of plane wave p_0 incidence (see Eq. (4.11) in the previous paragraph), we write the field in domain I as follows:

$$p_I = p_0 + \sum_{n=0}^{\infty} A_n \frac{H_n^{(1)}(kr)}{H_n'^{(1)}(ka)} \cos(n\theta) + \sum_{n=1}^{\infty} B_n \frac{H_n^{(1)}(kr)}{H_n'^{(1)}(ka)} \sin(n\theta). \quad (4.25)$$

We represent the field in domains II and III as the superposition of standing waves:

$$p_{II} = \sum_{n=0}^{\infty} C_n \frac{J_{\alpha_n}(kr)}{J'_{\alpha_n}(ka)} \cos(\alpha_n \theta), \quad (4.26)$$

$$p_{III} = \sum_{n=0}^{\infty} D_n \frac{J_{\beta_n}(kr)}{J'_{\beta_n}(ka)} \cos(\beta_n(\theta - 2\theta_1)), \quad (4.27)$$

where α_n and β_n are determined from the boundary conditions on the rigid boundaries of the wedge. Consequently

$$\alpha_n = \frac{\pi n}{2\theta_1}, \quad \beta_n = \frac{\pi n}{2(\pi - \theta_1)}, \quad n = 0, 1, 2, ...$$

If we consider the incidence of cylindrical wave (the source of cylindrical wave S is situated in domain II), we should use the known representation of the source field in wedge domain [17]:

$$p_0 = \frac{\pi i}{4\theta_1} \begin{cases} \sum_{n=0}^{\infty} \varepsilon_m J_{\alpha_n}(kr) H_{\alpha_n}^{(1)}(kR) \cos(\alpha_n\theta_0) \cos(\alpha_n\theta), & r < R, \\ \sum_{n=0}^{\infty} \varepsilon_n J_{\alpha_n}(kR) H_{\alpha_n}^{(1)}(kr) \cos(\alpha_n\theta_0) \cos(\alpha_n\theta), & R < r < a, \end{cases} \quad (4.28)$$

In this case, the source field (4.28) should be added to the right part of Eq. (4.26), and value p_0 should be excluded in Eq. (4.25).

We write the matching conditions on the boundary between the partial domains I and II, III (Fig. 4.13):

$$\frac{\partial p_I}{\partial r} = \begin{cases} \frac{\partial p_{II}}{\partial r}, & r = a, \quad \theta = [0, 2\theta_1], \\ \frac{\partial p_{III}}{\partial r}, & r = a, \quad \theta = [2\theta_1, 2\pi], \end{cases} \quad (4.29)$$

$$p_I = p_{II}, \quad r = a, \quad \theta = [0, 2\theta_1], \quad (4.30)$$

$$p_I = p_{III}, \quad r = a, \quad \theta = [2\theta_1, 2\pi]. \quad (4.31)$$

We perform the algebraization of functional Eq. (4.29) two times: the first time—we use the system of functions $\cos(n\theta)$, $n = 0, 1, 2, ...$, the second time—we use the system $\sin(n\theta)$, $n = 1, 2, ...$ both systems of functions are orthogonal on the interval $\theta =$

$[0, 2\pi]$. We perform the algebraization of functional Eq. (4.30) due to the orthogonality of the system of functions $\cos(\alpha_n \theta)$, $n = 0, 1, 2, ...$ on interval $\theta = [0, 2\theta_1]$. We use the system of functions $\cos(\beta_n(\theta - 2\theta_1))$, $n = 0, 1, 2, ...$ on the interval $\theta = [2\theta_1, 2\pi]$ for Eq. (4.31). We obtain the infinite system of linear algebraic equations of the second kind in unknown coefficients A_n, B_n, C_n, D_n. The infinite system of equations is solved by reduction method.

4.3.2 The Singularities of Plane Wave Reflection from the Corner

It was considered above that when angle is $2\theta_1 = 90°$, the corner has the property to reflect the incident wave in the opposite direction in relatively wide range of incident angle θ_0, (Fig. 4.13). Figure 4.14 illustrates it.

The difference in reflectance properties of the corner and the band is well seen here. The band efficiently reflects the waves when the directive beam of the incident wave is perpendicular to the surface of the band. The angles θ_0 area of effective reflection narrows rapidly with an increase of frequency. The band behaves like a mirror. The behavior of the corner is quite different in contrast to the band. The corner reflects the waves in the direction of their source in relatively wide range of angles θ_0. This property remains almost constant with an increase of frequency.

It is interesting to compare the structure of the field in the vicinity of the band and corner. Figure 4.15 shows that the field in the vicinity of the corner is more concentrated

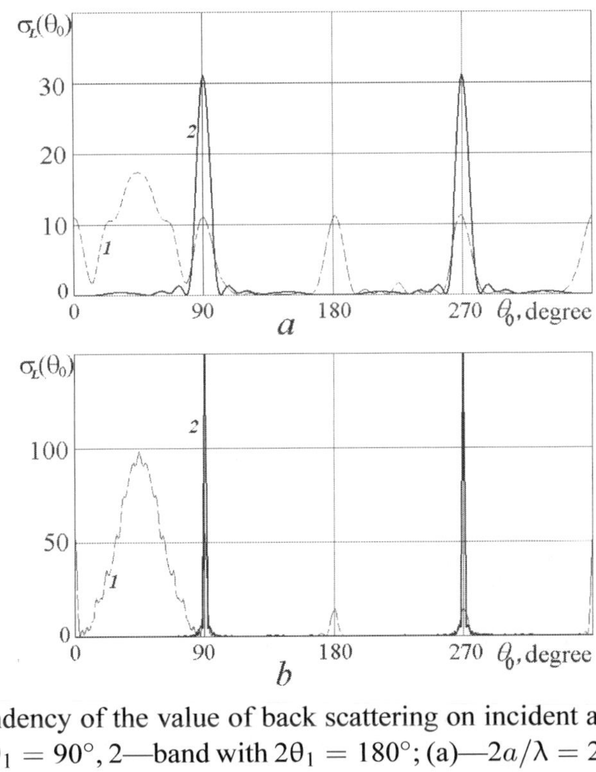

FIG. 4.14: Dependency of the value of back scattering on incident angle of plane wave: 1—corner with $2\theta_1 = 90°$, 2—band with $2\theta_1 = 180°$; (a)—$2a/\lambda = 2.4$, (b)—$2a/\lambda = 12$

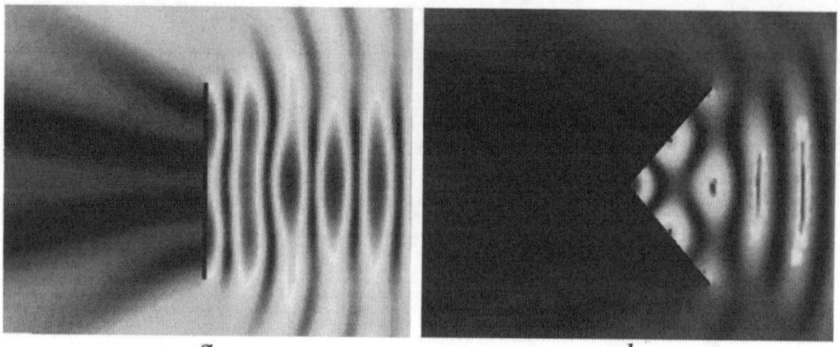

a b

FIG. 4.15: Field structure near the band (a) and the corner (b), $2a/\lambda = 2.4$

in the direction of incident wave, unlike the field in the vicinity of the band. This character of the corner field structure remains in the relatively wide range of angles θ_0.

4.3.3 The Characteristics of Cylindrical Wave Reflection from the Corner. Corner Antenna

We consider the case, when the source of cylindrical waves S is in the opening of the corner at distance R from point O, Fig. 4.13. Such system can be considered as corner antenna [3,4,7]. Let us examine its acoustical properties.

We consider the pressure variation in the far field depending on the source position relative to the origin of coordinates O, Fig. 4.16. It follows form Fig. 4.16 that the sound source should be placed near the point O for higher energetic efficiency of antenna. The directivity pattern almost does not change in the frequency band about octave. The width of directivity pattern $\theta_{0.7}$ (at the pressure level 0.7) is within the limits $60° \leq \theta_{0.7} \leq 70°$, and the level of back radiation is $\sigma \leq 0 < 0.19$.

Let us consider the change of directivity pattern depending on opening angle $2\theta_1$. Figure 4.17 shows the directivity patterns for the different opening angles. As it follows from these data the optimum opening angles are in the region $80° \leq 2\theta_1 \leq 100°$.

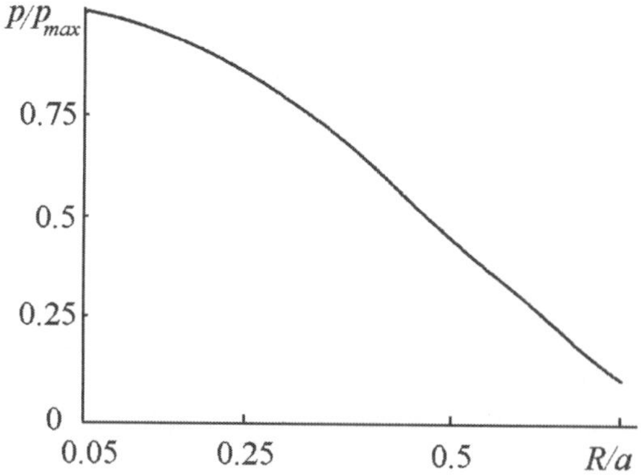

FIG. 4.16: Normalized pressure dependency, of corner antenna in the far field, on source position S, $2\theta_1 = 90°$, $a/\lambda = 0.5$

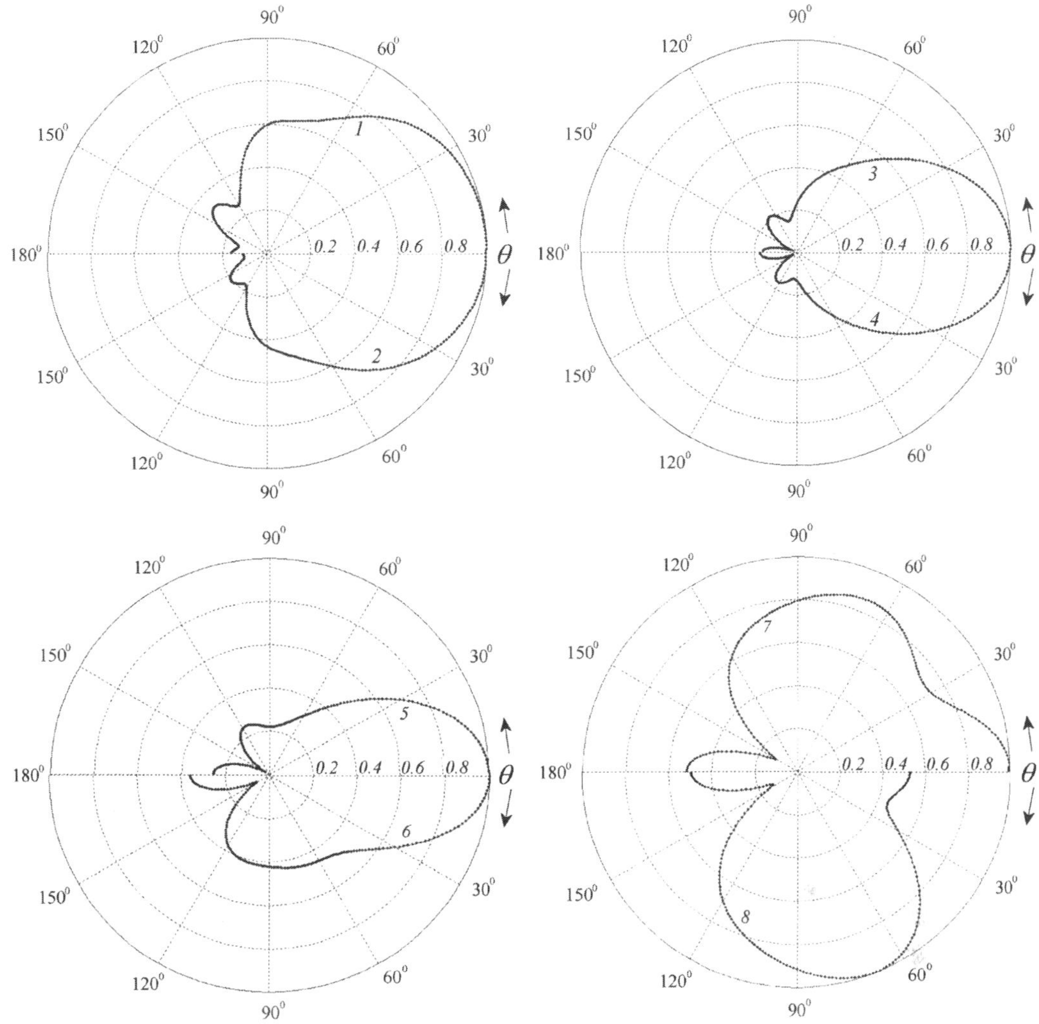

FIG. 4.17: The dependency of the change of shape of directivity pattern on opening angle $2\theta_1$ for $a/\lambda = 0.5$ and $R/a = 0.5$: 1–8 correspond to $2\theta_1 = 40°, 60°, 80°, 100°, 120°, 140°, 160°, 180°$

In addition, Fig. 4.18 shows the generalized data. These data was obtained on the basis of analysis of the bulk of numerical calculations. As it follows from these data (for the indicated source position), the optimum angles of corner opening are in the region $80° \leq 2\theta_1 \leq 100°$. The pressure in the far field is maximum, the level of back radiation is minimum, the width of the main lobe of directivity pattern is moderate. We should choose the angle of opening $2\theta_1 \approx 140°$ for high concentration of sound energy. The pressure on the directivity pattern axis decreases at about 25%, and the back level of radiation increases almost in 2 times. Consequently, there is no tangible benefit in energetic efficiency of antenna.

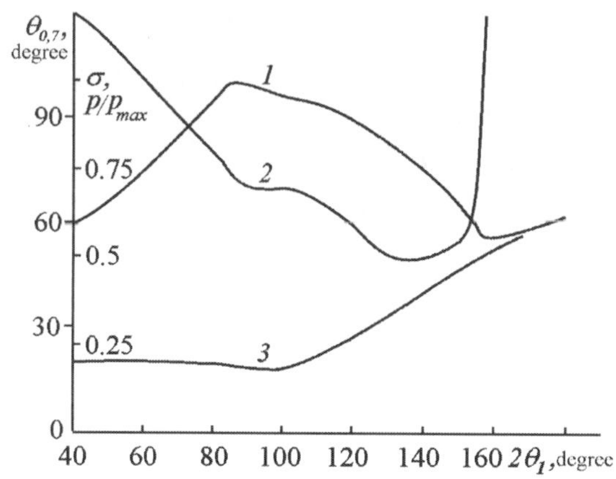

FIG. 4.18: The dependency of acoustic parameters of corner antenna on opening angle $2\theta_1$, $a/\lambda = 0.5$, $R/a = 0.5$: 1—relative pressure in the far field $p/p_{\max}$, 2—the width of the main lobe of directivity pattern $\theta_{0.7}$, 3—the level of back radiation σ

4.4 SOUND WAVES DIFFRACTION BY WEDGE

4.4.1 The Waves Diffraction by Classical (Sharp) Wedge

The problem of wave diffraction by wedge is a fundamental problem. A lot of physical theories of the interaction between waves and bodies are based on the results of this problem solving. The literature on this problem is quite extensive. The physical phenomena, connected with the influence of the wedge point on the incident waves is under investigation. The waves are of different nature – sound, elastic, electro-magnetic, and light waves. There are two reasons of interest in this problem. The first reason is theoretical (mathematical). Within the context of Euclidean geometry, it is impossible to determine the normal derivative on the point of wedge (its edge) (in acoustics – normal component of particle velocity). It was discussed in the first chapter. It leads to the ambiguity of required solution of diffraction problem. Besides, particle velocity tends to infinity when approaching the edge. Works [13,20] and the first chapter of this monograph give detailed information about the behavior of the velocities field near the edge of the wedge and about the mathematical methods of solving this problem.

The second reason of interest in diffraction of incident wave by wedge is the practical importance of this process. It is know that when the wave is incident on the wedge, the cylindrical wave appears on its edge. This wave interferes with the incident wave, and it also causes the waves appearance in the geometrical shadow zone. But the physical reason of the cylindrical wave appearance on the edge of the wedge was unknown for a long time. Huygens was possibly the first who gave the correct explanation of the reason of cylindrical wave appearance on the edge of the wedge. The idea of his theory is as follows. Assume that there is the sharp wedge with the acoustically rigid edges σ_1 and σ_2, the plane wave front p runs along the boundary σ_1, Fig. 4.19.

The edges are acoustically rigid, and the normal components of particle velocities on their surfaces are equal to zero. Consequently, the acoustic impedance of the edges

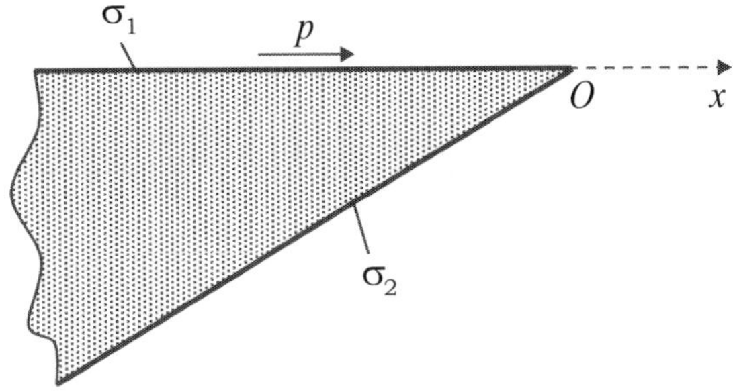

FIG. 4.19: Sharp wedge

surfaces is equal to infinity. Thus, while the wave front is propagating along the boundary σ_1, the sound pressure on the surface of this face remains constant. As soon as the wave front passes the point O, that is the edge of the wedge, the acoustical impedance decreases sharply. It leads to the stepwise fall of sound pressure and, as a result, to the appearance of cylindrical sound wave [25], which comes from the edge. This wave is often called the edge wave. Particle velocity is coordinate derivative of pressure. Consequently, there is a stepwise increasing of particle velocity in the vicinity of edge. Formally this velocity tends to infinity, when approaching the point O. Mathematically it means that function, which describes particle velocity, in point O has a peculiarity. The speed of particle velocity increase depends on the wedge sharpness. The sharper it is, the higher is the speed. The speed is maximum in the case, when the wedge becomes the semi-plane. We note that the described process of the wave formation does not break the law of conservation of energy. The energy of the edge wave appears due to the some part of energy of incident wave. An increase of particle velocity when infinitely approaching the edge does not mean that sound energy increases infinitely. The peculiarity of the function of particle velocity in point O is integratable for all the parameters of the wedge (this question is considered in [20]).

Now we mention briefly a historical background [17,21]. The problem of waves diffraction by wedge with ideal edges was solved for the first time by variable separation method of Poincare in 1892–1897 [22,23]. A little later, A. Sommerfeld [24] solved the problem of waves diffraction by semi-infinite plane screen by the method of so called ramified solutions. The researches paid a lot of attention to the question of the diffraction of waves of different nature by wedge. The works [21,26–29] are of special interest. In the last of these works the problem of sound diffraction by wedge with impedance edges was consided. The variable separation method could not be used.

In view of the above mentioned, let us start the presentation of solution of problem of cylindrical wave diffraction by wedge with ideal properties of edges. Methodically we will construct the solution in the same way, as it was done in [17,28,30].

The geometry of two-dimensional problem of cylindrical wave scattering on the wedge is shown in Fig. 4.20. The wave is created by harmonic point source S in the form of infinite long line, which is perpendicular to the figure plane and parallel to the edge of the wedge. The center O of polar coordinate system (r, θ) is on the wedge point with an exterior angle equal α. On the edges of the wedge then σ_1 and σ_2 have the

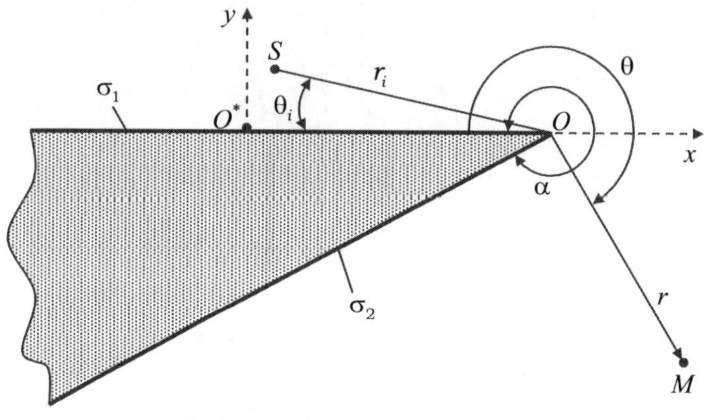

FIG. 4.20: Sharp wedge geometry

coordinates $\theta = 0$, $0 \leq r < \infty$ and $\theta = \alpha$, $0 \leq r < \infty$. The source S is in the point with coordinates $\mathbf{r}_i = (r_i, \theta_i)$, and the point of observation M is determined by vector $\mathbf{r} = (r, \theta)$.

Let us write the function of the source (Green function) for the free space in two dimensional case [17,30,31]

$$G(\mathbf{r}, \mathbf{r}_0) = \frac{i}{4} H_0^{(1)} (k |\mathbf{r}_0 - \mathbf{r}|) . \tag{4.32}$$

To write the function of the source with wedge (Fig. 4.20) it is necessary to solve the corresponding diffraction problem (the details of its solution are given in [17,26,31]). We omit the problem solution and give the final formulas:

1. if the surfaces σ_1 and σ_2 are acoustically rigid, then

$$G(\mathbf{r}, \mathbf{r}_i) = \frac{\pi i}{2\alpha} \sum_{m=0}^{\infty} \varepsilon_m J_{\nu_m}(kr_<) H_{\nu_m}^{(1)}(kr_>) \cos(\nu_m \theta) \cos(\nu_m \theta_i), \tag{4.33}$$

$$\nu_m = \frac{m\pi}{\alpha},$$

2. if the surfaces σ_1 and σ_2 are acoustically soft, then

$$G(\mathbf{r}, \mathbf{r}_0) = \frac{\pi i}{\alpha} \sum_{m=1}^{\infty} J_{\nu_m}(kr_<) H_{\nu_m}^{(1)}(kr_>) \sin(\nu_m \theta) \sin(\nu_m \theta_i), \tag{4.34}$$

$$\nu_m = \frac{m\pi}{\alpha},$$

3. if the surface σ_1 is acoustically rigid, and surface σ_2 is acoustically soft, then

$$G(\mathbf{r}, \mathbf{r}_0) = \frac{\pi i}{\alpha} \sum_{m=0}^{\infty} J_{\nu_m}(kr_<) H_{\nu_m}^{(1)}(kr_>) \cos(\nu_m \theta) \cos(\nu_m \theta_i), \tag{4.35}$$

$$\nu_m = \frac{(2m+1)\pi}{2\alpha},$$

$r_<$ is the smallest of two distances r and r_i, and $r_>$ is the largest of two distances r and r_i [31], $\varepsilon_0 = 1$ and $\varepsilon_m = 2$ when $m > 0$.

If we know the Green function, it is easy to determine the sound pressure, created by the source of cylindrical waves [17]. It is enough to multiply the Green function by value $-ik\rho c\upsilon_0 r^* \alpha/2\pi$, where r^*—radius of the source of cylindrical waves, and υ_0—particle velocity of its surface.

4.4.2 Plane Sound Wave Diffraction by Wedge

Let us consider the problem of plane wave diffraction by wedge. The source S should be placed far enough from the wedge [17]. Let tend the value $kr_>$ in the formulas given above to infinity and use the asymptotic representation of Hankel function for $kr_i = kr_> \gg 1$:

$$H^{(1)}_{\upsilon_m}(kr_i) \approx \sqrt{\frac{2}{\pi k r_i}} \exp\left(ikr_i - \frac{i\nu_m\pi}{2} - \frac{i\pi}{4}\right). \tag{4.36}$$

Besides, we normalize the sound pressure to the pressure on edge, which is created by point (linear) source in the free space. We use the Green function for it (4.34), where $r = 0$, $kr_i \gg 1$. Than

$$G(0, r_i) \approx \frac{i}{4}\sqrt{\frac{2}{\pi k r_i}} \exp\left(ikr_i - \frac{i\pi}{4}\right). \tag{4.37}$$

We divide the expressions (4.33)–(4.35) by the right hand side (4.37) and we use asymptotics (4.36). Then we write the full pressure sound field, which arises in the consequence of scattering of plane sound wave by wedge of unit amplitude:

1. acoustically rigid wedge

$$p = \frac{2\pi}{\alpha}\sum_{m=0}^{\infty} \varepsilon_m \exp\left(-\frac{i\nu_m\pi}{2}\right) J_{\nu_m}(kr)\cos(\nu_m\theta)\cos(\nu_m\theta_i), \tag{4.38}$$

$$\nu_m = \frac{m\pi}{\alpha},$$

2. acoustically soft wedge

$$p = \frac{4\pi}{\alpha}\sum_{m=1}^{\infty} \exp\left(-\frac{i\nu_m\pi}{2}\right) J_{\nu_m}(kr)\sin(\nu_m\theta)\sin(\nu_m\theta_i), \tag{4.39}$$

$$\nu_m = \frac{m\pi}{\alpha},$$

3. acoustically rigid face σ_1 and acoustically soft face σ_2

$$p = \frac{4\pi}{\alpha}\sum_{m=0}^{\infty} \exp\left(-\frac{i\nu_m\pi}{2}\right) J_{\nu_m}(kr)\cos(\nu_m\theta)\cos(\nu_m\theta_i), \tag{4.40}$$

$$\nu_m = \frac{(2m+1)\pi}{2\alpha}.$$

Thus, using Eqs. (4.38)–(4.40) we can investigate the plane wave diffraction $p_i = \exp(-ikr\cos(\theta - \theta_i))$. The plane wave front is parallel to the edge, and the direction of incidence is determined by angle θ_i.

4.4.3 The Analysis of Calculation Results for the Sharp Wedge

As it was mentioned above, the wave processes in the vicinity of the edge of the wedge are of the main interest. Consequently, we fix attention on these processes.

When calculating, it is assumed that the source S is situated on the edge σ_1 (i.e., $\theta_i = 0$) and the wave distance from the source to the edge is $r_i/\lambda = 1.3$ (here λ—the wave length in the acoustic medium surrounding the wedge). The additional Cartesian coordinate system xO^*y is introduced (see Fig. 4.20). The center O^* of this system is brought in coincidence with the point of the source S position. The number of the kept terms in the series (4.33)–(4.35), when calculating, is 100.

Figure 4.21 shows distributions of pressure and radial component of particle velocity along the O^*x axis for the wedge with acoustically rigid edges. The part of distance, equal O^*O, coincides with the surface of edge σ_1, and the part after the point O is beyond the wedge in acoustical medium.‡

Let us consider some peculiarities of the behavior of the values $\tilde{p}$ and $\tilde{\upsilon}$ in Fig. 4.21. As we can see, Huygens was right—there is a pressure-jump on the edge. The jump becomes more abrupt with an increase of the wedge sharpness. This jump causes edge cylindrical wave on the edge of the wedge. Edge wave in the illuminated zone interferes with the incident wave (the pressure oscillations appear on the surface of the face σ_1) and penetrates into the zone of geometrical shadow.

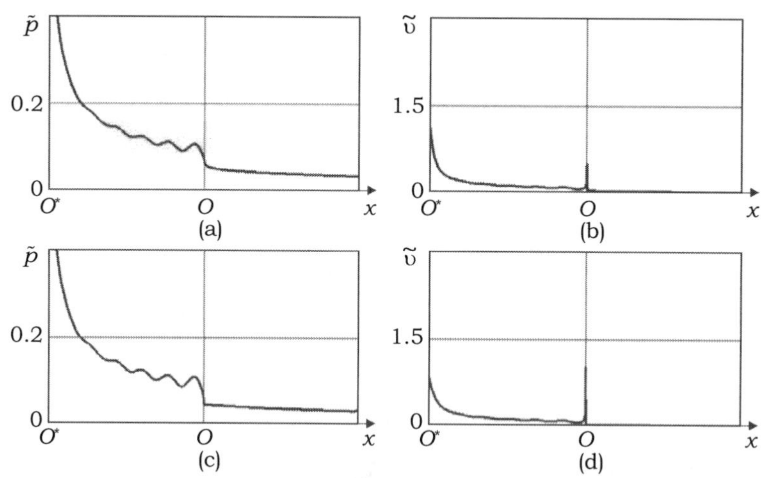

FIG. 4.21: Normalized pressure distribution $\tilde{p}$ and particle velocity $\tilde{\upsilon}$ (both edges are acoustically rigid) along the axis O^*x when $r_i/\lambda = 1.3$: (a,b)—$\alpha = 270°$, (c,d)—$\alpha = 360°$

‡Note that the interval of the axis O^*x beyond the edge σ_1 is (geometrically) a dividing line between the illuminated domain and the domain of shadow.

Pressure jump causes the appearance of peculiarity of particle velocity on the edge.[§] The sharper is edge the stronger is peculiarity (compare the levels of particle velocity on the edge in Figs. 4.21(b) and 4.21(d)). The comparison of these levels gives only the relative idea about the character of peculiarity on the edge. The calculations were performed for the finite number of the terms of series (4.33). If we want to estimate rigorously the character of these peculiarities, it is necessary to use the method of Maue [20], as it was done in the first chapter. In this case, when $\alpha = 270°$, the character of peculiarity of particle velocity υ when approaching the edge is as follows $\upsilon \approx A/r^{1/3}$, and when $\alpha = 360°$ it is $\upsilon \approx A/r^{1/2}$ respectively. The peculiarity is stronger for the semi-plane than for the direct angle; A is a constant here.

Let us consider now the behavior of pressure distributions of radial component of particle velocity along the axis O^*x in the case when the face of the wedge σ_1 is acoustically rigid, and the face σ_2 is acoustically soft. Figure 4.22 shows such data. As we can see, the pressure jumps near the edge are more sharp, than in the case, when the both faces are acoustically rigid. Consequently, the peculiarities of particle velocity on the edge are more pronounced. In the case, when $\alpha = 270°$, the character of particle velocity peculiarity is as follows $\upsilon \approx A/r^{2/3}$ when approaching the edge, and when $\alpha = 360°$ it is $\upsilon \approx A/r^{3/4}$, respectively.

The space distribution of acoustic pressure in the vicinity of the wedge is also of interest. Figure 4.23 shows such data for the right-angled wedge. The strong interference effect between the incident and edge waves is observed in the illuminated area. The interference is shown as the alternation of zones of high and low pressure. It is clearly seen how the edge wave flows into the shadow zone. When the faces have different acoustic properties, the pressure level on shadow zone is about 10 dB lower, than in the case, when both faces are acoustically rigid.

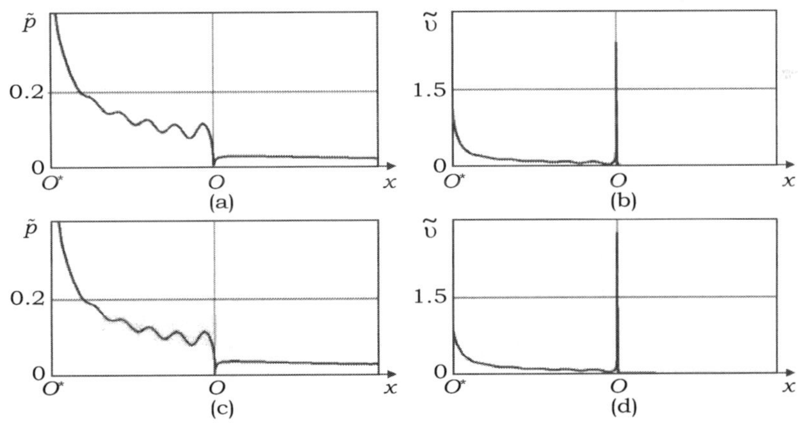

FIG. 4.22: The distribution of normalized pressure and particle velocity (one edge is acoustically rigid, the other—soft) along the axis O^*x when $r_1/\lambda = 1.3$: (a,b)—for $\alpha = 270°$, (c,d)—for $\alpha = 360°$

[§]The peculiarity, that is an infinite velocity increase, is the consequence of the neglect of viscosity of acoustic medium and nonlinearity.

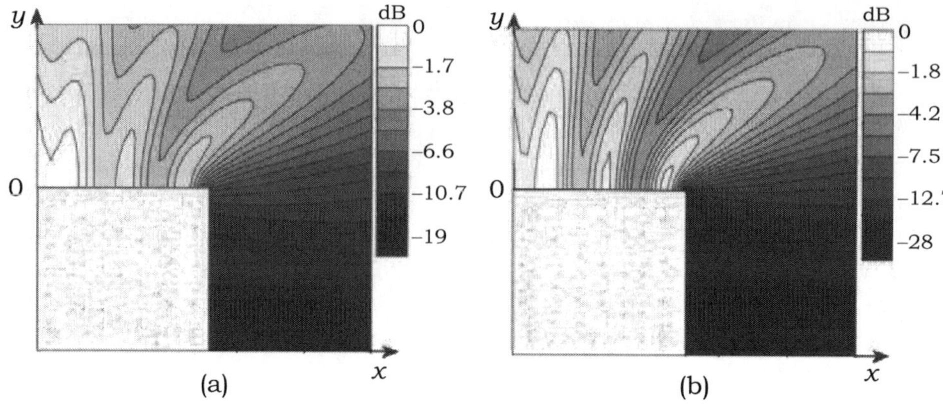

FIG. 4.23: Space distribution of pressure field when $\alpha = 270°$ and $r_1/\lambda = 1.3$: (a)—both faces of the wedge are acoustically rigid, (b)—face σ_1 is acoustically rigid, and face σ_2 acoustically soft

It is interesting to consider the situation when the wedge becomes a plane. Angle α becomes equal $180°$. Fig. 4.24 shows pressure space distribution for this case. When the faces of the wedge are acoustically rigid, there are no special wave processes. Cylindrical wave, excited by source S, is propagating in semi-plane and the level of its pressure drops gradually according to the law $1/\sqrt{r}$. But if the surface σ_1 is acoustically rigid, and σ_2 acoustically soft, the sound pressure on the rigid part of the surface is different from zero, and on the soft part—is equal to zero. That is why there is a pressure jump in the point O, and it is well seen in Fig. 4.24(b), and the peculiarity of particle velocity appears. The result of it is that the edge wave appears. This wave interferes with the incident wave, and creates the sequence of zones of high and low pressure over the surface σ_1. We have the zone of smooth and rapid drop of the pressure level down to zero over the surface σ_2, when approaching the surface σ_2.

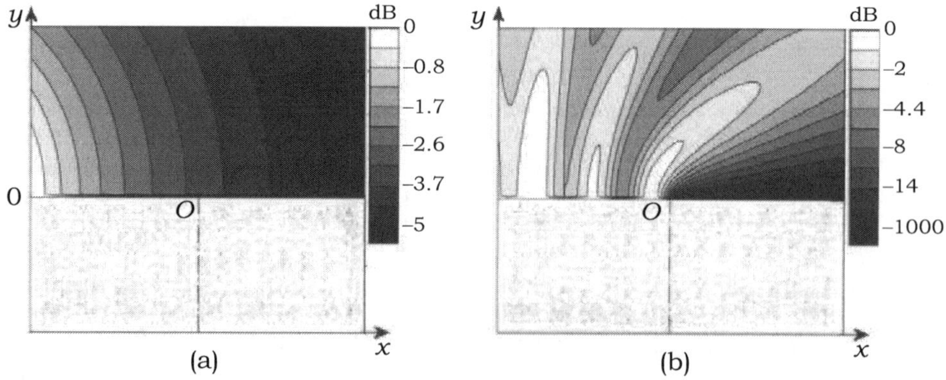

FIG. 4.24: Space distribution of the field pressure when $\alpha = 180°$ and $r_1/\lambda = 1.3$: (a)—surface σ_1 and σ_2 are acoustically rigid, (b)—surface σ_1 is acoustically rigid, and σ_2 acoustically soft

4.4.4 Waves Diffraction by Rounded Wedge

The diffraction of waves by shaped (classical) wedge was considered, and we found out that there is a pressure jump on its edge. The edge wave appears in the result of it. Let us consider now what happens if the sharp edge of the wedge is rounded as cylindrical surface. Figure 4.25 shows the geometry of such wedge and all the notations.

Let us consider the plane problem of cylindrical wave scattering by rounded wedge. All the surfaces of the wedge are acoustically rigid. Let us introduce two polar coordinate systems: $rO\theta$ and $r_1O_1\theta_1$ to construct the problem solution (the coordinates of some point M are indicated on these coordinate systems in Fig. 4.25). The radius of the wedge rounding is $a = O_1A = O_1B$. The angle of rounded wedge α is determined as the angle of the wedge without rounding. The sound source S have the coordinates $(r_i, \theta_i = 0)$, it is situated on wedge surface σ_1 and coincides with the origin of additional coordinate system xO^*y.

Let us designate the following domains to use the partial domains method. Exterior domain I is set in the coordinate system $rO\theta$: $b \le r < \infty$, $0 \le \theta \le \alpha$. It is the exterior of the circle with radius b, bounded by the faces of rounded wedge. Domain II is the interior of the circle with radius b excluding the zone of intersection with the circle of radius a with the point O_1 as a center.

The pressure in domain I is written as follows:

$$p_I = \frac{\pi i}{2\alpha}\sum_{n=0}^{\infty} \varepsilon_n H_{\xi_n}^{(1)}(kr_i)\, J_{\xi_n}(kr)\cos(\xi_n\theta) + \sum_{n=0}^{\infty} A_n \frac{H_{\xi_n}^{(1)}(kr)}{H_{\xi_n}^{(1)\prime}(kb)}\cos(\xi_n\theta), \quad (4.41)$$

where $\xi_n = n\pi/\alpha$. The first sum here describes the field of the source of the usual (sharp) wedge. The second sum determines the field of the scattered wave in the form of the set of cylindrical waves coming away from the wedge. The prime means derivative of function with respect to argument.

The pressure in domain II is represented as follows

$$p_{II} = p_{II}^{(1)} + p_{II}^{(2)}, \quad (4.42)$$

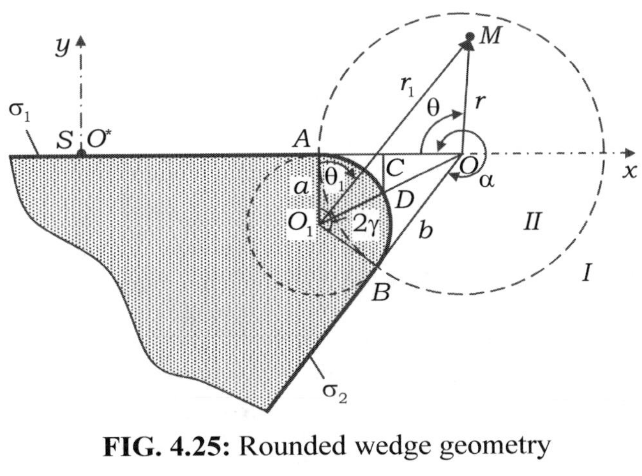

FIG. 4.25: Rounded wedge geometry

where

$$p_{II}^{(1)} = \sum_{n=0}^{\infty} B_n \frac{J_n(kr)}{J_n'(kb)} \cos(n\theta) + \sum_{n=1}^{\infty} \tilde{B}_n \frac{J_n(kr)}{J_n'(kb)} \sin(n\theta), \tag{4.43}$$

$$p_{II}^{(2)} = \sum_{n=0}^{\infty} C_n \frac{H_{\eta_n}^{(1)}(kr_1)}{H_{\eta_n}^{(1)\prime}(ka)} \cos(\eta_n \theta_1), \quad \eta_n = \frac{n\pi}{2\gamma}. \tag{4.44}$$

Expression (4.43) is the general solution for the interior of the circle of radius b with point O as the center. The functions $\cos(n\theta)$ and $\sin(n\theta)$ have the orthogonality property on the full circle ($0 \le \theta \le 2\pi$). But partial domain II is not a full circle. The zone of intersection with the circle of radius a and point O_1 as the center is excluded. Consequently, when writing the conditions of the fields matching on the boundaries of partial domains I and II ($r = b,\ 0 \le \theta \le \alpha$), we should add the boundary condition for solution $p_{II}^{(1)}$ on the nonphysical area ($r = b,\ \alpha \le \theta \le 2\pi$), (see below). The possibility of using such procedure was discussed in Chapter 1.

Solution (4.44) is general for the interior of dihedral domain BO_1A (the functions $\cos(\eta_n\theta_1)$ are orthogonal when the angle coordinate changes $0 \le \theta_1 \le 2\gamma$). That is why the solution $p_{II}^{(2)}$ provides the fulfillment of the boundary condition on the rigid surface of the wedge rounding (arc ADB).

The sound fields matching conditions on the boundary of domains I and II are as follows:

$$p_{II} = p_I, \quad r = b, \quad \theta = [0, \alpha], \tag{4.45}$$

$$\begin{cases} \dfrac{\partial p_{II}}{\partial r} = \dfrac{\partial p_I}{\partial r}, & r = b, \quad \theta = [0, \alpha], \\ \dfrac{\partial p_{II}^{(1)}}{\partial r} = F(\theta), & r = b, \quad \theta = [\alpha, 2\pi]. \end{cases} \tag{4.46}$$

It is to be recalled that it makes no difference whether we set pressure or radial velocity on the nonphysical part of the boundary ($r = b,\ \alpha \le \theta \le 2\pi$). The velocity was chosen in condition (4.46). Function $F(\theta)$ can be arbitrary. But it is desirable to decrease the field jumps on the ends of the region (points A and B). We choose the function $F(\theta)$ in the following form

$$F(\theta) = \frac{\partial G(b, \alpha; r_i, \theta_i)}{\partial r} + \left[\frac{\partial G(b, 0; r_i, \theta_i)}{\partial r} - \frac{\partial G(b, \alpha; r_i, \theta_i)}{\partial r}\right] \frac{\theta - \alpha}{2\pi - \alpha}, \tag{4.47}$$

G—the source function for the usual wedge, $\alpha \le \theta \le 2\pi$.

The boundary condition on the surface of wedge rounding is as follows:

$$\frac{\partial p_{II}}{\partial r_1} = 0, \quad r_1 = a, \quad \theta_1 = [0, 2\gamma]. \tag{4.48}$$

The algebraization of functional equations (4.45), (4.46) and boundary condition (4.48) is performed by standard method. The properties of orthogonality of the set of functions $\cos(\xi_n\theta)$ on the interval $0 \le \theta \le \alpha$ are used for condition (4.45) and the properties of orthogonality of functions $\cos(n\theta)$ and $\sin(n\theta)$ on the interval $0 \le \theta \le 2\pi$

are used for condition (4.46). The orthogonality of function $\cos(\eta_n\theta_1)$ on the interval $0 \leq \theta_1 \leq 2\gamma$ is used for condition (4.48). We obtain the infinite system of linear algebraic equations of the second kind in unknown coefficients A_n, B_n, $\tilde{B}_n$, C_n.

The number of the kept unknowns in the systems of linear algebraic equations is 100 when we perform the calculations. The matching conditions (4.46) were fulfilled with graphical accuracy, and boundary condition (4.45)—with relative accuracy not less than 1%.

4.4.5 Calculation Results Analysis for Rounded Wedge

The obtained above solution allows us to carry out the numerical analysis of sound field diffraction by wedge in the case when the edge is rounded. Figure 4.26 shows the normalized distributions of pressure and of radial component of particle velocity along the axis O^*x for rounded wedge with acoustically rigid boundaries and acoustically rigid surface of rounding. Figure 4.27 shows the space distributions of pressure in the vicinity of rounded edge. Let us compare the dependencies of distribution of pressure and radial component of particle velocity in Figs. 4.21(a), 4.21(b), 4.26(a), 4.26(b) and also the space distribution of pressure in Figs. 4.23(a) and 4.27(a). As we can see, when the radiuses of edge rounding are small, there are no special changes in the field of diffraction. Despite the rounding of the edge, the pressure jump is almost the same as in the case of sharp edge. Consequently, the peak particle velocity appears. The edge wave appears as a result of it. This wave interferes with incident wave and creates the succession of zones of high and low pressure over the surface σ_1 of the wedge [these zones are well seen in Fig. 4.27(a)]. It causes the ripple of the graphs of pressure and particle velocity on the intervals O^*A of O^*x axis [Figs. 4.26(a) and 4.26(b)]. The peculiarity of particle velocity cannot appear on the rounded edge of wedge from the mathematical point of

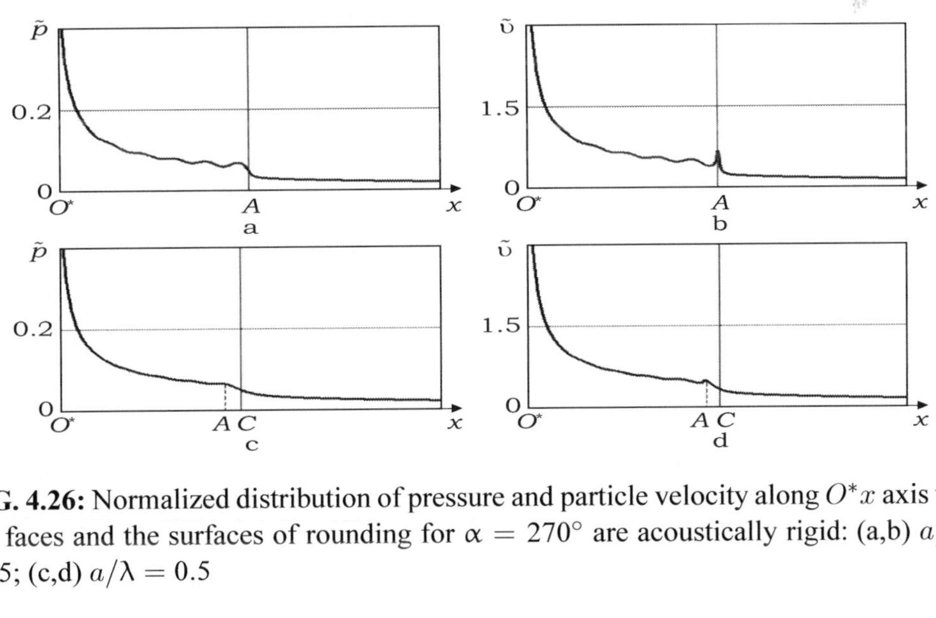

FIG. 4.26: Normalized distribution of pressure and particle velocity along O^*x axis when the faces and the surfaces of rounding for $\alpha = 270°$ are acoustically rigid: (a,b) $a/\lambda = 0.05$; (c,d) $a/\lambda = 0.5$

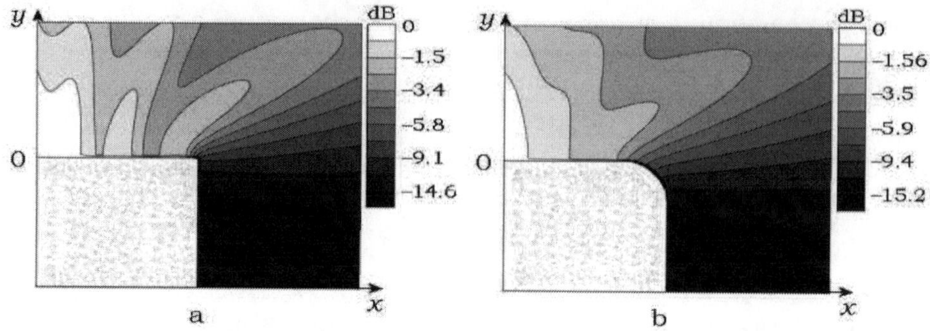

FIG. 4.27: Space distribution of pressure field with acoustically rigid faces and surface of rounding for $\alpha = 270^{\circ}$: (a) $a/\lambda = 0.05$; (b) $a/\lambda = 0.5$

view. But, as we can see, all the characteristic features of the field of diffraction remain almost the same.

The situation is quite different, when the radius of the edge rounding is comparable with the wave length. Let us consider the graphs of pressure and particle velocity in Figs. 4.26(c), 4.26(d) and space distribution of the pressure field in Fig. 4.27(c). There is no pressure jump in this case, and consequently, the peak of particle velocity does not occur. The edge wave is almost absent. This is proved by the absence of interference zones of high and low pressure over the surface σ_1 of wedge [Fig. 4.27(b)] and by the absence of ripple on the graphs of pressure and particle velocity on the intervals O^*A of axis O^*x [Figs. 4.26(c) and 4.26(d)].

It should be noted in conclusion that the results given in this chapter are based on the investigations performed by the authors in monograph [13] and articles [14,15,16].

REFERENCES

1. Gutin, L.D., Concerning the theory of sound parabolic concentrator, *Proc. Electrical Industry of Low Current*, vol. **9**, pp. 9–25, 1935. [in Russian]

2. Rozenberg, L.D., *Focusing Sound Systems*, Moscow, Leningrad: The USSR Academy of Science Printing House, p. 111, 1949. [in Russian]

3. Kyun, R., *Microwave Antennas*, Leningrad: Sudostroenie, p. 517, 1967. [in Russian]

4. Ayzenberg, G.Z., Yampolskiy, V.G., and Tereshin, O.I., *USW Antennas*, Part 1, 2, Moscow: Svyaz, p. 381, 1977. [in Russian]

5. Feld, Y.N. and Ansryan, A.K., Scalar wave diffraction by parabolic mirror, *Radiotechnichs Eletronics*, vol. **22**, no. 1, pp. 1–9, 1977. [in Russian]

6. Kanevskiy, I.N., *Sound and Ultrasonic Waves Focusing*, Moscow: Nauka, p. 336, 1977. [in Russian]

7. Fradin, A.Z., *Ultra-High Frequencies Antennas*, Moscow: Sovetskoe Radio, p. 647, 1957. [in Russian]

8. Rzhevkin, S.N., *A Course of Lectures on the Theory of Sound*, Moscow: MSU Printing House, p. 335, 1960. [in Russian]

9. Isakovich, M.A., *General Acoustics*, Moscow: Nauka, p. 495, 1973. [in Russian]
10. Shenderov, E.L., *Wave Problems of Hydro-Acoustics*, Leningrad: Sudostroenie, p. 347, 1972. [in Russian]
11. Ivanov, E.A., *Electromagnetic Waves Diffraction by Two Bodies*, Minsk: Nauka i Tehnika, p. 584, 1968. [in Russian]
12. Bobber, R., *Hydro-Acoustic Measurements*, Moscow: Mir, p. 362, 1974. [in Russian]
13. Grinchenko, V.T. and Vovk, I.V., *Wave Problems of Sound Scattering by Elastic Shells*, Kiev: Naukova dumka, p. 240, 1986. [in Russian]
14. Vovk, I.V., Sound waves diffraction by parabolic reflector, *J. Acoust.*, vol. **25**, no. 5, pp. 670–674, 1979. [in Russian]
15. Vovk, I.V. and Gulega, L.G., Sound field of spherical reflectora with grating of radiators in focal area, *J. Acoust.*, vol. **36**, no. 1, pp. 17–20, 1990. [in Russian]
16. Grinchenko, V.T. and Matsypura, V.T., Sound scattering by finite wedge objects, *Acustychnyi visnyk*, vol. **6**, no. 2, pp. 23–33, 2003. [in Russian]
17. Shenderov, E.L., *Sound Scattering and Radiation*, Leningrad: Sudostroenie, p. 304, 1989. [in Russian]
18. Beranek, L., *Acoustic Measurements*, Moscow: Foreign Literature Printing House, p. 626, 1952. [in Russian]
19. Schults, F.V., Ruckgaber, G.M., Schindler, J.K., and Rogers, C.C., The theoretical and numerical determination of the finite cone, *Proc. IEEE*, vol. **53**, no. 8, pp. 1212–1213, 1965.
20. Hönl, H., Maue, A., and Westpfahl, K., *Diffraction Theory*, Moscow: Mir, p. 428, 1964.
21. Babich, V.M., Lyalinov, M.A., and Grikurov, V.E., *Sommerfeld-Malyuzhintsa Method in Diffraction Theory*, St. Petersburg State University, SPb., p. 101, 2003. [in Russian]
22. Poincaré, J.H., Sur la polarization par diffraction, *Acta Math.*, vol. **16**, pp. 297–339, 1892.
23. Poincaré, J.H., Sur la polarization par diffraction, *Acta Math.*, vol. **20**, pp. 313–355, 1897.
24. Sommerfeld, A., Mathematische theorie der diffraction, *Math. Ann.*, vol. **47**, pp. 317–374, 1896.
25. Vovk, I.V. and Grinchenko, V.T., *The Sound Born by a Flow*, Kiev: Naukova Dumka, p. 221, 2010. [in Russian]
26. Petrashen, G.I., Nikolaev, B.G., and Kouzov, D.P., Concerning the method of series in diffraction theory of waves from plane corner domains, dynamic problems of elasticity theory, vol. **32**, pp. 5–165, 1958. (Scientific notes LSU; series Mathematical Science; no. 246). [in Russian]
27. Bowman, J. and Senior, T.B.A., Diffraction of a dipole field by a perfectly condacting half-plane, *Radio Sci.*, vol. **2**, no. 11, pp. 1339–1345, 1967.
28. Kending, R.P. and Hayek, S.I., Diffraction by a hard-soft barrier, *J. Acoust. Sci. Am.*, vol. **70**, no. 4, pp. 1156–1165, 1981.
29. Malyuzhinets, G.D., Excitation, reflection and radiating of surface waves on the wedge with given faces impedance, *Report USSP Academy of Science*, vol. **121**, no. 3, pp. 436–439, 1959. [in Russian]
30. Grinchenko, V.T., Vovk, I.V., and Matsypura, V.T., *Fundamentals of Acoustics*, Kiev: Naukova dumka, p. 639, 2007. [in Russian]

31. Felsen, L. and Markuvits, N., *Waves Radiation and Scattering*, Moscow: Mir, vol. **2**, p. 555, 1978.

CHAPTER 5

Acoustic Properties of Sound Barriers

"What is this, Holmes?" I cried. "This is beyond anything which I could have imagined."

Arthur Conan Doyle, "The Adventure of the resident patient"

Unfortunately, the progress in science and technology is accompanied with negative developments. For example, the noise pollution of environment. The reason of it is the increasing of traffic speed and density, and also the increasing of intensity of work of different industrial machines and so on.

It is well known that the noise with level higher than 65 dB can cause complete or partial hearing loss of a person [1]. At present, the offices and residential premises are affected by traffic noise, which exceeds the permissible noise level. The noise levels in hospitals, schools and pre-schools are also unacceptable. Absent-mindedness when performing intellectual work, increased fatigability, irritability, and headache are the most evident consequences of human exposure to noise impact. The consequences can be even more serious. According to present knowledge, municipal noise increases the incidence of neuroses, different autonomic disturbances, hypertension and other diseases, caused by the chronic-fatigue of brain. It should be noted that the noise reaction of a definite person is very subjective. It depends on the acoustic properties of the source, the place, the exposure time and the state of nervous system. For example, noise level 70–80 dB(A) is sometimes not perceived as irritatory agent when a person is in a subway car (the tired people can even sleep in subway). But an exterior noise of the same level penetrating in a living room is unbearable.

In the developed countries, a lot of attention is paid to the investigation of cities acoustic pollution and to the elaboration of its reduction measures. This is evident from the fact that there is a continuous stream of publications in top scientific and technical journals (see, for example, [2–5] and bibliography in them).

Noise reduction with the aid of acoustic barriers (acoustic baffles) holds a special place. Barriers are placed between the noise source and the area, which should be protected against noise. The examples of such areas are the residential buildings, the sidewalks near the traffic arteries, and the place of work on the factory floor. The barriers are popular because they are relatively cheap and simple-to-use. Figure 5.1 shows the photographs of real barriers.

Such barriers are made of non-transparent materials (concrete, wood, metal), and also from transparent plastics [6,7]. As we can see, they can have different design.

The described barriers can protect from noise due to the effect of acoustic shielding. It means that the acoustical shadow appears around a protected object. Within the scope

ISBN: 978–1–56700–459–5 © 2018 by Begell House Inc.

FIG. 5.1: Examples of sound barriers

of ray acoustics (geometrical acoustics) theory [8], it is assumed that noise does not penetrate into the acoustical shadow zone. Acoustical shadow zone is separated from illuminated zone of space by semi-infinite right line, coming from the source through upper edge of sound barrier (Fig. 5.2). At first glance, it seems that it is enough to select the corresponding height of acoustic barrier to protect some object (all the object should be in geometric shadow) and the noise will be removed. But the results of ray theory application are satisfactory only when the sound wave length is much less than the barrier height and we can neglect a diffraction effect.

In most cases, the lion's share of sound energy of traffic and industrial noise lies in the range of relatively law frequencies. As an example, Fig. 5.3 shows averaged experimental noise spectrums of traffic flows. And we can see, that the maximum of noise energy is in domain 50, ..., 200 Hz, which is equivalent of wave length 1, 5, ..., 6 m.

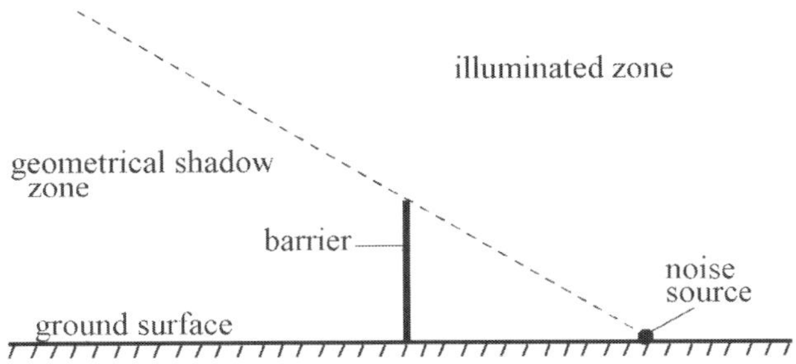

FIG. 5.2: Noise source allocation scheme and acoustic barrier

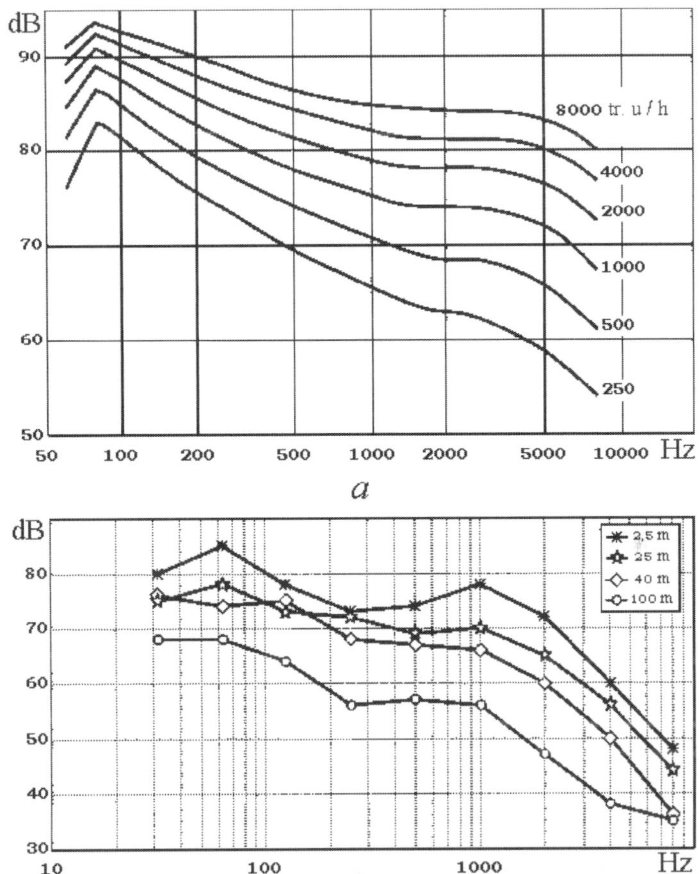

FIG. 5.3: Noise averaged spectrums of traffic flows; (a)—for the given traffic intensity—transport units per hour (tr. u/h) [9]; (b)—for the given distance from highway side (measurements performed by Institute of hydromechanics NAS in 2007 year on Pobedy Boulevard in Kiev with transport density 7000 tr. u/h)

The height of acoustic barriers built near traffic arteries range from 3 to 8 m. When studying the sound scattering by bodies, comparable with the length of sound wave, it is necessary to use the rigorous methods for statement and solution of corresponding diffraction problem [10,11]. In the contrary case, the results of scattered field estimation can be false.

The bulk of publications, dedicated to the estimation of sound fields scattered by barriers, is performed on the base of using the approximate approaches (ray acoustics methods, Keller method and other asymptotical methods [9,12]). The specified methods allow us to obtain good fields estimations behind the barrier (in the zone of its acoustic shadow) for the cases, when the barrier height exceeds the length of incident sound wave.

Diffraction problems, concerned with sound barriers models, are considered in this chapter on the base of partial domains method. The solutions are obtained, which allow us to analyze the sound fields scattered by barriers in all the practically interesting frequency range.

5.1 CLASSICAL SOUND BARRIER

5.1.1 Physical and Mathematical Models of Barrier

Classical barriers are the solid vertical walls made of rigid material; the height of such barriers is much more than their thickness. The barriers are mounted on the sides of the highways as a continuous wall. The length of this wall is much more than its height. Assume the traffic flow is heavy (more than two thousand transport units per hour). In this case, the traffic noise level almost does not depend on a separate transport unit noise level.

In view of the foregoing, let us consider the following ideal physical model of classical barrier, situated along the highway (Fig. 5.4).

Assume that an infinite acoustically hard surface simulates the ground. Acoustically hard thin barrier is at point O. This barrier is infinite and perpendicular to the figure plane. The height of barrier r_h, and its inclination angle about the ground θ_0. The effective height of barrier $h = r_h \sin\theta_0$ (see Fig. 5.4). Linear harmonic sound source S with frequency ω is on the right hand, parallel to the barrier at distance b from the barrier and at a height y_S from the ground. The source is an infinite thread. The observation point is denoted with P. All the semi-space is filled with ideal acoustic medium with density ρ and sound velocity c. These parameters correspond to the air.

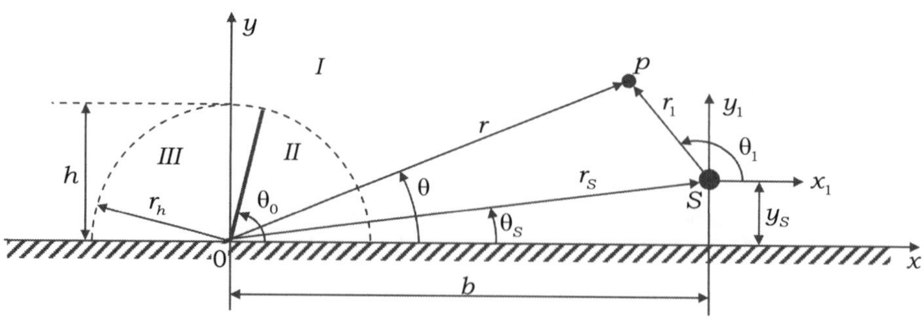

FIG. 5.4: Physical model geometry

The physical model, described mathematically, is equivalent to a plane problem. In this model the parameters of sound field are independent of one coordinate (in our case this coordinate is perpendicular to the figure plane). The normal component of sound field particle velocity on the ground and barrier is equal to zero.

This model corresponds to the practical situations and allows us to construct rigorous analytical solution [13].

5.1.2 Analytical Solution Construction

To construct the solution of the problem, let us introduce the polar coordinate system (r, θ) with point O as center (Fig. 5.4). Let us partition all the sound field existence domain into three domains: domain I is determined by formulas $r \geq r_h$, $0 \leq \theta \leq \pi$; domain II occupies the right sector of semicircle with radius r_h: $r \leq r_h$, $0 \leq \theta \leq \theta_0$; domain III is the left sector of semicircle: $r \leq r_h$, $\theta_0 \leq \theta \leq \pi$.

The described above partial domains were defined in such a way as to construct the problem solution. We can construct the general solution of Helmholtz equation for such domains. The solution of initial boundary problem is reduced to the fulfillment of matching conditions at the boundaries of partial domains. We do not write time factor $\exp(-i\omega t)$.

Let us place center O_1 of the second polar coordinate system (r_1, θ_1) in the location of source S, Fig. 5.4. The pressure field of linear source with the amplitude equal to unity is determined by expression [14]: $p_0 = H_0^{(1)}(kr_1)$, where $H_0^{(1)}(kr_1)$—Hunkel function of the first kind of zero order, $k = \omega/c$—wave number, $\omega = 2\pi f$, f—frequency. If $r_S = \sqrt{b^2 + y_S^2} > r_h$, the sound source is in domain I, and if $r_S < r_h$—in domain II.

Let $r_S > r_h$. The pressure field in domain I then should be written as follows

$$p_I = H_0^{(1)}(kr_1) + \sum_{n=0}^{\infty} A_n \frac{H_n^{(1)}(kr)}{H_n'^{(1)}(kr_h)} \cos(n\theta), \tag{5.1}$$

where angle functions $\cos(n\theta)$ are chosen in such a way as to automatically fulfill boundary conditions on rigid surface $\theta = 0$ and $\theta = \pi$; the prime on the Hankel function means argument derivative. The totality of arbitrary elements A_n allows us to fulfill matching conditions at the boundaries with domains II and III.

Let us represent the pressure field in domain II as the superposition of standing waves:

$$p_{II} = \sum_{n=0}^{\infty} B_n \frac{J_{\alpha_n}(kr)}{J'_{\alpha_n}(kh)} \cos(\alpha_n\theta), \quad \alpha_n = \frac{n\pi}{\theta_0}. \tag{5.2}$$

The angle functions here $\cos(\alpha_n\theta)$ are chosen according to the boundary conditions on the rigid surfaces of the plane and screen (when $\theta = 0$ and $\theta = \theta_0$). If the condition $r_S < r_h$ is fulfilled, then the expression for the field of source $H_0^{(1)}(kr_1)$ should be transferred from Eq. (5.1) into the right-hand side of Eq. (5.2).

We write the pressure field for domain III as follows:

$$p_{III} = \sum_{n=0}^{\infty} C_n \frac{J_{\beta_n}(kr)}{J'_{\beta_n}(kh)} \cos(\beta_n(\theta - \theta_0)), \quad \beta_n = \frac{n\pi}{\pi - \theta_0}. \tag{5.3}$$

Standard notations for Bessel and Hankel functions are used in Eqs. (5.1)–(5.3).

The sound source can be in domain I, and in domain II. The linear source field $H_0^{(1)}(kr_1)$ should be written in coordinate system (r, θ). For this purpose, we will use the Green function for an infinite wedge with acoustically rigid boundaries (the field of linear source differs from Green function only in constant factor [14]):

$$G(r, \theta; r_S, \theta_S) = \frac{\pi i}{2\alpha} \sum_{n=0}^{\infty} \varepsilon_n J_{\nu_n}(kr_<) H_{\nu_n}^{(1)}(kr_>) \cos(\nu_n \theta_S) \cos(\nu_n \theta),$$
$$\nu_n = \frac{n\pi}{\alpha}, \quad \varepsilon_n = \begin{cases} 1, & n = 0 \\ 2, & n > 0 \end{cases}, \tag{5.4}$$

where α—opening angle of wedge, $r_<$—is the least distance from r and r_S, $r_>$—the largest distance from two indicated distances.

When the source is spaced in domain I, the opening angle of wedge corresponds to the plane and is equal π. In such a situation the expression for pressure field p_{0I}, created by linear source, is as follows

$$p_{0I} = \frac{i}{2} \begin{cases} \sum_{n=0}^{\infty} \varepsilon_n J_n(kr) H_n^{(1)}(kr_S) \cos(n\theta_S) \cos(n\theta), & r_h < r < r_S, \\ \sum_{n=0}^{\infty} \varepsilon_n J_n(kr_S) H_n^{(1)}(kr) \cos(n\theta_S) \cos(n\theta), & r > r_S. \end{cases} \tag{5.5}$$

According to (5.1), the field in domain I

$$p_I = p_{0I} + \sum_{n=0}^{\infty} A_n \frac{H_n^{(1)}(kr)}{H_n'^{(1)}(kh)} \cos(n\theta). \tag{5.6}$$

If the source is situated in domain II, the opening angle of the wedge is equal θ_0 and the expression for the pressure field of the source is as follows:

$$p_{0II} = \frac{\pi i}{2\theta_0} \begin{cases} \sum_{n=0}^{\infty} \varepsilon_m J_{\nu_n}(kr) H_{\nu_n}^{(1)}(kr_S) \cos(\nu_n \theta_S) \cos(\nu_n \theta), & r < r_S, \\ \sum_{n=0}^{\infty} \varepsilon_n J_{\nu_n}(kr_S) H_{\nu_n}^{(1)}(kr) \cos(\nu_n \theta_S) \cos(\nu_n \theta), & r_S < r < r_h, \end{cases} \tag{5.7}$$

where $\nu_n = \alpha_n = n\pi/\theta_0$. Taking into account expression (5.2), we have

$$p_{II} = p_{0II} + \sum_{n=0}^{\infty} B_n \frac{J_{\alpha_n}(kr)}{J'_{\alpha_n}(kh)} \cos(\alpha_n \theta). \tag{5.8}$$

Let us write the system of functional equations, which defines the conditions of sound field continuity at the boundaries of domains I and II, III:

$$p_I = \begin{cases} p_{II}, & r = r_h, \quad \theta = [0, \theta_0], \\ p_{III}, & r = r_h, \quad \theta = [\theta_0, \pi], \end{cases} \tag{5.9}$$

$$\frac{\partial p_I}{\partial r} = \frac{\partial p_{II}}{\partial r}, \quad r = r_h, \quad \theta = [0, \theta_0], \tag{5.10}$$

$$\frac{\partial p_I}{\partial r} = \frac{\partial p_{III}}{\partial r}, \quad r = r_h, \quad \theta = [\theta_0, \pi]. \tag{5.11}$$

Taking into account the orthogonality of the corresponding sets of functions, we will algebraize the functional system of Eqs. (5.9)–(5.11). We obtain an infinite system of linear algebraic equations in unknown coefficients A_n, B_n, C_n. We do not cite the final form of this system because of its awkwardness.

We turn our attention now to the solution of system of algebraic equations and the estimation of accuracy of numerical results. The solution of infinite system of equations was considered in the previous chapters. Such algorithms are effective due to the calculation of the singularities in the vicinity of angular points (in our case it is an upper edge of a barrier). It allows us to obtain the quantitative estimations of sound fields characteristics in the domains arbitrarily close to angular points. If we want to find out the field properties in the points distant from the angles, the accurate results can be achieved using simple reduction method. We will proceed exactly this way.

Let us consider the accuracy of fulfillment of matching conditions at the boundaries of partial domains. To improve the accuracy of sound field estimation, we should increase the number of kept unknown complex coefficients when solving the algebraic system of equations by reduction method. It is especially important for the high frequency domain of noise spectrum. The numerical experiments were conducted, which allow us to estimate the degree of fulfillment of sound fields matching conditions for the wide range of values of geometric parameters of barrier and frequency. We can obtain the results, which can be used in practice, if the total number of kept unknown complex coefficients is 120 $\div$ 270. The relative accuracy of pressure matching and particle velocity matching will be no worse than 0.01, except for the points in the immediate neighborhood of barrier edge.

5.1.3 Numerical Results Analysis

Let us begin with the analysis of sound field distribution around the barrier. This is the most illustrative and informative property of sound field. Figures 5.5–5.8 shows such data, calculated for the different values of frequency and for the different distances from the source to the barrier $h = 5.5$ m high. The barrier is perpendicular to the ground ($\theta_0 = 90°$, Figs. 5.5 and 5.6), or it is inclined (Figs. 5.7 and 5.8). The source S is situated at the ground level ($y_S = 0$, Fig. 5.4).

Each figure shows the adjacent to the barrier fragment of space distribution of difference in levels of sound pressures in decibels (i.e., pressure relative value)

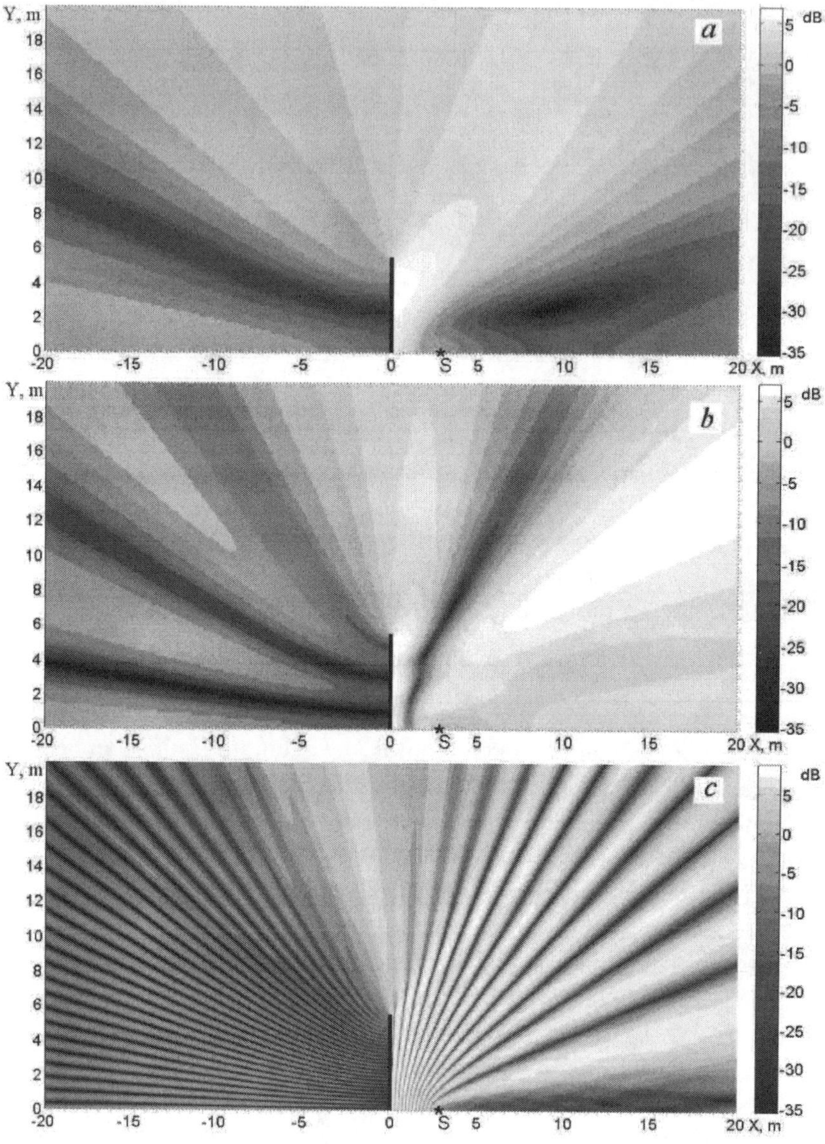

FIG. 5.5: Value ΔL distribution around the barrier; $h = 5.5$ m, $b = 2.5$ m, $\theta_0 = 90°$, $y_S = 0$; (a)—$f = 34$ Hz; (b)—$f = 85$ Hz; (c)—$f = 850$ Hz

$$\Delta L = 20 \lg \frac{p}{p_0}, \tag{5.12}$$

here p—pressure in the observation point when there is a barrier, p_0—pressure in the observation point without a barrier. The coordinates of the considered domains on Ox and Oy axis are given in meters: 20 m in horizontal direction in front of the barrier and behind the barrier, 20 m up from the ground.

The dynamic range of ΔL variation in decibels is shown in figures with the aid of a scale, which is situated to the right of each figure. The negative values of ΔL are plotted

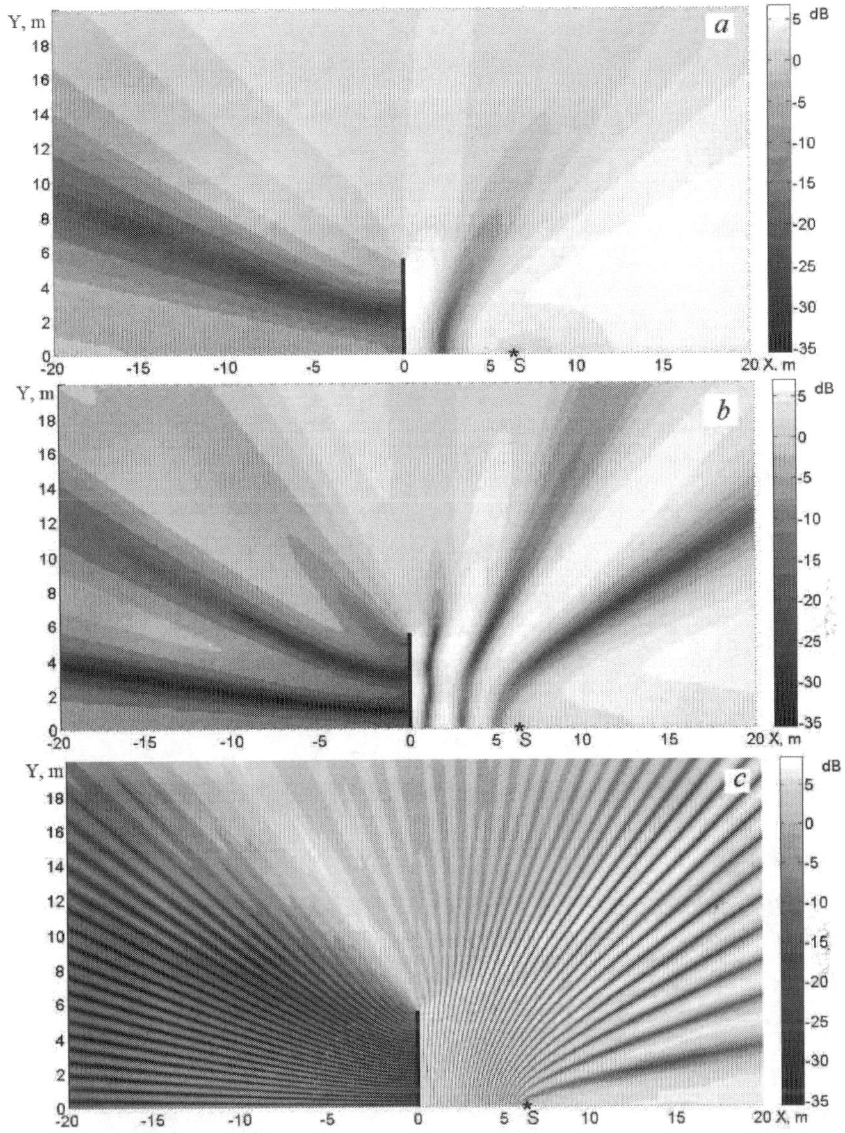

FIG. 5.6: Value ΔL distribution around the barrier; $h = 5.5$ m, $b = 6$ m, $\theta_0 = 90°$, $y_S = 0$; (a)—$f = 34$ Hz; (b)—$f = 85$ Hz; (c)—$f = 850$ Hz

down from 0, that corresponds to sound level reduction caused by the barrier, and the values plotted upwards correspond to the sound level increasing in the given point of space, caused by the presence of the barrier.

The analysis of data on sound field distribution around barrier allows us to determine the general regularities of the system source-barrier:

1. the whole sound shadow depth behind the barrier increases with an increase of frequency; it should be expected due to the increase of barrier wave height;

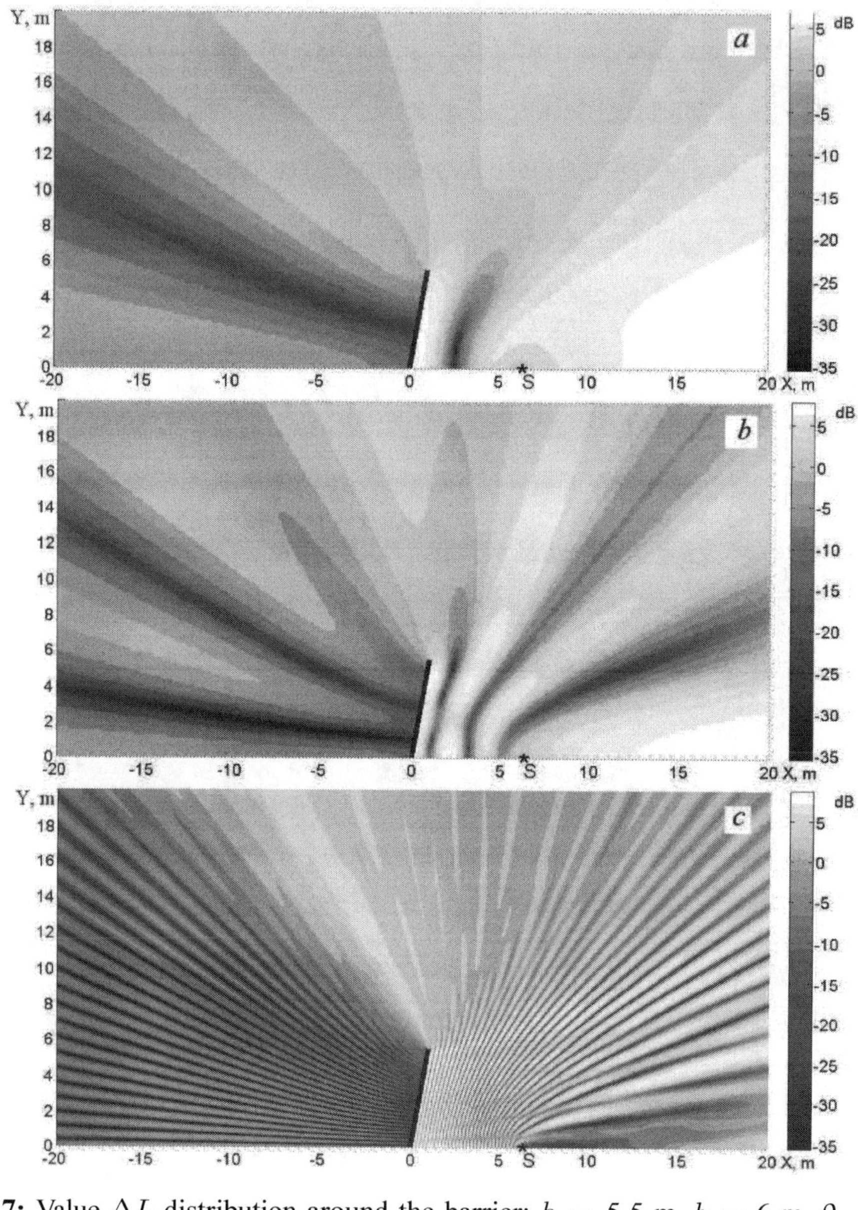

FIG. 5.7: Value ΔL distribution around the barrier; $h = 5.5$ m, $b = 6$ m, $\theta_0 = 80°$; (a)—$f = 34$ Hz; (b)—$h = 85$ Hz; (c)—$f = 850$ Hz

2. the total field nonuniformity behind the barrier increases with an increase of frequency; this is because of the linear source field interference with diffraction field on the barrier edge;

3. wave interference before the barrier becomes more noticeable when moving the source away from the barrier; this is because of the interaction of direct waves from the source and the waves reflected from the barrier in the opposite direction;

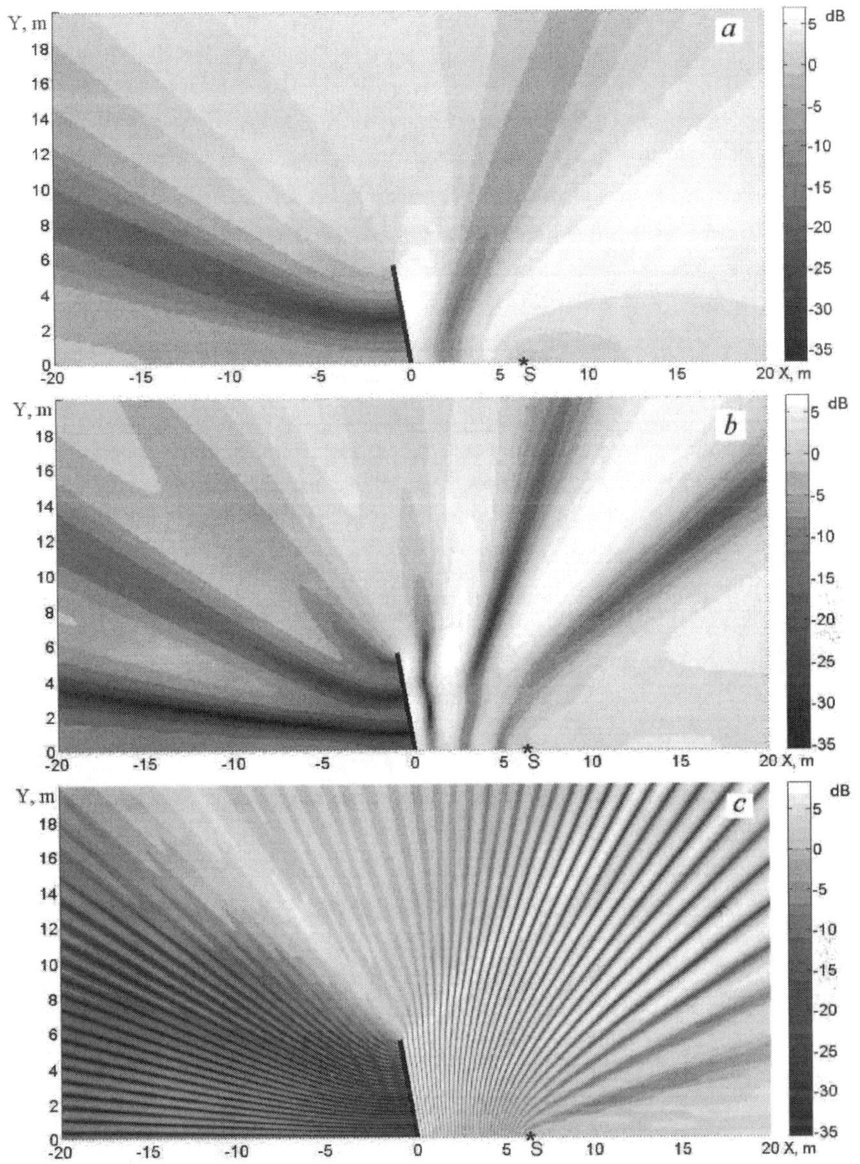

FIG. 5.8: Value ΔL distribution around the barrier; $h = 5.5$ m, $b = 6$ m, $\theta_0 = 100°$, $y_S = 0$; (a)—$f = 34$ Hz; (b)—$f = 85$ Hz; (c)—$f = 850$ Hz

4. when analyzing the influence of the angle of slope of the barrier about the ground, it may be noted, that this parameter influences on the size of sound shadow domain (it increases when the barrier bends to the sound source, it decreases when the barrier bends in the opposite side from the source). The angle of slope influences on the depth of the sound shadow to a lesser extent. The middle level of ΔL value in the domain of geometrical shadow remains almost unchangeable for all three variants of barrier slope.

Pressure space distribution in the far region of the source is described by the directivity pattern. The directivity pattern is defined as the angular distribution of pressure amplitude normalized to maximum. The "source-barrier" system can be considered as radiating system in this case. Expression (5.6) is computational for the field in domain I, taking into account the presence or absence of the sound source in domain I. The asymptotic representation of Hankel function is used $H_n^{(1)}(kr)$ when $kr \gg 1$.

Figures 5.9 and 5.10 show directivity patterns at frequencies 34 and 85 Hz; the distance between the source and barrier 2.5 m [figures (a)], 6 m [figures (b)].

The height of barrier is the parameter of the curves. As one should expect, the noise-protective properties of the barrier weaken with a decrease of barrier height. The decrease of barrier height from 5.5 m to 4 m influences the field distribution in the far field to a lesser extent than the further decrease of the height to 2 m.

It is of interest to compare the results, obtained using the given algorithm, with the results in work [3]. The geometry of problem considered in work [3], is shown in Fig. 5.11(a). The approximate solution of wave diffraction of point source S on the screen of finite height is obtained here. The screen is placed on a rigid surface. The point of observation P is in the far field of radiating system "source-barrier."

Figure 5.11(b) shows two frequency characteristics of the noise reduction value ΔL, computed using two methods for fixed point P of acoustic shadow zone. Curve 1

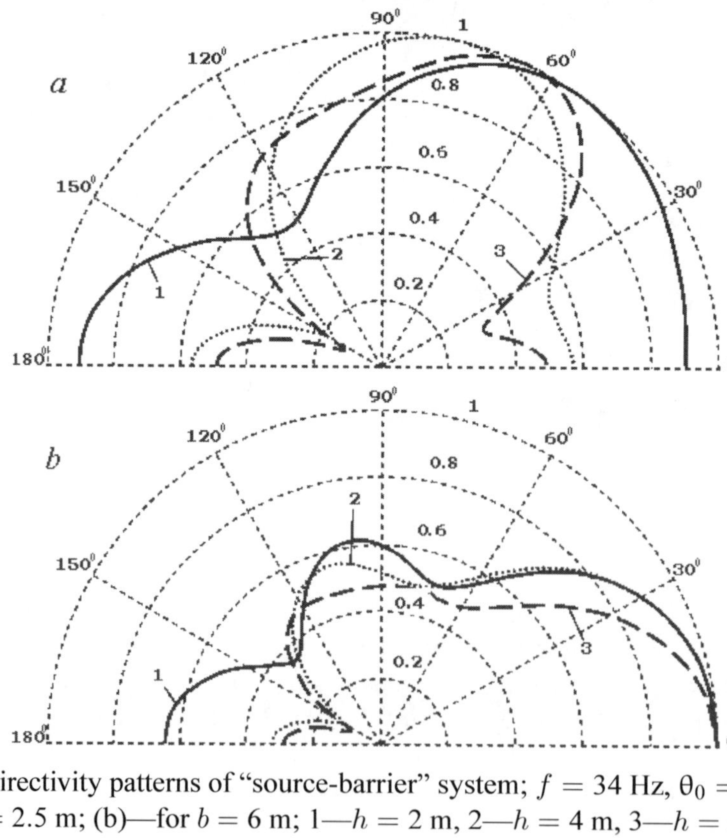

FIG. 5.9: Directivity patterns of "source-barrier" system; $f = 34$ Hz, $\theta_0 = 90°$, $y_S = 0$: (a)—for $b = 2.5$ m; (b)—for $b = 6$ m; 1—$h = 2$ m, 2—$h = 4$ m, 3—$h = 5.5$ m

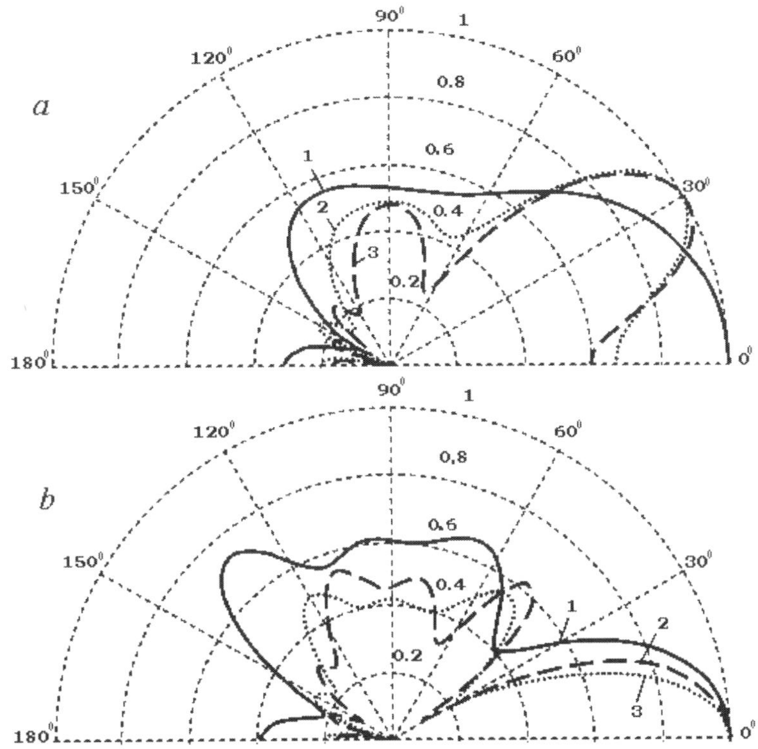

FIG. 5.10: The directivity patterns of system "source-barrier;" $f = 85$ Hz, $\theta_0 = 90°$, $y_S = 0$: (a)—for $b = 2.5$ m; (b)—for $b = 6$ m; 1—$h = 2$ m, 2—$h = 4$ m, 3—$h = 5.5$ m

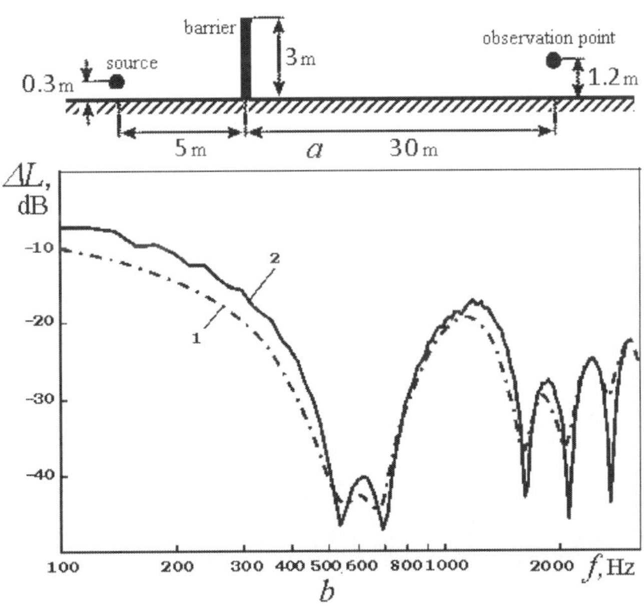

FIG. 5.11: Problem geometry (a); (b)—frequency dependencies of value ΔL; 1—point source [3], 2—linear source

corresponds to the data in work [3] given the point sound source, and curve 2—to the algorithm for linear harmonic sound source described above. As we can see, there is a satisfactory fit between the both curves, while the sources are of various kinds. A little divergence of the curves can be explained by the fact the point sound source is used in one case, and linear source in the other. The linear source has a smaller decrease of pressure amplitude per distance unit. It is for this reason the sound levels of the linear source behind the barrier are larger than of the point source (3 dB bigger for the frequencies from 100 to 300 Hz), and interference minimums are deeper.

In conclusion of this paragraph, we provide an example of estimation of classical barrier acoustical properties. Let us assume that it is necessary to protect the pedestrians, walking on a sidewalk near a highway, from traffic noise. It was agreed that a concrete barrier of four-meter height should be built between the sidewalk and the highway. Let us estimate (at some characteristic frequencies) the efficiency of such barrier. Let us estimate how the barrier can decrease the level of traffic noise in the sidewalk area. We assume that the pedestrians are two meters apart from the barrier, when they are walking on the sidewalk, and the sound source is at the ground level.

Figure 5.12 shows the levels of sound pressure (for the point of observation without a barrier), calculated along Oy axis at the distance 2 m from the barrier in the shadow zone. We assume that the height of a person does not exceed 2 m. It follows from the graphs, that barrier can reduce the traffic noise level in the sidewalk area in the amount for at least 12 dB in the frequency range from 34 to 850 Hz.

The given regularities allow us to draw a conclusion. The barrier efficiency of noise level reduction depends on the source noise spectrum, and on mutual arrangement of the source, the barrier, and the area protected from noise. Barriers efficiency depends also on the form and the physical properties of their surfaces (these questions are considered further). There are a lot of parameters of the problem to solve and the field character

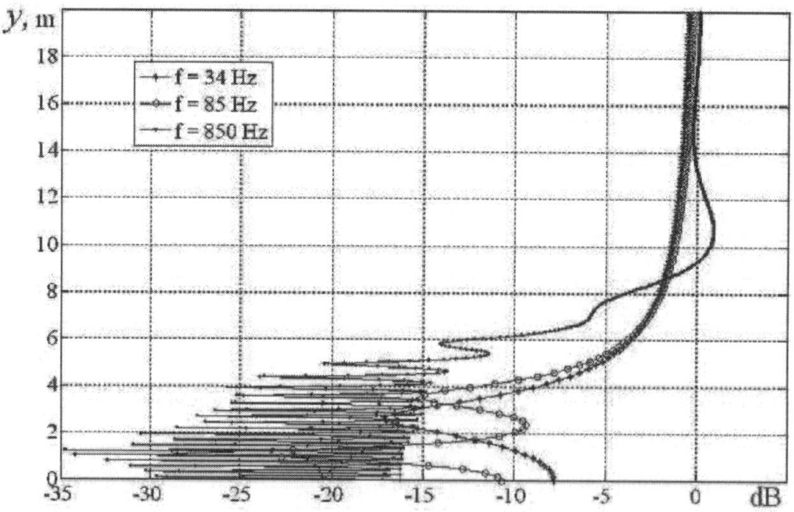

FIG. 5.12: Vertical distribution of sound pressure levels 2 m apart from the barrier in its shadow zone: $h = 4$ m, $b = 2.5$ m, $\theta_0 = 90°$, $y_S = 0$

behind the barrier is rather complex. Consequently, it is possible to predict the efficiency of the barriers basing on the strict solution of the problems of sound dissipation. We should solve the problems for specific purposes, with specific conditions and barrier location.

5.2 ACOUSTIC PROPERTIES OF V-BARRIER

5.2.1 Basic Idea

Sound barriers in the form of vertical rigid wall of specified height are widely used to protect residential areas from traffic noise due to their simple design and relative cheapness. The barrier height is the main parameter influencing the barrier efficiency. The higher barrier is the higher is its efficiency, all other parameters being equal. But this solution is irrational from the stand point of economy and aesthetics. Consequently, we are looking for the ways how to increase the barriers efficiency without increasing their height [4,15]. One of the possible constructions of the barrier with advanced efficiency is given bellow. And theoretical analysis of its acoustic properties estimation is given.

Let's get back to the traditional barrier in the form of a vertical wall, and we recall its basic limitations. Figure 5.13(a) shows such a barrier and the noise source S, from which it is necessary to protect the zone at the left, behind the barrier. Let us consider the given situation within the framework of geometrical theory of diffraction [12]. According to this theory, all the semi-space above the surface level of the road can be divided into illuminated area and shadow area behind the barrier. The line separating these areas lies on semi-infinite line, coming through the source and the edge of the barrier. The boundary [Fig. 5.13(a)] begins near the barrier edge. The sound field in the illuminated area is determined by five components:

1. the direct rays from the source,
2. the rays reflected from the barrier,
3. the rays reflected from the road surface,
4. the rays reflected from the road and from the barrier, and also
5. the ray's fan, which appears when the direct ray falls on the barrier edge.

The sound field in geometric shadow area is defined by the rays scattered by the barrier edge. These rays are the main reason of sound field appearance in shadow area behind the barrier.

One of the ways to weaken the intensity of rays scattered on the edge of the barrier, is to create some additional field in the zone of the barrier edge. This field should be equal in intensity to the field created by diffraction rays fan on the edge of the barrier, but it should be different from this field in 180° phase. These fields may give total intensity close to zero, when being added in the barrier edge area. It weakens the field in the barrier

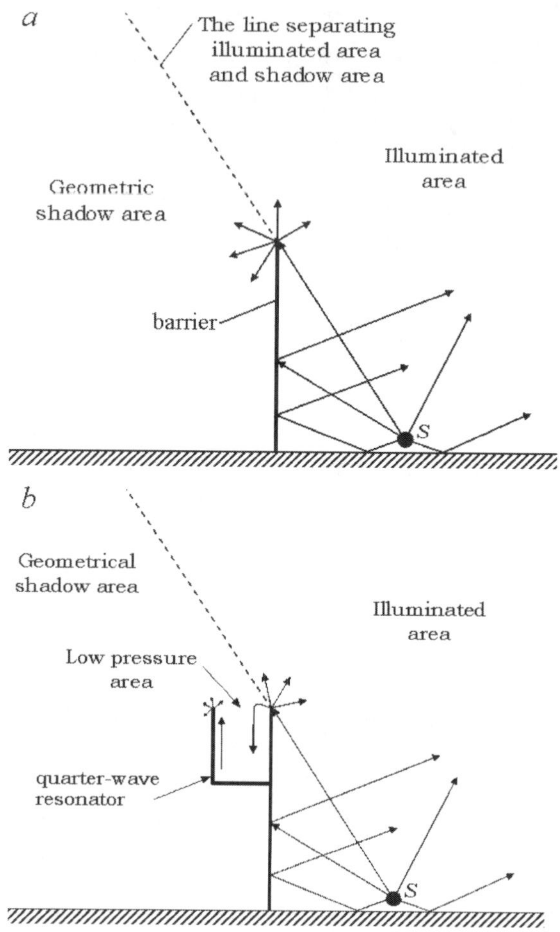

FIG. 5.13: The scheme of the barrier in the form of vertical wall (a) and the barrier with resonator near its edge (b)

shadow area. A similar idea was investigated experimentally in work [4], but there was not a rigorous analytical solution of a problem.

To implement this idea, we place a quarter-wave resonator in the form of a glass with the rigid walls and bottom near the barrier edge [Fig. 5.13(b)]. When the sound wave is incident on the inlet of resonator, two sound rays will propagate inside of it: direct, moving from the inlet to the bottom and the ray, reflected from the bottom and moving in opposite direction. A standing wave appears along the resonator axis because of the interference of these waves. If the resonator depth is equal to quarter wavelength, the maximum sound pressure area is near the resonator bottom, and minimum sound pressure area is near its inlet. Consequently, the level of noise, “flowing” behind the barrier, will be reduced.

Let us consider the structural scheme of the barrier, with a quarter-wave resonator in the area of its edge. We will show that the resonator efficiency can be higher than the efficiency of traditional barrier in the form of a simple wall of the same height.

5.2.2 Physical Barrier Model

The proposed construction of sound barrier with quarter-wave resonator near the edge is represented in Fig. 5.14. It contains two inclined to each other rigid panels 1. The bottom part of the space between them is filled with acoustically hard material 2 up to a definite level, and the top part is empty, and it is open to the surrounding space. Consequently, the upper part of the space between the walls is a peculiar kind of resonator. The depth of resonator can be changed due to the choice of material filling the bottom part of the space between the walls. Such barrier is called V-barrier.

We consider that the location of V-barrier and the traffic flow characteristics are the same as, for a classical barrier. Thus, as in the previous problem, we can confine ourselves to the consideration of two-dimensional problem in the plane perpendicular to the barrier. Then we can write the physical model of V-barrier, located along the highway (Fig. 5.14) as follows.

Let the earth surface be represented as an infinite acoustically hard surface. The V-barrier is located in the point O. The barrier consists of two rigid walls inclined to each other at an angle δ. The height of the walls r_h and inclination angle relative to the earth's surface θ_0 (Fig. 5.15). The filling level between the walls is equal r_d. The height of the walls r_h is related to the height of V-barrier h by expression $r_h = h/\sin\theta_0$. The linear

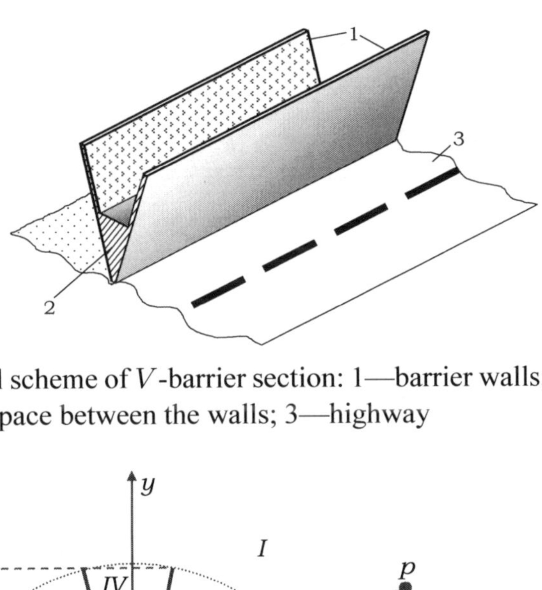

FIG. 5.14: Structural scheme of V-barrier section: 1—barrier walls; 2—acoustically hard material, filling the space between the walls; 3—highway

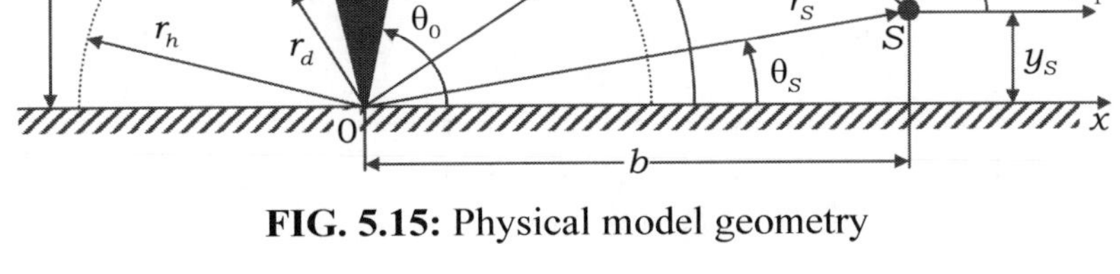

FIG. 5.15: Physical model geometry

harmonic sound source S is to the right of the barrier at the height y_S at the distance b from barrier. The point of observation is denoted by P. This model allows us to construct rigorous analytical problem solution [16].

5.2.3 Analytical Solution Construction

Let us introduce the polar coordinate system (r, θ) with the center in the point O. Let us partition all the domain of sound field existence into four different domains, according to partial domains method. Domain I is on the exterior of semicircle with radius r_h, it is bounded with the coordinates $r \geq r_h$, $0 \leq \theta \leq \pi$. The other three domains are the adjoining sectors inside the circle with radius r_h. Thus, domain II, situated between the ground and the right wall of the barrier, is given by the following inequality $r \leq r_h$, $0 \leq \theta \leq \theta_0$. Domain III is between the ground and the left wall of the barrier, that is $r \leq r_h$, $\theta_0 + \delta \leq \theta \leq \pi$. And domain IV is inside the resonator: $r_d \leq r \leq r_h$, $\theta_0 \leq \theta \leq \theta_0 + \delta$. All the model surfaces (Fig. 5.15) are ideally rigid.

The sound source can be in domain I ($r_S > r_h$), or in domain II ($r_S < r_h$). We should use the Green function for an infinite wedge with acoustically hard boundaries [Eq. (5.4)] to write the sound field of linear source.

Let $r_S > r_h$. The sound field in domain I should then be written as Eq. (5.6)

$$p_I = p_{0I} + \sum_{n=0}^{\infty} A_n \frac{H_n^{(1)}(kr)}{H_n'^{(1)}(kh)} \cos(n\theta), \tag{5.13}$$

The aggregate of arbitrary coefficients A_n allows us to fulfill the matching conditions at the boundary with domains II, III, and IV. Let us represent the pressure field in domain II as the superposition of standing waves:

$$p_{II} = \sum_{n=0}^{\infty} B_n \frac{J_{\alpha_n}(kr)}{J'_{\alpha_n}(kh_h)} \cos(\alpha_n \theta), \quad \alpha_n = \frac{n\pi}{\theta_0}. \tag{5.14}$$

The angle functions here $\cos(\alpha_n \theta)$ are chosen according to the boundary conditions on the rigid surfaces of the ground and the barrier ($\theta = 0$ and $\theta = \theta_0$). The sequence of coefficients B_n provides for the fulfillment of matching conditions at the boundary with domain I. Likewise, the pressure field in domain III is as follows:

$$p_{III} = \sum_{n=0}^{\infty} C_n \frac{J_{\beta_n}(kr)}{J'_{\beta_n}(kh)} \cos(\beta_n(\theta - \theta_0 - \delta)), \quad \beta_n = \frac{n\pi}{\pi - \theta_0 - \delta}. \tag{5.15}$$

Sound field in domain IV should be written as follows:

$$p_{IV} = \sum_{n=0}^{\infty} \left(D_n \frac{J_{\gamma_n}(kr)}{J'_{\gamma_n}(kr_h)} - C_n \frac{N_{\gamma_n}(kr)}{N'_{\gamma_n}(kr_d)} \right) \cos(\gamma_n(\theta - \theta_0)), \quad \gamma_n = \frac{n\pi}{\delta}. \tag{5.16}$$

From the boundary condition on the resonator bottom $\partial p_{IV}/\partial r = 0$, $r = r_d$, $\theta_0 \leq \theta \leq \theta_0 + \delta$, we find the relation between coefficients C_n and D_n: $C_n = -D_n(J'_{\gamma_n}(kr_d)/J'_{\gamma_n}(kr_h))$. Then Eq. (5.16) takes the form of

$$p_{IV} = \sum_{n=0}^{\infty} D_n \left(\frac{J_{\Upsilon_n}(kr)}{J'_{\Upsilon_n}(kr_h)} - \frac{J'_{\Upsilon_n}(kr_d)}{J'_{\Upsilon_n}(kr_h)} \frac{N_{\Upsilon_n}(kr)}{N'_{\Upsilon_n}(kr_d)} \right) \cos\left(\Upsilon_n (\theta - \theta_0)\right). \tag{5.17}$$

We write the system of functional equations determining the continuity conditions of the sound field at the boundaries of domains I and II, III, IV.

$$p_I = \begin{cases} p_{II}, & r = r_h, \quad 0 \le \theta \le \theta_0, \\ p_{IV}, & r = r_h, \quad \theta_0 \le \theta \le \theta_0 + \delta, \\ p_{III}, & r = r_h, \quad \theta_0 + \delta \le \theta \le \pi, \end{cases} \tag{5.18}$$

$$\frac{\partial p_I}{\partial r} = \frac{\partial p_{II}}{\partial r}, \quad r = r_h, \quad 0 \le \theta \le \theta_0, \tag{5.19}$$

$$\frac{\partial p_I}{\partial r} = \frac{\partial p_{IV}}{\partial r}, \quad r = r_h, \quad \theta_0 \le \theta \le \theta_0 + \delta, \tag{5.20}$$

$$\frac{\partial p_I}{\partial r} = \frac{\partial p_{III}}{\partial r}, \quad r = r_h, \quad \theta_0 + \delta \le \theta \le \pi. \tag{5.21}$$

We algebrize Eqs. (5.18)–(5.21) using orthogonality properties of the corresponding sets of angle functions. Consequently, we obtain an infinite system of linear algebraic equations of the second kind in unknown coefficients A_n, B_n, C_n, D_n. When solving the infinite system, we will follow recommendations given in Section 5.1.

5.2.4 Numerical Results Analysis

Let us perform a comparative analysis of the space distribution of pressure sound fields for two types of barriers: classical barrier in the form of simple wall and V-barrier. Figures 5.16 and 5.17 shows the data, calculated for different frequencies, when the effective barriers height is $h = 4$ m. The resonator dimensions for each frequency are selected individually, according to the following rules:

1. the depth of the cavity between the walls is about quarter-wavelength
$$r_h - r_d \approx \lambda/4,$$
2. the distance between the upper edges of the walls is small in relation to the wavelength
$$2r_h \sin(\delta/2) < \lambda/3.$$

As we can see, the barrier and the resonator have the noise level in the shadow zone lower than the noise level of a traditional barrier at all the frequencies under consideration. In particular, the field's levels near the ground behind the V-barrier are lower, than behind the traditional barrier about 6 dB (for the frequencies 34 Hz and 85 Hz) and about 10 dB for the frequency 850 Hz. Thus, the calculation data shows, that V-barrier is more effective noise protection due to the resonator.

Let us consider in more detail, what is happening inside the resonator in the area adjacent to its inlet (throat), Fig. 5.17(a). The pressure near the resonator bottom is high

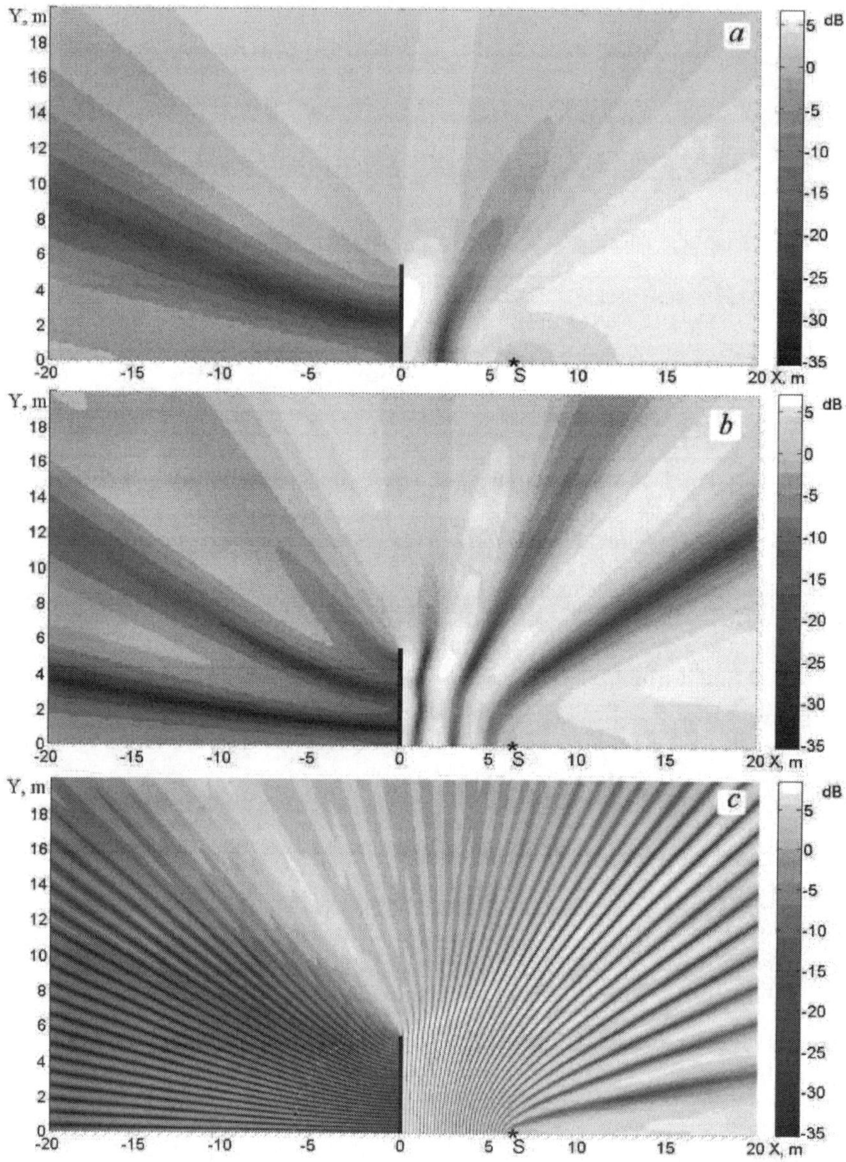

FIG. 5.16: The value of ΔL distribution around a classical barrier: $h = 4$ m, $b = 6$ m, $\theta_0 = 90°$, $y_S = 0$; (a)—$f = 34$ Hz; (b)—$f = 85$ Hz; (c)—$f = 850$ Hz

because of the field distribution here (about +8 dB). The pressure near the inlet of resonator is low (about –10 dB). The ideal situation for sound protection is when this zone of low pressure (like a "lid") covers completely the inlet of resonator and is parallel to the ground. It would be like this, if the source of sound was on Oy axis above the resonator. But if the sound source is alongside of the barrier (and the resonator), the low pressure zone near the inlet of resonator is above the right wall (on the line connecting the source with the right edge of resonator).

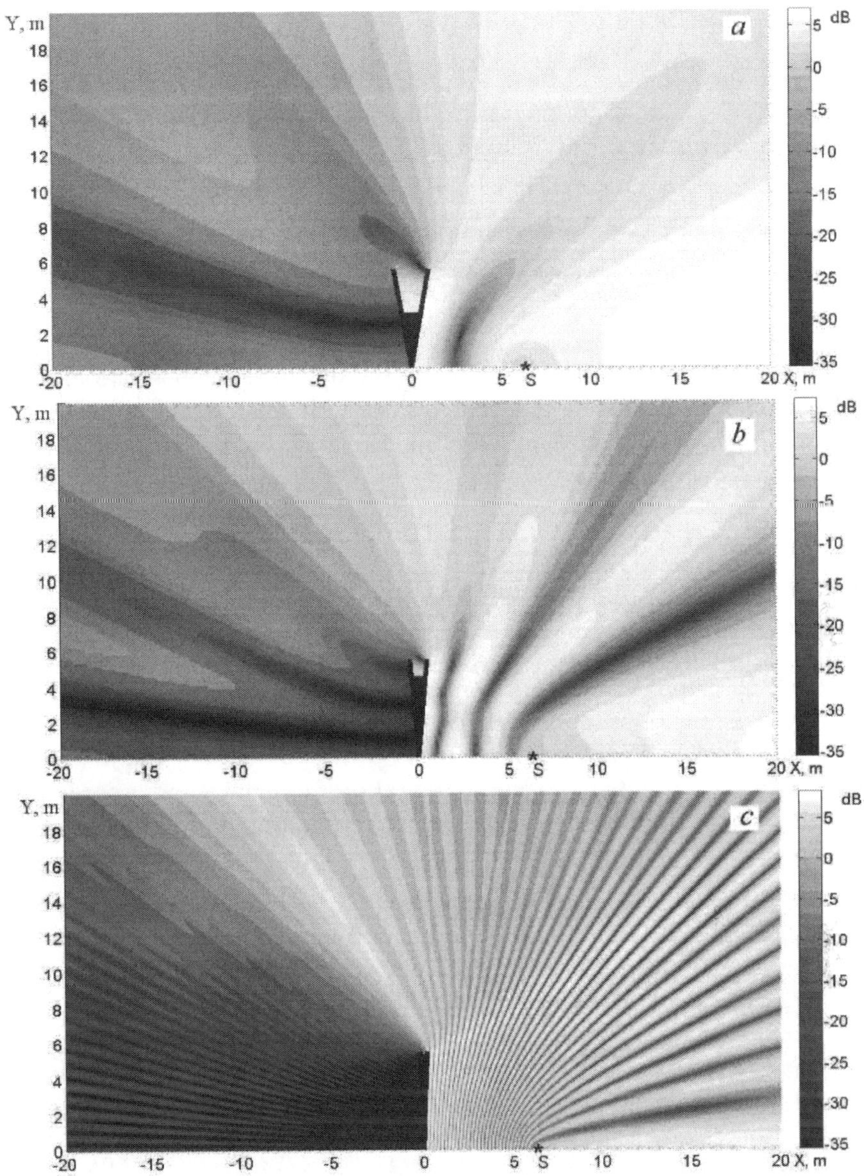

FIG. 5.17: Value ΔL distribution around V-barrier: $h = 4$ m, $b = 6$ m, $\theta_0 = 90°$, $y_S = 0$; (a)—$f = 34$ Hz; (b)—$f = 85$ Hz; (c)—$f = 850$ Hz

To analyze the barrier with resonator efficiency, we have to analyze the influence of its parameters, such as the opening angle of resonator δ, the angle of slope θ_0 and the resonator depth. As it follows from the numerical analysis, we have found some important regularities:

1. the V-barrier efficiency is higher than the efficiency of traditional barrier in the form of a wall in almost all the frequency range under consideration;

2. the most efficient barrier is when its symmetry axis coincides with axis Oy;

3. the optimum resonator aperture width is about 0.2λ, λ—wave length;

4. optimum resonator depth is about 0.25λ.

The following recommendations may be useful in practice:

1. the barrier under consideration is especially efficient with relatively narrow-range noise source, or the source with several dominant frequency components;

2. it makes sense to include the resonator into the barrier construction, if the problem frequencies are relatively low (barrier wave dimension $kh \leq 10, ..., 15$). There is no difference in efficiency between traditional and V-barrier at higher frequencies.

5.3 INTEGRAL CHARACTERISTICS OF CLASSICAL AND V-BARRIER

5.3.1 Estimation Method of Integral Characteristics of the Acoustic Barriers

Sound pressure variation value ΔL is calculated to analyze the sound field's properties in the vicinity of the barrier. It gives the comprehensive information about the local values of sound pressure levels in any point of space. It should be noted that the local values of sound pressure depend on the barrier configuration, frequency and noise source position relative to the barrier. Such detailed comparative evaluation of different barriers ability to protect from noise is rather labor-intensive task. We also need some generalized estimations of noise-protective properties of the barriers to design the barriers. Such estimations allow us to choose the barrier design without a detailed comparison of sound field's local values.

It is worthwhile to use the energetic characteristics of sound field such as: full power radiated by source W_0 (with barrier) and sound field power W_D, penetrating into the zone of barrier geometric shadow due to diffraction [17]. Figure 5.18 shows the barrier with all the parameters, which are necessary to find the opening angle of geometric shadow ϕ.

We take the value

$$G = \frac{W_D}{W_0}, \tag{5.22}$$

as an integral space criterion for the estimation noise-protective properties of the barrier. This value is an energy coefficient of sound transmission into barrier shadow zone. Plane problem of sound diffraction, medium loss free condition, is under consideration. The average power of the source per unit length in a period can be calculated by arc L integration. This arc is of arbitrary radius r, and it encloses the barrier and source S:

$$W_0 = \int_0^{\pi} I(r, \theta) d\theta, \tag{5.23}$$

the intensity I in each point of the arc L is expressed as sound pressure p in this point and radial (normal to the arc) particle velocity component v_r: $I = 0.5\mathrm{Re}\,(pv_r^*)$, the asterisk means complex conjugation.

We can find the value W_D by calculating the power flow, through the arc L_W with radius r_W (Fig. 5.18). This arc is situated in shadow zone. It is bounded by the ground and by the boundary line of geometric shadow SP, coming from the source through the barrier edge:

$$W_D = \int\limits_{\pi-\phi}^{\pi} I\,(r_W, \theta)d\theta. \tag{5.24}$$

According to the problem geometry of V-barrier (Fig. 5.18), the opening angle of integration arc ϕ is expressed through the angle of geometric shadow α, barrier height h and arc radius r_W as follows:

$$\phi = \alpha + \arcsin\left(\frac{h}{\sin\theta_0}\frac{\sin(\alpha+\theta_0)}{r_W}\right), \tag{5.25}$$

and for classical barrier ($\delta = 0$, $\theta_0 = 90°$)

$$\phi = \alpha + \arcsin\left(h\frac{\cos\alpha}{r_W}\right). \tag{5.26}$$

The behavior of the value of ϕ results from the last two formulas when r_W increases infinitely: $\lim\limits_{r_W\to\infty}\phi = \alpha$. The power level of sound field, coming through the arc L_W, does not changes for rather large distances ($r_W \gg \lambda$, $r_W \gg h, r_S$) (in this case the integration arc is the part of the common wave front, formed by system "source-barrier-ground").

Let us introduce the integrated space-frequency criterion to make a quantitative description of frequency dependency of barrier efficiency

$$\Psi = \int\limits_{f_1}^{f_2} G\,(f)\,df, \tag{5.27}$$

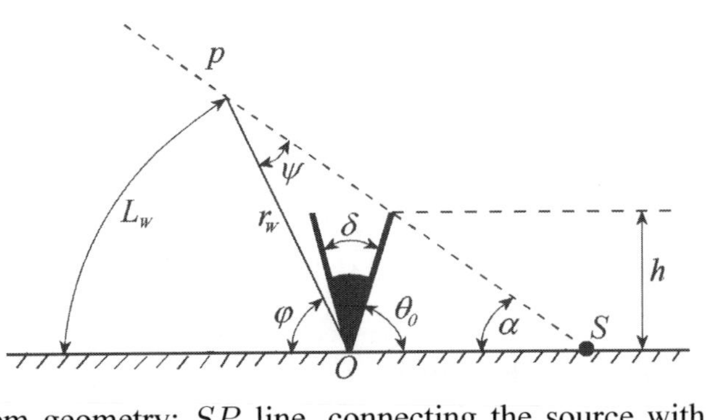

FIG. 5.18: Problem geometry: SP line, connecting the source with the barrier edge, separates the shadow zone from the illuminated zone

where $G(f)$—frequency dependency of space integrated criterion, f_1 and f_2—the boundaries of frequency band in which the integration is carried out. This criterion allows us to describe the efficiency of sound barrier, taking into account the frequency dependencies of barrier. It makes easier the comparative analysis of efficiency of sound barriers of varying design.

5.3.2 Numerical Results Analysis

We turn our attention now to the calculation of space criterion of barrier efficiency G. We begin with studying its dependence on the integration arc radius r_W. It allows us to find the value of radius, starting with which, the value of $G(r_W)$ can be considered as constant value. Figure 5.19 shows the corresponding calculations for classical barrier

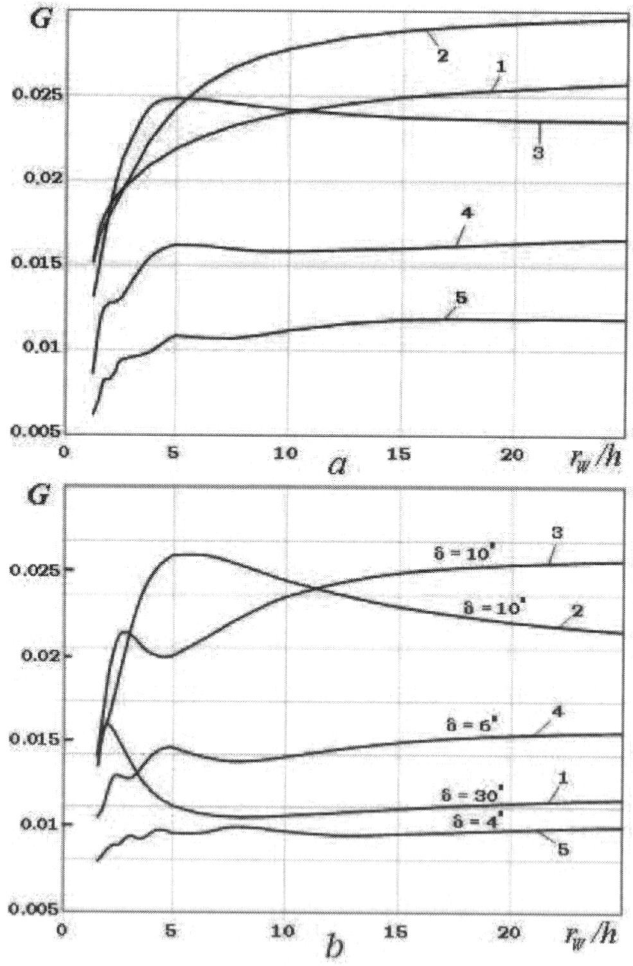

FIG. 5.19: Space criterion G dependency on normalized radius r_w/h; (a)—classical barrier, (b)—V-barrier; 1—$f = 34$ Hz, 2—$f = 85$ Hz, 3—$f = 125$ Hz, 4—$f = 250$ Hz, 5—$f = 500$ Hz

[Fig. 5.19(a)] and V-barrier [Fig. 5.19(b)]. Every figure shows the graphs for five different frequencies. The value of integration radius r_W is plotted along the abscissa axis. It is normalized to the efficient barrier height h. The height of both barriers is $h = 4$ m. The depth and opening angle of V-barrier for each curve are tuned to the frequency under consideration [opening angle δ is shown in Fig. 5.19(b)]. The source is 6 m apart from the barrier and it is situated on the ground.

The graphs behavior is the same in both figures. The data for the curves are variable when $r_W/h \leq 5$. The curves are almost parallel to abscissa axis for the large values of r_W/h. We note that V-barrier is much more efficient than classical (6 times larger and more).

Figure 5.20 shows the dependencies of the value of G on frequency for the both types of the barriers under investigation. We turn our attention first to curves 1 and 2. Curve 1 corresponds to classical barrier. Curve 2 is constructed for V-barrier. The resonator depth here is tuned to each frequency in order to correspond to quarter sound wave length. Opening angle $\delta = 10°$ is unchanged. It is optimal for resonator frequency 96 Hz. In real life it is impossible to realize such tuning. But it is interesting to investigate this situation. From the graphs shown in Fig. 5.20 it is seen that the value of G depends on frequency for classical barrier, and for V-barrier. The curves oscillations in both cases are caused by interferential phenomena, arising because of the interaction of direct wave from the source with the wave scattered by the barrier.

We turn our attention now to the frequency domain from 25 to 34 Hz (Fig. 5.20). We can see that the curve corresponding to the ideal V-barrier, which can be tuned to the current frequency, is on the higher level than for classical barrier. This points to the

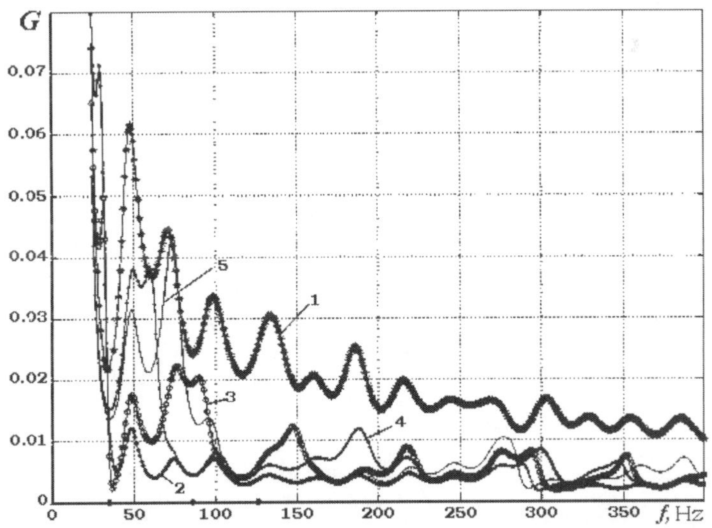

FIG. 5.20: Frequency dependency of space criterion G: $r_w = 20$ m, $h = 4$ m, $b = 6$ m, $y_s = 0$, $\delta = 10°$, $\theta_0 = 85°$; 1—classical barrier, 2—V-barrier, the resonator is tuned to the current frequency, 3—resonator is tuned to frequency 34 Hz, 4—resonator is tuned to frequency 85 Hz, 5—resonator is tuned to frequency 125 Hz

fact that V-barrier is less efficient than classical barrier in this frequency domain. It is related to the fact that for the frequencies less than 34 Hz the opening angle of resonator $\delta = 10°$ is too small and the sound wave "passes over" the resonator. When the wave distance between the top edges of resonator is so small (in this case it is about 0.07λ, instead of the recommended 0.2λ) all the construction behaves as the barrier with rather broad continuous top edge. The sound-protective properties of such barrier are worse than the properties of classical thin screen [2].

Curve 2 in Fig. 5.20 is constructed for V-barrier. The depth of its resonator varies with the frequency under consideration. To make a V-barrier with such frequency characteristic is almost impossible. Consequently, we will consider a more real V-barrier with fixed resonator dimensions. Curves 3, 4, 5 in Fig. 5.20 correspond to the situations, when the resonator of V-barrier is tuned to frequencies 34, 85, and 125 Hz. Such frequencies are chosen not coincidentally. Frequency 34 Hz is of interest, because it is one of the lowest frequencies, in the domain under consideration. 85 Hz and 125 Hz correspond to maximum levels in transport noise spectrum (see Fig. 5.3). Space-frequency criterion Ψ was calculated for each of these curves. The values of this criterion are given in Table 5.1.

As we might expect, the efficiency of barrier with fixed resonator tuning is lower than the efficiency of barrier with the resonator, which can be tuned to the frequency of noise source. Nevertheless, its efficiency is higher, than the efficiency of classical barrier. The V-barrier is the most efficient at the frequencies to which it is tuned. The tonal noise or narrow band is a rare phenomena. The efficiency of the barrier for such noise is maximum. When we choose the geometrical dimensions of resonator for broadband noise, we should focus on its dominant frequency components. The V-barrier tuned to frequency 85 Hz is the most efficient for the spectrums of transport noise given in Fig. 5.3 (Fig. 5.22, curve 3). The V-barrier, tuned to frequency 125 Hz, is well suited for protection from noise with dominant frequency components higher than 100 Hz. The V-barrier, tuned to frequency 34 Hz is valid for sound protection for the frequencies lower than 70 Hz. The analysis of value of space-frequency criterion shows, that the relation between the values of Ψ for the different barriers changes depending on the frequency domain. When integrating in domain from 100 to 400 Hz, we obtain the set of values of Ψ, which correspond to the graphs in Fig. 5.20. The efficiency of tunable V-barrier is five times larger than the efficiency of classical barrier. The efficiency of V-barrier, tuned to 34 Hz or to 85 Hz—is 3,5 times larger, and the efficiency of barrier tuned to 125 Hz—4 times larger than the efficiency of classical barrier.

TABLE 5.1: Integral criterion of barier efficiency

Frequency domain, Hz	**Space-frequency criterion Ψ values**				
	Classical barrier	**V-barrier, $\delta = 10°$, $\theta_0 = 85°$**			
		Tuneable	**34 Hz**	**85 Hz**	**125 Hz**
25–400	7.96	2.13	2.85	2.90	3.13
50–400	7.00	1.44	2.29	2.28	2.56
100–400	5.18	1.15	1.50	1.48	1.39

Basing on the conducted analysis, we may say that the integral criterions, proposed for the estimation of the barrier acoustic properties, (space criterion G and space-frequency criterion Ψ) are the clear and valid characteristic of their sound-protective properties. They can be used to analyze the different designs of the barriers to estimate their efficiency.

5.4 THE EFFECT OF SOUND BARRIER SURFACE PROPERTIES ON ITS EFFICIENCY

The design of classical barrier is simple. It is a thin vertical wall made of acoustically rigid material. We need to increase the height of the barrier in order to increase the shadow depth and its sizes behind the barrier. As it was mentioned above, such approach cannot be used in many circumstances. The ways to increase the efficiency of barriers, which are not higher than 4–6 meters, are looked for. The V-barrier was considered above as one of the possible problem solutions. The materials with special properties, used to improve the noise protective properties of the barriers, are considered in this paragraph [6,18].

As an example, Fig. 5.21 shows the photos of the barriers. These photos were made by one of the authors in the autumn of 2009 in Austria and Italy. It may be assumed that the surfaces of all three barriers possess some sound—absorbing properties. The barriers have different design. Figure 5.21(a) shows the barrier surface made of the panels of interlaced wooden ribbons. Such panels are often used for the treatment of walls and ceilings of concert halls. Figure 5.21(b) shows the barrier in the form of bookcases partially filled with sand. It is fair to assume that these bookcases are the quarter-wave resonators. Figure 5.21(c) shows the barrier in the form of the set of horizontal panels with pinpointing perforations. There is a rigid wall from backside of these panels at a distance.* It is so called resonant absorber [6].

There are a lot of different designs of barriers. It means that there is no optimal barrier design which meets several criteria, namely: acoustic efficiency, aesthetics and not very high price.

Let us add the following property to the model described in the Section 5.1.1 (see Fig. 5.4) in order to study the acoustic properties of barriers with the walls having sound absorption properties. The illuminated surface of the barrier is an impedance surface. It allows us to model the reflectance and absorption properties of real constructions. We assume that the sound source is on the ground and barrier is perpendicular to the ground.

The problem can be solved by partial domains method. Solution construction corresponds to Section 5.1.2. We write the pressure field in domain II as follows

$$p_{II} = \sum_{n=0}^{\infty} B_n \frac{J_{2n}(kr)}{J'_{2n}(kh)} \cos(2n\theta) + \sum_{n=0}^{\infty} D_n \cos\left(\frac{n\pi z}{h}\right) \exp(i\gamma_n x). \quad (5.28)$$

*The barrier would be absolutely sound-transparent without a rigid wall.

FIG. 5.21: Real barriers, used on the highways in Austria and Italy; (a) barrier made of interlaced thin wooden ribbons, (b) barrier in the form of a bookcase, partially field with sand, (c) barrier made of horizontal perforated panels

The first sum here is the totality of particular solutions of Helmholtz equations in the domain in the form of quadrant with radius h. Consequently, with due choice of coefficients B_n, this sum allows us to fulfill the arbitrary matching conditions of the fields on the interface of domains I and II. Angle functions $\cos(2n\theta)$ are chosen on the presumption that the ground and the illuminated side of barrier are acoustically rigid.

The illuminated side of barrier is characterized by some impedance Z. There is the second sum in Eq. (5.28) in order to fulfill the boundary condition on this surface. This sum is represented as the set of normal waves of plane-parallel waveguide with rigid boundary surfaces. The width of waveguide is h. Consequently, we can fulfill the impedance boundary condition on the illuminated side of the barrier by means of D_n coefficients selection.

Let us write the system of functional equations. It shows the conditions of continuity of sound field on the interface of domains I, II, III and boundary condition on the illuminated barrier surface:

$$\frac{\partial p_I}{\partial r} = \begin{cases} \dfrac{\partial p_{II}}{\partial r}, & r = h, \quad \theta = [0, \pi/2], \\ \dfrac{\partial p_{III}}{\partial r}, & r = h, \quad \theta = [\pi/2, \pi], \end{cases} \tag{5.29}$$

$$p_I = p_{II}, \quad r = h, \theta = [0, \pi/2], \tag{5.30}$$

$$\frac{1}{i\omega\rho}\frac{\partial p_{II}}{\partial r} = \begin{cases} 0, & x = 0, \quad y = [0, h_1], \\ \dfrac{p_{II}}{Z}, & x = 0, \quad y = [h_1, h], \end{cases} \tag{5.31}$$

$$p_I = p_{III}, \quad r = h, \theta = [\pi/2, \pi]. \tag{5.32}$$

Than we bring it to the infinite system of linear algebraic equations of the second kind in unknown coefficients A_n, B_n, D_n, C_n. We can obtain the satisfactory results for practical purposes if the total amount of unknown complex coefficient is about 130.

Let us set the following geometric parameters of barrier model (Fig. 5.4): barrier height $h = 4$ m, the distance from the barrier to the sound source $b = 6$ m, the radius of the arc where the energy flow is calculated $r_w = 20$ m. It is to be recalled that $y_s = 0$, $\theta_0 = 90°$. Let us analyze the noise-protective properties of the barrier. We start with the situation when the illuminated side of the barrier has the properties of acoustically ideal surface (Fig. 5.22). Curve 1, in this figure, corresponds to acoustically rigid surface of the illuminated barrier side, curve 2—acoustically soft surface, curve 3—absorbing surface. For these three variants, the value of specific (per unit area) and normalized to the value of ρc impedance of the illuminated barrier surface is $Z = \infty$, $Z = 0$, $Z = 1$, respectively.

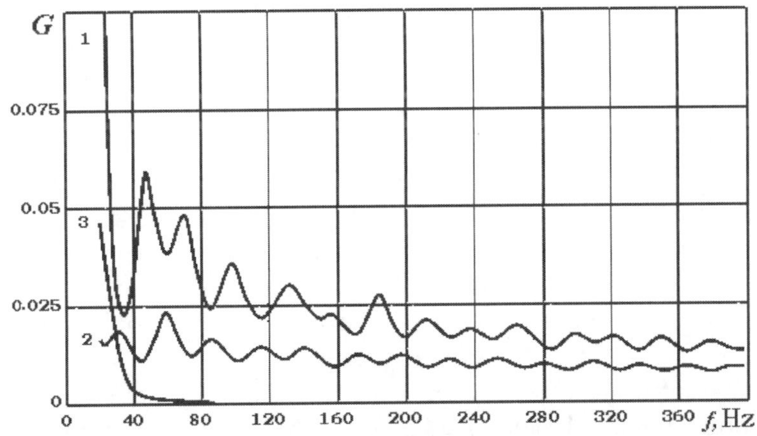

FIG. 5.22: Energy coefficient G frequency dependency for the case of acoustically ideal surface of illuminated barrier side: 1—acoustically rigid, 2—acoustically soft, 3—totally absorbing

Curve 1 in Fig. 5.22 shows the potentials of classical barrier with the acoustically rigid surfaces. We can indicate three sections on this curve: the first—very-low frequencies domain to 30 Hz, the second—low frequencies from 30 Hz to 120 Hz, the third—middle and high frequencies (above 120 Hz). On the first section, the curve rises, with an increase of frequency. The sound-protective properties of barrier reduce here. Such barrier properties result from the small wave height of the barrier. On the second section, the value of coefficient G decreases rapidly, when there are strong oscillations, with an increase of frequency. On the third section, curve 1 oscillates and approaches to some value of coefficient G with an increase of frequency. The value of the coefficient is determined according to the geometrical theory of diffraction.

Curve 3 in Fig. 5.22 shows maximum possible values of energy coefficient G for the barrier with an absorbing surface. As we can see, the barrier becomes almost sound non-transparent for the frequencies above 30 Hz. Curve 3 (as well as curve 1) rises steeply up on the section of very-low frequencies (lower than 30 Hz). And we can observe the improvement of sound-protective properties in this frequency range in comparison with the barrier with acoustically rigid surfaces.

The run of the curve 2, which corresponds to the acoustically soft surface of the illuminated barrier side, is different from curves 1 and 3. Curve 2 is between these curves. The value of G coefficient is approximately constant in the domain of very-low frequencies. The barrier has such properties due to the fact that the phase of the wave reflected from acoustically soft wall is shifted 180° with respect to the incident wave. The interaction between them leads to the fact that the pressure on the barrier surface becomes equal to zero, and the pressure in immediate proximity to the barrier decreases. The general efficiency improvement is the result of it. But it is impossible to create an acoustically soft surface in the air environment in the wide range of frequencies. That is why curve 2 is rather a guide, than the model of real barriers.

Let us turn attention to curve 3 in Fig. 5.22. It is of interest to find out, how the value of coefficient G changes, if the ideally absorbing material is only on some section of the surface of illuminated barrier side. The rest of the surface is acoustically rigid.

The calculations given in Fig. 5.23 provide an answer for such question. Here curve 1 corresponds to the case when the barrier surface is covered completely, curves 2, 3, 4—when it is covered partially. The absorber is on the upper part of the barrier surface, starting from its top edge ($y = h$). Curve 2 corresponds to the case when 50% of the surface of illuminated barrier side is covered, curve 3—25% and curve 4—10%. As we can see, when 50% of the surface is covered (curve 2), the noise-protective properties are improved in almost all the considered frequency range (curve 1 in Fig. 5.22). When 25% of the surface is covered (curve 3) the noise-protective properties are improved in the domain of relatively high frequencies (above 120 Hz). As to the case when 10% of the surface is covered (curve 4), the effect is slight.

The curves represented in Figs. 5.22 and 5.23 allow us to conclude: the noise-protective barrier properties can be improved considerably, when an absorber covers not less than 25–50% of the surface of barriers illuminated side.

The sound-protective properties of the barriers with an absorber were considered above, regardless of sound frequency. Such study is useful for the modeling of absorbing

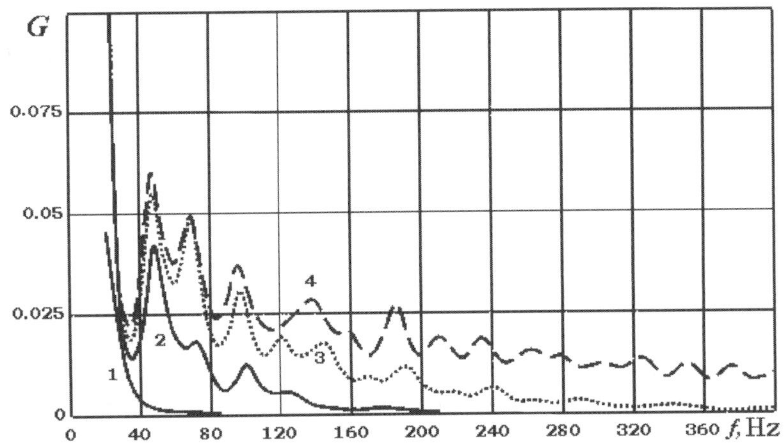

FIG. 5.23: Frequency dependency of energy coefficient G for the cases, when the ideally absorbing material is on some part of the surface of illuminated barrier side, starting from its top edge ($y = h$): 1—100% area of surface, 2—50%, 3—25%, 4—10%

coatings close to the real constructions (see. Fig. 5.21). Two of such models are shown in Fig. 5.24. We turn now to the first of them [Fig. 5.24(a)], which is the grating of quarter-wave resonators [19,20]. It is fair to assume that this model corresponds to a real barrier, which is shown in Fig. 5.21(b). Specific and normalized to the value ρc impedance of such construction is as follows [11]:

$$Z = \frac{r}{\rho c} + i\text{ctg}\left(\frac{\omega b}{c}\right), \tag{5.33}$$

where the coefficient r determines the sound energy loses inside this construction, b—resonator depth. Note that expression (5.33) makes sense in the domain of frequencies, for which the dimension d [Fig. 5.24(a)] is less than half sound wave length.

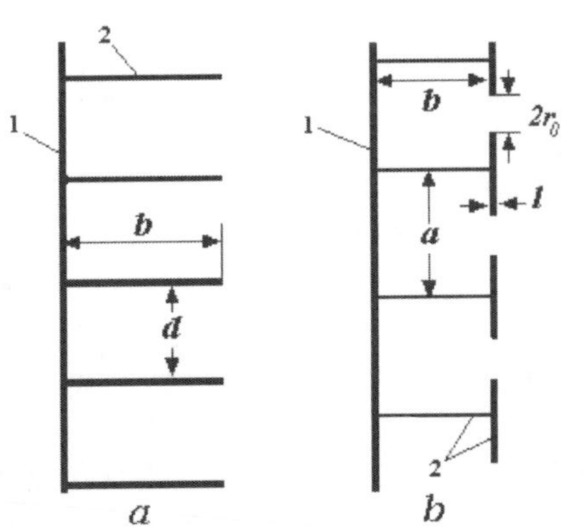

FIG. 5.24: Absorbing barrier cover constructions; (1—acoustically rigid barrier surface, 2—resonator walls): a—quarter-wave resonators grating, b—Helmholtz resonator grating

The value of b is equal to quarter sound wave length at the frequency of the main resonance of construction [Fig. 5.24(a)]. The pressure in the plane of inlet section of resonators is almost equal to zero (when $r = 0$ it is equal to zero). This construction simulates the situation close to the wave incidence on the acoustically soft surface.

The frequency dependencies of energy coefficient G are shown in Fig. 5.25 for the cases, when absorbing cover is made in the form of quarter-wave resonators [Fig. 5.24(a)]. Curve 1 in the figure corresponds to the acoustically rigid screen surface, curves 2 and 3 correspond to the construction in the form of the resonator grating, which covers all the illuminated side of the barrier. Resonator depth is $b = 0.25$ m for curve 2, $b = 0.50$ m for curve 3. The resonance frequencies corresponding to such dimensions are approximately equal 330 Hz and 165 Hz. The normalized coefficient is set equal $r/(\rho c) = 0.1$. As we can see in Fig. 5.25, each of the quarter-wave resonators constructions reduces the value of coefficient G in rather wide frequency range. The domain of these frequencies depends on resonator depth b.

We consider now the second model of absorbing cover in the form of the grating of Helmholtz resonators [Fig. 5.24(b)]. The grating covers all the surface of the illuminated barrier side. It is fair to assume that this model corresponds to a real barrier. The photo of this barrier is shown in Fig. 5.21(c). The parameters of a particular Helmholtz resonator depend on its geometrical dimensions: depth b and width a of resonator chamber, with diameter $2r_0$ and length l of Helmholtz resonator throat (we assume that, the resonator throat cross section is a circle, the area of which is $\sigma = \pi r_0^2$, and the cross section of chamber—a square with an area $S = a^2$). The natural frequency of Helmholtz resonator in the air [21,22]

$$f_0 \approx \frac{93 r_0}{a\sqrt{b\,(l + \Delta l)}} \tag{5.34}$$

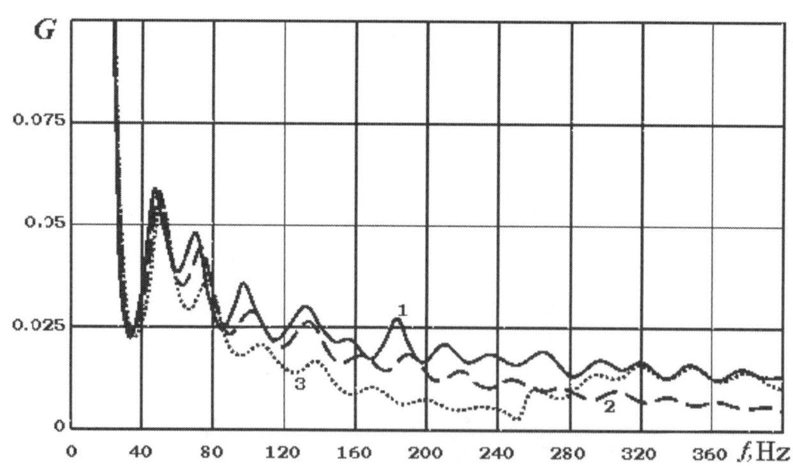

FIG. 5.25: Frequency dependency of energy coefficient G for the cases, when all the surface of illuminated barrier side is made in the form of a grating of quarter-wave resonators (loss coefficient $r/(\rho c) = 0, 1$): 1—barrier surface is acoustically rigid, 2—$b = 0.25$ m, 3— $b = 0.5$ m

The impedance of Helmholtz resonator (in fractions ρc and calculated per unit area of resonator chamber base)

$$Z = \frac{n\tilde{r}}{\rho c} + i\left(\mathrm{ctg}\left(\frac{\omega b}{c}\right) - \frac{\omega n}{c}(l + \Delta l)\right), \tag{5.35}$$

here $\tilde{r}$—friction coefficient in the constructional elements of Helmholtz resonator (it is usually small net or fabric, near the resonator throat, [22]), $n = S/\sigma$. The value $\Delta l = \pi r_0/2$ is determined by the added mass at the ends of resonator throat.

We obtain the barrier models with different frequency characteristics of energy coefficient G by changing the geometric and physical parameters of resonator construction. Figure 5.26 shows three calculation variants (curves 2, 3, and 4). Resonator throat diameter $2r_0$, chamber width a and friction coefficient $\tilde{r}$ are constant. Variable parameters are the resonator chamber depth b and throat length l. The chamber depth b increases, and the throat length l decreases with an increase of the curves 2, 3, 4 number. It leads to a decrease of equivalent elasticity and Helmholtz resonator mass. And the values of natural resonator frequencies change little, but its Q factor changes substantially.

As we can see in Fig. 5.26, for curves 2, 3, 4, there is a substantial coefficient G reduction for the range of frequencies in the vicinity of Helmholtz resonator natural frequency. Curve 2 corresponds to the narrowest band construction, and curve 4—construction with the widest range of the effective noise protection. According to the calculations, the extending of range of effective noise protection may cause the essential nonuniformity in coefficient G frequency characteristic. Curve 4 is an example. If we know the frequency band of intense noise, the grating of resonators should be used with the resonance frequency in the middle of this frequency band.

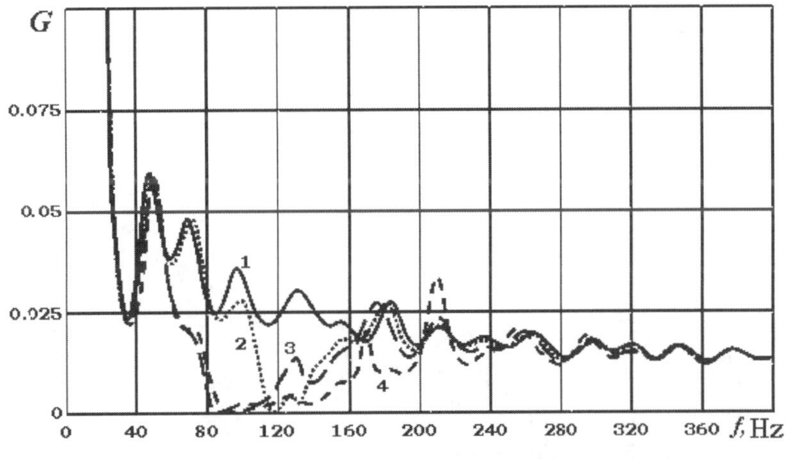

FIG. 5.26: Frequency dependency of energy coefficient G for the cases, when the whole surface of illuminated barrier side is covered with the grating of Helmholtz resonators $a = 0.05$ m, $2r_0 = 4 \times 10^{-3}$ m, $\tilde{r} = 2$; 1—barrier surface is acoustically rigid, 2—$b = 0.08$ m, $l = 10 \times 10^{-3}$ m, $f_0 \approx 115$ Hz, 3—$b = 0.32$ m, $l = 2.5 \times 10^{-3}$ m, $f_0 \approx 88$ Hz, 4—$b = 0.40$ m, $l = 0.5 \times 10^{-3}$ m, $f_0 \approx 97$ Hz

5.5 NOISE-PROTECTIVE PROPERTIES OF THE BARRIERS, SITUATED ALONG THE BOTH SIDES OF A HIGHWAY

The sources of literature analysis (see, for example, [6,23] and bibliography in them) show, that, as a rule, the cases are considered, when there is a barrier only on the one wayside. There are a lot of situations in practice, when it is necessary to protect the area adjacent to the both sides of highway from transport noise and the barriers are along the both waysides. Figure 5.27 shows the photo, made by one of the authors in Italy. As we can see, sound-protective barriers are along the both waysides of highway. It is fair to assume that if the distance between barriers is divisible by some part of wave length, the resonance phenomena may appear. These phenomena can decrease the efficiency. This presumption is to a certain degree proved by work [24]. The fact of sharp acoustic pressure fluctuations in the zone between the barriers is proved experimentally (using the model) in this work.

5.5.1 Physical and Mathematical Models

Let us consider the following physical model of two barriers system [25], Fig. 5.28. Assume that two infinite (along the direction perpendicular to the figure plane) barriers (2 and 3) with height h are on the infinite acoustically rigid surface 1, modeling the earth surface, in points $x = 0$ and $x = a$. The surface between the barriers $0 < x < a$, $y = 0$ simulates the highway. The sound source S, in the form of infinite pulsating band, is on this surface and it is parallel to the barriers. The position of the sound source is specified by coordinates $x = a_1$, $x = a_2$. The observation point is denoted by M. The semi-space $y > 0$ is filled with air with density ρ and sound velocity c. The barriers surface exposed to the source S (illuminated barriers surface), is characterized by complex conductivity Y. The barrier surface in the shade is always acoustically rigid. From a mathematical standpoint, the described physical model is equivalent to a plane problem.

FIG. 5.27: The highway section with sound-protective barriers along its both sides

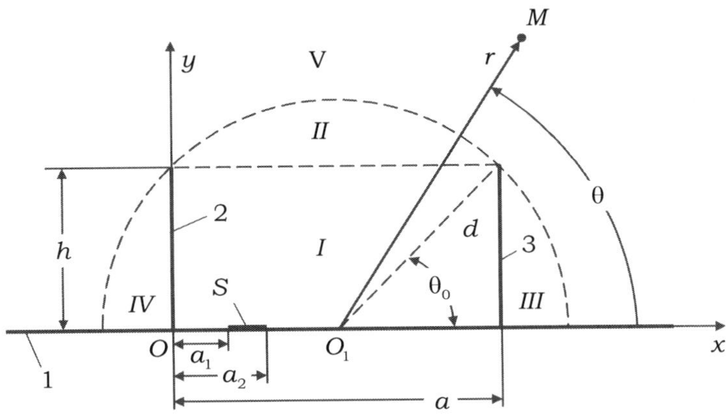

FIG. 5.28: Mathematical model geometry of two barriers system: 1—ground, 2, 3—barriers

5.5.2 Analytical Solution Construction

Let us introduce Cartesian coordinate system (x, y) with point O as center and polar coordinate system (r, θ) with point O_1 as center (Fig. 5.28). As it follows from Fig. 5.28, the relation between the coordinates of point M in the indicated coordinate systems is as follows:

$$x = r\cos\theta + b, \quad y = r\sin\theta, \quad b = OO_1, \tag{5.36}$$

$$r = \sqrt{(x-b)^2 + y^2}, \quad \cos\theta = \frac{x-b}{r}. \tag{5.37}$$

We partition all the domain of sound field existence into five domains according to the partial domains method:

1. domain I is the space between the barriers $0 \le x \le a$, $0 \le y \le h$;
2. domain II is bounded by the surfaces $(0 \le x \le a, 0 \le y \le h)$ and $(r = d, \theta_0 \le \theta \le \pi - \theta_0)$;
3. domain III is bounded by the surfaces $(a \le x \le d + b, y = 0)$, $(x = a, 0 \le y \le h)$, and $(r = d, 0 \le \theta \le \theta_0)$;
4. domain IV is bounded by the surfaces $(-(d - b) \le x \le 0, y = 0)$, $(x = 0, 0 \le y \le h)$, and $(r = d, \pi - \theta_0 \le \theta \le \pi)$;
5. domain V is the exterior of semicircle with radius $d = \sqrt{b^2 + h^2}$, i.e., $r \ge d$, $0 \le \theta \le \pi$.

The field in domain I should be represented in such form, as to fulfill the conditions: 1) boundary conditions on the surface of the road $(0 \le x \le a, y = 0)$, 2) impedance conditions on the barriers surfaces, 3) matching conditions of the sound fields on the surface $(0 \leq x \leq a, y = h)$.

The conditions indicated above can be fulfilled, if the field in domain I is represented as superposition of normal waves of two parallel-plate waveguides with rigid surfaces of width a and h, respectively. Thus,

$$\begin{aligned} p_I &= \sum_{n=0}^{\infty} A_n \cos\left(\beta_n^{(1)} x\right) \exp\left(i\gamma_n^{(1)} y\right) \\ &+ \sum_{n=0}^{\infty} A_n^{(1)} \cos\left(\beta_n^{(1)} x\right) \exp\left(-i\gamma_n^{(1)} (y-h)\right) \\ &+ \sum_{n=0}^{\infty} A_n^{(2)} \cos\left(\beta_n^{(2)} y\right) \exp\left(i\gamma_n^{(2)} x\right) \\ &+ \sum_{n=0}^{\infty} A_n^{(3)} \cos\left(\beta_n^{(2)} y\right) \exp\left(-i\gamma_n^{(2)} (x-a)\right), \end{aligned} \tag{5.38}$$

where

$$\beta_n^{(1)} = \frac{n\pi}{a}, \quad \gamma_n^{(1)} = \sqrt{k^2 - \left(\beta_n^{(1)}\right)^2}, \quad \beta_n^{(2)} = \frac{n\pi}{h},$$
$$\gamma_n^{(2)} = \sqrt{k^2 - \left(\beta_n^{(2)}\right)^2}, \quad k = \frac{\omega}{c}. \tag{5.39}$$

Boundary condition on the road surface $(0 \leq x \leq a, y = 0)$ is as follows:

$$\left.\frac{1}{i\omega\rho}\frac{\partial p_{II}}{\partial y}\right|_{y=0} = V(x), \tag{5.40}$$

where

$$V(x) = \begin{cases} \upsilon_0, & a_1 \leq x \leq a_2, \\ 0, & 0 \leq x < a_1 \cup a_2 < x \leq a \end{cases}$$

the function, specifying the distribution of sound field particle velocity amplitude along the surface of highway. We substitute (5.38) into boundary condition (5.40), and we find the relation between the coefficients A_n and $A_n^{(1)}$:

$$A_n = A_n^{(1)} \exp\left(i\gamma_n^{(1)} h\right) + \frac{\rho c V_n}{a\delta_n \gamma_n^{(1)}}, \tag{5.41}$$

where $\delta_0 = 1$, $\delta_n = 0,5$ when $n > 0$, and coefficients

$$V_n = \int_0^a V(x) \cos\left(\beta_n^{(1)} x\right) dx. \tag{5.42}$$

With allowance for (5.41), expression (5.38) for the field in domain I is as follows:

$$
\begin{aligned}
p_I = &\sum_{n=0}^{\infty} A_n^{(1)} \cos\left(\beta_n^{(1)} x\right)\left[\exp\left(i\gamma_n^{(1)}(y+h)\right)+\exp\left(-i\gamma_n^{(1)}(y-h)\right)\right] \\
&+\sum_{n=0}^{\infty} \frac{\rho c V_n}{a\delta_n \gamma_n^{(1)}} \cos\left(\beta_n^{(1)} x\right)\exp\left(-i\gamma_n^{(1)} y\right) \\
&+\sum_{n=0}^{\infty} A_n^{(2)} \cos\left(\beta_n^{(2)} y\right)\exp\left(i\gamma_n^{(2)} x\right) \\
&+\sum_{n=0}^{\infty} A_n^{(3)} \cos\left(\beta_n^{(2)} y\right)\exp\left(-i\gamma_n^{(2)}(x-a)\right).
\end{aligned}
\tag{5.43}
$$

Consequently, we can fulfill the impedance conditions on the barrier surfaces due to the choice of $A_n^{(2)}$ and $A_n^{(3)}$ coefficients. The set of coefficients $A_n^{(1)}$ allows us to fulfill the matching conditions of sound fields on the boundaries of domains I and II.

Pressure field in domain II we write as follows

$$
\begin{aligned}
p_{II} = &\sum_{n=0}^{\infty} A_n^{(4)} \cos\left(\beta_n^{(1)} x\right)\exp\left(i\gamma_n^{(1)}(y-h)\right) \\
&+\sum_{n=0}^{\infty} A_n^{(5)} \frac{J_{\beta_n^{(3)}}(kr)}{J'_{\beta_n^{(3)}}(kd)} \cos\left(\beta_n^{(3)}(\theta-\theta_0)\right),
\end{aligned}
\tag{5.44}
$$

where

$$
\beta_n^{(3)} = \frac{n\pi}{\pi - 2\theta_0}. \tag{5.45}
$$

The first sum here is the superposition of normal waves of parallel-plate waveguide of width h. It provides the sound fields matching on the boundary of domains I and II. The second sum is the set of particular solutions of Helmholtz equation in the domain in the form of the sector of a circle of radius d. This sum allows us to fulfill the arbitrary conditions of fields matching on the interface of domains II and V.

The expressions for the fields in domains III and IV also have the structure similar to Eq. (5.44):

$$
\begin{aligned}
p_{III} = &\sum_{n=0}^{\infty} A_n^{(6)} \cos\left(\beta_n^{(2)} y\right)\exp\left(i\gamma_n^{(2)}(x-a)\right) \\
&+\sum_{n=0}^{\infty} A_n^{(7)} \frac{J_{\beta_n^{(4)}}(kr)}{J'_{\beta_n^{(4)}}(kd)} \cos\left(\beta_n^{(4)}\theta\right),
\end{aligned}
\tag{5.46}
$$

$$
\begin{aligned}
p_{IV} = &\sum_{n=0}^{\infty} A_n^{(8)} \cos\left(\beta_n^{(2)} y\right)\exp\left(-i\gamma_n^{(2)} x\right) \\
&+\sum_{n=0}^{\infty} A_n^{(9)} \frac{J_{\beta_n^{(4)}}(kr)}{J'_{\beta_n^{(4)}}(kd)} \cos\left(\beta_n^{(4)}(\theta-(\pi-\theta_0))\right),
\end{aligned}
\tag{5.47}
$$

where

$$\beta_n^{(4)} = \frac{n\pi}{\theta_0}. \tag{5.48}$$

And the field in domain V is as follows:

$$p_V = \sum_{n=0}^{\infty} A_n^{(10)} \frac{H_n^{(1)}(kr)}{H_n'^{(1)}(kd)} \cos(n\theta). \tag{5.49}$$

Let us write the system of functional equations. It determines the continuity conditions of sound field on the interface of domains I, II, III, IV, V and boundary conditions of barriers surfaces:

$$p_I = p_{II}, \quad x = [0, a], \quad y = h, \tag{5.50}$$

$$\frac{\partial p_I}{\partial y} = \frac{\partial p_{II}}{\partial y}, \quad x = [0, a], \quad y = h, \tag{5.51}$$

$$\frac{1}{i\omega\rho}\frac{\partial p_I}{\partial x} = -Yp_I, \quad x = 0, \quad y = [0, h], \tag{5.52}$$

$$\frac{1}{i\omega\rho}\frac{\partial p_I}{\partial x} = Yp_I, \quad x = a, \quad y = [0, h], \tag{5.53}$$

$$\frac{\partial p_{III}}{\partial x} = 0, \quad x = a, \quad y = [0, h], \tag{5.54}$$

$$\frac{\partial p_{IV}}{\partial x} = 0, \quad x = 0, \quad y = [0, h], \tag{5.55}$$

$$p_{II} = p_V, \quad r = d, \quad \theta = [\theta_0, \pi - \theta_0], \tag{5.56}$$

$$p_{III} = p_V, \quad r = d, \quad \theta = [0, \theta_0], \tag{5.57}$$

$$p_{IV} = p_V, \quad r = d, \quad \theta = [\pi - \theta_0, \pi], \tag{5.58}$$

$$\frac{\partial p_V}{\partial r} = \begin{cases} \dfrac{\partial p_{III}}{\partial r}, & r = d, \quad \theta = [0, \theta_0], \\ \dfrac{\partial p_{II}}{\partial r}, & r = d, \quad \theta = [\theta_0, \pi - \theta_0], \\ \dfrac{\partial p_{IV}}{\partial r}, & r = d, \quad \theta = [\pi - \theta_0, \pi]. \end{cases} \tag{5.59}$$

The standard procedure of functional system algebraization (5.50)–(5.59) leads to the infinite system of linear algebraic equations of the second kind in unknown coefficients $A_n^{(1)}, \ldots, A_n^{(10)}$. As in the problems considered above, the field characteristics away from barriers edges are of interest. And simple reduction method can be used for solution of infinite system of equations [10,11]. As a rule, the number of unknowns in algebraic equations system is determined experimentally. We omit this standard analysis [10,11]. Note that we can obtain the results, which can be used in practice, if the total number of unknown complex coefficients is about 300.

Along with the amplitude of sound field distribution in the vicinity of barriers, the integral estimations of noise-protective barriers properties are of special interest. We use value (5.22) $G = W_D/W_0$ as an integral criterion for noise-protective barrier properties estimation. It shows the part of sound energy, radiated by the source, which is in the zone of barriers geometric shadow. The expression for full power of the source per unit length is as follows:

$$W_0 = \frac{1}{2}\int_0^a \mathrm{Re}\left[p\left(x, y=0\right)\right] V\left(x\right) dx. \qquad (5.60)$$

The value of W_D is defined as sound field power flow, which goes through the arcs l_1 and l_2 of radius r_w, situated in the zone of the barriers geometric shadow (see Fig. 5.29). The arcs l_1 and l_2 are bounded by ground from one side, and by rays 1 and 2 from the other side. These rays go through the barrier edges. The position of rays 1 and 2 is determined by angles ψ_1 and ψ_2. The values of angles ψ_1, ψ_2 and radius r_w allows us to find the range of variation of angle θ of polar coordinate system along the arcs l_1 and l_2 (see Fig. 5.29). Consequently, we can find the sought-for value W_D as follows:

$$W_D = \int_{l_1} I(l_1)\, dl + \int_{l_2} I(l_2)\, dl. \qquad (5.61)$$

Note that when the value of radius $r_w (r_w \gg \lambda, r_w \gg h)$ is rather large the level of power of sound field, going through the arcs l_1 and l_2, almost does not change. In this case, the integration arc is the part of wave front, formed by the system "source-barriers-ground."

5.5.3 Calculation Data Analysis

We set the following geometric parameters of the considered mathematical model (Fig. 5.28): barrier height $h = 4$ m, the width of highway $a = 12$ m, the band width, simulating the noise source S is defined by the coordinates $a_1 = 1$ m and $a_2 = 3$ m. The

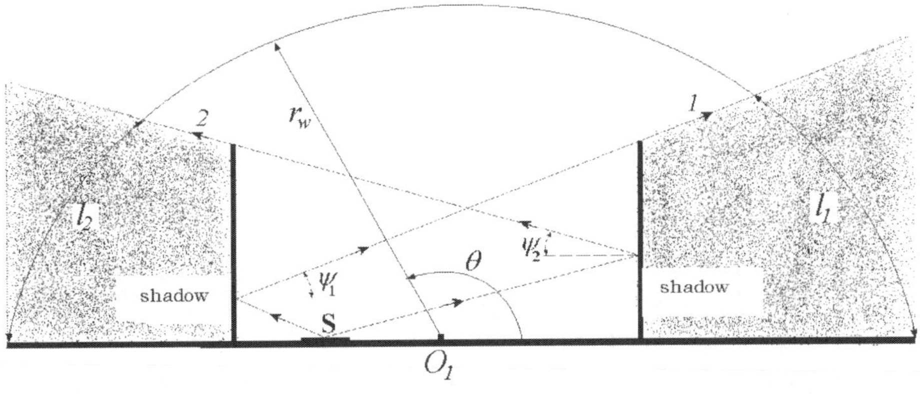

FIG. 5.29: Geometric shadow of the source behind the barriers

radius of arc, on which the power flow is calculated $r_w = 2d \approx 7.2$ m. The amplitude of sound source velocity $\upsilon_0 = 1$ m/s.

Figure 5.30 shows frequency dependencies of energy coefficient G for different acoustic properties of the illuminated barrier sides: curve 1 corresponds to acoustically rigid surface, curve 2—absorbing surface, curve 3—the grating of Helmholtz resonators covers the surface of illuminated barriers sides (the grating of resonators description was given in Section 5.4). We suppose that the acoustic properties of two barriers are the same and constant over the whole surface.

The parameters of a specific Helmholtz resonator depend on its geometric dimensions: the depth b_1 and width b_2 of resonator chamber, diameter $2r_0$ and length l of Helmholtz resonator throat (we assume that the cross section of resonator throat is a circle with area $\sigma_1 = \pi r_0^2$, and the cross section of chamber is a square $\sigma_2 = b_1 b_2$ in area). If we know these values, we can calculate the natural Helmholtz resonator frequency and impedance (in fractions of ρc per unit area of resonator chamber base $\sigma_2 = b_1 b_2$, $n = \sigma_2/\sigma_1$) using the Eqs. (5.34) and (5.35). Curve 3 in Fig. 5.30 was obtained for the following parameters of Helmholtz resonator: $b_1 = 0.40$ m, $b_2 = 0.05$ m, $2r_0 = 4 \times 10^{-3}$ m, $l = 0.5 \times 10^{-3}$ m, $\tilde{r} = 2$. The natural frequency of Helmholtz resonator $f_0 \approx 97$ Hz for the given parameters. For curves 1 and 2, the value of specific (per unit area) and normalized to the value of ρc impedance of illuminated barrier surface is $Z = \infty$ and $Z = 1$, respectively.

Curve 1 in Fig. 5.30 shows the potentials of classical barrier with acoustically rigid surfaces. Four frequency domains can be marked on the given curve: the first—very-low frequencies domain to 30 Hz, the second—low frequencies from 30 Hz to 120 Hz, the third—medium frequencies from 120 Hz to 300 Hz, the fourth—high frequencies (above 300 Hz). In the first domain, the curve goes up with a decrease of frequency. The noise-protective barrier properties worsen. It is due to the small wave height of the barrier. In the second domain, we can see a series of abrupt and relatively large spikes of G coefficient. But its middle level decreases rapidly with an increase of frequency. There are also the

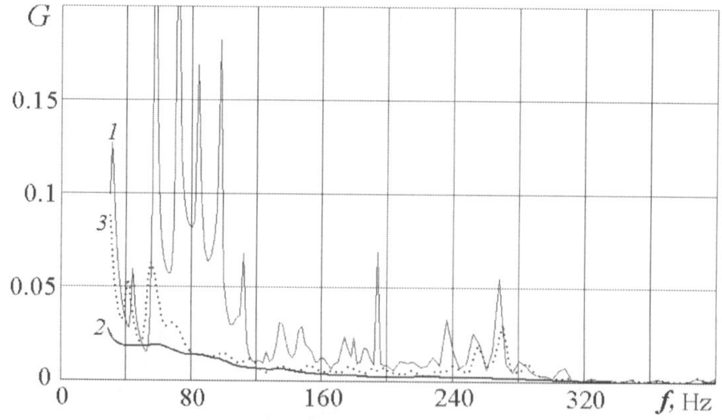

FIG. 5.30: Frequency dependency of energy coefficient G for three variants of surface of illuminated barrier side: 1—acoustically rigid, 2—completely absorbing, 3—grating of Helmholtz resonators (resonance sound absorbers)

spikes in the third domain, but the general level is lower than in the second domain. In the fourth domain, curve 1 asymptotically approaches zero with further increase in frequency.

Curve 2 in Fig. 5.30 lies lower than curve 1 and it has not the spikes in the entire considered frequency domain. It shows the minimum possible levels of energy coefficient G, which can be reached, using the barriers with completely absorbing surfaces. In the domain of very-low frequencies (below 30 Hz), curve 2 comes up, but not so sharply, as curve 1 and it causes a significant improvement of noise-protective properties, in comparison to the barrier with acoustically rigid surfaces.

The run of curve 3, which corresponds to the case when the surface of the illuminated barrier side is covered with the grating of Helmholtz resonators, differs from the curves 1 and 2. As we can see, curve 3 is close to curve 1in the domain of low frequencies, up to 55 Hz. Thus, the barrier with the grating of Helmholtz resonators has much in common with the barrier with acoustically rigid surfaces in the domain where the frequencies are below the resonance frequency of Helmholtz resonators. But in the domain of resonators resonance and up to 250 Hz we can see a significant improvement of noise-protective barrier properties. This is particularly significant, because there is an irregularity of coefficient G for the barriers with acoustically rigid walls in this frequency domain. The resonators efficiency decreases for the frequencies higher than 250 Hz, i.e., for the frequencies which are higher than the frequency of resonators resonance. The curves 1 and 3 are almost the same. Thus, if we know the frequency band of intense noise, the gratings of resonators can be used, the resonance frequency of which is in the middle of this band.

Figure 5.30 shows the irregularity of curve 1 (especially at frequencies from 50 Hz to 120 Hz). We can assume that there are the resonance phenomena in partial domain I (Fig. 5.28). They may be caused by standing waves between barriers 2 and 3. We will show that it is possible.

Taking into account that noise source S is loaded on partial domain I, we should estimate its radiation specific impedance. Taking into account the normalization to ρc, its calculation formula is as follows:

$$Z' = R' + iX' = \frac{1}{\rho c\,(a_2 - a_1)\,\upsilon_0} \int_{a_1}^{a_2} p\,(x, y = 0)dx. \tag{5.62}$$

Figure 5.31 shows the frequency dependency of real R' (curve 1) and imaginary X' (curve 2) parts of radiation impedance of source S. As we can see, 1 tends to unity with an increase of frequency, and curve 2 tends to zero. Such tendency is typical for any sound source. The positions of splashes along the frequency axis for curves 1 in Figs. 5.30 and 5.31 almost coincide. The sharp fluctuations of imaginary part X' (curve 2) of radiation impedance correspond to the real part R' splashes. The negative values X' here change to positive values and vice versa. Such behavior of the value of X' points out that we deal with a complex (multiresonant) sound source in the form of a system "radiating band—space between the barriers". The sequence of resonances and anti-resonances is typical [26]. It is to be recalled that for the chosen time dependency $\exp(-i\omega t)$, the values of $X' < 0$ correspond to mass impedance, and the values of $X' > 0$—correspond to elastic

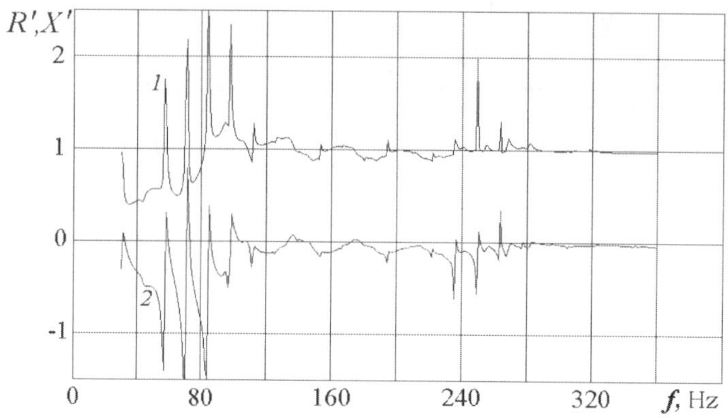

FIG. 5.31: Frequency dependency of real R' (curve 1) and imaginary X' (curve 2) parts of radiation impedance of sound source S

impedance. Such behavior of R' and X' points at the fact that there are the resonance phenomena in partial domain I and the connected with it splashes of radiated acoustic power of source S and coefficient G.

Space distribution of sound field at resonance frequencies is of interest. Figure 5.32 (for the resonance frequencies) shows the pressure amplitude distribution normalized to maximum values between the barriers and in the vicinity of the highway with acoustically rigid barriers. For Figs. 5.3(a)–5.3(e) these frequencies are as follows: 31, 57, 71, 84, and 249 Hz. These are the frequencies at which curve 1 (Fig. 5.30) have sharp spikes. As we can see, there is clearly defined standing wave along Ox axis for the indicated frequencies of sound source in the space between the barriers. Such distribution is typical for the pressure amplitude field of appropriate modes of parallel-plate waveguide with rigid boundaries. Function $\cos{(n\pi x/a)}$ determines pressure distribution in the cross section of n-th mode of parallel-plate waveguide of width a with the rigid boundaries. Consequently, the standing waves shown in Figs. 5.32(a)–5.32(e) are typical for the modes enumerated $n = 2, 4, 5, 6, 9$, respectively. Figure 5.32(f) shows the situation, when there is no space resonance. The frequency of 64 Hz is between the resonance frequencies here.

Thus, the resonance phenomena appear for the model under investigation (Fig. 5.28) at certain frequencies of sound source, in the space between the barriers. We can observe the space resonance, when the source energy is in one of the waveguides modes. Consequently, there is no interference effect between the modes and, as a consequence, there is a sharp increase of energy flow in the waveguide.

One of the effective ways to remove the space resonance phenomena is to use the barriers with the absorbing walls. The space distributions of pressure amplitude, represented in Figs. 5.33 and 5.34, illustrates it. Here Fig. 5.33 corresponds to the situation with the ideally absorbing illuminated barrier sides, and Fig. 5.34—corresponds to the case when the barriers surface is covered with the grating of Helmholtz resonators.

It is easy to note the difference between the space structures of sound field, when comparing these graphs with Figs. 5.32(b), 5.32(c), and 5.32(e). The periodicity in

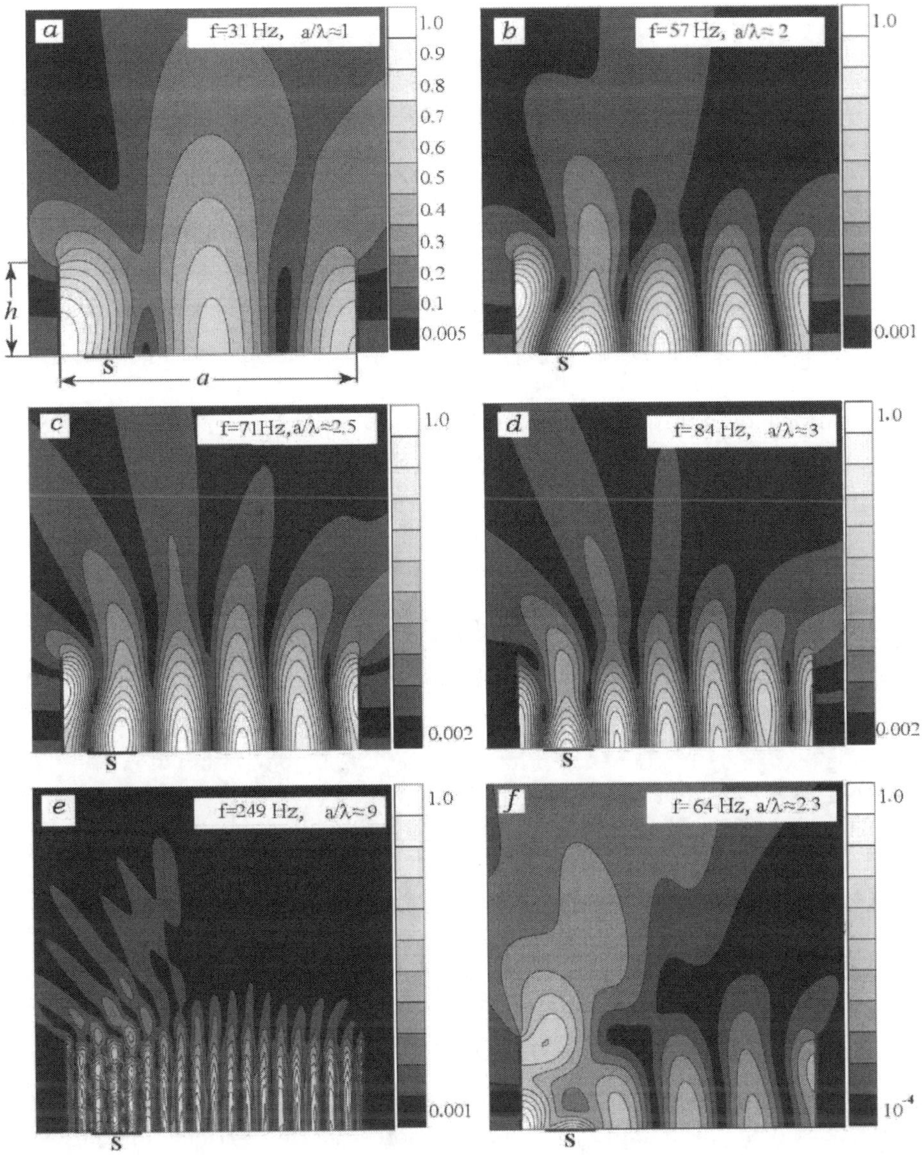

FIG. 5.32: Pressure amplitude distribution between the barriers and in the vicinity of highway; the barriers surfaces are acoustically rigid ($Y = 0$): (a)—$f = 31$ Hz, (b)—$f = 57$ Hz, (c)—$f = 71$ Hz, (d)—$f = 84$ Hz, (e)—$f = 249$ Hz, (f)—$f = 64$ Hz

pressure amplitude variation in Fig. 5.34 is not very noticeable. The resonance field structure is absent in Fig. 5.33, which corresponds to the completely absorbing barrier surfaces.

The resonance phenomena reduction, which appear in the space between the barriers, can be gained by covering the barrier walls with sound-absorbing covers. As may be supposed, the slope of the barriers (may be small) away from the highway is a simpler way.

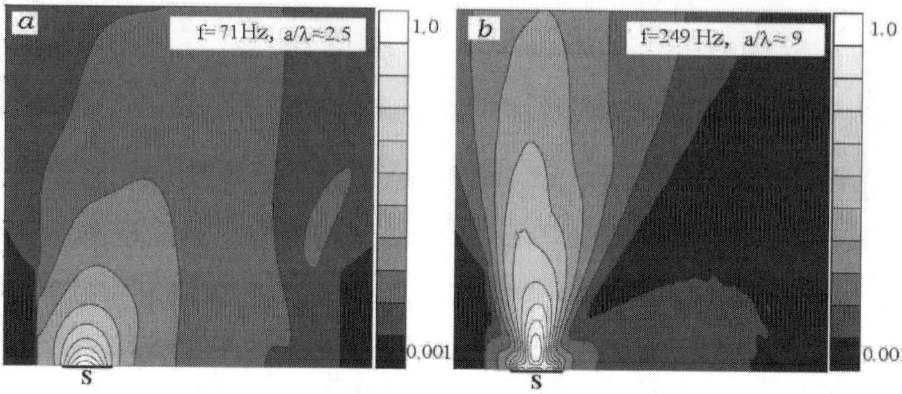

FIG. 5.33: Pressure amplitude distribution between the barriers and in the vicinity of highway; ideally absorbing illuminated barriers surfaces ($Y = 1$): (a)—$f = 71$ Hz, (b)—$f = 249$ Hz

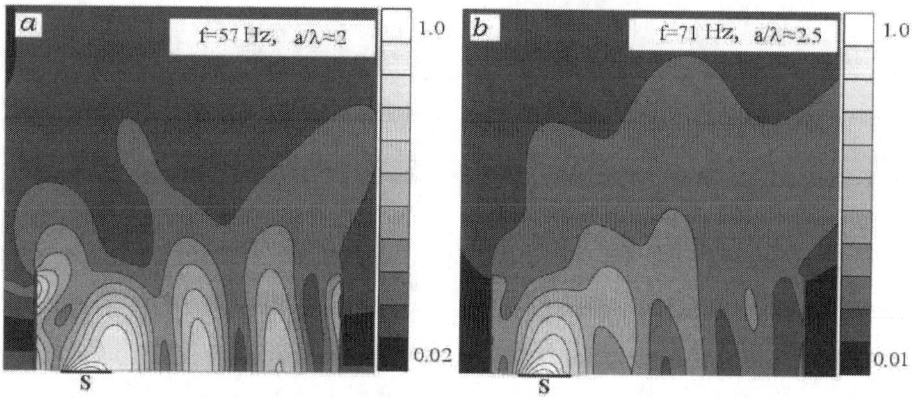

FIG. 5.34: Pressure amplitude distribution between the barriers and in the vicinity of highway; illuminated barriers surfaces covered with the grating of Helmholtz resonators ($Y = 1/\bar{Z}$): (a)—$f = 57$ Hz, (b)—$f = 71$ Hz

If we look closely at photo in Fig. 5.27, we can note, that the barriers are gently sloped. In theory, the slope should weaken the resonance phenomena in the space between the barriers. To make the quantitative estimation of the slope influence on the coefficient G and sound field distribution between the barriers, it is necessary to develop the corresponding physical and mathematical models and to perform all the necessary calculations. Let us show how it can be done.

5.5.4 Sloped Barriers

If the barriers are sloped, the mathematical model geometry (Fig. 5.28) changes and takes the form as in Fig. 5.35. Let us introduce one more polar coordinate system (R, ψ) with point O_2 as centre, to analyze this problem. Barriers slope is defined by angle ψ_0. The

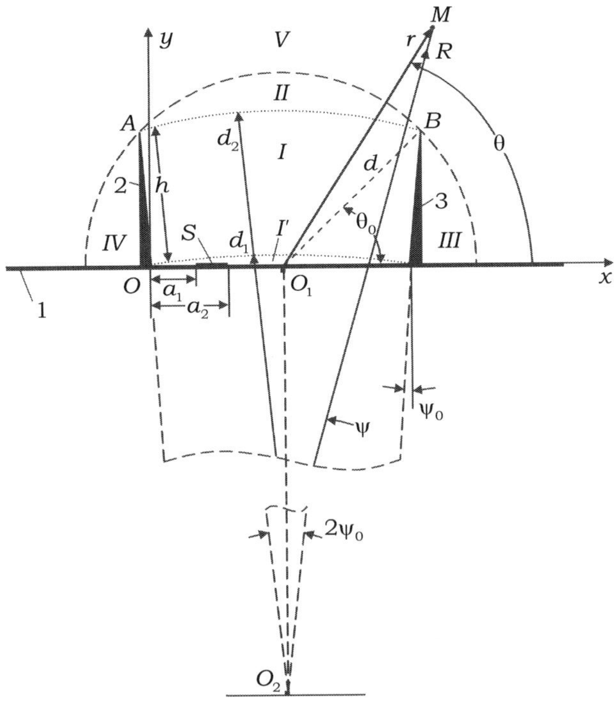

FIG. 5.35: Geometry of barriers with slope system: 1—ground, 2 and 3—barriers

slope angle ψ_0 is rather small in real constructions. To simplify the mathematical model it is necessary to represent the sloped barrier as the construction with triangular profile. In Fig. 5.35 such barriers 1 and 2 are shown as blacken triangles.

The new problem geometry leads to the change of partial domains form and to the occurrence of one more partial domain. This domain is denoted as I' in Fig. 5.35. Thus, we partition the whole space of sound field existence into six domains:

1. domain I' is bounded by the surfaces $(0 \le x \le a, y = 0)$ and $(R = d_1, \pi/2 - \psi_0 \le \psi \le \pi/2 + \psi_0)$, where $d_1 = O_2O = b/\sin\psi_0$, $b = OO_1$;

2. domain I is bounded by the barriers and the surfaces $(R = d_1, \pi/2 - \psi_0 \le \psi \le \pi/2+\psi_0)$ and $(R = d_2, \pi/2-\psi_0 \le \psi \le \pi/2+\psi_0)$, where $d_2 = O_2A = h+d_1$;

3. domain II is bounded by the surfaces $(R = d_2, \pi/2 - \psi_0 \le \psi \le \pi/2 + \psi_0)$ and $(r = d, \theta_0 \le \theta \le \pi - \theta_0)$, where $d = \sqrt{(h\cos\psi_0)^2 + (b + h\sin\psi_0)^2}$;

4. domain III is bounded by the surfaces $(a + h\sin\psi_0 \le x \le d + b, y = 0)$, $(x = a + h\sin\psi_0, 0 \le y \le h\cos\psi_0)$, and $(r = d, 0 \le \theta \le \theta_0)$;

5. domain IV is bounded by the surfaces $(-(d - b) \le x \le -h\sin\psi_0, y = 0)$, $(x = -h\sin\psi_0, 0 \le y \le h\cos\psi_0)$, and $(r = d, \pi - \theta_0 \le \theta \le \pi)$;

6. domain V is the exterior of semicircle of radius d, i.e., $r \ge d, 0 \le \theta \le \pi$.

Let us write the expressions for the sound fields in the indicated partial domains.

$$p_{I'} = \sum_{n=0}^{\infty} B_n^{(1)} \cos(\alpha_n x) \exp(i\eta_n y) + \sum_{n=0}^{\infty} B_n^{(2)} \frac{J_{\beta_n^{(1)}}(kR)}{J'_{\beta_n^{(1)}}(kd_1)} \cos\left(\beta_n^{(1)}(\psi - \psi_{00})\right),$$

where $\alpha_n = n\pi/a$, $\eta_n = \sqrt{k^2 - (\alpha_n)^2}$, $\psi_{00} = \pi/2 - \psi_0$, $\beta_n^{(1)} = n\pi/2\psi_0$;

$$p_I = \sum_{n=0}^{\infty} A_n^{(1)} \frac{H^{(1)}_{\beta_n^{(1)}}(kR)}{H'^{(1)}_{\beta_n^{(1)}}(kd_1)} \cos\left(\beta_n^{(1)}(\psi - \psi_{00})\right) + \sum_{n=0}^{\infty} A_n^{(2)} \frac{J_{\beta_n^{(1)}}(kR)}{J'_{\beta_n^{(1)}}(kd_2)} \cos\left(\beta_n^{(1)}(\psi - \psi_{00})\right),$$

$$p_{II} = \sum_{n=0}^{\infty} A_n^{(3)} \frac{H^{(1)}_{\beta_n^{(1)}}(kR)}{H'^{(1)}_{\beta_n^{(1)}}(kd_2)} \cos\left(\beta_n^{(1)}(\psi - \psi_{00})\right) + \sum_{n=0}^{\infty} A_n^{(4)} \frac{J_{\beta_n^{(3)}}(kr)}{J'_{\beta_n^{(3)}}(kd)} \cos\left(\beta_n^{(3)}(\theta - \theta_0)\right),$$

where $\beta_n^{(3)} = n\pi/(\pi - 2\theta_0)$;

$$p_{III} = \sum_{n=0}^{\infty} A_n^{(5)} \cos\left(\beta_n^{(2)} y\right) \exp\left(i\gamma_n^{(2)}(x - (a + h\sin\psi_0))\right) + \sum_{n=0}^{\infty} A_n^{(6)} \frac{J_{\beta_n^{(4)}}(kr)}{J'_{\beta_n^{(4)}}(kd)} \cos\left(\beta_n^{(4)}\theta\right),$$

$$p_{IV} = \sum_{n=0}^{\infty} A_n^{(7)} \cos\left(\beta_n^{(2)} y\right) \exp\left(-i\gamma_n^{(2)}(x + h\sin\psi_0)\right) + \sum_{n=0}^{\infty} A_n^{(8)} \frac{J_{\beta_n^{(4)}}(kr)}{J'_{\beta_n^{(4)}}(kd)} \cos\left(\beta_n^{(4)}(\theta - (\pi - \theta_0))\right),$$

where $\beta_n^{(2)} = n\pi/h\cos\psi_0$, $\gamma_n^{(2)} = \sqrt{k^2 - \left(\beta_n^{(2)}\right)^2}$, $\beta_n^{(4)} = n\pi/\theta_0$;

$$p_V = \sum_{n=0}^{\infty} A_n^{(9)} \frac{H_n^{(1)}(kr)}{H_n'^{(1)}(kd)} \cos(n\theta).$$

We omit the following analytic transformations: the sound fields matching conditions at the boundaries of partial domains and analytical solution construction. It is not different from the data given above. We turn our attention now to the numerical results analysis. We need to find out how the barrier slope helps to improve their noise–protective properties?

Figure 5.36 shows the calculated G coefficient values depending on the frequency for two cases: without barriers slope and with their slope equal to ten degrees. Curve 1 (barriers without a slope) shows an irregularity in the frequency band from 40 to 120 Hz and there are the big sharp splashes. These irregularities are weakened to a certain extend when the barrier is sloped (curve 2). The value of the splashes near the frequencies 57 Hz and 71 Hz decreases, and the splashes almost disappear in the vicinity of frequencies 84 Hz and 98 Hz. The estimates suggest that, curve 2 irregularity decreases considerably at the higher frequencies comparing to curve 1. Thus, the slope of the barriers improves the noise–protective properties of the barriers.

5.6 SOUND-PROTECTIVE BARRIERS PROPERTIES, SITUATED ALONG THE STREETS

Intra-urban traffic is the most widely-spread and intense noise source in the cities [24,27–30]. The traffic flow contribution into acoustic contamination of the streets is from 60% to 80% depending on the traffic area covering (asphalt or block stone), the rate of traffic, the concentration of traffic [30]. The people living near the traffic areas are affected by high noise level. In the towns, the noise protection of sideways and windows in the buildings, situated along the streets, is different from the noise protection outside the town. Residential accommodations outside the towns can be situated rather far from highways. Consequently, the problem of sound protection in a town requires detailed consideration [24,28–30]. The articles [24,28] deserve special attention. These articles describe the model experiments, which were conducted to estimate the possibility to reduce the transport noise level using classical rigid barriers, situated along the sideways. Unfortunately, there is no absolute answer in these articles for a question whether it is advisable to mount the barriers along the streets in the towns. It is shown in these articles, that frequency dependencies of sound pressure of traffic noise have oscillating character with the differential in pressure, up to 60 dB.

The analysis of the barriers effectiveness in an urban setting is conducted in the given paragraph, basing on the constructed physical and mathematical models [31].

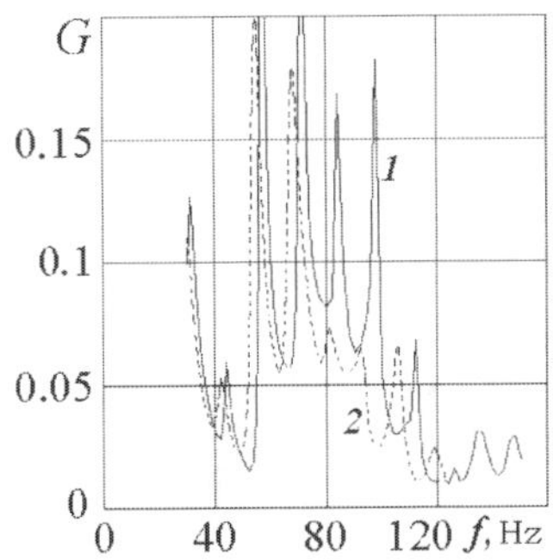

FIG. 5.36: Energy coefficient G dependency on frequency: 1—barriers without a slope (dependency curve in Fig. 5.30), 2—sloped barriers (angle $\psi_0 = 10°$)

5.6.1 Physical and Mathematical Models

Let us consider the physical model of typical urban street with close development. The sound barriers are situated along the both sides of the street Fig. 5.37. In this figure, acoustically rigid surface $y = 0$, $x = [0, a + b + c]$ simulates the surfaces of sidewalk and traffic area. The street is bounded with buildings. The faces of these buildings are described as acoustically rigid surfaces $x = 0$, $y = [0, H] \cup x = [-\infty, 0]$, $y = H$ and $x = a + b + c$, $y = [0, H] \cup x = [a + b + c, \infty]$, $y = H$. The models geometry does not depend on the direction, which is perpendicular to the figure plane. Two barriers of height h are in the planes $x = b$ and $x = a + b$. The surfaces between the buildings and the barriers $x = [0, b]$, $y = 0$ and $x = [a + b, a + b + c]$, $y = 0$ simulates the sidewalk. The surface between the barriers $x = [b, a + b]$, $y = 0$ is the traffic area. There is a sound source S as an infinite pulsing band of width $a_2 - a_1$ on this surface. The source simulates the noise of traffic flow. The semi-space $y > 0$ is filled with area with density ρ and sound velocity c. The barriers surfaces, faced to the source S (illuminated barriers surface), are characterized by complex conductivity Y. The shade side of the barriers is always acoustically rigid. It should be reminded that the conductivity of ideal acoustically rigid surface is equal to zero ($Y = 0$). And the conductivity of ideally absorbing surface (normalized to the value ρc) $Y = 1$.

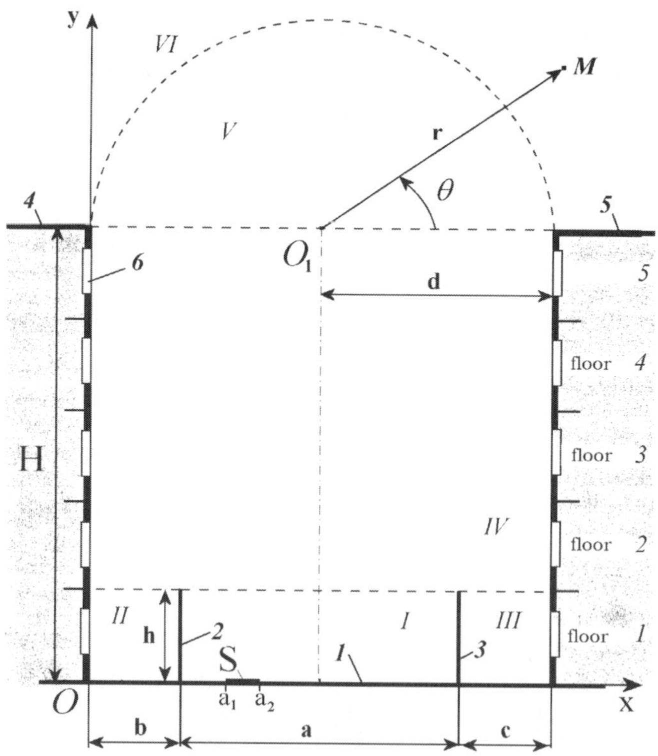

FIG. 5.37: Model geometry: 1—sidewalk surface and traffic area surface, 2, 3—barriers, 4, 5—buildings, 6—windows

5.6.2 Analytical Solution Construction

Let us introduce the Cartesian coordinate system (x, y) with point O as center and polar coordinate system (r, θ) with point O_1 as center (Fig. 5.37). As it follows from the figure, the relation between the point M coordinates in the given coordinate systems is as follows:

$$x = r\cos\theta + d, \quad y = r\sin\theta + H, \tag{5.63}$$

$$r = \sqrt{(x-d)^2 + (y-H)^2}, \quad \cos\theta = (x-d)/r, \tag{5.64}$$

where $d = b + a/2$.

We partition all the field existence domain into six domains (Fig. 5.37):

1. domain I is the space between barriers 2 and 3 $b \le x \le a+b, 0 \le y \le h$;
2. domain II is the space between building 4 and barrier 2 to the left of the traffic area $0 \le x \le b, 0 \le y \le h$;
3. domain III is the space between building 5 and barrier 3 to the right of the traffic area $a+b \le x \le a+b+c, 0 \le y \le h$;
4. domain IV is the space between the buildings $0 \le x \le a+b+c, h \le y \le H$;
5. domain V is bounded by surfaces $0 \le x \le a+b+c$, $y = H$ and $r = d$, $0 \le \theta \le \pi$;
6. domain VI $r \ge d, 0 \le \theta \le \pi$.

The field in domain I, is represented in the form that fulfills the following conditions: boundary conditions on the traffic area surface $b \le x \le a+b$, $y = 0$, impedance conditions on the barriers surfaces, sound fields matching conditions on the surface $b \le x \le a+b, y = H$. These conditions can be fulfilled, if the field in domain I is represented as superposition of normal waves of two parallel-plate waveguides with the rigid surfaces of width a and h, respectively. Consequently,

$$\begin{aligned} p_I = &\sum_{n=0}^{\infty} A_n \cos\left(\beta_n^{(1)}(x-b)\right)\exp\left(i\gamma_n^{(1)}y\right) \\ &+\sum_{n=0}^{\infty} A_n^{(1)} \cos\left(\beta_n^{(1)}(x-b)\right)\exp\left(-i\gamma_n^{(1)}(y-h)\right) \\ &+\sum_{n=0}^{\infty} A_n^{(2)} \cos\left(\beta_n^{(2)}y\right)\exp\left(i\gamma_n^{(2)}(x-b)\right) \\ &+\sum_{n=0}^{\infty} A_n^{(3)} \cos\left(\beta_n^{(2)}y\right)\exp\left(-i\gamma_n^{(2)}(x-a-b)\right), \end{aligned} \tag{5.65}$$

where

$$\beta_n^{(1)} = \frac{n\pi}{a}, \quad \gamma_n^{(1)} = \sqrt{k^2 - \left(\beta_n^{(1)}\right)^2}, \quad \beta_n^{(2)} = \frac{n\pi}{h},$$
$$\gamma_n^{(2)} = \sqrt{k^2 - \left(\beta_n^{(2)}\right)^2}, \quad k = \frac{\omega}{c}. \tag{5.66}$$

Boundary condition on the traffic area surface ($b \le x \le a + b, y = 0$) is as follows:

$$\frac{1}{i\omega\rho} \frac{\partial p_I}{\partial y}\bigg|_{y=0} = V(x), \tag{5.67}$$

where

$$V(x) = \begin{cases} \upsilon_0, & a_1 \le x \le a_2, \\ 0, & b \le x < a_1 \cup a_2 < x \le b + a \end{cases}$$

the function, setting the distribution of particle velocity amplitude on the source S surface and traffic area surface. Substituting (5.65) into boundary condition (5.67), we find the relation between coefficients A_n and $A_n^{(1)}$:

$$A_n = A_n^{(1)} \exp\left(i\gamma_n^{(1)} h\right) + \frac{\rho c V_n}{a\delta_n \gamma_n^{(1)}}, \tag{5.68}$$

where $\delta_0 = 1$, $\delta_n = 0, 5$ when $n > 0$, and coefficients

$$V_n = \int_0^a V(x) \cos\left(\beta_n^{(1)} x\right) dx.$$

Taking into account Eq. (5.68), expression (5.65) for the field in domain I changes to

$$\begin{aligned} p_I = &\sum_{n=0}^{\infty} A_n^{(1)} \cos\left(\beta_n^{(1)} (x - b)\right)\left[\exp\left(i\gamma_n^{(1)} (y + h)\right) + \exp\left(-i\gamma_n^{(1)} (y - h)\right)\right] \\ &+ \sum_{n=0}^{\infty} \frac{\rho c V_n}{a\delta_n \gamma_n^{(1)}} \cos\left(\beta_n^{(1)} (x - b)\right) \exp\left(i\gamma_n^{(1)} y\right) \\ &+ \sum_{n=0}^{\infty} A_n^{(2)} \cos\left(\beta_n^{(2)} y\right) \exp\left(i\gamma_n^{(2)} (x - b)\right) \\ &+ \sum_{n=0}^{\infty} A_n^{(3)} \cos\left(\beta_n^{(2)} y\right) \exp\left(-i\gamma_n^{(2)} (x - a - b)\right). \end{aligned} \tag{5.69}$$

We can fulfill the impedance conditions on the barriers surfaces due to the choice of coefficients $A_n^{(2)}$ and $A_n^{(3)}$. The set of coefficients $A_n^{(1)}$ allows us to satisfy the matching conditions of sound fields on the boundaries of domains I and II. We write the pressure field in domain II as follows:

$$p_{II} = \sum_{n=0}^{\infty} A_n^{(4)} \cos\left(\beta_n^{(3)} x\right) \exp\left(i\gamma_n^{(3)} y\right) + \sum_{n=0}^{\infty} \tilde{A}_n^{(4)} \cos\left(\beta_n^{(3)} x\right) \exp\left(-i\gamma_n^{(3)} (y-h)\right), \tag{5.70}$$

where $\beta_n^{(3)} = n\pi/b$, $\gamma_n^{(3)} = \sqrt{k^2 - \left(\beta_n^{(3)}\right)^2}$. Boundary condition:

$$\frac{\partial p_{II}}{\partial y} = 0, \quad x = [0, b], \quad y = 0, \tag{5.71}$$

we find the relation $\tilde{A}_n^{(4)} = A_n^{(4)} \exp\left(-i\gamma_n^{(3)} h\right)$. Taking into account this equality, we write expression (5.70) as follows:

$$p_{II} = \sum_{n=0}^{\infty} A_n^{(4)} \cos\left(\beta_n^{(3)} x\right) \frac{\exp\left(i\gamma_n^{(3)} (y+h)\right) + \exp\left(-i\gamma_n^{(3)} (y-h)\right)}{1 + \exp\left(2i\gamma_n^{(3)} h\right)}. \tag{5.72}$$

Likewise the field in domain III can be written as follows:

$$p_{III} = \sum_{n=0}^{\infty} A_n^{(5)} \cos\left(\beta_n^{(3)} (x - a - b)\right) \frac{\exp\left(i\gamma_n^{(3)}(y+h)\right) + \exp\left(-i\gamma_n^{(3)}(y-h)\right)}{1 + \exp\left(2i\gamma_n^{(3)} h\right)}. \tag{5.73}$$

The field in domain IV, is similar to expression (5.70), and is as follows:

$$p_{IV} = \sum_{n=0}^{\infty} A_n^{(6)} \cos\left(\beta_n^{(4)} x\right) \exp\left(i\gamma_n^{(4)} (y-h)\right) + \sum_{n=0}^{\infty} \tilde{A}_n^{(7)} \cos\left(\beta_n^{(4)} x\right) \exp\left(-i\gamma_n^{(4)} (y-H)\right), \tag{5.74}$$

where $\beta_n^{(4)} = n\pi/(a+b+c)$, $\gamma_n^{(4)} = \sqrt{k^2 - \left(\beta_n^{(4)}\right)^2}$.

We write the field in domain V as follows:

$$p_V = \sum_{n=0}^{\infty} A_n^{(8)} \cos\left(\beta_n^{(4)} y\right) \exp\left(i\gamma_n^{(4)} (x-H)\right) + \sum_{n=0}^{\infty} A_n^{(9)} \frac{J_n(kr)}{J_n'(kd)} \cos(n\theta). \tag{5.75}$$

The first sum here is the superposition of the normal waves of parallel-plate waveguide of width $(a+b+c)$, it provides sound fields matching on the boundaries of domains IV and V. The second sum is the totality of particular solutions of Helmholtz equation in the semi-circular domain of radius d. This sum allows us to fulfill the fields matching conditions at the interface of domains V and VI. The field in domain VI is as follows:

$$p_{VI} = \sum_{n=0}^{\infty} A_n^{(10)} \frac{H_n^{(1)}(kr)}{H_n'^{(1)}(kd)} \cos(n\theta). \tag{5.76}$$

We write the system of functional equations. It determines the sound field continuity conditions at the interface of domains I, II, III, IV, V, VI and boundary conditions on the barriers surfaces:

$$p_I = p_{IV}, \quad x = [b,\ a+b], \quad y = h, \tag{5.77}$$

$$p_{II} = p_{IV}, \quad x = [0, b], \quad y = h, \tag{5.78}$$

$$p_{III} = p_{IV}, \quad x = [a+b, a+b+c], \quad y = h, \tag{5.79}$$

$$\frac{\partial p_{IV}}{\partial y} = \begin{cases} \dfrac{\partial p_{II}}{\partial y}, & x = [b, a+b], \quad y = h, \\ \dfrac{\partial p_{I}}{\partial y}, & x = [0, b], \quad y = h, \\ \dfrac{\partial p_{III}}{\partial y}, & x = [a+b, a+b+c], \quad y = h, \end{cases} \tag{5.80}$$

$$p_{IV} = p_V, \quad x = [0, a+b+c], \quad y = H, \tag{5.81}$$

$$\frac{\partial p_{IV}}{\partial y} = \frac{\partial p_V}{\partial y}, \quad x = [0, a+b+c], \quad y = H, \tag{5.82}$$

$$p_V = p_{VI}, \quad r = d, \quad \theta = [0, \pi], \tag{5.83}$$

$$\frac{\partial p_V}{\partial r} = \frac{\partial p_{VI}}{\partial r}, \quad r = d, \quad \theta = [0, \pi], \tag{5.84}$$

$$\frac{1}{i\omega\rho}\frac{\partial p_I}{\partial x} = -Yp_I, \quad x = b, \quad y = [0, h], \tag{5.85}$$

$$\frac{1}{i\omega\rho}\frac{\partial p_I}{\partial x} = Yp_I, \quad x = a+b, \quad y = [0, h], \tag{5.86}$$

Then we obtain an infinite system of linear algebraic equations of the second kind in unknown coefficients $A_n^{(1)}$, ..., $A_n^{(10)}$. We solve this system by the simple reduction method. The field characteristics away from the barriers edges are of main interest in our problem. Field properties at a distance from barrier edges are of interest in our problem. The number of unknowns in the system of algebraic equations are determined by experiment. The results appropriate for practical application can be obtained in this problem if the total amount of unknown complex coefficients is 350–600 for the frequency range from 30 Hz to 1000 Hz.

5.6.3 Calculated Data Analysis

Let us set the following geometric parameters of the considered model (Fig. 5.37): barrier height $h = 4$ m, traffic area width $a = 12$ m, sidewalks width at the left and at the right of the sidewalk is $b = c = 4$ m, the band, simulating the sound source S, is specified by the coordinates $a_1 = 5$ m and $a_2 = 7$ m. Velocity amplitude of sound source $\upsilon_0 = 1$ m/s. The height of five-storied building is $H = 19$ m. The height of windows is taken equal to 2 m. The bottom levels of the windows relative to the surface of sidewalk are equal to 2; 5.5; 9; 12.5; 16 m, respectively. Air properties: density $\rho = 1.27$ kg/m^3, sound velocity $c = 333.3$ m/s.

Let us see how the levels of sound pressures change in some characteristic points of space between the buildings, depending on whether the barriers are absent or present (barriers surface is acoustically rigid). As a criterion, we use the ratio from work [24], namely

$$\Delta L = 20 \lg \left(\frac{|p_1(x_j, y_j)|}{|p_2(x_j, y_j)|} \right) \text{ dB}, \tag{5.87}$$

where $|p_1|$—pressure amplitude in the point with coordinates x_j, y_j without the barriers, $|p_2|$—pressure in the same point with barriers.

We should simplify the original problem, and exclude domains I, II, and III (see Fig. 5.37), when performing the calculations for the case without barriers. But it can be omitted. It is enough to assume that the barriers height $h \to 0$ to satisfy the conditions $h \neq 0$ and $h/\lambda \ll 1$, where λ—the wave length. We compared the data of calculations performed for such two models. The comparison showed, that when $h/\lambda \leq 0.1$ the results agree up to fourth decimal.

Figure 5.38 shows the frequency dependencies of value ΔL for two points on the face of the left building (see Fig. 5.37). When we compare these dependencies with similar dependencies from work [24], it may be noted that there is a good qualitative and quantitative correspondence. The main feature of these dependencies is their oscillating character with the pressure drops, reaching 50–60 dB. If we average these dependencies (even visually) we can see that the barriers help to reduce the pressure on some intervals of frequency domains (about 5–10 dB) in the points under consideration.

It is not possible to estimate the integrated efficiency of barriers, using expression (5.87), because it is of local character. We assume that the sound protection of the windows in buildings is the most important, because the noise penetrates into the offices and apartments through the windows. We introduce some parameter to estimate the noise effect on the windows surface. We introduce this parameter as the mean sound pressure on the windows surface, divide by volume velocity of the source

$$J = \frac{1}{S_0' \upsilon_0 (a_2 - a_1)} \int_{S'} |p| \, dS', \tag{5.88}$$

the integration here is performed on windows surface S' (computation is performed per unit windows width for plane problem), S_0'—windows area, a $\upsilon_0(a_2 - a_1)$—volume particle velocity of the source. The dependency of J parameter on frequency for the windows on different floors is of primary interest.

At first, we consider the case when there are no barriers. Figures 5.39(a)–5.39(e) shows frequency dependencies J for the windows from the ground-floor [Fig. 5.39(a)] to the fifth floor [Fig. 5.39(e)] of the building to the left of the barriers, and in

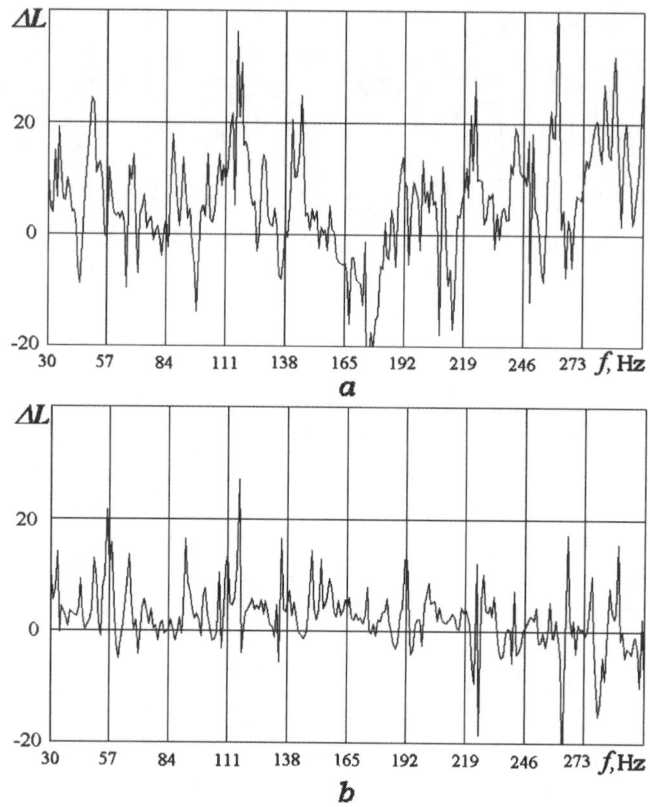

FIG. 5.38: Frequency dependency of value ΔL in the points with coordinates: (a)—$x = 0$ m, $y = 3$ m; (b)—$x = 0$ m, $y = 10$ m

Figs. 5.39(f)–5.39(k), respectively for the building to the right of the barriers. As we can see, frequency dependencies of J parameter have a lot of sharp relatively narrow-band spikes of large amplitude against relatively low (on the average) parameter J value. The character of frequency dependency of J vary only slightly from one floor to another. It ought to be noted that there is some reduction of spike values of function J for the fourth and the fifth floors. The value of spikes, which are beyond the diagram boundaries and their frequencies are indicated near the spikes in fractional form. The upper number indicates the frequency, and the bottom number indicates the maximum value of spike amplitude.

What is the reason of appearance of such sharp spikes in J parameter frequency dependency? To answer this question, we will we perform the computations of pressure field in the space between the buildings at the frequencies, corresponding to some of these spikes. Figures 5.40(a)–5.40(d) shows these data. It follows from their analysis:

1. the spikes in frequency dependencies of value of J are caused by resonance phenomena; these phenomena appear in the area, bounded by acoustically rigid faces of buildings and by the surface of traffic area and of sidewalks; visually the field distribution is as the pulled bands vertically alternating;

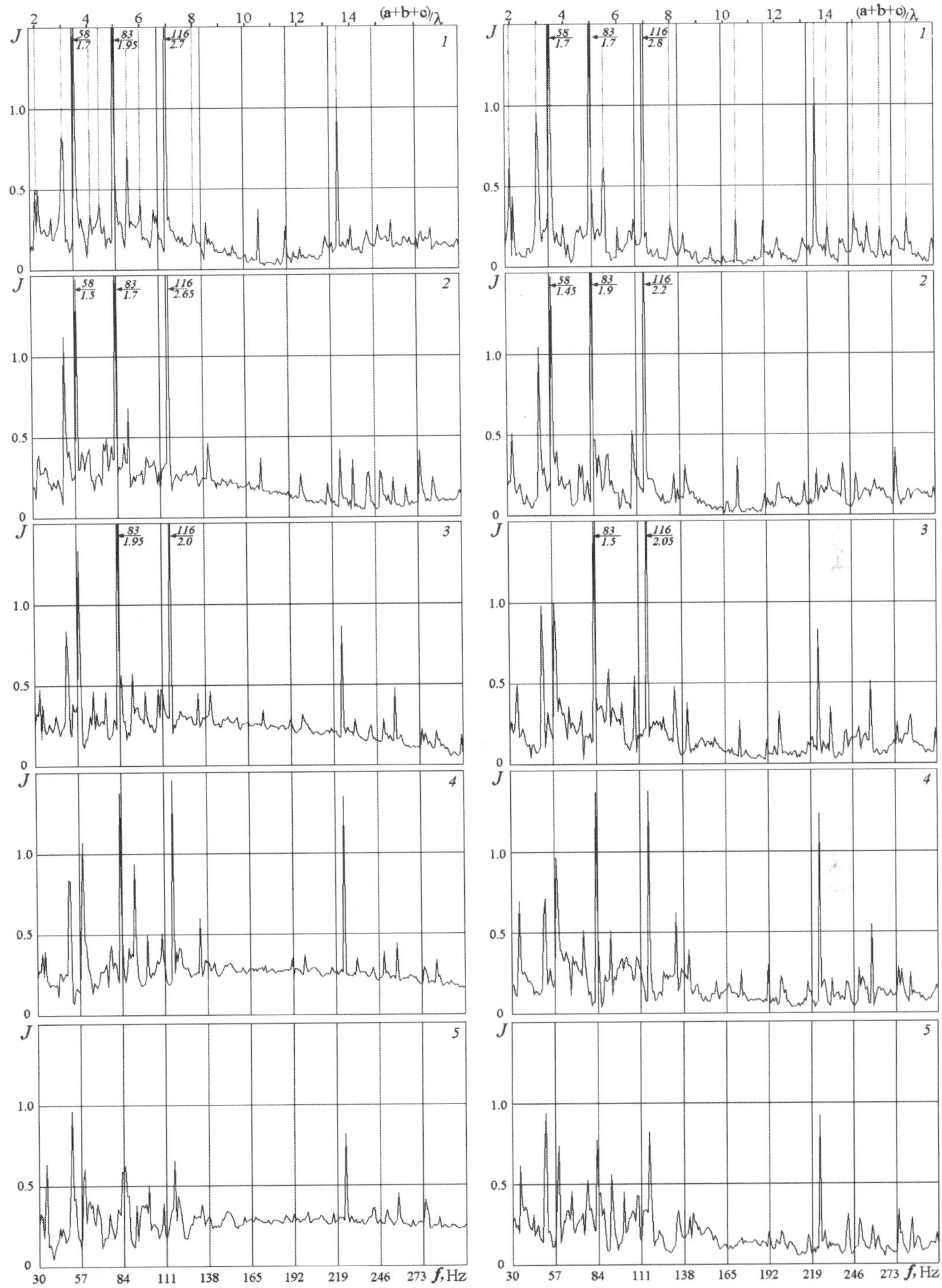

FIG. 5.39: Frequency dependency of J parameter without barriers: the left column—the building to the left of the traffic area, right column—the building to the right of the traffic area; the floors are numbered from the top down

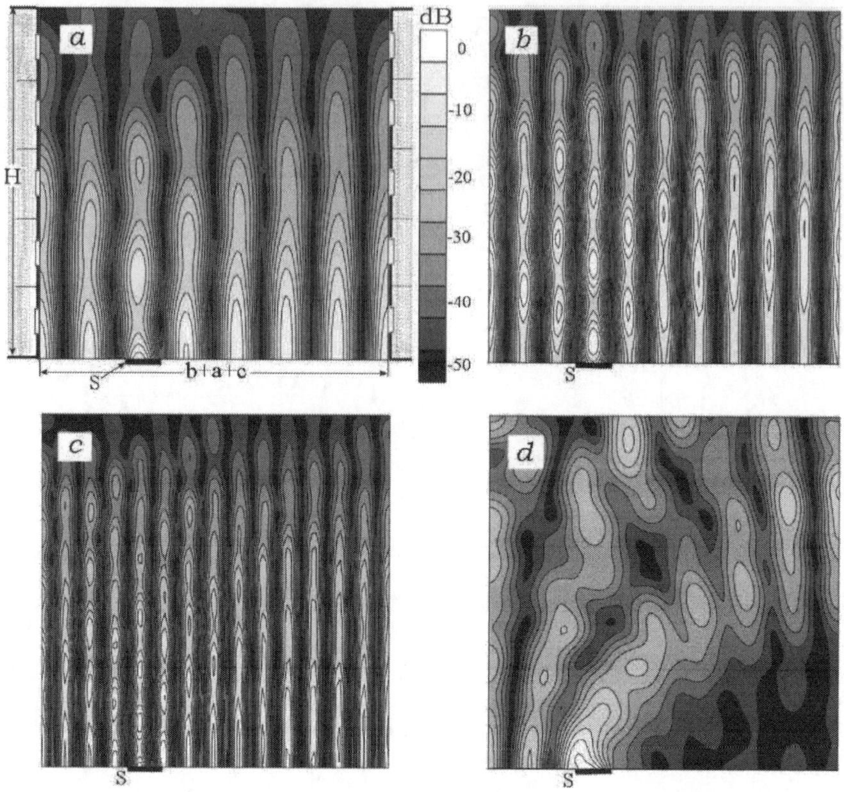

FIG. 5.40: Pressure amplitude field in the space between the buildings, when the barriers are absent at frequencies: (a)—58 Hz, (b)—83 Hz, (c)—116 Hz, (d)—70 Hz

2. the resonances may appear in the cases when the ratio of distance between the buildings to the wave length is $n/2$, where $n = 1, 2, 3, \ldots$ (the scale of values $(b + a + c)/\lambda$ is given in Figs. 5.39(a) and 5.39(f), at the top of the graphs, here λ—wave length in the air); the surfaces of buildings faces are acoustically rigid, and for this reason, the pressure loops appear on them, i.e., the zones of maximum pressure value [21];

3. the area between the buildings is an open resonator; consequently, the resonance field structure "gets fuzzy" near the upper (open) part of the resonator; the pressure drops, resulting in the weakening of the spike values of function J for the fourth and the fifth floors (it was mentioned above).

It should be mentioned that not all the potentially-enable resonances can be realized to the full extent. Their efficiency depends largely on the source S wave dimensions and its position on the traffic area. In other words, the source is in good agreement with the area volume between the faces of the buildings only at some resonance frequencies. Consequently, on the curves in Fig. 5.39 we can see not only the large spikes (with relatively good agreement), but also a lot of small spikes, with poor agreement.

Is it of interest for you to observe the field distribution on the non-resonance frequencies? For this purpose we turn to Fig. 5.40(d), where the pressure field distribution is shown between the buildings faces at frequency 70 Hz. There are no resonance spikes here. As we can see, the field distribution is of rather "fuzzy" character. Mean value of the pressure amplitude decreases gradually, when moving from the source to the upper floors.

Let us consider now the situation, when there are the barriers with acoustically rigid surfaces along the street (Fig. 5.37). Figure 5.41 shows frequency dependencies of J parameter for the same calculations variants, as in Fig. 5.39.

When comparing the graphs in Figs. 5.39 and 5.41, it may be noted that:

1. the mean level of parameter J for the lower floors, and especially for the first floor, decreases considerably (about 4–6 dB); but its frequency dependency has a lot of spikes;

2. the levels of the most spike values of parameter J decrease, at least, starting from the second floor, and above;

3. resonance spike at frequency 83 Hz almost does not change its high level.

Thus, the application of acoustically rigid barriers leads to some decrease of mean level of sound pressure on the windows of the lower floors and to decrease spikes values of the pressure on the upper floors. Nevertheless, there is only little decrease of the value of parameter J. The question arises, why the parameter J almost does not change? We assume that the application of the barriers creates the conditions for the additional series of resonance frequencies appearance. The additional spikes appear in the frequency dependency of parameter J. The barriers are situated parallel to each other and parallel to the buildings faces. It leads to the formation of other three open resonators—one between the barriers and two between the barriers and the faces surfaces. To be sure in this fact, we turn to Fig. 5.42. We start with Fig. 5.42(a), which shows the pressure field distribution at frequency 45 Hz. We can see here the resonance phenomena between the barriers and the building faces. The condition $b/\lambda = c/\lambda = 1/2$, is fulfilled at this frequency, and we can see a pronounced spike, Fig. 5.41(a), in frequency dependency. Now we turn attention to the fact that the resonance is more pronounced between the left barrier and the left face of the building, than between the right barrier and the right face. It is due to the fact that the source is closer to the left barrier.

The resonances can appear at higher frequencies, when the following condition is fulfilled $b/\lambda = c/\lambda = m/2$, where $m = 1, 2, 3, \ldots$. Figures 5.41(a) and 5.41(e) shows the additional scales of values a/λ, b/λ, $(a + b + c)\,/\lambda$ in the upper part of the graphs. These scales allow us to estimate the frequencies of potentially-enable resonances between the barriers and the faces of the buildings.

Figure 5.42(b) illustrates the resonance between the barriers at frequency 58 Hz. It corresponds to the wave dimension $a/\lambda = 2$, (general condition of the resonance appearance between the barriers is as follows $a/\lambda = m/2$, where $m = 1, 2, 3, \ldots$).

Figure 5.42(c) is very interesting. Pressure field distribution is shown here, when the resonance phenomena appear simultaneously at one frequency (in this case at frequency

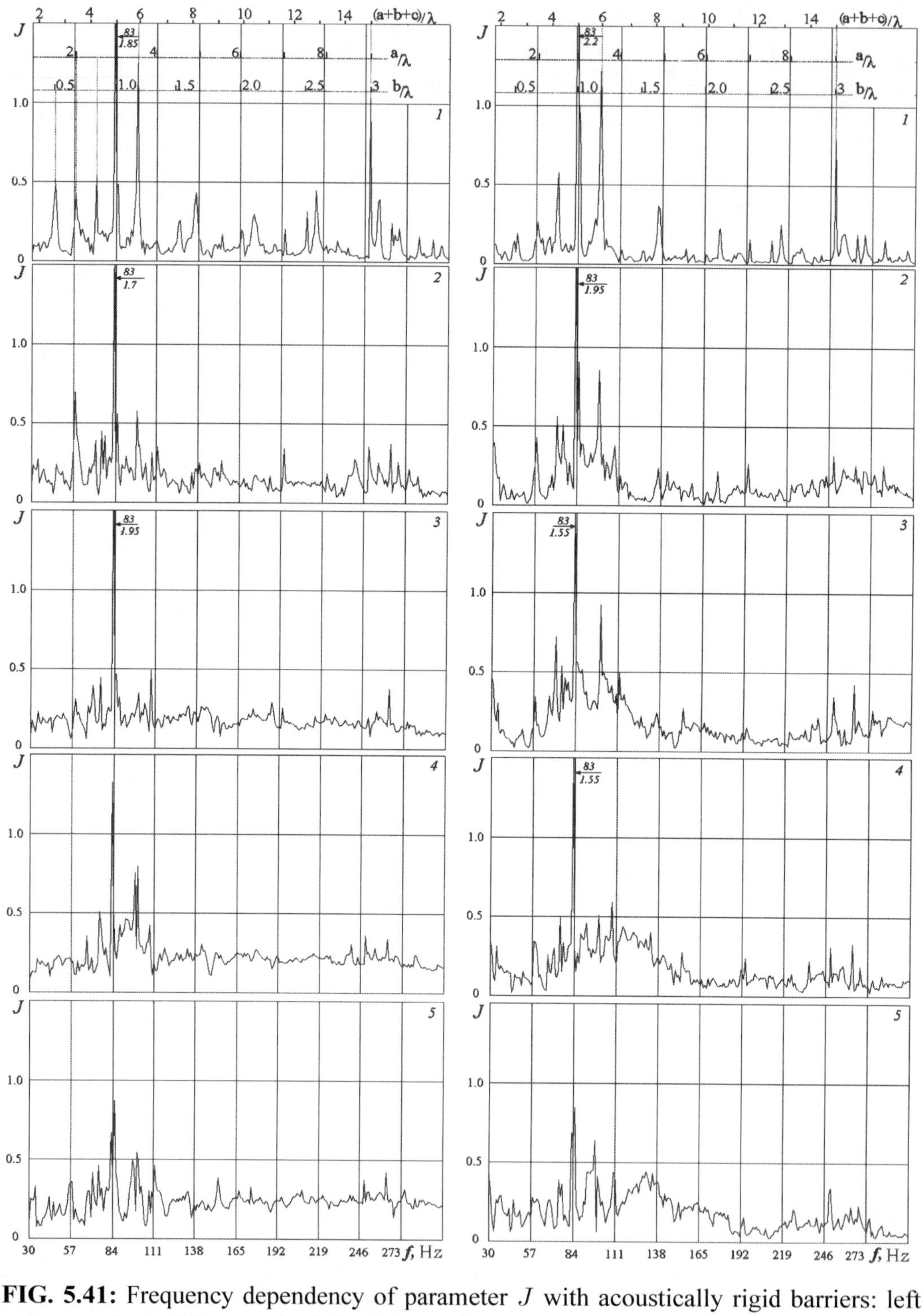

FIG. 5.41: Frequency dependency of parameter J with acoustically rigid barriers: left column—the building to the left of the traffic area, right column—the building to the right of the traffic area; 1–5—floors numbering (from the top down)

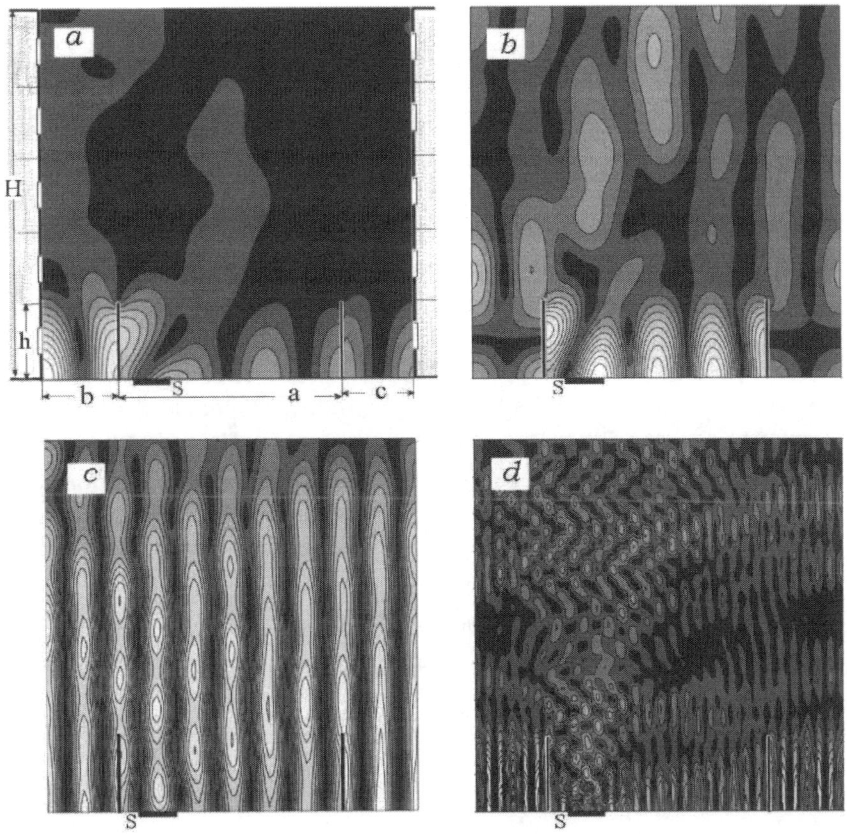

FIG. 5.42: Pressure amplitude field in the space between the buildings. When the frequency of the sound source is (barriers are acoustically rigid): (a)—45 Hz, (b)—58 Hz, (c)—83 Hz, (d)—249 Hz

83 Hz), in all the domains (between the buildings, between the barriers and between the barriers and buildings). It corresponds to the case, when $(b + a + c)/\lambda = 5$, $a/\lambda = 3$ and $b/\lambda = c/\lambda = 1$. It seems that the sound field "ignores" the barrier [compare Figs. 5.40(b) and 5.42(c)]. Such "global" resonance occupies all the area between the buildings. That is why this resonance dominates on all the floors (see Fig. 5.41) of both buildings. The situation is similar at the frequency 249 Hz [see Fig. 5.42(d)], where there is also a "global" resonance. But it is not so efficient here, as at frequency 83 Hz. This is due to the fact that at frequency 249 Hz the wave size of sound source $(a_2 - a_1)/\lambda = 1.54$, i.e., it is greater than wave length. The source becomes directional and it is unable to insonify uniformly all the space between the buildings. Figure 5.42(e) shows that the relatively narrow cone of acoustic energy comes from the source. It expands and "gets fuzzy." That is why the spike is clearly seen only in the area of the first floor, at frequency 249 Hz. The spike level above the first floor is rather small.

In summary we can conclude: the application of barriers with acoustically rigid surfaces on the urban streets to protect the buildings and sidewalks from traffic noise may be ineffective because of the resonances between the buildings, between the barriers and between the barriers and the faces of the buildings. It is fair to assume that the barriers with sound absorbing walls, can protect the sidewalks and windows in the buildings more efficiently.

Figure 5.43 shows frequency dependencies of function J for the barriers with ideally absorbing surfaces ($Y = 1$) from the traffic way side.

Comparing the graphs in Fig. 5.43 with the graphs in Figs. 5.39 and 5.41 it is easy to verify that the level of all the resonance spikes decreases and the mean level of parameter J decreases in the entire frequency band under consideration. There are the isolated resonance spikes, but their level is no more than 0.5–0.3. If we turn to Fig. 5.44, which shows the structure of pressure amplitude field for the barriers with ideally absorbing walls, we can see, that the "global" resonances disappeared. There are only the resonances between the barriers and the faces of buildings of relatively low levels. The walls of the barriers, faced to the buildings, should also be absorbing to remove the resonances.

Figure 5.45 shows frequency dependencies of parameter J for the barriers, covered with the grating of Helmholtz resonators. We select the following parameters of Helmholtz resonator: $b_1 = 0.40$ m, $b_2 = 0.05$ m, $2r_0 = 4 \times 10^{-3}$ m, $l = 0.5 \times 10^{-3}$ m, $\tilde{r} = 2$. The natural frequency of Helmholtz resonator $f_0 \approx 97$ Hz.

As we can see, the frequency dependencies of parameter J for the barriers with the walls in the form of the gratings of Helmholtz resonators are little different from the barriers with the ideally absorbing walls. Compare the graphs in Figs. 5.43 and 5.45. It is fair to say that, the barriers with the absorbing walls can be implemented in practice. They can ensure effective noise protection on the modern urban streets. We note finally,

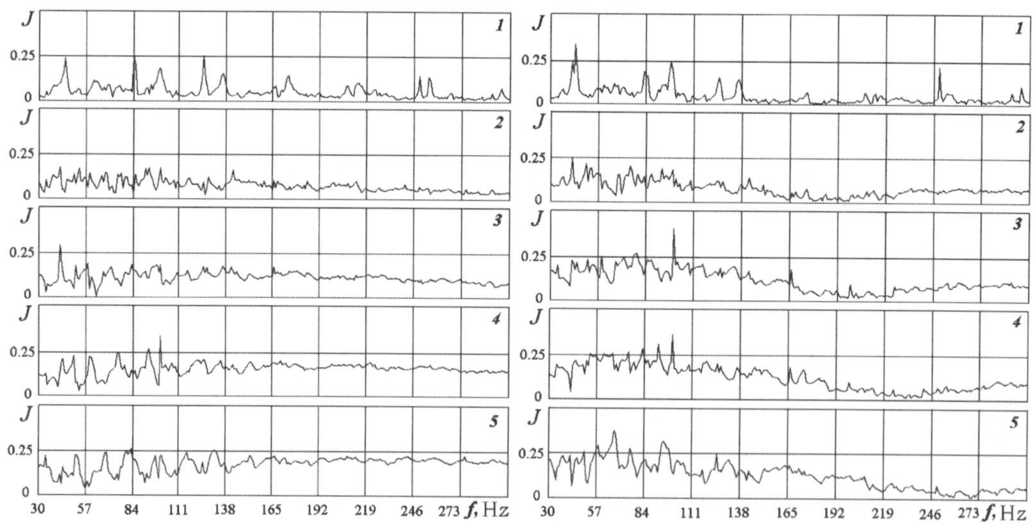

FIG. 5.43: Frequency dependency of parameter J with barriers, the surfaces of which are faced to traffic area, and absorb ideally (the surfaces of barriers, faced to the buildings, acoustically rigid): the left column—the building to the left of the traffic way, the right column—the building to the right of the traffic way, the floors are numbered from the top down

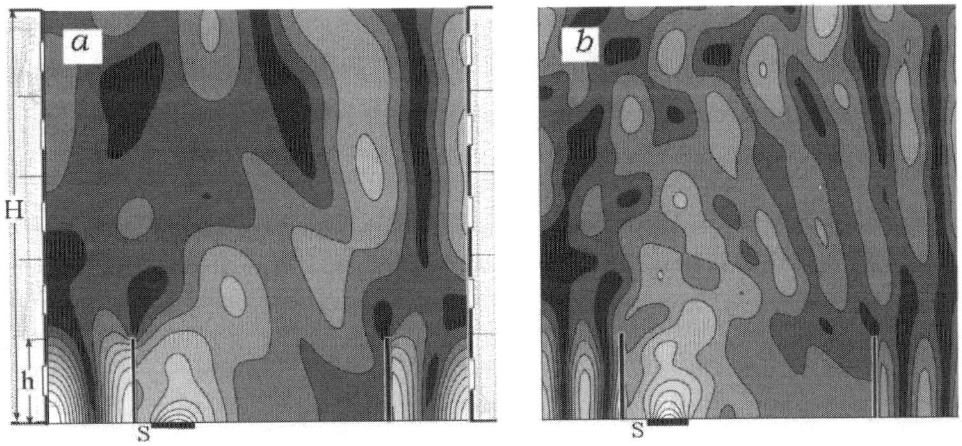

FIG. 5.44: Pressure amplitude field in the space between the buildings when sound source frequency is: (a)—45 Hz, (b)—84 Hz (barriers surfaces, faced to the traffic way, are ideally absorbing, and the surfaces of the barriers, faced to the buildings—are acoustically rigid)

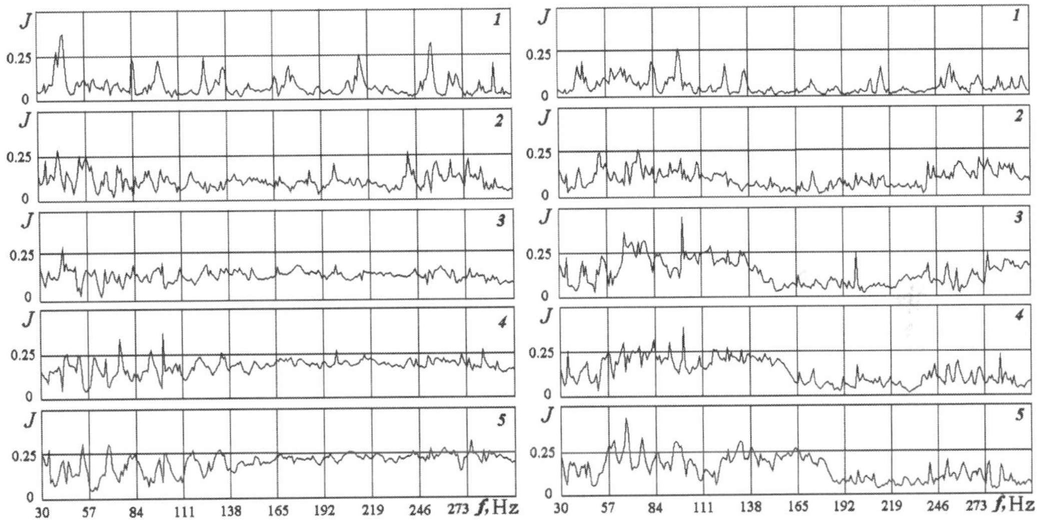

FIG. 5.45: Frequency dependency of parameter J when the surfaces of the barriers, faced to the traffic area, are covered with the grating of resonant absorbers (the barriers surfaces faced to the buildings are acoustically rigid): the left column—the buildings to the left of the traffic area, the right column—the buildings to the right of the traffic area, the floors are numbered from top down

that the considered problem can easily be used in more complicated case, when all the walls of the barriers absorb sound energy.

The results given in Chapter 5, are based on works [13,16–18,25,31], which were published.

REFERENCES

1. Osipov, L.G., Bobilev, V.I., and Borisov, L.A., et. al., *Sound Insulation and Acoustic Absorption*, Moscow: AST and Astrel Publishing House, p. 450, 2004. [in Russian]
2. Kurze, U.J., Noise reduction by barriers, *J. Acoust. Soc. Am.*, vol. **55**, no. 3, pp. 504–518, 1974.
3. Isei, T., Embleton, T.F.W., and Piercy, J.E., Noise reduction by barriers on finite impedance ground, *J. Acoust. Soc. Am.*, vol. **67**, no. 1, pp. 46–58, 1980.
4. Okubo, T. and Fujiwara, K., Efficiency of noise barrier on the ground with an acoustically soft cylindrical edge for practical use, *J. Acoust. Soc. Am.*, vol. **105**, no. 6, pp. 3326–3335, 1999.
5. Okubo, T. and Yamamoto, K., Procedures for determining the acoustic efficiency of edge-modified noise barriers, *Appl. Acoust.*, vol. **68**, no. 7, pp. 797–819, 2007.
6. Osipov, L.G. and Judin, E.Y., *Noise Reduction in the Buildings and Residential Areas* (Ed.), Moscow: Stroyizdat, p. 558, 1987. [in Russian]
7. Asdrubali, F. and Pispola, G., Properties of transparent sound-absorbing panels for use in noise barriers, *J. Acoust. Soc. Am.*, vol. **121**, no. 1, pp. 214–221, 2007.
8. Li, K.M., Knok, M.P., and Law, M.K., A ray model for hard parallel noise barriers in high-rise cities, *J. Acoust. Soc. Am.*, vol. **123**, no. 1, pp. 121–132, 2008.
9. Osipov, L.G., Prutkov, B.G., Shishkin, I.A., and Karagodina, I.L., *City-Planning Methods of Noise Elimination*, Moscow: Stroyizdat, p. 215, 1975. [in Russian]
10. Grinchenko, V.T. and Vovk, I.V., *Wave Problems of Scattering by Elastic Shells*, Kiev: Naukova dumka, p. 240, 1986. [in Russian]
11. Grinchenko, V.T., Vovk, I.V., and Matsypura, V.T., *Basic Acoustics*, Kiev: Naukova dumka, p. 640, 2007. [in Russian]
12. Borovikov, V.A. and Kinber, B.E., *Geometrical Theory of Diffraction*, Moscow: Svyaz, p. 248, 1978. [in Russian]
13. Vovk, I.V., Konchenko, T.A., and Matsypura, V.T., Concerning one rigorous method of estimation of acoustical properties of noise barriers, *Acustychnyi Visnyk*, vol. **7**, no. 4, pp. 21–27, 2004. [in Russian]
14. Shenderov, E.L., *Sound Radiation and Scattering*, Leningrad: Sudostroyenie, p. 304, 1989. [in Russian]
15. Menounou, P. and Ho You, J., Experimental study of the diffracted sound field around jagged edge noise barriers, *J. Acoust. Soc. Am.*, vol. **116**, no. 5, pp. 2842–2854, 2004.
16. Vovk, I.V., Matsypura, V.T., and Sotnikova, T.A., Concerning one method of noise barriers effectiveness improvement, *Acustychnyi visnyk*, vol. **9**, no. 2, pp. 17–26, 2006. [in Russian]
17. Vovk, I.V. and Sotnikova, T.A., Integral acoustic characteristics of V-noise barrier, *Acustychnyi visnyk*, vol. **10**, no. 3, pp. 25–29, 2007. [in Russian]
18. Vovk, I.V. and Matsypura, V.T., The influence of the properties of the noise barrier surface on its efficiency, *Acustychnyi visnyk*, vol. **13**, no. 1, pp. 3–10, 2010. [in Russian]
19. Sevrugova, N.V., Sound field near the plane periodic structure with impedance bands, *J. Acoust.*, vol. **27**, no. 1, pp. 138–142, 1981. [in Russian]

20. Shestopalov, V.P., Litvinenko, L.N., Masalov, S.A., and Sologub, V.G., *Waves Diffraction by the Gratings*, Kharkov: Kharkov University Printing House, p. 287, 1973. [in Russian]

21. Rzhevkin, S.N., *Course of Lectures on Sound Theory*, Moscow: Moscow State University Printing House, p. 336, 1960. [in Russian]

22. Furduev, V.V., *Electroacoustics*, Moscow, Leningrad: OGIZ, p. 516, 1948. [in Russian]

23. May, D.N. and Osman, M.M., Highway noise barriers: New shapes, *J. Sound Vib.*, vol. **71**, pp. 73–101, 1980.

24. Li, K.M., Kwok, M.P., and Law, M.K., A ray model for hard parallel noise barriers in high-rise cities, *J. Acoust. Soc. Am.*, vol. **123**, no. 1, pp. 121–132, 2008.

25. Vovk, I.V. and Matsypura, V.T., Sound protective properties of the barriers, situated along the both sides of traffic area, *Acustychnyi Visnyk*, vol. **13**, no. 4, pp. 3–14, 2010. [in Russian]

26. Skudrzyk, E., *Simple and Complex Vibratory Systems*, Pennsylvania: Pennsylvania State University Press, p. 514, 1968.

27. Kotzen, B. and English, C., *Environmental Noise Barriers*, London, New York: Spon Press, p. 257, 2009.

28. Li, K.M. and Tang, S.H., The predicted barrier effect in the proximity of tall buildings, *Acoust. Soc. Am.*, vol. **114**, pp. 821–832, 2003.

29. Osipov, L.G., (Ed.), *Sound Protection in City-Planning*, Moscow: Stroyizdat, p. 96, 1993. [in Russian]

30. Ivanov, N.I., *Engineering Acoustics, Theory and Practice of Noise Elimination*, Moscow: Logos, p. 424, 2008. [in Russian]

31. Vovk, I.V., Grinchenko, V.T., and Matsypura, V.T., Sound protective properties of barriers, situated along the urban streets, *Acustychnyi Visnyk*, vol. **15**, no. 2, pp. 3–16, 2012. [in Russian]

CHAPTER 6

Soung Propagation in the Human Respiratory System

"Breathe deeper: you're excited"

Ilf and Petrov, "The Twelve Chairs"

6.1 THE PHENOMENON OF RESPIRATION SOUNDS

The human body, throughout its life, continually generates various sounds, including sounds that occur during breathing. People from ancient times were constantly interested in the causes of these sounds and attempted to use them to assess the condition of the respiratory system. For example, it is known that the physicians of ancient Greece and among them, the "father of medicine", Hippocrates (460–377 BC AD) resorted to the practice of auscultation of breath sounds [1,2]. Hippocrates even described a variety of sound phenomena that occur in during various respiratory tract diseases. For example, he was well familiar (in modern terminology) with crackles, which he identified with the sounds of a boiling vinegar. Physicians of ancient China resorted to auscultation of internal organs, which is indisputably evidenced [2]. It is surprising that then, for centuries, listening to breath sounds, as a diagnostic technique, was virtually forgotten. The honour of re-opening an important diagnostic significance of breath sounds, but now as a method of physical science-based studies of the human body, belongs to the famous French clinician Rene Laennec (1781–1826), who in 1818 published the classic treatise "On Mediate Auscultation or Treatise on the Diagnosis of the Diseases of the Lungs and Heart" [3]. We also owe to him the invention of the stethoscope, a wooden tube with bells at its ends, the first physical device for listening to the sounds produced by the body. In his treatise, Laennec formulated some verbal models* of respiratory sounds, offered their classification and gave a qualitative description of their possible causes. Later on, many generations of medical scientists continued and deepened the study started Laennec. What are the main results achieved in this area over the past almost 200 years?

First of all, the Laennec stethoscope was improved, giving birth to microstethoscope.† Verbal models of virtually all types of breath sounds were formulated, both in norm (i.e., for a healthy individual) and in pathology (i.e., for a sick individual). These models are qualitative, descriptive, based on the principle of similarity with a certain

*Qualitative, descriptive models.

†Microstethoscope, unlike a stethoscope, is a binaural device with flexible acoustic line.

ISBN: 978–1–56700–459–5 © 2018 by Begell House Inc.

sound, occurring in nature or in everyday life. For example, in some pathologies of the respiratory system, the sounds that occur are called moist wheezes in medicine [4]. These wheezes in medicine are divided into large, medium and small bubbling wheezes and are associated with the sounds of bursting bubbles of different calibres in a glass of champagne. Or, for example, dry wheezes, which can be buzzing, humming, whistling—a clear association with sounds emitted by various insects as they fly. Thus, using verbal models all respiratory sounds get classified. Further, parts of the surface of the thorax where auscultation of these or those sounds was best were identified. Finally, some qualitative relationships between the disease type and the nature of resulting sounds were established, at least statistically.

We now consider main drawbacks of the auscultation of respiratory sounds using a classic microstethoscope to assess the condition of the respiratory system. First of all, this is a significant subjective assessment of the nature and type of emerging sounds. Indeed, the energy of the thorax surface vibrations, caused by breathing is very low and therefore the sound pressure level, reaching the ears of a doctor through the tubes of microstethoscope, is also very low. Therefore, a physician's ability to hear the characteristics of certain sounds will substantially depend on his hearing acuity as well as on the acoustic properties‡ of a microstethoscope. Furthermore, physicians need a good auditory memory, given that to effectively and efficiently classify noises and sounds they need to "keep" in mind not only images of a lot of these sounds, but also of their different variations. Obviously, not every doctor has this ability. Another important factor to assess the feasibility of auscultation relates to physiological features of sound perception by the human ear. It is well known that the frequency range perceived by the ear is limited and extends from approximately 50–70 Hz to 15–18 kHz. In addition, the greatest hearing acuity is in the vicinity of 1000 Hz, and as it approaches the low and high frequencies the hearing acuity deteriorates substantially [5]. Finally, the background noise and all kinds of interferences that are almost always present during the chest auscultation will roughen the hearing threshold, which does not contribute to identifying subtle features of sounds.

It is these deficiencies of classic auscultation on the one hand and the rapid development of electronics and computer technology over the past 20–25 years on the other, that stimulated the search for and creation of electronic registration of the sounds of human activity, excluding in turn subjective judgment. A significant progress on this matter has been achieved to date. Please refer to e.g., [6,7] and an extensive bibliography therein.

At the same time, it became obvious that without an understanding of the physical causes of the sound sources during respiration, without determining their spatial localization and without detecting peculiarities of sound propagation in the chest from these sources to the surface of the chest (where they are registered) it is impossible to reliably establish correspondences between the traditional verbal models of sounds and specific physical changes that occur in the organs of the respiratory system. Unfortunately, for obvious reasons, it is very difficult to experimentally investigate these problems *in vivo*.

‡Unfortunately, acoustic properties of microstethoscopes are not standardized and to a large extent depend on the manufacturer; besides their amplitude—frequency characteristics are very indented and in the frequency range of 50–2000 Hz are relatively uneven, from—10 dB to +15 dB [4].

Therefore, at this time, high hopes lie with mathematical and physical modelling of the processes of sound generation and propagation in the human respiratory system, and some progress has already been made here [8–17].

In conclusion, it should be said that a lot of work related to improvements to methods of respiratory sound registration and classification, as well as to mathematical and physical modelling of processes of sound generation and propagation in the human respiratory system, has been continuously carried out for many years in laboratories in the US, Canada, Israel, Japan, France and other leading countries in Europe as well as in Ukraine, in particular at the Institute of Hydromechanics of the National Academy of Sciences of Ukraine. It should also be noted that in 1976 The International lung sounds association (ILSA), an international society whose aim is to coordinate research on the problems associated with the sounds of breathing, was created. ILSA organizes annual international conferences, details of which, including place and timing, can be found on the official website http://www.ilsa.us.

6.2 THE STRUCTURE OF THE RESPIRATORY SYSTEM AND AERODYNAMIC EFFECTS THEREIN THAT GENERATE BREATH SOUNDS

First, we briefly look at the anatomical features of the human respiratory system structure and its function. As is well known [18], the main function of the respiratory system is to provide a gas exchange, supply of oxygen to the blood and removal of carbon dioxide therefrom. This is achieved through breathing, that is, cyclical (inhale and exhale) run of the airdown numerous respiratory tracts. The upper respiratory tract, the nasal cavity, mouth and pharynx provide the connection with the environment. Next is the larynx, which performs two functions: one is a subsequent airway and another one is a voice generating system. Then immediately follows the bronchial tree, which consists of the trachea and branching bronchi of various calibres, bronchioles, alveolar ducts and alveoli[§] connected in series [Fig. 6.1(a)].

Alveolar ducts and alveoli form the lung parenchyma[¶] and serve as respiratory departments with thin walls, where the gas exchange takes place during breathing. The entire respiratory system is placed inside the chest and it is surrounded by several layers of biological tissue, including the muscle layer, ribs, the fat layer and the skin. Inside, between the right and the left lungs, lies the mediastinum, where the heart, the oesophagus, and a number of other vital organs are placed.

We begin our analysis of the air propagation in the respiratory system from the larynx, Fig. 6.2. Due to the increase of the chest cavity volume as a result of rising ribs and lowering diaphragm, the phase of inspiration begins, and the air passing through the nose and mouth, travels to the vestibule of the larynx. The geometry of the latter is rather complicated, however, in the first approximation, it can be presented in the form of a tube converging towards the vocal cords. From the aerodynamics standpoint the vestibule of

§The alveoli are about 0.3 mm in diameter, the total amount of alveoli for an adult is $\sim$ 600 mill.

¶The parenchyma is of biological tissue which actually forms the lungs.

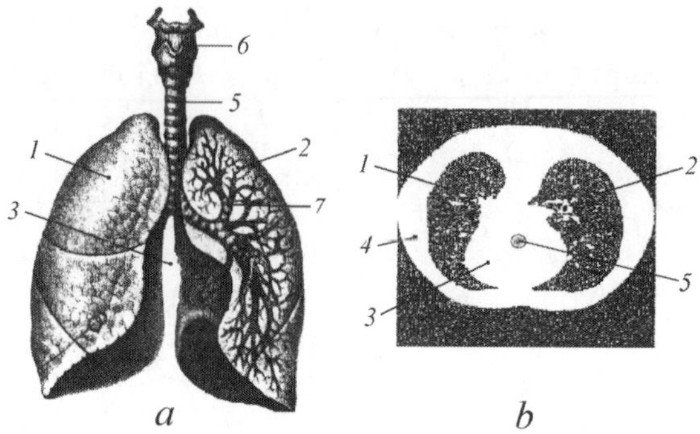

FIG. 6.1: The respiratory system (without the upper respiratory tract) [18]; (a)—the view in the frontal plane (the left lung is dissected by a vertical plane), (b)—the horizontal section of the chestat the level slightly above the bifurcation of the trachea into the main bronchi (CT); 1—the right lung, 2—the left lung, 3—the mediastinum, 4—musculo-rib and fat layers, 5—the trachea, 6—the larynx, 7—the bronchial tree

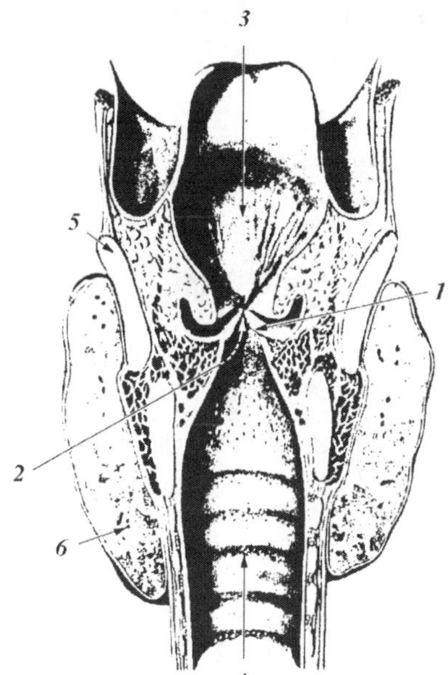

FIG. 6.2: The front section of the cavity of the larynx through the middle portion of the vocal cords [19]; 1—the vocal cords, 2—the glottis, 3—the vestibule of the larynx, 4—the trachea, 5—thyroid cartilage, 6—thyroid

the larynx can be viewed as confusor, i.e., a rather gradually tapering part that provides a transition from a large section of the initial portion of the vestibule of the larynx to the small cross section of the glottis.

The analysis in [8] shows that the ratio of the cross sectional area of the larynx to the beginning of the vestibule area of the glottis in the inspiratory phase averages about 5–6, whereas the taper angle of the larynx vestibule is about 60 degrees. From aero hydromechanics it is well known that the airflow will accelerate 5–6 times in such a channel. Further, the flow passes through the glottis, the shape of which largely depends on the individual; however, the most common one has the shape close to an isosceles triangle.

After passing through the glottis, the flow enters the lower cavity of the larynx, in the first approximation an expanding tube (a diffuser in aerodynamics) connecting the glottis and the trachea. As a result, the airflow slows down 3–5 times in this part [8]. Then the flow moves through the trachea, whose shape resembles a cylindrical tube, and into the bronchi.

It is appropriate to consider here the structure of the bronchial tree in more detail [Fig. 6.1(a) and Table 6.1]. As is well known [9,18], the bronchial tree begins with the trachea, also called the windpipe. The trachea, in turn, divides into two main bronchi, the left one and the right one, each then respectively entering the left and the right lungs. The main bronchi branch into two secondary, or lobar, bronchi, each of which in turn branches into two bronchi and so on. Thus, the division of airways toward the periphery of the bronchial tree is dichotomous,[||] and the branching points of the bronchi are called bifurcations. The branching airways (top to bottom) are called generations and are denoted by the letter z. Thus, for the trachea $z = 0$, for the main bronchi $z = 1$ and so on. According to [9], the bronchial tree (together with the trachea) has 24 generations, and the number of airways in each generation equals approximately 2^z. Therefore, from the aero hydromechanics standpoint, the bronchial tree is essentially a type of airway channel with numerous branches, well-known in physics and engineering. It is important to note that the total cross-sectional area of the respiratory tract in each subsequent generation grows exponentially (Table 6.1).

It is apparent from the above analysis that the narrowest part of the bronchial tree is the glottis, whose area is, on average, five times less than the cross sectional area of the trachea. This fact clearly shows that the highest airflow velocity during respiration will be in the glottis. Considering that, when breathing with an average intensity, the airflow rate is about 0.5–1.0 litters/sec, the flow velocity in the glottis may reach 10–20 m/s. This corresponds to Reynolds numbers of about 5000–10,000. Therefore, it is in the area of the glottis that the conditions for intense turbulence in the airflow during respiration, and, as a result, for pressure pulsations on the trachea walls are created. As the airflow moves to the periphery of the bronchial tree, its velocity falls rapidly, because the total cross-sectional area of the airways in each succeeding generation grows exponentially (Table 6.1).

Pressure pulsations on the trachea walls cause oscillations of the trachea walls, which in turn excite sound waves with frequencies from about 70 Hz to 1500 Hz in the air inside the trachea [1,2,6,8,22,23] and in the biological tissues surrounding the trachea. This physical mechanism for converting energy of the gas or liquid flow into the sound energy is well known [13,21]. Hence, the surface oscillations of the thorax, caused by pressure pulsations of the turbulent flow in the trachea will have two components. The first component exists due to wave propagation by air, i.e., through the bronchial tree,

[||] Dichotomy, from Greek διχοτομεω means divided into two parts.

TABLE 6.1: The overall architecture of the bronchial tree, its airways and some parameters

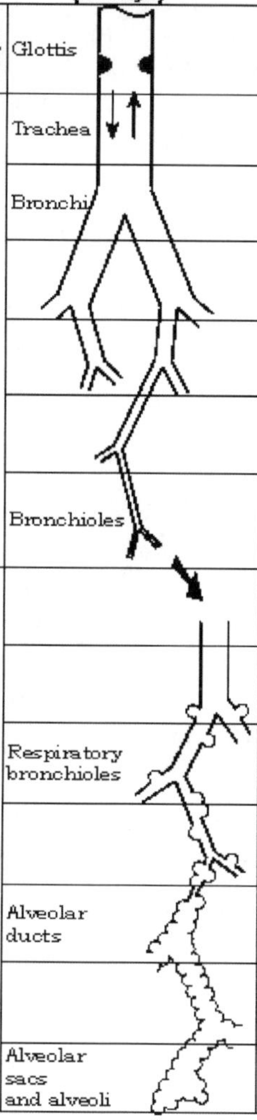

Zone	Anatomical structure (airways)	z	Number of branches	Diameter, cm	Length, cm	Total cross-section, cm^2	Relative velocity of airflow	Reynolds number, Re
Larynx	Glottis	–	–	~0.82	0.3	0.54	~5	~5000
Conductive zone	Trachea	0	1	1.675	12	2.54	~1	~1000
	Bronchi	1	2	1.136	4.78	2.33		
		2	4	0.705	1.91	2.13		
		3	8	0.436	0.76	2.0	1.26	
		4	16	0.34	1.27	2.48		
	Bronchioles	↓						
Transient and respiratory zone		17	131072	0.038	0.141	300		
		18						
	Respiratory bronchioles	19						
		20	1048576	0.032	0.08	1600		
	Alveolar ducts	21						
		22						
	Alveolar sacs and alveoli	23	8388608	0.029	0.05	11800	0.0002	~0.2

causing wall oscillations of all its airways and therefore re-radiation of sound to the surrounding biological tissues, including the lungs, and further through the biological tissue surrounding the lungs, to the surface of the thorax. The second component is due to wave propagation in biological tissues surrounding the trachea. They first penetrate through the mediastinum, and then through the lungs and the bio-tissues surrounding the lungs, to the surface of the thorax [1,2,7].

In addition to the examined mechanism of sound generation, initiated by the turbulent airflow in the area of the glottis, there is another natural mechanism of sound generation, which is human speech. During speech vocal cords vibrate, as a result the cross section of the glottis continuously changes causing pulsed air flow in the trachea and the upper respiratory tract [20]. These oscillations in turn cause oscillations of the trachea walls and of the walls of the further airways. Sound waves that occur as a result of these oscillations will be propagating in the thorax in two ways, as already described above. The frequency range of speech signals lies between 80 Hz and 1000 Hz [13]. Note that the evaluation of the quality of voice transmission of simple signals, such as vowel sound "e" (or the numbers "ninety-nine" or "sixty-six" are traditionally mentioned) through the respiratory system is also widely used in medicine for the diagnosis of her condition, and this method is called bronchophony [1,2,6].

We reviewed the physical mechanisms of sound generation in the respiratory system, assuming that all the airways are open, i.e., that the airflow can move there freely during respiration. This situation is most common in people with a healthy respiratory system, i.e., "the norm." However, in certain diseases (pathology) other physical mechanisms of sound generation may occur, different form the ones described above. However, we will not consider them here.

6.3 THE BRONCHIAL TREE AS A BRANCHING ACOUSTIC WAVEGUIDE

Above we have shown that during respiration and speech, the sound source is localized at the beginning of the trachea. Then the sound starts propagating along the bronchial tree and the biological tissues surrounding the trachea. In this section we will pay attention to the features of sound propagation of the bronchial tree as the acoustic waveguide, and try to model this process.

Indeed, from the perspective of acoustics, the bronchial tree can be represented as the air-filled branching waveguide with elastic walls, surrounded by the biological tissue with specific acoustic properties. Given the dichotomous branching pattern of the airways toward the periphery of the bronchial tree, consider a typical (repeated several times at different scales**) fragment of the bronchial tree and its geometric characteristics [10] (Fig. 6.3).

First, the diameter d and the length l of the daughter airways are almost always less than the diameter and the length of their parent, i.e., the conditions $d(z+1) < d(z)$ and $l(z+1) < l(z)$ hold (Table 6.2). Secondly, the cross-sectional area of the parent (only for the first three generations) is greater than the total cross-sectional area of their 2 daughter airways, i.e., their ratio.

However, beginning with $z > 2$, the value $\alpha(z) < 1$. Thirdly, although the angle of furcating ϕ varies within a relatively wide range, it is on average about 30° [24]. Fourth, the connection between parent and daughter airways is very smooth and has an average radius of curvature of $R(z) \approx 6d(z+1)$ [24]. Fifth, the airway wall thickness δ is small and is less than 10% of its diameter [25]. Furthermore, it should be noted that from the 17th generation onwards ($z \geq 17$) airway walls (respiratory bronchioles) are not solid but have holes (perforations). Through these holes the internal volume of the airways is connected with the internal cavities of the alveoli. The alveoli thinnest walls contain blood capillary vessels, where, in fact, the gas exchange takes place during respiration.

**The bronchial tree has fractal structure.

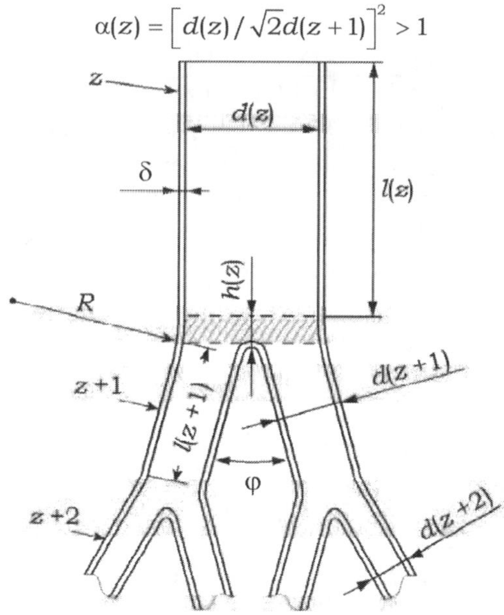

FIG. 6.3: Bifurcation fragment of the bronchial tree

The transverse dimensions of the alveoli, as already mentioned, are extremely small and are about 300 microns. The degree of wall perforation is characterized by the parameter $\eta(z)$, indicating which part of the wall surface is perforated. Thus, the magnitude of $\eta(z)$ for the bronchioles of the 17th, 18th, and 19th generations equals to 12, 25, and 50%, respectively. Beginning from the 20th generation (Table 6.2), the walls already present a thin net of elastic collagen and individual muscle fibres and for them $\eta(z) = 100\%$. Finally, bronchioles of the 23rd generation (alveolar sacs) are structurally identical to the alveolar passages, but instead of branching out, close at the end by a group of alveoli. It should be stated that the alveoli and alveolar pathways in fact form the foundation for the lung biological tissue, the parenchyma. The density of the parenchyma is $\rho_p \approx 0.3 \times 10^3$ kg/m^3, and the speed of sound propagation there is $c_p \approx 30$–50 m/s. Furthermore, the parenchyma has the capacity to absorb sound energy, which increases with frequency [16].

The parenchyma almost entirely surrounds the airways, starting from the lobe bronchi, i.e., from second generation. The trachea and main bronchi are surrounded by biological tissues of the mediastinum, the mean densities of which equal to $\rho_c \approx 10^3$ kg/m^3, and the speed of the sound is close to $c_c \approx 1.5 \times 10^3$ m/s.

Earlier, our attention was focused on the architecture of the bronchial tree and on geometric parameters of its elements (airways). We now focus on the mechanical properties of the airway walls. Using the results presented in [26,27], which provide data about the mechanical properties of the airway walls, we estimated average values of the unit masses $M(z)$, elasticity $K(z)$, and losses $R(z)$ of the airway walls for each generation up to the 18th (the walls of subsequent generations are completely perforated and do not represent any obstacle to sound oscillations). The results of these estimates are shown in

TABLE 6.2: Geometric and mechanical parameters of airways

z	$d(z)$, cm	$l(z)$, cm	$\alpha(z)$	$M(z)$, kg/m^2	$K(z)$, kg/m^{2}s$^2\times10^6$	$R(z)$, kg/m^{2}s$\times10^3$	$f_0(z)$, Hz	Anatomical structure
0	1.675	12.0	1.080	7.800	1.67	8.10	73	trachea
1	1.136	4.78	1.300	1.300	1.97	3.80	195	main bronchi
2	0.705	1.91	1.300	1.340	2.44	4.60	215	lobe bronchi
3	0.436	0.76	0.830	0.210	3.57	6.80	650	—"—
4	0.340	1.27	0.822	0.135	4.54	8.40	910	segmental bronchioles
5	0.266	1.07	0.800	0.110	4.40	9.10	960	—"—
6	0.211	0.90	0.780	0.090	5.36	11.40	1200	—"—
7	0.169	0.76	0.754	0.090	6.82	14.00	1370	—"—
8	0.138	0.64	0.736	0.060	6.40	13.80	1660	—"—
9	0.114	0.54	0.726	0.060	7.80	17.00	1830	—"—
10	0.095	0.46	0.710	0.043	9.20	20.00	2240	—"—
11	0.080	0.39	0.680	0.040	10.90	23.80	2600	terminal bronchioles
12	0.069	0.33	0.660	0.038	8.50	17.00	2350	—"—
13	0.060	0.27	0.660	0.044	9.75	19.50	2350	—"—
14	0.052	0.23	0.620	0.032	10.80	21.60	2900	—"—
15	0.047	0.20	0.625	0.024	12.10	24.20	3550	—"—
16	0.042	0.16	0.614	0.027	13.30	26.70	3500	—"—
17	0.038	0.14	0.556	0.021	14.70	30.00	–	respiratory bronchioles
18	0.036	0.12	0.590	0.023	16.00	32.00	—	—"—
19	0.033	0.10	0.535	—	—	—	—	—"—
20	0.032	0.08	0.570	—	—	—	—	alveolar ducts
21	0.030	0.07	0.500	—	—	—	—	—"—
22	0.030	0.06	0.646	—	—	—	—	—"—
23	0.029	0.05	—	—	—	—	—	alveolar sacs

Table 6.2. Using these data, we also estimated the resonance frequencies of the radial oscillations of the airways $f_0(z)$ (for a hypothetical case, when they are in a vacuum).

Analysing these data, it is easy to note that as z increases, the unit mass of the walls decreases rapidly. This is quite natural given the rapid decrease of the diameter and the thickness of the airway walls. However, the elasticity of the walls increases, apparently due to an increase in their curvature. As a result, the resonant frequency increases as z increases.

We now analyse wave dimensions of the airways. As mentioned above, the main portion of the sound energy of noises and sounds generated in the trachea, lies in the frequency range of 70 Hz to 1500 Hz. Given this and assuming the speed of sound in air is 330 m/s, it is easy to define a range of wavelengths suitable for the study of sound propagation in the bronchial tree. It will be 22 cm $\leq \lambda \leq$ 4700 cm. Considering the data in Table 6.2, it can be concluded that in the whole range of frequencies the wave diameter $d(z)/\lambda$ of the airway of any generation is very small and does not exceed 0.08. It is also easy to see that starting from the second generation $z \geq 2$ wavelengths of the airways are $l(z)/\lambda \leq 0.09$, i.e., also very small. The only exceptions are the trachea and the main bronchi, the wavelengths of which are much larger, and are $l(0)/\lambda \leq 0.54$ and $l(1)/\lambda \leq 0.22$, respectively. Thus, only the length of the trachea is comparable to half of the wavelength, and even then only in the high-frequency part of the frequency range under consideration.

6.4 MATHEMATICAL MODELLING OF THE SOUND PROPAGATION IN THE BRONCHIAL TREE

Given the above analysis, we can conclude that, in terms of acoustics, the bronchial tree represents a waveguide of a very complex geometry with numerous irregularities in the form of bifurcations, changes in cross-sections and changes in the mechanical properties of the walls. It is well known from the theory of acoustic waveguides [28–34] that the presence of even single irregularities in the waveguide can substantially complicate the propagation of sound waves there. It is then quite difficult to solve the problem of sound propagation in such a waveguide taking into account all its peculiarities. Therefore, we introduce and justify the hypothesis and some assumptions to simplify the solution of this problem.

6.4.1 Adopted Hypotheses and Assumptions

First of all, we consider what role a waveguide bifurcation can play. Given the quantitative results presented in [32], and taking into account the fact that for all generations of airways, the condition $d(z)/\lambda \ll 1$ holds and that $\phi \approx 30^\circ$, it is easy to see that the impact of bifurcations on the process of wave propagation in the bronchial tree is infinitesimal and can be neglected. This conclusion is fully consistent with the findings obtained in [33,35] when analysing characteristics of sound propagation in narrow waveguides and channels.

We now turn to the assessment of the role of cross-sectional jumps that occur at the junction of parent and daughter airways (in Fig. 6.3. the area is highlighted by hatching). Generally speaking, the transition of a round cross-section of the parent in two round cross-sections of the daughterairways occurs according the scenario associated with a complex cross-section configuration transformation described in detail in [24]. However, it is more important to us that the wave dimensions of the junction area are very small: $d(z)/\lambda \ll 1$ and $h(z) \ll 1$. This allows us to replace the complex geometry of the junction area with a simpler one in order to relatively easy access the impact of the junction on the process of sound propagation.

6.4.2 Acoustic Model of Junctions (Bifurcations) of Airways

We assume that the junction area is a short (compared to the diameter) section of the tube with different diameters of the inlet and output openings (Fig. 6.4). Inlet diameter $d(z)$ equals to the diameter of the parent airway, and the output diameter $\sqrt{2}d(z+1)$ is calculated assuming that the total cross-sectional area of two daughter airways equals $2\pi\left[d(z+1)\right]^2/4$.

Using this approach we formally accounted for the value of the cross-section jump that is observed in the airway junction (bifurcation) area of the bronchial tree. To simplify further estimates we also assumed that the surfaces of the part of the tube that models airway junctions are acoustically rigid. It is apparent that the lower the height of the tube's part $h(z)$, the more this assumption shall hold.

6.4.3 Sound Fields in Junctions

Taking into account the above assumptions, we estimate the sound field in the bifurcation and the possible changes in acoustic characteristics caused by the cross-section jump. Assume that the narrow tube is filled with air having the wave resistance ρc, at the inlet of the tube (in the cross-section $x = 0$, see Fig. 6.4) we have oscillating velocity $\tilde{V}(z)$, and the bottom tube opening [the cross-section $x = h(z)$] is loaded on the unit impedance $Z(z+1)$. Besides, to be specific, we assume that $d(z) < \sqrt{2}d(z+1)$. Then, taking into account the adopted hypothesis and assumptions, and in a one-dimensional model framework, velocity potential Φ_z inside the bifurcation of the generation z can be presented as

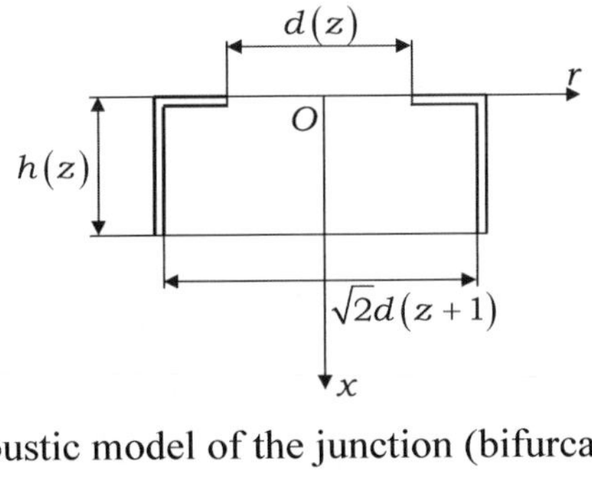

FIG. 6.4: Acoustic model of the junction (bifurcation) of airways

$$\Phi_z = A\cos kx + B\sin kx, \tag{6.1}$$

where $k = \omega/c$, A and B are unknown complex coefficients. The boundary conditions when $x = 0$ are

$$-\frac{\partial \Phi_z}{\partial x} = \begin{cases} \tilde{V}(z), & 0 \leq r \leq d(z)/2, \\ 0, & d(z)/2 \leq r \leq \sqrt{2}d(z+1)/2, \end{cases} \tag{6.2}$$

and when $x = h(z)$ are

$$p_z = -\frac{\partial \Phi_z}{\partial x} Z(z+1), \quad 0 \leq r \leq \sqrt{2}d(z+1)/2, \tag{6.3}$$

where $p_z = -i\omega\rho\Phi_z$.

Condition (6.2) enables us to determine coefficient B in the Eq. (6.1):

$$B = -\alpha(z)\tilde{V}(z)/k. \tag{6.4}$$

Condition (6.3) allows us to determine coefficient A. Taking into account that $h(z)/\lambda \ll 1$ and omitting basic interim calculations, we get

$$A = \frac{\alpha(z)\tilde{V}(z)\left[Z(z+1) - i\rho c k h(z)\right]}{k\left[i\rho c + Z(z+1)kh(z)\right]}. \tag{6.5}$$

Let us now determine the input unit impedance of the bifurcation $\tilde{Z}(z)$, i.e., the impedance at the surface $x = 0$. Given (6.1), (6.4), (6.5) and using common definition of the impedance [36,37], we finally get

$$\tilde{Z}(z) = i\alpha(z)\frac{Z(z+1) - i\omega\rho h(z)}{i + Z(z+1)\omega h(z)/\rho c^2}. \tag{6.6}$$

Thus, a cross-section jump in the bifurcation area plays the role of an acoustic transformer, and the input impedance of bifurcation $\tilde{Z}(z)$ is determined by four parameters: the transformation coefficient $\alpha(z)$, the output impedance $Z(z+1)$, unit mass of the air volume in the bifurcation area $\rho h(z)$ and its unit elasticity $\rho c^2/h(z)$.

It is interesting to see how input impedance of the bifurcation changes when the output impedance reaches limits. For example, if $Z(z+1) \to \infty$, i.e., the lower wall of the bifurcation is absolutely rigid, then $Z(z) \approx i\alpha(z)\rho c^2/\omega h(z)$ and the input impedance is entirely determined by the elasticity of the bifurcation volume. If the lower wall is absolutely soft $Z(z+1) \to 0$, then the input impedance is determined only by the mass of the air volume $Z(z) \approx -i\alpha(z)\omega\rho h(z)$. Finally, Eq. (6.6) can be simplified further as $h(z) \to 0$. In the limit we get

$$\tilde{Z}(z) \to \alpha(z)Z(z+1). \tag{6.7}$$

This last equation can prove to be useful when studying sound propagation in airways of large orders.

The solution obtained above allows us to determine the oscillating velocity of the surroundings $V(z+1)$ at the lower orifice of the bifurcation. Simple calculations show the following relation

$$V(z+1) = i\alpha(z)\tilde{V}(z)\rho c\left[i\rho c + Z(z+1)kh(z)\right]^{-1}, \tag{6.8}$$

and as $h(z) \to 0$

$$V(z+1) \to \alpha(z)\tilde{V}(z). \tag{6.9}$$

To conclude, it should be noted that Eqs. (6.6)–(6.9) also hold when $d(z) > \sqrt{2}d(z+1)$.

6.4.4 Acoustic Model of the Airways

Above, we have presented an acoustic model of the airway junction (bifurcation) zone and offered its mathematical description. Now we turn to the acoustic model of the main elements of the bronchial tree, that of the airways (i.e., trachea, bronchi, bronchiole, etc.). Figure 6.5 presents one of the possible models of a single airway. The main element of the model is the part of the tube filled with air; the walls of this part are characterised by the mechanical impedance $\Psi(z) = -i\omega M(z) + iK(z)/\omega + R(z)$.

Suppose that the input (upper) end of the tube is characterised by an oscillating velocity $V(z)$, induced by the wave which propagated from an airway of the preceding generation, whereas the output (bottom) end of the tube is loaded on the input impedance $\tilde{Z}(z)$ of the bifurcation, which we have already determined before. We shall also assume that the tube walls, from the outside, are loaded on the infinite ($0 \le x \le l(z)$, $d(z)/2 \le r \le \infty$) layer of the medium with some wave resistance determined by the location of an airway in the bronchial tree. It should be noted here that in reality airways are surrounded by finite domain of a medium, however extremely complex geometry of the chest cavity

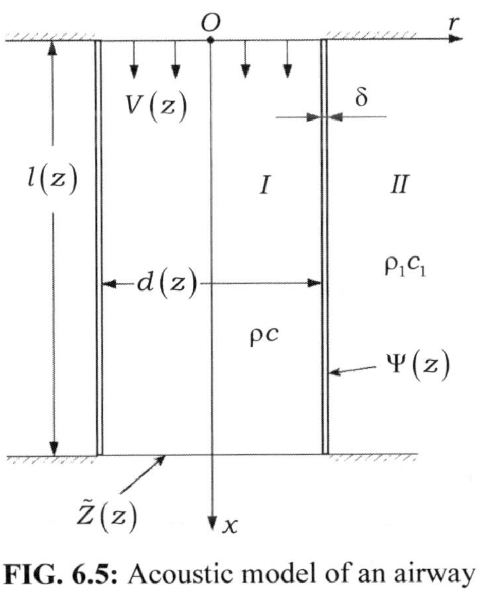

FIG. 6.5: Acoustic model of an airway

in general and of the bronchial tree in particular does not allow us so far to offer any adequate acoustic model. At the same time, as mentioned before, the parenchyma substantially absorbs sound energy which explains the fact that there are almost no waves reflected from its walls [16,17]. Therefore we can justify proceeding with modelling the medium surrounding airways by an infinite layer.

6.4.5 Sound Fields in Airways

The adopted acoustic model of a single airway allows us to move to the mathematical modelling of the problem. Following the method of partial domains [38,39], we divide the entire area of sound field (Fig. 6.5) into two subdomains—the inner one (domain I), determined by coordinates $0 \le x \le l(z)$, $0 \le r \le d(z)/2$ and the outer one (domain II): $0 \le x \le l(z)$, $d(z)/2 \le r \le \infty$. The field in the first subdomain can be presented as:

$$\Phi_1(z) = \sum_{n=0}^{\infty} [A_n \cos k_n x + B_n \sin k_n x] J_0(\gamma_n r) + \sum_{m=0}^{\infty} C_m \cos \alpha_m x F_0(k_m r), \quad (6.10)$$

where $k_n = \sqrt{k^2 - \gamma_n^2}$, $k_m = \sqrt{k^2 - \alpha_m^2}$; $k = \omega/c$, $\omega = 2\pi f$, f is the frequency, c is sound velocity in the air; $\alpha_m = \pi m/l$, γ_m are the roots of the equation $\gamma_m J_1(\gamma_m d(z)/2) = 0$;

$$F_0(k_m r) = \begin{cases} J_0(k_m r), & k > \alpha_m, \\ I_0(k_m r), & k < \alpha_m. \end{cases}$$

The field in the second subdomain can be presented as

$$\Phi_2(z) = \sum_{q=0}^{\infty} D_q \cos \alpha_q x Q_0(k_q r), \quad (6.11)$$

where $k_q = \sqrt{k_1^2 - \alpha_q^2}$, $\alpha_q = \pi q/l$, $k_1 = \omega/c_1$, c_1 is the sound velocity in the second subdomain;

$$Q_0(k_q r) = \begin{cases} H_0^{(1)}(k_q r), & k_1 > \alpha_q, \\ K_0(k_q r), & k_1 < \alpha_q. \end{cases}$$

We have adopted the following notations for cylindrical functions: $J_0()$, $I_0()$, $H_0^{(1)}()$, $K_0()$ are the Bessel functions of the real and imaginary arguments; the Hankel function of the first kind and the Macdonald function respectively. Below, the upper index in the form of apostrophe used with these functions denotes the first full derivative.

Below we set boundary conditions and the conditions of pairing of field components taking into account adopted hypothesise and assumptions:

$$-\frac{\partial \Phi_1(z)}{\partial x} = V(z), \quad x = 0, \quad 0 \le r \le d(z)/2, \quad (6.12)$$

$$p_1(z) = -\tilde{Z}(z)\frac{\partial \Phi_1(z)}{\partial x}, \quad x = l, \quad 0 \le r \le d(z)/2, \tag{6.13}$$

$$p_1(z) - p_2(z) = -\Psi(z)\frac{\partial \Phi_2(z)}{\partial r}, \quad 0 \le x \le l(z), \quad r = d(z)/2, \tag{6.14}$$

$$\frac{\partial \Phi_1(z)}{\partial r} = \frac{\partial \Phi_2(z)}{\partial r}, \quad 0 \le x \le l(z), \quad r = d(z)/2. \tag{6.15}$$

Here $p_1(z) = -i\omega\rho\Phi_1(z); p_2(z) = -i\omega\rho_1\Phi_2(z)$ are sound pressures in subdomains; ρ and ρ_1 are densities of mediums in the first and second subdomains respectively.

In accordance with the method of partial domains, functional Eqs. (6.12)–(6.15) can be transformed in the infinite system of linear algebraic equations based on the completeness and orthogonality of functions in series (6.10) and (6.11), namely, $J_0(\gamma_n r)$, $\cos \alpha_n x$, $\cos \alpha_q x$. We shall skip this rather standard procedure.

Thus we have shown a formal approach [since we do not know neither velocity $V(z)$, nor impedance $\tilde{Z}(z)$] to determine sound fields in airways and bifurcation of any generation z. Next we present how, using this approach, we can numerically estimate sound fields in the entire bronchial tree.

6.4.6 An Algorithm for Numerical Evaluation of Sound Fields in the Bronchial Tree

We begin with the periphery of the bronchial tree and consider the airway of the last, 23rd generation. As we mentioned before, its walls are completely perforated and as such essentially present a boundary (a cylindrical surface) between the two mediums: air and the parenchyma. The end of this airway can be loaded in two different ways. If the end is located inside the lung, then acoustically it will be loaded on the parenchyma wave resistance $Z_p = \rho_p c_p$. If the end is located on the outer surface of the lung which is in contact with the inner surface of the thorax, then its acoustical impedance will be presented by the unit impedance of the thorax $Z_g = -i\omega M_g + iK_g/\omega + R_g$.

Analysis of the literature gives the following averaged values of the thorax impedance components: unit mass $M_g \approx 10$ kg/m^2, unit elasticity $K_g \approx 5 \cdot 10^5$ kg/m^{2}s^2 and unit losses $R_g \approx 10^4$ kg/m^{2}s. Hence, we know possible acoustic loads of the end of the 23rd airway. We shall discuss specific choices of these loads later as we analyse numerical results.

Knowing the load at the end, unit oscillating velocity at the entrance of the 23rd airway and other parameters of the latter according to the table, we can determine field $\Phi_1(23)$ and input impedance $Z(23)$ using (6.10)–(6.15). Next, using (6.6) and (6.7) we can determine output impedance $\tilde{Z}(22)$ of the airway of the 22nd generation. Repeating the procedure, we estimate input impedance $Z(22)$ of the 22nd airway, then output impedance $\tilde{Z}(21)$ of the 21st airway and so on. Using this procedure continuously (we shall hereinafter call it a "reverse procedure," given the sequence of impedance evaluation from the periphery of the bronchial tree to its start—the trachea), we determine input and output impedances for all airways. The last to estimate is the output impedance $\tilde{Z}(0)$ of the trachea.

Now that we know output impedance of the trachea and using data from Table 6.2, we can estimate all energy and local characteristics of the sound field of the trachea, including power of sound oscillations $W(0)$ at the trachea entrance, power of sound oscillation $W_2(0)$ that travelled to the domain II, power of sound oscillations $\tilde{W}(0)$ that propagated to the main bronchi as well as values of pressure and oscillating velocity of any point of the bronchial tree. To achieve this, we need to set oscillating velocity $V(0)$ at the trachea entrance [e.g., in Eq. (6.12) setting $V(0) = 1$ m/s], solve system (6.12)–(6.15) and calculate series (6.10) and (6.11).

Once we determine all field characteristics in the trachea, including oscillating velocity $\tilde{V}(0)$ at the exit point, we can estimate oscillating velocity $V(1)$ at entry points to main bronchi using Eqs. (6.8) and (6.9). Similar to how the output impedance of the main bronchi was determined before, we can now estimate energy and local characteristics of the sound field in the main bronchi. We use the same procedure we used to estimate the field in the trachea (hereinafter referred to as the "direct procedure" given the sequence of moving from the trachea to the periphery of the bronchial tree). Applying the procedure to subsequent airways, we estimate, for each of them, all energy and local characteristics and essentially solve the problem (under the adopted assumptions) of sound propagation in the bronchial tree taking into account sound energy penetration through airway walls.

6.5 ANALYSIS OF NUMERICAL RESULTS

6.5.1 Input Impedance of the Bronchial Tree

Why is it so important to know such an exotic, at first glance, from the point of a human's physiology, acoustic characteristic as the input impedance of the bronchial tree? To answer this not so easy question, we need to refer to some initial explanations. As we have pointed before, the respiratory tract (as other human organs) processes a tremendous physical, geometric and physiological complexity, which in itself represents a substantial barrier for theoretical and experimental study of its acoustic properties. At the same time, it is also not clear today how to conduct non-invasive experimental studies of the acoustic field inside the respiratory tract with the aim to accumulate data about the differences of acoustic processes there in normal and pathology states. For that very reason it is only possible to discuss acoustic processes inside the respiratory tract on a qualitative level, based on the results of registering the modulus of the amplitude of oscillations of the thorax surface. In such a situation, as we have already mentioned, many researches focused on building acoustic and mathematical models that enable one to describe, at least in first approximation, the process of sound propagation in the respiratory tract. However, to evaluate relevancy and effectiveness of such models need to be tested; in other words, it is necessary to qualitatively and quantitatively compare acoustic characteristics obtained through these models with corresponding experimental data. Unfortunately, it is not enough to have data on amplitude of oscillations of the thorax surface to test workable theoretical models aimed at assessment of sound fields inside the respiratory tract.

Such seemingly deadlock situation was solved after the original methodology of non-invasive direct measurement of input acoustic impedance of the bronchial tree in a human being *in vivo* was proposed and implemented in 1976 [40]. Later, in [26] and [27] with the help of this methodology, a volume of experimental data was accumulated, sufficient, in the first approximation, for more or less reliable testing of newly developed acoustic and mathematical models of the respiratory tract, and first of all models of the bronchial tree.†† It is due to the abovementioned reasons that it seems important and interesting to start researching acoustic characteristics of the bronchial tree using comparative analysis of existing experimental data [27] and calculated frequency dependencies of its input impedance that we obtained.

We now turn to Fig. 6.6, which shows calculated frequency dependencies of the imaginary and real components of the input acoustic impedance of the bronchial tree as well as same experimental dependencies obtained through measurements conducted on healthy volunteers. We chose only three experimental dependencies out of seven presented in [27], namely those that provide visual representation of existing dispersals of these characteristics caused by natural variability of geometric and physical parameters of airways, and quite possibly in the first place by dispersals of the longitude dimension of the trachea.

Comparison of calculated and experimental curves (Fig. 6.6) apparently shows that all their qualitative features coincide. At the same time, it should be noted that the calculated curves are shifted with respect to the experimental ones in the direction of lower frequencies (approximately by 30%). As a result, we observe slightly worse quantitative congruence of calculated and experimental curves. Yet, the results of general comparative analysis of the curves show that proposed models capture well main features of the input impedance of the bronchial tree. This, on the one hand, proves that the hypotheses and assumptions chosen for modelling are correct, and, on the other hand, allows us to use

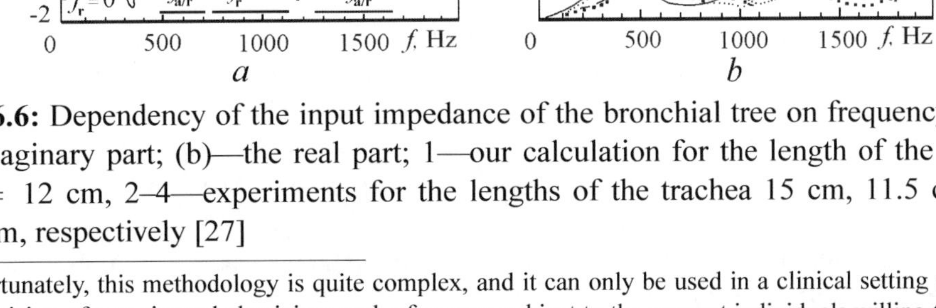

FIG. 6.6: Dependency of the input impedance of the bronchial tree on frequency; (a)—the imaginary part; (b)—the real part; 1—our calculation for the length of the trachea $l(0) = 12$ cm, 2–4—experiments for the lengths of the trachea 15 cm, 11.5 cm, and 10.3 cm, respectively [27]

††Unfortunately, this methodology is quite complex, and it can only be used in a clinical setting under the supervision of experienced physicians and, of course, subject to the consent individuals willing to participate in such scientific experiments. Therefore, the amount of quantitative data received to date is relatively small.

models for estimates of other acoustic characteristics, which cannot be obtained experimentally.

Let's look at curves in Fig. 6.6 in more detail. As can be seen, behaviour of frequency dependencies of the imaginary component of the input impedance $X(0)$ is determined by a series of resonances occurring in the bronchial tree. This is directly indicated by the curves crossing the frequency axis. It follows from this unambiguously that the bronchial tree behaves as a typical acoustical waveguide of a finite length. Indeed, the most distinctive feature of finite waveguides is the sequential alternation of resonance and anti-resonance with the increase of frequency [36]. Figure 6.6 suggests that at low frequencies input impedance is of mass character (curves describing magnitude $X(0)$ are below the frequency axis), while at $f \to 0$ both components of impedance tend to zero. As is known [41], for any multi-resonance acoustical or mechanical system the point $f = 0$ can formally be considered as first resonance $f_r^{(1)}$ of the system. This first formal resonance is followed by a real one anti-resonance $f_{a/r}^{(1)}$. After that the impedance becomes elastic (the curves are above the frequency axis). Then the second resonance $f_r^{(2)}$ occurs, where the impedance changes sign again and has mass character. This resonance is followed by second anti-resonance $f_{a/r}^{(2)}$ and so on. Figure 6.6(a) in its bottom part shows wide bold horizontal lines which denote experimental values of resonance and anti-resonance frequencies obtained from measurements of the input impedance performed on seven volunteers [27]. The spread in experimental values of resonance and anti-resonance frequencies is rather big and is determined by the spread in geometric and physical parameters of the bronchial tree and the respiratory tract as a whole. In particular, the spread in values for $f_{a/r}^{(1)}$ is about 38%, for $f_r^{(2)}$ is about 40%, and for $f_{a/r}^{(2)}$ is about 30%.

When studying any multi-resonance acoustical systems, such characteristic as the ratio of subsequent resonance frequencies to the previous ones is also of interest. As follows from Fig. 6.6, in our case the calculated ratio is $f_{a/r}^{(2)}/f_{a/r}^{(1)} = 2.85$, while for the experimental one (averaged over seven volunteers) is $f_{a/r}^{(2)}/f_{a/r}^{(1)} \approx 2.4$. Thus, the difference between the values of these characteristics obtained theoretically and experimentally is only about 15%, which is thought to be a good agreement in acoustics.

Considering the real part of the input impedance of the bronchial tree [Fig. 6.6(b)], it is easy to see that this characteristic takes minimum values in the areas close to resonance frequencies, and maximum (peak) values in the areas close to the anti-resonance frequencies. This property is characteristic of all acoustical waveguides of a finite length.

We used an average length of the trachea of 12 cm to calculate data in Fig. 6.6. It is, however, well known that dimensions of all airways are characterized by rather high variability. For instance, the already cited publication [27] presents results of trachea measurements for seven people and the trachea length varies from 9.5 to 15 cm. Oscillations of resonance and anti-resonance frequencies were observed. These oscillations were determined by the points of transition through zero of the imaginary part of the input impedance of the bronchial tree [see Fig. 6.6(a), curves 2–4]. In particular, within the mentioned spreads of trachea length the ratio of the maximum value of frequency of the first resonance to its minimum value is about 1.4. Taking this into account we calculated input impedance of the bronchial tree for two dimensions of trachea, 9.5 and

15 cm, all other dimensions of all the other airways being equal. It appeared that the ratio of the maximum value of the first resonance to its minimum value was 1.3, which agrees with the experimental ratio. Thus the assumption that the spread of resonance and anti-resonance frequencies observed in experiments [27] is determined mainly by the spread in longitude dimensions of the trachea was correct.

Summarizng the comparative analysis of theoretical and experimental data it can be concluded that the developed theoretical model of the bronchial tree rather adequately describes impedance characteristics of a real bronchial tree and can hence serve as a good basis for qualitative and quantitative estimates of other important acoustic characteristics of the bronchial tree.

6.5.2 Sound Energy Flow Distribution in the Bronchial Tree

The sound generated in the trachea due to oscillations of vocal cords and/or pressure pulsations of the turbulised airflow during respiration travels along the bronchial tree as along the wave guide, gradually losing its energy. There are two main reasons that cause sound degeneration. First, losses in viscoelastic walls of the airways, whose oscillations are caused by the sound waves propagating inside the bronchial tree, and secondly, losses from re-radiation of sound by the oscillating walls of the airways in the biotissues (parenchyma and mediastinum) surrounding the bronchial tree.

Thus, one of the important and interesting problems about the sound propagation in the bronchial tree lies in estimating the portions of sound energy traveling in the air inside the tree, the energy that is absorbed by the airway walls and the energy radiated in the surrounding bio-tissues. It should be said that this problem has been of interest to researchers for a long time; however, it has not been solved so far neither experimentally, nor theoretically. We have already discussed above difficulties associated with the experimental estimate of sound fields in the bronchial tree. As to the theoretical estimates, they have not really been done in either earlier [42] or later [27] works, and here is why. In [42], when developing the acoustic model of the bronchial tree the airway walls were assumed to be absolutely rigid thus eliminating any energy losses in the walls and energy penetration through the walls. In other words it was assumed that the entire sound energy generated in the trachea without any losses reaches alveoli and only there radiates in the parenchyma, causing oscillations of the thorax. In later works [27], oscillations of the walls were taken into consideration; however, the fact that they can re-radiate sound energy in the biotissues surrounding bronchial tree was completely excluded.

Taking into account the above and wide application of our acoustic model we further analyse obtained with its help quantitative estimates of sound energy flow distribution in the bronchial tree. Figure 6.7 presents frequency dependencies of total (across all generations) relative portions of the sound energy $\sigma_3 = \left[\sum_{z=0}^{23} W_3(z,f)\right]/W_0$ lost in viscoelastic walls of all the airways and radiated by $\sigma_2 = \left[\sum_{z=0}^{23} W_2(z,f)\right]/W_0$ all airway walls in biotissues surrounding the bronchial tree. Here W_0 is the sound energy injected at the entrance of the trachea; and $W_3(z,f)$ and $W_2(z,f)$ represent the energy

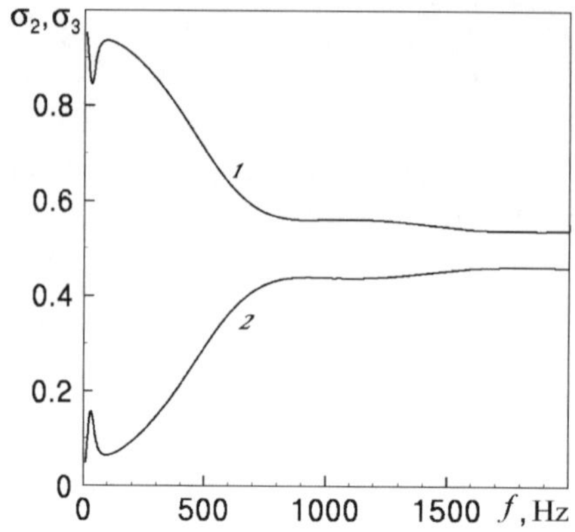

FIG. 6.7: Dependencies of the relative proportions of sound energy on frequency; 1—the total portion of the energy lost in the viscoelastic walls of all the airways, 2—the total portion of the energy emitted by the walls of all the airways in the bio-tissues surrounding the bronchial tree

lost in the walls and the energy radiated by the airway walls of z generation in the surrounding biotissues, respectively. As seen, the total portion of the acoustic energy lost in the airway walls is high especially at low frequencies ($f < 200$ Hz), where losses reach almost 90%. As the frequency increases, energy losses in the walls decrease and from $f \approx 800$ Hz stabilize at 55%. The total portion of energy that penetrates in the biotissues surrounding the bronchial tree is small at low frequencies and makes about 10%. As frequency increases, this portion also increases and above 800 Hz stabilizes at about 45%. We were also interested in the total portion of energy that reaches airways of the last 23rd generation and from there radiates in the parenchyma. Estimates showed that the magnitude of this energy is infinitesimally small and is less than $10^{-6}W_0$. This estimate gives ground to affirm that conclusions in [27] stating that the entire sound energy generated in the trachea reaches the end of the bronchial tree without losses and only there radiates in the parenchyma are fundamentally incorrect.

Results presented in Fig. 6.7 show a general picture of the acoustic energy flow distribution in the bronchial tree but cannot obviously address the question of how the energy propagates along the airways of each generation. In particular, it does not address questions of how quickly the energy dampens traveling down the bronchial tree, in what generation the energy absorption by the walls is the highest, and in what generation the energy is re-radiated in the biotissues surrounding the bronchial tree. To answer these questions we estimated the balance of energies in each generation. The results of these estimates are depicted in Figs. 6.8–6.10.

We first look at Fig. 6.8, which shows the dynamics of the energy decay $\bar{W}_1 = 10\lg W_1(z,f)/W_0$ as it propagates down the bronchial tree. Here $W_1(z,f)$ is the energy flowing into airways of z generation. As seen, at low frequencies ($f < 100$ Hz) almost 90% of the energy is lost in the cervical and thoracic regions of the trachea, another 7–9% is lost in the main bronchi and the remaining 1–3% of the energy is lost in small airways. Thus, at low frequencies sound generated at the trachea entrance dampens quickly and

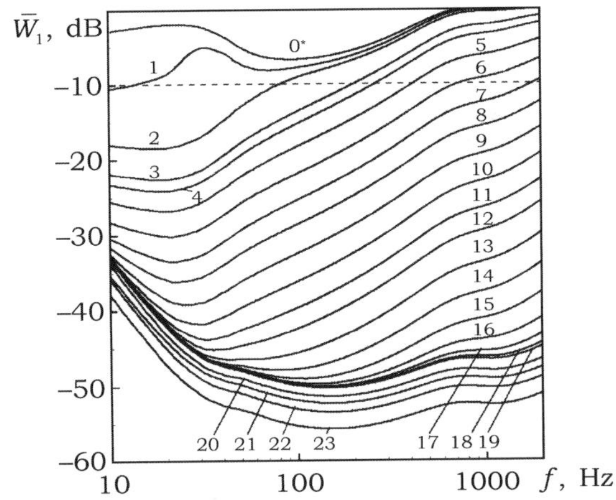

FIG. 6.8: The relative proportion of sound energy, flowing into the airway of the generation number z. The curve number corresponds to the generation number

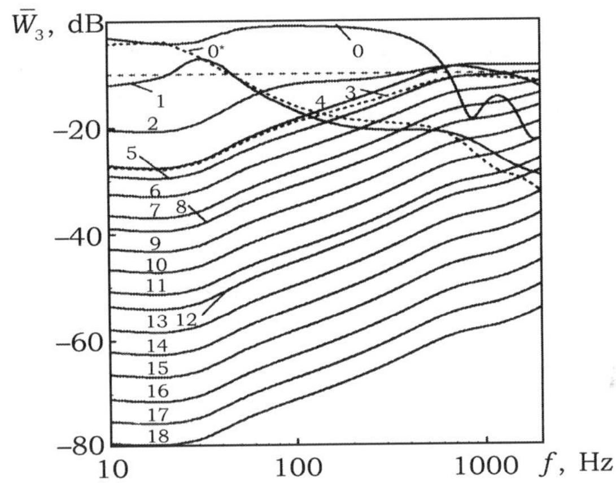

FIG. 6.9: The relative proportion of sound energy, lost in the walls of airways in each generation z. The curve number corresponds to the generation number

penetrates in the bronchial tree only as far as the first generation. At higher frequencies energy does not decay as quickly, and in the range of $f > 500$ Hz it can relatively effectively reach 3rd–5th generation.‡‡ Taking this into account it can be said that, in spite

‡‡Note that the curve numbers (except for the trachea) in Figs. 6.8–6.10 are identical to the generation numbers of the airways. With regard to the trachea, in contrast to the classical definition of its number as the generation $z = 0$, we have assigned it two numbers and here's why. As is known, anatomically trachea is divided into two segments, the cervical and thoracic. It is obvious that the outer walls of these two segments of the trachea are acoustically loaded on different media. The walls of the cervical portion are loaded essentially on the air environment, and the walls of the thoracic area on the biological tissue of the mediastinum. It is precisely because of these circumstances, that we have assigned the trachea two different generation numbers. For the portion of the cervical part, we have given the classic number $z = 0$, whereas for the thoracic segment we have given the number $z = 0^*$.

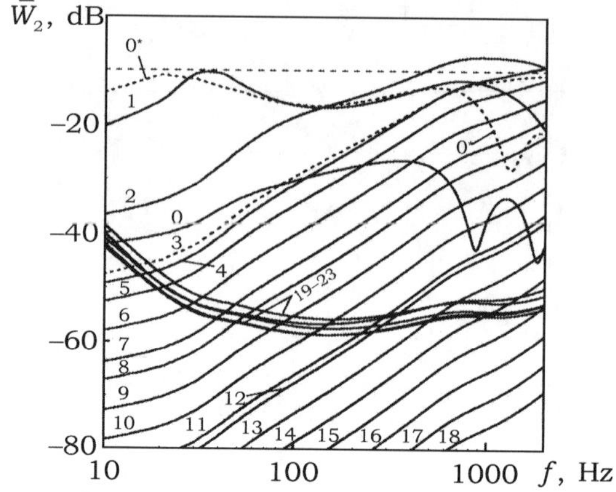

FIG. 6.10: The relative portion of sound energy, that travels through the walls of airways in each generation z. The curve number corresponds to the generation number

of expectations, the bronchial tree does not represent an effective acoustic waveguide, since the sound energy propagating through it experiences high decay due to the losses in the viscoelastic airway walls and re-radiation of the energy in the surrounding biotissues.

We now estimate energy losses $\bar{W}_3 = 10 \lg W_3(z, f)/W_0$ in airway walls of each generation and energy $\bar{W}_2 = 10 \lg W_2(z, f)/W_0$ re-radiated by them in the surrounding biotissues (Figs. 6.9 and 6.10). As follows from Fig. 6.9 at frequencies up to 50 Hz the main portion (almost 90%) of energy is absorbed by the trachea walls and partially by the main bronchi walls. In the range of 50 to 500 Hz big portion of energy is absorbed by the walls of the cervical part of trachea. Above 500 Hz losses are determined by contributions of the waveguides of the 2nd–8th generations. Remarkably, in this frequency range losses in the trachea walls and the main bronchi are relatively low. Note that on Fig. 6.9 there are no curves corresponding to generations $z > 18$. This is related to the fact that beginning with 19th generation airways have essentially no walls.

The share of energy re-radiated in the biotissues surrounding the bronchial tree, as follows from Fig. 6.10, is small. At frequencies up to 100 Hz energy radiation in biotissues is explained mainly by the oscillations of the walls of the thoracic part of trachea and main bronchi, while at higher frequencies, (100 to 500 Hz) it is explainedby oscillations of the lobar bronchi walls and partially of the main bronchi walls. At frequencies above 500 Hz the energy caused by the oscillations of the airway walls of 3rd–5th generations starts playing a significant role. Contribution from the wall oscillations of the airways of higher generations is minor.

Recapping data analysis presented in Fig. 6.10 it is worth paying attention to behaviour of several curves. In particular, we note abnormally low level of energy re-radiated by the walls of the cervical part of the trachea. This fact can be easily explained. First, the walls of the cervical part of the trachea re-radiate energy in the air, surrounding the neck. Thus this gives almost no additional input to the energy injected in the biotissues of the thorax. Secondly, wave resistance of the air is much less than that of the parenchyma and the mediastinum. Consequently, acoustic load of the walls of the cervical part of the trachea is very low and as a result the level of re-radiated energy is low. Abnormally high (compared to the airways of 11th–18th generations) level of energy radiated in the parenchyma by the airways of 19th–23rd generation is also noticeable. This feature also has an easy explanation. The airways of these generations have no walls and thus no additional losses. That is why the energy traveling in the parenchyma through these last airways is higher than at the airways of 11th–18th generation, although absolute values of the energy for all the small airways are very low.

It is worthwhile introducing summarising data about acoustic energy flow distribution in the bronchial tree obtained on the basis of results presented in Figs. 6.8–6.10. These summarising data are in essence relative integral (over frequency) levels of energy, derived from the following equation:

$$\tilde{W}_i(z) = 10\lg \frac{\int_{10}^{2000} W_i(z, f)df}{\int_{10}^{2000} W_0 df},$$

where numbers $i = 1, 2, 3$ correspond to energies, entering airways, re-radiated in the surrounding biotissues and absorbed by the viscoelastic walls. Figure 6.11 shows how integral levels of all three parts of energy change as the energy flow moves inside the bronchial tree.

6.6 MATHEMATICAL MODELLING OF THE SOUND PROPAGATION IN THE ORGANS OF THE THORAX

We have formulated an acoustic and mathematical model of the bronchial tree and examined in detail the features of the sound propagation therein, initiated by the vibrations of

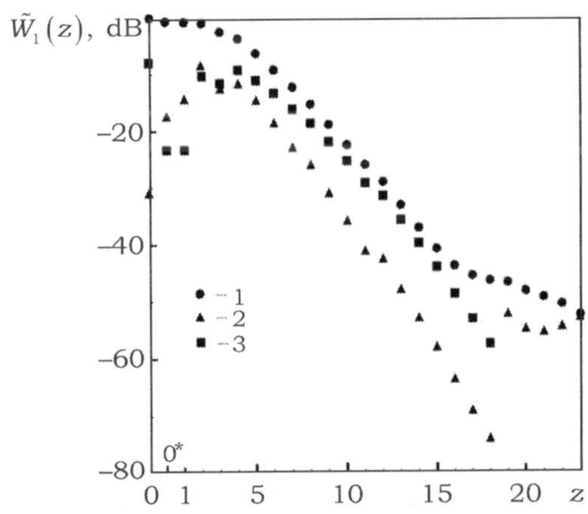

FIG. 6.11: Integral levels of sound energy as a function of the generation number z; $i = 1$: energy flowing into the respiratory tract; $i = 2$: energy transmitted through the wall into biological tissues; $i = 3$: energy absorbed by the walls

internal surfaces of the trachea walls. We now formulate appropriate models of internal organs of the thorax and consider the features of sound propagation there initiated by oscillations of external surfaces of the trachea walls. We first consider already existing models of the thorax. In 1989, a simple physical model was proposed in [16]. The model consists of two elements, a tube, simulating the trachea, and an annular layer of viscous liquid surrounding the tube, simulating the lung parenchyma. Axisymmetric vibrations of the pipe wall produced cylindrical sound waves in the ring layer and oscillations of its outer surface. Assuming that the outer diameter of the annular layer is substantially less than the length of the sound wave in the layer, the authors (based on the theory of electromechanical analogies [41]) replaced the original physical three-dimensional model with the equivalent electric scheme with lumped parameters. Using this primitive model, the authors estimated the transfer function during the signal propagation from the trachea to the outer surface of the parenchyma, and showed that this function tends to decrease as the frequency increases, which in general agrees with experimental data.

In 1995 another paper appeared [17], which offered a more realistic (and more complex) model of the thorax, which comprised of the trachea, annular layer of the lung parenchyma, musculo-rib annular layer and the outer muscular fat layer. The sound source (the same as in [16]) was axisymmetric oscillations of the trachea walls. Accounting in the model for a larger number of organs in the thorax allowed obtaining more realistic transfer functions forthe signal propagation from the trachea to the outside surface of the thorax. In addition, this model enabled to estimate the frequency dependency of the mechanical impedance of the thorax surface, which allowed developing requirements for the acoustic sensor registering oscillations of the thorax.

As we noted above, the thorax contains not only lungs but other organs as well. They all differ from each other not only in the configuration, volume and location of the thorax, but also in density, speed of sound propagation within them, and the magnitude of its dampening. Moreover, even a glance at the illustrations in the anatomical atlas [19] and Fig. 6.1(b) assure that the thorax is not an axisymmetric object in the geometric sense. First, the chest has two lungs (right and left), and secondly, there is the mediastinum between the trachea and the lungs. The models of the thorax used in [16,17] are axisymmetric: the lungs are modelled by a solid cylindrical layer and the mediastinum is absent altogether. Therefore, it is obvious that these models are far from reality and require substantial improvements.

6.6.1 Acoustic and Mathematical Models of the Thorax

We take the model developed in [17] as the basis for acoustic model, and outline below a number of substantial changes. First, let us look at anatomical images of the lungs and note some of the important features of their structure (see Fig. 6.1). Figure 6.1(a) depicts a general view of the lungs and the respiratory tract (the larynx, the trachea, the bronchi and the remaining portion of the bronchial tree). Figure 6.1(b) gives a real tomographic image of the cross-section of the lungs [43]. As is well known [18,19], and can be clearly seen from the Figures presented, the lungs are a paired organ, i.e., divided into the right and the left lung, and their three-dimensional structure is extremely complex. However, if we consider only the horizontal (middle) section [Fig. 6.1(b)], then in first approximation the shape of each lung looks like a certain part (segment) of the cylindrical layer. Further, in between the lungs and the trachea there is an area (which, in the first approximation, has a cylindrical shape), called the mediastinum, containing a whole range of bodies, including numerous blood vessels, lymph nodes, nerve trunks, and so on. Around the middle of the mediastinum, there is the trachea, from where the main bronchi branch off.

Thus, the vibrations of the outer surfaces of the trachea walls caused by the airflow in the trachea or during speech generate sound waves, first and foremost in the mediastinum. Then they penetrate the lungs and the areas between the lungs from the back and the chest sides [see Fig. 6.1(b)]. Finally, after passing through the musculo-rib layer and the fat layer they reach the surface of the skin of the chest, causing its oscillations.

The results of this brief anatomical analysis make us believe that a more realistic study of the sound propagation in the thorax would require a more complex model of the thorax than the one developed in [17]. This can be achieved by introducing the mediastinum area and four other areas occupied by the right and the left lungs and the areas between the lungs from the back and chest sides into the model (see Fig. 6.12).

Let us look at the proposed physical model in more detail. As in [17], this is a two-dimensional model, i.e., its properties are determined in polar coordinates (r, ψ). The model has the shape of a circle and the main sound source is oscillating trachea walls. On the outer surface of the trachea (at $r = r_0$) the sound pressure is given as p_0. The region I, bounded by coordinates $r_0 \leq r \leq r_1, 0 \leq \psi \leq 2\pi$ simulates the mediastinum. Segments of cylindrical layers II and IV (shaded areas on Fig. 6.12), defined by the coordinates $r_1 \leq r \leq r_2, 2\pi - \psi_0 \leq \psi \leq \psi_0$ and $r_1 \leq r \leq r_2, \pi - \psi_0 \leq \psi \leq \pi + \psi_0$, respectively, simulate the left and the right lungs. Segments of cylindrical layers III

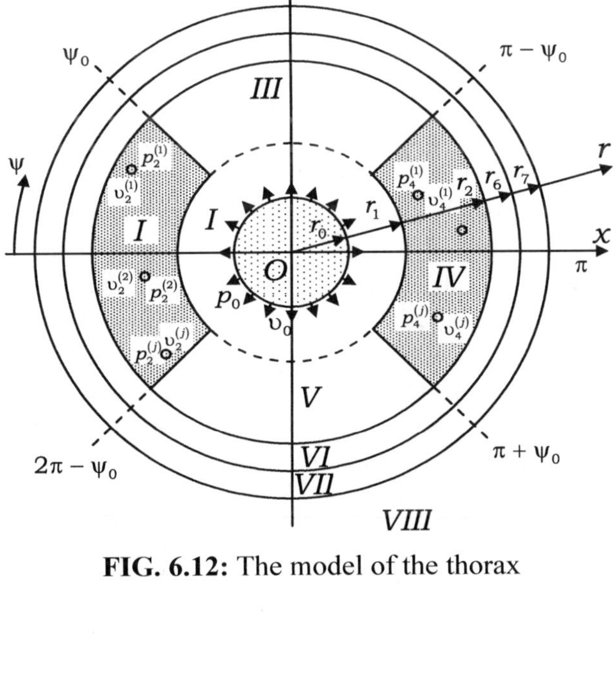

FIG. 6.12: The model of the thorax

and V, with boundary coordinates $r_1 \leq r \leq r_2, \psi_0 \leq \psi \leq \pi - \psi_0$ and $r_1 \leq r \leq r_2, \pi + \psi_0 \leq \psi \leq 2\pi - \psi_0$, respectively, model interpulmonary areas. The circular cylindrical layer VI, with boundary coordinates $r_2 \leq r \leq r_6$, $0 \leq \psi \leq 2\pi$ models the intercostals layer, and the circular layer VII, with boundary coordinates $r_6 \leq r \leq r_7$, $0 \leq \psi \leq 2\pi$ simulates the fat layer and the surface of the skin of the chest. In addition, we introduced the external area VIII into the model with coordinates $r \geq r_7, 0 \leq \psi \leq 2\pi$ simulating airspace surrounding the thorax.

As we discussed above, the vibrations of the external surfaces of the walls of the trachea are not the only source of acoustic noise, which can be detected on the chest surface. Fluctuations in the inner surfaces of the walls of the trachea also generate noises, but inside the trachea. They propagate along the bronchial tree, penetrate into the lungs due to the oscillations of the bronchi and bronchiole walls of different generations and further reach the surface of the thorax. Such sound sources are local in nature. Given small wave size of the bronchi and bronchioles, they can be considered as point sources. Therefore, to examine the propagation of sound in the thorax, generated by such source type, we introduced, in the areas simulating lungs, point sources $p_2^{(1)}$, $p_2^{(2)}$, ..., $p_2^{(j)}$ in domain II, and $p_4^{(1)}$, $p_4^{(2)}$, ..., $p_4^{(j)}$ in domain IV, where j is the number of the source (Fig. 6.12). Obviously we need to state their location coordinates when accounting for such sources in the model.

Assume that each of the three-dimensional domains I–VIII of the adopted physical model is filled with acoustic medium of density ρ_m and sound propagation velocity $c_m = \overline{c_m}(1 - i\mu_m)$, where μ_m is the coefficient of attenuation and m is the domain number.

Now we are left with the last open question. What acoustic properties should be given to the boundaries, separating domains of lungs II and IV and interpulmonary areas III and V, i.e., surfaces $r_1 \leq r \leq r_2$ at $\psi = \psi_0, \pi - \psi_0, \pi + \psi_0, 2\pi - \psi_0$? The answer may be simple, if we recall that the characteristic impedance of the lung parenchyma is much smaller than the wave resistance of the muscle-fat biotissue and organs filled with blood and lymph. Indeed, for example, according to studies [16,17] the average characteristic impedance of the lung parenchyma is about 10^4 kg m^{-2}s^{-1}, and the other biotissues have a characteristic impedance of 1.5×10^6 kg m^{-2}s^{-1} and above. Thus, the impedance of the lung biotissues is two orders of the magnitude lower than that of other tissues. This assessment allows us to use the following assumptions: from the side of the lungs, the borders can be considered acoustically rigid, whereas on the interpulmonary side, the borders are acoustically soft.

Now that the physical model of the thorax has all the necessary properties, we can formulate its equivalent mathematical model and proceed with solving the problem. So, we have a two-dimensional problem of sound radiation in cylindrical coordinates. We consider the sources to be harmonic with frequency ω. The entire domain of the sound field is divided into a number of sub-domains adjoining each other, each of which is filled with an acoustic medium with known values of density, sound velocity and sound attenuation in the medium. The coordinates of harmonic sound sources and distribution on the surface of the sound field are given. We need to determine the fields of pressure and oscillating velocities in each sub-domain at certain frequencies. This is a classic multi-layer problem of sound radiation in which we need to solve the Helmholtz equation in cylindrical coordinates with the following boundary conditions.

On the surface of trachea[§§]

$$p_I = p_0; \quad r = r_0, \quad 0 \le \psi \le 2\pi, \tag{6.16}$$

where p_0 is the pressure amplitude of the sound source on the surface, simulating the trachea, and p_I is the pressure field in domain I. From inside the lung regions II and IV, we write the boundary condition for acoustically rigid surfaces

$$\begin{aligned} \frac{\partial p_{II}}{\partial \psi} &= 0, \quad r_1 \le r \le r_2, \quad \psi = \psi_0, \quad \psi = 2\pi - \psi_0, \\ \frac{\partial p_{IV}}{\partial \psi} &= 0, \quad r_1 \le r \le r_2, \quad \psi = \pi - \psi_0, \quad \psi = \pi + \psi_0, \end{aligned} \tag{6.17}$$

where p_{II} and p_{IV} are the pressures in the domains II and IV. From the side of interpulmonary areas III and V, we write the boundary condition for acoustically soft surfaces

$$\begin{aligned} p_{III} &= 0, \quad r_1 \le r \le r_2, \quad \psi = \pi + \psi_0, \quad \psi = 2\pi - \psi_0, \\ p_V &= 0, \quad r_1 \le r \le r_2, \quad \psi = \psi_0, \quad \psi = \pi - \psi_0, \end{aligned} \tag{6.18}$$

where p_{III} and p_V are the pressures in the domains III and V. With regard to the point sources in the lung areas, we discuss below how to introduce them.

6.6.2 Solution of the Problem

We apply the method of partial domains [37–39] to solve the above stated boundary problem. Initially, assume that the problem is symmetric relative to the axis (Fig. 6.12), i.e., it is sufficient to consider only the area $y \ge 0$ where the angular coordinate is within the range $0 \le \psi \le \pi$. Such an assumption, on the one hand, produces virtually no loss of generality of the investigated problem and, on the other hand, facilitates the construction of the computational algorithm.

In each of the partial domains we present the sound field as a superposition of the known partial solutions of the Helmholtz equation for the given domain. The pressure field in region I can be given as

$$p_I = \sum_{n=0}^{\infty} A_n^{(1)} \frac{J_n(k_1 r)}{(J_n(k_1 r_1))'} \cos(n\psi) + \sum_{n=0}^{\infty} B_n^{(1)} \frac{Y_n(k_1 r)}{(Y_n(k_1 r_0))'} \cos(n\psi), \tag{6.19}$$

where $k_1 = \omega / c_1$ is the wave number, and $J_n(k_1 r)$ and $Y_n(k_1 r)$ are Bessel and Neumann functions respectively. Due to the presence in Eq. (6.19) of two sequences of arbitrary values $A_n^{(1)}$ and $B_n^{(1)}$, it is possible to satisfy the required conditions at the boundary surfaces of domain I, which are circles with radii r_0 and r_1 (Fig. 6.12).

[§§] We can set an equivalent condition $\upsilon_I = \upsilon_0; r = r_0, 0 \le \psi \le 2\pi$.

Point sources can be present in domain II. Domain II in itself resembles a hard wedge, indicated by rays $\psi = 0$ and $\psi = \psi_0$. We define the pressure field of the point source in the domain II through Green's function for a wedge with rigid boundaries [37]

$$G_j^{(2)}(r,\psi,r_j,\psi_j) = p_2^{(j)}\frac{\pi i}{2\psi_0}\sum_{n=0}^{\infty}\varepsilon_n J_{\alpha_n}(k_2r_<)\,H_{\alpha_n}^{(1)}(k_2r_>)\cos(\alpha_n\psi)\cos(\alpha_n\psi_j) \quad (6.20)$$

Here $k_2 = \omega/c_2$ is the wavenumber, r and ψ are the coordinates of an arbitrary point in the domain II; r_j and ψ_j are the coordinates of the source, which is assigned number j; $\varepsilon_0 = 1$, $\varepsilon_n = 2$ at $n > 0$; $H_{\alpha_n}^{(1)}(k_2r_>)$ is the Hankel function of the first kind. Values $r_<$ and $r_>$ determine the smaller and the greater of distances r and r_j.

Numbers α_n, $n = 0, 1, 2, \ldots$ are determined by the boundary conditions on flat surfaces of the domain II. Since these surfaces are deemed to be acoustically rigid (6.17), then $\alpha_n = n\pi/\psi_0$. Then the field in the domain II can be expressed as

$$\begin{aligned} p_{II} = &\sum_{n=0}^{\infty} A_n^{(2)}\frac{J_{\alpha_n}(k_2r)}{(J_{\alpha_n}(k_2r_2))'}\cos(\alpha_n\psi) + \sum_{n=0}^{\infty} B_n^{(2)}\frac{Y_{\alpha_n}(k_2r)}{(Y_{\alpha_n}(k_2r_1))'}\cos(\alpha_n\psi) \\ &+\sum_{j=1}^{J_2} G_j^{(2)}(r,\psi,r_j,\psi_j), \end{aligned} \quad (6.21)$$

where J_2 is the number of point sources in the domain II ($y \geq 0$).

The field in the domain III can be presented as

$$\begin{aligned} p_{III} = &\sum_{n=0}^{\infty} A_n^{(3)}\frac{J_{\beta_n}(k_1r)}{(J_{\beta_n}(k_1r_2))'}\sin(\beta_n(\psi-\psi_0)) \\ &+\sum_{n=0}^{\infty} B_n^{(3)}\frac{Y_{\beta_n}(k_1r)}{(Y_{\beta_n}(k_1r_1))'}\sin(\beta_n(\psi-\psi_0)). \end{aligned} \quad (6.22)$$

Here the numbers β_n, $n = 0, 1, 2, \ldots$ are determined from the boundary conditions on flat surfaces in the domain III. In the domain III, they are deemed acoustically soft (6.18), hence $\beta_n = (n+1)\pi/(\pi - 2\psi_0)$.

Point sources may also be present in the domain IV. By analogy with formula (6), we shall present the field in the domain IV as

$$\begin{aligned} p_{IV} = &\sum_{n=0}^{\infty} A_n^{(4)}\frac{J_{\alpha_n}(k_4r)}{(J_{\alpha_n}(k_4r_2))'}\cos(\alpha_n(\psi-\pi)) \\ &+\sum_{n=0}^{\infty} B_n^{(4)}\frac{Y_{\alpha_n}(k_4r)}{(Y_{\alpha_n}(k_4r_1))'}\cos(\alpha_n(\psi-\pi)) + \sum_{j=1}^{J_4} G_j^{(4)}(r,\psi,r_j,\psi_j), \end{aligned} \quad (6.23)$$

where $k_4 = \omega/c_4$ is the wave number, and J_4 is the number of point sources in the region IV ($y \geq 0$). The Green function of the point source in IV is as follows:

$$G_j^{(4)}(r,\psi,r_j,\psi_j) = p_4^{(i)}\frac{\pi i}{2\psi_0}\sum_{n=0}^{\infty}\varepsilon_n J_{\alpha_n}(k_2 r_<)H_{\alpha_n}^{(1)}(k_2 r_>) \times \cos(\alpha_n(\psi-\pi))\cos(\alpha_n(\psi_j-\pi)). \tag{6.24}$$

Fields in the domains VI and VII are written in the form similar to equation (6.19):

$$p_{VI} = \sum_{n=0}^{\infty} A_n^{(6)}\frac{J_n(k_6 r)}{(J_n(k_6 r_6))'}\cos(n\psi) + \sum_{n=0}^{\infty} B_n^{(6)}\frac{Y_n(k_6 r)}{(Y_n(k_6 r_2))'}\cos(n\psi), \tag{6.25}$$

$$p_{VII} = \sum_{n=0}^{\infty} A_n^{(7)}\frac{J_n(k_7 r)}{(J_n(k_7 r_7))'}\cos(n\psi) + \sum_{n=0}^{\infty} B_n^{(7)}\frac{Y_n(k_7 r)}{(Y_n(k_7 r_6))'}\cos(n\psi), \tag{6.26}$$

where $k_6 = \omega/c_6$, $k_7 = \omega/c_7$.

The field in the domain VIII, which manifests an exterior with respect to the cylindrical surface of the radius r_7, can be presented as a superposition of cylindrical waves

$$p_{VIII} = \sum_{n=0}^{\infty} A_n^{(8)}\frac{H_n^{(1)}(k_8 r)}{\left(H_n^{(1)}(k_8 r_7)\right)'}\cos(n\psi) \tag{6.27}$$

where $k_8 = \omega/c_8$.

We form a system of functional equations, which defines the conditions of continuity of the sound field at the boundaries between domains I, II, III, IV, VI, VII, and VIII:

$$p_I = p_0, \quad r = r_0, \quad \psi = [0, 2\pi], \tag{6.28}$$

$$\frac{1}{i\omega\rho_1}\frac{\partial p_I}{\partial r} = \begin{cases} \dfrac{1}{i\omega\rho_2}\dfrac{\partial p_{II}}{\partial r}, & r = r_1, \quad \psi = [0, \psi_0], \\ \dfrac{1}{i\omega\rho_3}\dfrac{\partial p_{III}}{\partial r}, & r = r_1, \quad \psi = [\psi_0, \pi-\psi_0], \\ \dfrac{1}{i\omega\rho_4}\dfrac{\partial p_{IV}}{\partial r}, & r = r_1, \quad \psi = [\pi-\psi_0, \pi], \end{cases} \tag{6.29}$$

$$p_I = p_{II}, \quad r = r_1, \quad \psi = [0, \psi_0], \tag{6.30}$$

$$p_I = p_{III}, \quad r = r_1, \quad \psi = [\psi_0, \pi-\psi_0], \tag{6.31}$$

$$p_I = p_{IV}, \quad r = r_1, \quad \psi = [\pi-\psi_0, \pi], \tag{6.32}$$

$$\frac{1}{i\omega\rho_6}\frac{\partial p_{VI}}{\partial r} = \begin{cases} \dfrac{1}{i\omega\rho_2}\dfrac{\partial p_{II}}{\partial r}, & r = r_2, \quad \psi = [0, \psi_0], \\ \dfrac{1}{i\omega\rho_3}\dfrac{\partial p_{III}}{\partial r}, & r = r_2, \quad \psi = [\psi_0, \pi-\psi_0], \\ \dfrac{1}{i\omega\rho_4}\dfrac{\partial p_{IV}}{\partial r}, & r = r_2, \quad \psi = [\pi-\psi_0, \pi], \end{cases} \tag{6.33}$$

$$p_{VI} = p_{II}, \quad r = r_2, \quad \psi = [0, \psi_0], \tag{6.34}$$

$$p_{VI} = p_{III}, \quad r = r_2, \quad \psi = [\psi_0, \pi - \psi_0], \tag{6.35}$$

$$p_{VI} = p_{IV}, \quad r = r_2, \quad \psi = [\pi - \psi_0, \pi], \tag{6.36}$$

$$\frac{1}{i\omega\rho_7}\frac{\partial p_{VII}}{\partial r} = \frac{1}{i\omega\rho_6}\frac{\partial p_{VI}}{\partial r}, \quad r = r_6, \quad \psi = [0, \pi], \tag{6.37}$$

$$p_{VII} = p_{VI}, \quad r = r_6, \quad \psi = [0, \pi], \tag{6.38}$$

$$\frac{1}{i\omega\rho_8}\frac{\partial p_{VIII}}{\partial r} = \frac{1}{i\omega\rho_7}\frac{\partial p_{VII}}{\partial r}, \quad r = r_7, \quad \psi = [0, \pi], \tag{6.39}$$

$$p_{VIII} = p_{VII}, \quad r = r_7, \quad \psi = [0, \pi]. \tag{6.40}$$

After substituting equations (6.19)–(6.27) in (6.28)–(6.40), we carry out a standard procedure of algebraisation of the obtained functional equations. This procedure leads to an infinite system of linear algebraic equations of the second kind for the unknown coefficients $A_n^{(g)}$, $g = 1, 2, 3, 4, 6, 7, 8$; $B_n^{(h)}$, $h = 1, 2, 3, 4, 6, 7$; $n = 0, 1, 2, ..., \infty$.

As is well known [37,38], when solving this type of problems with the method of partial domains, the number of unknowns retained in the resulting system of algebraic equations is determined empirically by analysing the accuracy of the law of conservation of energy and quality of boundary conditions and conjugation conditions at the boundaries of partial domains. Of great importance is also the frequency range where the derived estimates may be of interest. In our case, the most interesting frequency range is about 80 Hz to 1500 Hz, because breath sounds in this range show main diagnostic features characterizing the state of human respiratory system in healthy condition and in pathology [1,2,6,7,16,23].

With regard to the quality of the boundary conditions (6.16)–(6.18), they hold due to the possibility of obtaining a direct connection between the coefficients $A_n^{(1)}$ and $B_n^{(1)}$ in (6.19) and the appropriate choice of variables α_n and β_n in (6.21) and (6.22). As for the quality of the conditions of conjugation (6.29)–(6.40), as test numerical experiments have shown, quite acceptable results can be achieved if at low frequencies in the corresponding algebraic system we retain at least 130 unknowns, and at high frequency at least 520.

6.6.3 Analysis of the Results of Numerical Calculations

When carrying out numerical calculations, the results of which are presented below, the physical dimensions of the model elements in Fig. 6.12 were as follows: $r_0 = 0.01$ m, $r_1 = 0.07$ m, $r_2 = 0.14$ m, $r_6 = 0.15$ m, and the parameters of biological tissues, in each of the areas, were taken the same as in [17], namely: $r_0 = 0.01$ m, $r_1 = 0.07$ m, $r_2 = 0.14$ m, $r_6 = 0.15$ m, $r_7 = 0.16$ m, density $\rho_1 = 1000$, $\rho_2 = \rho_4 = 300$, $\rho_3 = \rho_5 = 1000$, $\rho_6 = 2000$, $\rho_7 = 1000$, and $\rho_8 = 1.22$ (all density values are in kg/m^3); the sound speed $c_1 = 1500(1 - i \cdot 0.2)$, $c_2 = c_4 = 30(1 - i \cdot 0.25)$, $c_3 = c_5 = 1500(1 - i \cdot 0.2)$, $c_6 = 2500(1 - i \cdot 0.3)$, $c_7 = 1500(1 - i \cdot 0.2)$, and $c_8 = 340$ (all speeds are in m/s). Note that the choice of complex values of the sound speed allows one to simulate the absorption of sound in the human tissues. In addition, during the first stages of calculations we assumed the absence of any point source in areas II and IV. We shall take into account their contribution to the overall acoustic picture of the thorax, and in particular to the oscillating velocity of the thorax in the paragraphs to follow.

6.6.3.1 The Acoustic Properties of the Trachea as the Sound Source

As mentioned above, the trachea has been considered as the main source of sound in the model of the thorax, whose walls vibrate under the influence of pulsation of the air flow pressure inside the trachea during breathing. Therefore, to solve the problem of sound propagation in the model of the thorax, we specified sound pressure p_0 on the outer surface of the trachea $r = r_0$, whose value does not depend on the frequency f or the angle ψ. Since our model, in contrast to the model developed in [17], is not axially symmetric, it can be assumed that the oscillating velocity on the surface $r = r_0$ will depend both on ψ and on f. To test this assumption, we turn to the graphs shown in Fig. 6.13. It can be clearly seen that, despite the absence of axial symmetry of the model, both the oscillating speed and its phase are almost independent of the angle ψ in the entire range of frequencies under consideration. At the same time, oscillating speed decreases rapidly as the frequency increases, whereas its phase is relatively weakly dependent on the frequency and lies in the range of 78 to 90 degrees. The latter circumstance indicates that the trachea as the source of sound is rather inefficient.

Let us look in more detail at the dependency of oscillating speed on the surface of the trachea on the frequency (see Fig. 6.14). As can be seen, the magnitude of velocity decreases rapidly as the frequency increases. An additional insight into the nature of this dependency showed that speed is inversely proportional to the frequency.

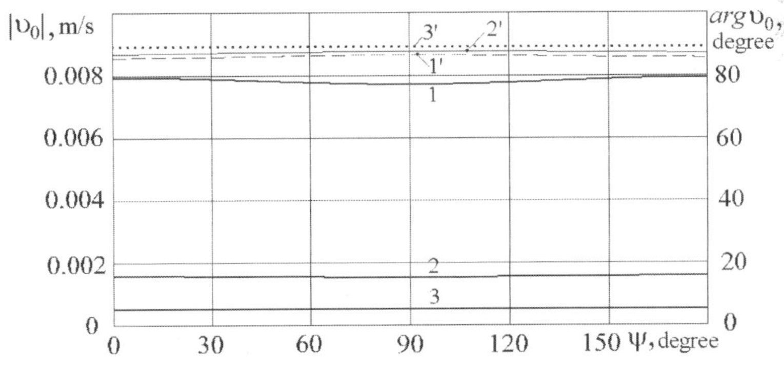

FIG. 6.13: Dependency of the oscillatory speed on the trachea surface and its phase on frequency;[¶¶] 1, 2, 3—oscillatory speed, for the frequencies of 100, 500, and 1500 Hz, respectively; 1′, 2′, 3′—phase, for frequencies of 100, 500, and 1500 Hz, respectively

[¶¶]Though the ranges were selected randomly, but so that, the first Eigen frequencies of the walls were to be in the range from ~ 50 to ~ 250 Hz. Any other information about real parameters range of the F_0, $\bar{\rho}$, L, and ϕ are absent today.

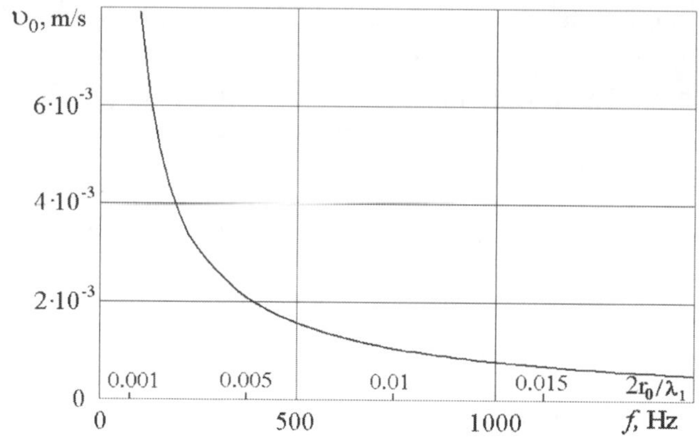

FIG. 6.14: Dependency of the oscillatory speed on the trachea surface and its phase on frequency; here the abscissa axis shows not only frequency values but also the wave diameter of the trachea, where $\lambda_1 = \mathrm{Re}c_1/f$

The discovered characteristics of oscillating speed and phase dependency on frequency (taking into account that trachea diameter does not exceed 0.02) the frequency range under consideration) suggest that the real part of the impedance of the radiation of the trachea, as the acoustic sound source, will be very small, while the imaginary part will be inertial in nature. Figure 6.14 confirms the validity of this statement. Indeed, the dimensionless (normalized to $2\pi r_0 \rho_1 \mathrm{Re}c_1$) real part of the impedance of the radiation in this frequency range does not exceed 2.5×10^{-3}, while dimensionless imaginary part of the impedance of the radiation increases linearly as the frequency increases.

It should be noted that the observed properties of the radiation impedance of the trachea are well known in acoustics [36–38] and are specific to small wave size sources. The linear dependency of the imaginary part of the impedance of the radiation is usually presented as $\mathrm{Im}Z = \omega M_a$, where M_a is the added (co-oscillating) mass of the environment to the source surface. We made simple calculations and estimated the added mass of the mediastinum medium to the trachea surface. As Fig. 6.15 suggests, it does not depend on the frequency and is approximately 1.27 kg per unit length of the trachea.

Thus, the trachea as the sound source is rather inefficient and this is a direct consequence of its small wave diameter in this frequency range.

6.6.3.2 Features of the Distribution of the Oscillating Speed on the Thorax Surface

We have shown above that despite the absence of axial symmetry of our model, oscillation velocity $\upsilon_0(r_0)$ on the surface of the trachea is practically independent of the angle ψ. It is now important to establish whether such property holds for oscillation velocity $\upsilon(r_7)$ on the surface of the chest $r = r_7$ (see Fig. 6.12). This is an essential information, because it is on the surface of the chest that speed sensors (or pressure sensors) are located for breath sound registration [2.3].

Figure 6.16 shows dependency graphs for oscillatory speed $|\upsilon(r_7)|$ on the surface of the thorax for three different frequencies. It is evident that this speed depends both on

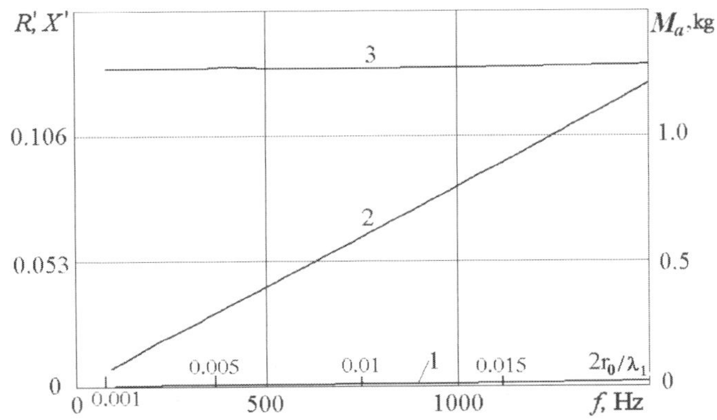

FIG. 6.15: The dimensionless impedance of the trachea radiation; 1—the real part R', 2—the imaginary part X', 3—associated mass

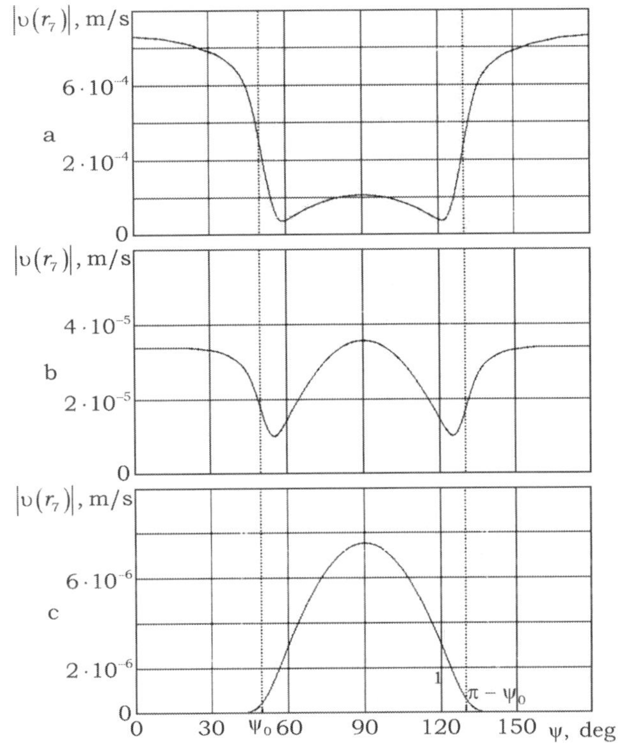

FIG. 6.16: Dependency of the oscillating speed on the thorax surface on angle ψ; (a,b,c)—for 100 Hz, 300 Hz, and 1500 Hz, respectively

angle ψ and on frequency. At the low end of the frequency range under consideration [see Fig. 6.16(a)] the value $|\upsilon(r_7)|$, detected above the lungs, is almost 4 times greater than that measured over the interpulmonary area. At mid frequencies [see Fig. 6.16(b)] oscillating speeds above lungs roughly equal those over the interpulmonary area. And at the high end of the range [see Fig. 6.16(c)] oscillating speed above lungs is almost two orders of magnitude lower than that over the interpulmonary area.

We showed by way of example how the oscillatory velocity redistributes along the surface of the thorax only on three fixed frequencies. Analysis of the obtained results suggests that the dependency of this redistribution on frequency is complex. It is therefore desirable to have a generalized (integrated) estimate of the distribution of the oscillatory speed $|\upsilon(r_7)|$ on the surface of the thorax on frequency. With this in mind, we have introduced two normalized integral parameters α_1 and α_2 defined as

$$\alpha_1 = \frac{2}{\pi r_0 |\upsilon_0(r_0, f)|} \int_0^{\psi_0} |\upsilon(r_7, f, \psi)| \, r_7 d\psi \tag{6.41}$$

$$\alpha_2 = \frac{2}{\pi r_0 |\upsilon_0(r_0, f)|} \int_{\psi_0}^{\pi/2} |\upsilon(r_7, f, \psi)| \, r_7 d\psi \tag{6.42}$$

Parameter α_1 represents an integral estimate of the oscillatory speed on the surface of the thorax above lungs and essentially characterizes the sound transparency of lungs by speed. Parameter α_2 gives an integral estimate of the oscillatory speed above the interpulmonary area and also characterizes its sound transparency by speed. Note that formulas (6.41) and (6.42) use the symmetry of the model in norm with respect to the axes Ox, Oy and the independence of the oscillatory speed on the surface of the trachea from the angle ψ, that we established above.

Figure 6.17 shows dependencies of integral parameters α_1 and α_2 on frequency. It is evident that the sound transparency of the interpulmonary area (curve 2) remains largely unchanged in the whole range of frequencies and is at 0.03–0.04 (or about –30 dB relative to the level $\alpha_1 = \alpha_2 = 1$).

The situation is different for the sound transparency of the lungs (curve 1). First, at around 70 Hz, we observe that sound transparency reaches its peak value. This peak

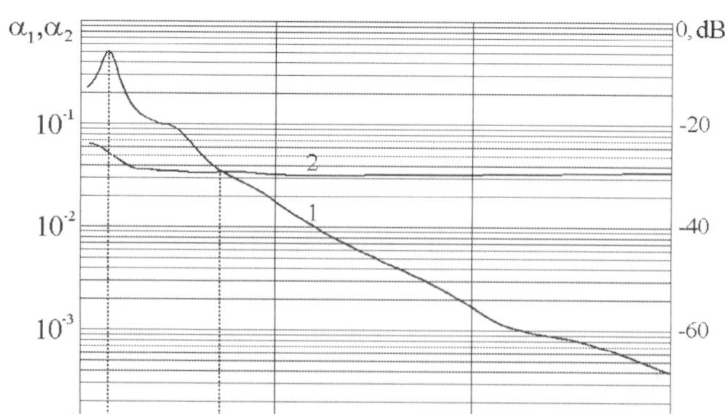

FIG. 6.17: Dependencies of integral parameters on frequency; 1—α_1, 2—α_2

has a clear resonance nature and its level is about 20 dB higher than that of curve 2. Secondly, the sound transparency of the lungs has a prominent tendency to decrease as the frequency increases. Thus, at low frequencies (up to about 300 Hz), the sound transparency of the lungs is higher than that of the interpulmonary area. At about 360 Hz, sound transparency of the lungs becomes equal to that of the interpulmonary area, and at higher frequencies, sound transparency of the lungs diminishes rapidly and at 1500 Hz becomes lower than that of the interpulmonary area by almost 40 dB.

Why then sound transparency of the lungs and of the interpulmonary area show such different dependency on frequency? We shall try to address this question. The loss factors that we use in the models of the lung medium and of the interpulmonary medium are 0.25 and 0.2, respectively, hence presenting only a 20 percent difference.

First of all, we recall the differences between the structures of the lungs and of the interpulmonary area (Fig. 6.12). If we observe physical properties of the bio-tissues, from the mediastinum to the surface of the chest in the interpulmonary area (where $\psi_0 \leq \psi \leq \pi - \psi_0$), we can see that all these biological tissues are characterized by high speed of propagation of the sound in them (not below $c = 1500$ m/s) and the magnitude of their wave resistance is not that different from each other. Given that the distance from the trachea to the chest surface is $r_7 - r_0 = 0.15$ m, it is easy to determine the distance measured in the wavelength of sound at the extreme frequencies: at $f = 100$ Hz, $(r_7 - r_0)/\lambda = (r_7 - r_0)f/c = 0.01$ and at 1500 Hz, $(r_7 - r_0)/\lambda = 0.15$. Thus, in the whole range of frequencies the distance from the trachea to the chest surface is substantially smaller than the wavelength. Therefore, it is difficult to expect that in this situation the sound transparency of the interpulmonary area can vary that much with frequency. This confirms the shape of the curve 2 in Fig. 6.17.

Now we look at the physical properties of bio-tissues, starting from the mediastinum to the chest surface in lung sectors (i.e., in sectors with $0 \leq \psi \leq \psi_0$ and $\pi - \psi_0 \leq \psi \leq \pi$). It is obvious that there is a significant difference in the physical properties of biological tissue of lung and its surrounding biological tissues. The speed of sound in the lung parenchyma is at least 50 times lower than in the surrounding biological tissues, and the characteristic impedanceof the lung parenchyma is about 150 times lower than that of the surrounding bio-tissues. If we look at the wave thickness of the lung, it turns out that at 100 Hz, $(r_2 - r_1)f/c = 0.23$ and at 1500 Hz, $(r_2 - r_1)f/c = 3.5$. Thus, at low frequencies of the frequency range considered, the wave thickness of the lung is rather small, while at the upper frequency it is rather large. Where can this situation lead? Let us first consider low frequencies.

As is well known in acoustics [38], if the layer of the small wave thickness, at the same time having a low impedance, lies between layers with high impedance, its physical properties are almost entirely determined by its elasticity S. In our case, for the parenchyma layer, $S = \rho_2 \mathrm{Re} c_2^2/(r_2 - r_1) = 3.95 \times 10^6$ kg/m^{2}s^2. The pressure will pass through this layer unchanged, while the speeds on its surfaces will obviously differ. Two thin layers, which simulate the intercostals and the fat layers (Fig. 6.12: annular regions VI and VII, respectively), are located on the outside of the lung ($r \geq r_2$). Considering the physical properties of these layers mentioned above, their total wave thickness at 100 Hz is not more than 0.0013, and their impedance is not less than 1.5×10^6 kg/m^{2}s.

Most importantly, these two layers are located between the lungs and the air medium surrounding the thorax, both of which have substantially lower impedances. Then, in accordance with [38], the physical properties of these two layers are almost entirely determined by the surface weight $M = \rho_6(r_6 - r_2) + \rho_7(r_7 - r_6)$. Oscillating speed will be transmitted unchanged through such layers, while pressures on their surfaces will differ. Given the accepted physical properties of these layers, their total surface mass will be about 30 kg/m^2.

To verify the validity of these statements we turn to Fig. 6.18, which shows the calculated dependences of pressure (absolute value) and oscillating speed on the coordinate r at 100 Hz. If we look at how pressure changes in the direction $\psi = 0$, that is in the direction passing through the centre of the lungs [Fig. 6.18(a), curve 1], it is easy to note that throughout the thickness of the lungs from r_1 to r_2, the pressure indeed remains almost constant. The speed of intercostal and fat layers from r_2 to r_7 in the direction $\psi = 0$ also remains constant [Fig. 6.18(b), curve 1]. For comparison, the same figure shows dependencies of the pressure and speed on the coordinate r in the direction $\psi = 90°$, that is in the direction which passes through the centre of intercostals area. As expected, here, both pressure and speed decrease monotonically, not following any specific patterns. The sharp drop in pressure in the area from r_2 to r_7 [Fig. 6.18(a)] is due to the fact that the characteristic impedance of the air surrounding the chest is almost four orders of the magnitude less than characteristic impedances of intercostals and fat layers.

It now becomes obvious that at low frequencies the system of the lungs and the adjacent intercostals and fat layers can be presented in the first approximation, as a spring with elasticity S connected in series with the mass M, i.e., this system is a simple series resonant circuit. The frequency of this circuit is easily determined and equals 61 Hz. Going back to curve 2 in Fig. 6.17, we can easily see that the resonant frequency calculated using simple methodology is in good agreement with the more rigorous calculation (the error does not exceed 9%).

Based on this analysis we can conclude that at low frequencies relatively high acoustic permeability in the areas where the lungs are located is determined by the resonant

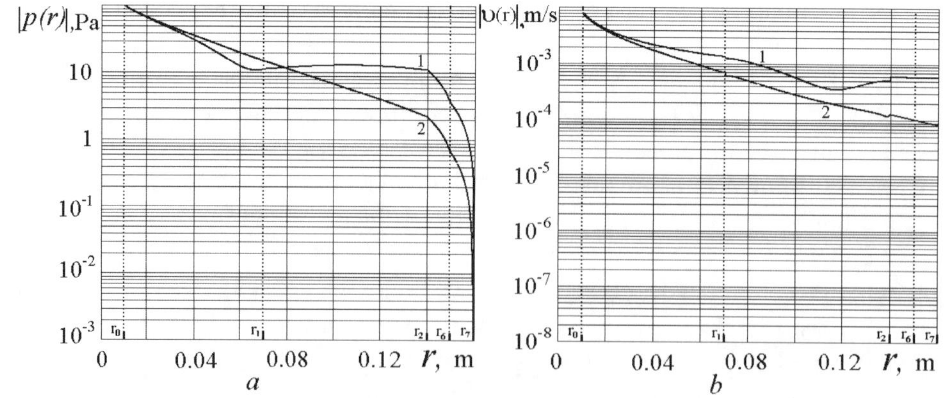

FIG. 6.18: Dependency of pressure (a) and oscillatory speed (b) on coordinate r at 100 Hz; curves 1, 2—in the directions $\psi = 0$ and $\psi = 90°$, respectively

phenomenon between the elasticity of the lung parenchyma and the mass of the adjacent intercostal and fat layers, i.e., we can talk about some peculiar resonance of the chest.

It should be said that the discovered resonance of the chest is not something unexpected. It has long been known in medical practice (see e.g., [44]), and it is usually detected in the frequency range from 40 to 90 Hz during percussion[||||] of the thorax. It is very important that our model adequately identifies this property of a real human chest.

We now turn back to Fig. 6.17 and try to understand why as the frequency increases, parameter α_1, unlike α_2, decreases so rapidly. To do this, we consider the simplest one-dimensional case, propagation of a plane sound wave in the direction of the coordinate x. Let the expression for sound pressure have the form

$$p = e^{ikx} \tag{6.43}$$

where $k = \omega/c$ and $c = c_0(1 - i\mu)$; c_0 is the speed of propagation of the sound wave, and μ is the loss factor. Then

$$k = \frac{\omega}{c_0(1 - i\mu)} = k_0\frac{1 + i\mu}{1 + \mu^2} \tag{6.44}$$

where $k_0 = \omega/c_0$.

Suppose, as in our model, $\mu \leq 0.3$. Then in the expression (6.44) the value of the denominator in the first approximation can be neglected and set

$$k \approx k_0(1 + i\mu) \tag{6.45}$$

Taking into account (6.45), the expression for pressure (6.43) takes the form

$$p \approx e^{ixk_0(1+i\mu)} = e^{ik_0x}e^{-k_0\mu x} = \frac{e^{ik_0x}}{e^{k_0\mu x}} \tag{6.46}$$

It is obvious that the numerator in the last equation describes a plane traveling wave [compare with (6.43)], and the denominator indicates how fast the wave dampens as it propagates along the axis x. If $k_0 = \omega/c_0 = 2\pi f/c_0 = 2\pi/\lambda_0$, the denominator in (6.46) can be written as

$$\delta(\mu, x/\lambda_0) = e^{2\pi\mu(xf/c_0)} = e^{2\pi\mu(x/\lambda_0)} \tag{6.47}$$

From the latter expression it follows that the attenuation of a sound wave depends not only on the value μ, but also on the frequency or wave size of the distance travelled x/λ_0. Of course, in case of cylindrical waves, which we deal with in our model, the situation is somewhat more complicated, because pressure will fall not only due to the presence of the dampening coefficient, but also due to the expansion of the wave front. However, in this case, this is not material.

[||||]Percussion is a diagnostic method for determining the condition of the lungs using a special light tapping on the chest and listening occurring while the sound effects [1,2].

Taking into account equation (6.47), we provide, in the first approximation, a comparative assessment of wave attenuation for the waves that travelled through the interpulmonary area and the lungs. We first consider the interpulmonary area, whose size from the trachea to the chest is $r_7 - r_1 = 0.15$ m. Since the sound wave propagation speed here is not less than 1500 m/s, and $\mu \leq 0.3$, the value $\delta(\mu, (r_7 - r_1)/\lambda_0)$ at 100 Hz will be $\sim$1.01, and at 1500 Hz it will be 1.19. Thus, the wave dampening in the interpulmonary area will be negligible, as follows from the behaviour of curve 2 in Fig. 6.17. We now estimate the attenuation of waves in the lungs, given that their thickness is $r_2 - r_1 = 0.07$ m, the speed of wave propagation in the parenchyma is 30 m/s, and $\mu = 0.25$. With these physical parameters, the value $\delta(\mu, (r_2 - r_1)/\lambda_0)$ at 100 Hz will be about 1.4, and at 1500 Hz it will already reach about 240. That's why curve 1 drops rapidly in Fig. 6.17, and this drop is primarily explained by the low speed of sound wave propagation in the parenchyma and, as a result, by large ratios $(r_2 - r_1)/\lambda_0$. By the way, these trivial calculations allow us to immediately give an estimate of the parameter α_1 reduction as the frequency changes from 100 Hz to 1500 Hz. It is obvious that the value of α_1 at 1500 Hz will be $240/1.4 = 170$ times (or by –45 dB) less than its value at 100 Hz in time. The exact calculation derived from curve 1, Fig. 6.17 gives us the value of approximately –50 dB, which differs from an estimate by only 5 dB.

Hence, the lungs, from the acoustic point of view, act like a rather effective filter of low frequencies.

In conclusion, we present the distribution pattern of the sound fields inside the chest on the two extreme frequencies—100 Hz (Fig. 6.19) and 1500 Hz (Fig. 6.20). Unfortunately, the range of variation of pressure and velocity here is of several orders of magnitude. Therefore, the data in Figs. 6.19 and 6.20 are presented in the logarithmic scale. As can be seen, the values of sound pressure levels are presented as a scale of shades of grey. A change in colour from white to black corresponds to the reduction of sound pressure level at 100 Hz by four orders of magnitude, and at 1500 Hz by five orders of magnitude. Thin white lines in Figs. 6.19(a) and 6.20(a) represent isobars.

With respect to the oscillatory speed distribution [Figs. 6.19(b) and 6.20(b)], the change in length of arrows from the longest to the shortest at 100 Hz corresponds to the speed reduction by one order of magnitude, and at 1500 Hz by five orders of magnitude.

As expected, the sound field distribution inside the chest as a whole is rather complicated. We first consider the pressure pattern distribution at 100 Hz, Fig. 6.19(a). It is clearly seen that throughout the entire thickness of the lungs, that is from r_1 to r_2, pressure level remains virtually unchanged. We discussed earlier the reasons behind such effect. The level of pressure in the interpulmonary area in general is significantly lower than in the lungs and tends to decline sharply as we approach the boundaries $\psi = \psi_0, r_1 \leq r \leq r_2$ and $\psi = \pi - \psi_0, r_1 \leq r \leq r_2$, separating the interpulmonary area and the lungs. The reason behind this effect is explained by the boundary conditions for acoustic softness ($p = 0$) accepted in our model for these boundaries from the interpulmonary area. We should note a certain complication of the nature of the distribution of the pressure field and of the velocity field in the musculo-rib and fat layers in the vicinity of surfaces $\psi_0, r_2 \leq r \leq r_7$ and $\pi - \psi_0, r_2 \leq r \leq r_7$. Let us examine this in more detail.

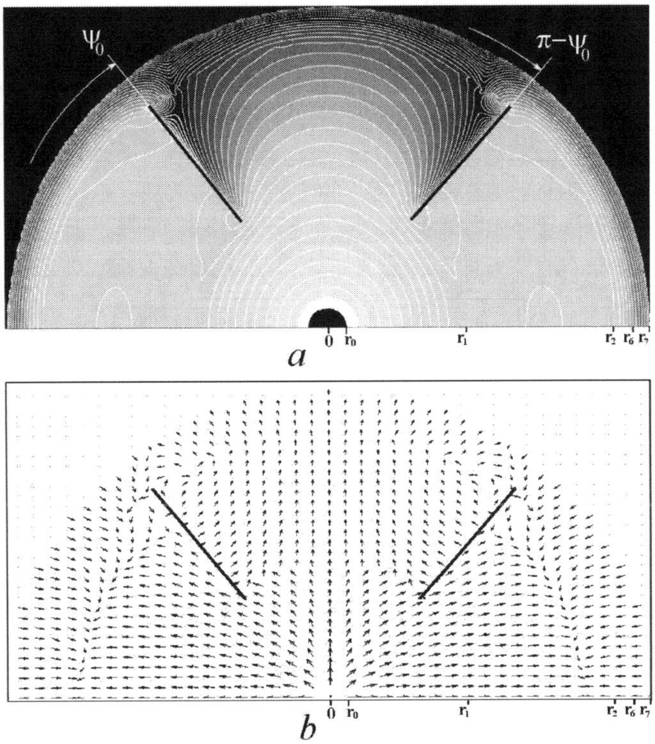

FIG. 6.19: Sound field distribution inside the thorax at 100 Hz; a—pressure field, b—velocity field

If we analyse the behaviour of the oscillating velocities in the musculo-rib and fat layers, it is possible to detect that the direction of velocities there over the lungs and over the musculo-rib area differ by 180°. This phase shift is caused by the significant difference between the sound propagation velocities in the lungs and in the interpulmonary area. Such phase shift inevitably leads to periodic oscillatory motion ("the overflow") of the medium in the musculo-rib and fat layers along the coordinate ψ through the surfaces $\psi_0,\ r_2 \le r \le r_7$ and $\pi - \psi_0, r_2 \le r \le r_7$. It is this "overflow" that can be seen in Fig. 6.19(a).

We now consider the features of the sound field distribution inside the chest at 1500 Hz (Fig. 6.20). As expected, the picture is a little more complicated than in Fig. 6.19. It can be clearly seen that the lion's share of sound energy now passes through interpulmonary area, while only insignificant portion of it passes through the lungs. Since the length of the sound wave in the parenchyma at this frequency is only 0.02 m, 3.5 lengths of a sound wave would fit into the thickness of the lungs. That's why, in Fig. 6.20(b), we can observe the change of phases of the oscillating velocities at 180°. Also, as at 100 Hz, the directions of oscillating velocities in the musculo-rib and fat layers over the lungs and over the interpulomonary area differ by 180°, which leads to a periodic oscillatory motion (periodic "overflow") of the medium through surfaces $\psi_0, r_2 \le r \le r_7$ and $\pi - \psi_0, r_2 \le r \le r_7$.

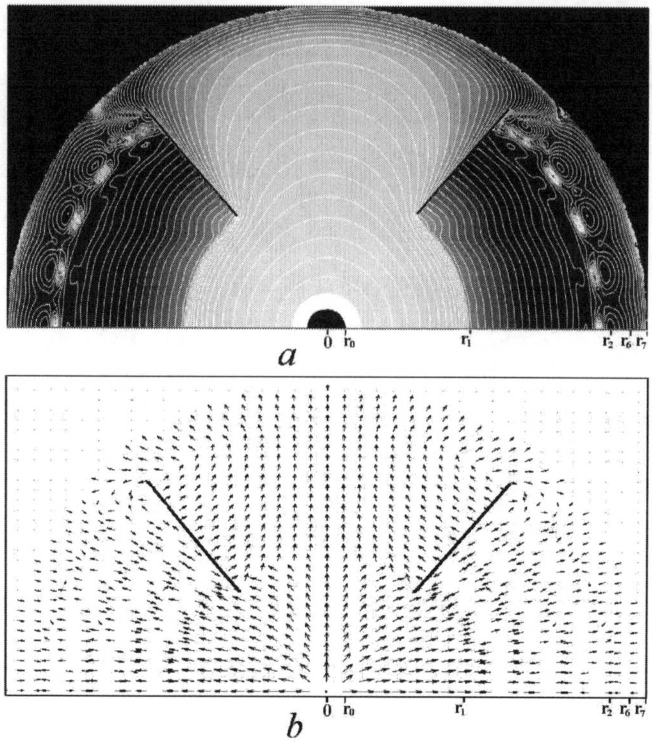

FIG. 6.20: Sound field distribution inside the thorax at 1500 Hz; (a)—pressure field, (b)—velocity field

6.7 A GENERALIZED MODEL OF SOUND PROPAGATION FROM THE TRACHEA TO THE CHEST

We recall that the sound in the respiratory system of a healthy person can be a result of either oscillations of the trachea walls under the influence of pressure fluctuations caused by unsteady air flow in the area of the glottis during breathing, or vibrations of the vocal cords (and with them the walls of the trachea) during the speech. There are two main ways of sound propagation from the trachea to the surface of the chest. The first way is through the bronchial tree into the lungs, then through an rib-muscular and fat layers to the surface of the chest. This path is often referred to as an "air" way because the first part of the passage happens through the air, filling the bronchial tree. The second way is through the tracheal wall, mediastinum, the lungs, rib-muscle and fat layers and on to the surface of the chest. This path is often referred to as a "structural" way, meaning that it passes through the biological tissues of organs (structures) inside the chest. Although these two ways of sound propagation seem obvious, they were first mentioned in 1956 in the paper [47]. In 2010, using of the method of sensing the lungs with complex acoustic signals and their subsequent special analysis, these two ways were confirmed experimentally [48]. In Section 6.6, we examined the mathematical model describing the "structural" way of sound propagation from the outer surface of the trachea wall through the mediastinum, the lungs, rib-muscle and fat layers to the surface of the chest. In Section 6.4 we considered mathematical model describing the "air" way of sound propagation, albeit only the portion that runs along the bronchial tree itself. The remainder of the path, through the lungs, rib-muscle and fat layers to the surface of the chest, was not considered. Therefore it is not possible to provide complete and detailed picture of sound propagation in the human thorax and to assess the contribution of each of the two paths of sound propagation in the oscillating speed on the surface of the thorax. Namely, this speed (or acceleration) is the main parameter available for experimental evaluation in vivo, and it is recorded both in various scientific studies related to acoustics of the respiratory system and in diagnostics of the lung diseases using traditional microstethophone.

Therefore, below we shall show how one can combine mathematical models discussed in Sections 6.4 and 6.6, in a generalized model that simultaneously takes into account both the "air" and the "structural" ways of sound propagation, allowing to obtain quantitative data describing the contribution of each path in the overall level of the oscillating speed on the surface of the thorax.

6.7.1 The Method and Its Implementation

Let us consider the image of the model in Fig. 6.12 and recall the structure of the model and adopted notations. In the centre lies the trachea of radius r_0, whose wall makes pulsating harmonic oscillations with an amplitude υ_0 (or their pressure p_0 is given). Region I, a cylindrical layer, bounded by radii r_0 and r_1, models the mediastinum. Shaded areas II and IV model the lungs, while regions III and V model the inter-lung areas. Cylindrical layer VI, bounded by radii r_2 and r_6, simulates the edge-muscular layer, and a cylindrical layer VII, bounded by radii r_6 and r_7, models the fat layer. Finally, region VIII ($r \geq r_7$) models the airspace surrounding the chest (the thorax).

Pulmonary areas II and IV may contain additional sound sources, characterized by some sound pressures $p_{2,4}^{(j)}$ and oscillating velocities $\upsilon_{2,4}^{(i)}$, which we shall use below. Every region of the adopted model is characterized by its density, speed of sound and the loss factor specified in Section 6.6.3.

We now consider the model of the bronchial tree (Section 6.4). There, we considered a model in the shape of a branching acoustic waveguide, consisting of resilient tubes, whose geometrical and physical characteristics are consistent with the real bronchi in each generation. The model warranted that the outer walls of each element of the bronchial tree are surrounded by bio-tissues whose physical properties corresponded to the location of each element within the thorax. So it was believed that the trachea and the main bronchi are surrounded by biological tissues of the mediastinum and the remaining bronchi and bronchioles are surrounded by the lung parenchyma. The developed algorithm allowed us to estimate both oscillating velocities of the walls of all the elements of the bronchial tree, and the proportions of the sound energy that passed through the wall into biological tissues and that was absorbed by the walls.

Thus, the quantitative data obtained using this model of the oscillating speeds of the bronchus and bronchiole walls can be used as input data for variables $\upsilon_2^{(j)}$ and $\upsilon_4^{(j)}$

in the model in Fig. 6.12. Knowing these velocities and taking into account that the diameter of the elements of the bronchial tree is significantly smaller than the wavelength propagating in the parenchyma, we can easily estimate the sound pressure from the walls of the bronchi and bronchioles in the lung parenchyma, using a known expression [49]

$$p_{2,4}^{(j)} = -ik_{2,4}\rho_{2,4}c_{2,4}Q_{2,4}^{(j)}G_{(j)}^{(2,4)}(r, \psi, r_{(j)}, \psi_{(j)}) \tag{6.48}$$

where indices 2, 4 indicate the number of the domain occupied by the left or the right lung, (j) is the number of the source in the lungs (Fig. 6.12), $k_{2,4}$, $\rho_{2,4}$, $c_{2,4}$ are the wave number, the density of the parenchyma and the speed of sound in the parenchyma of the right or the left lung, $Q_{2,4}^{(j)} = \upsilon_{2,4}^{(j)} 2\pi r_{2,4}^{(j)}$ is the volume velocity created by the j bronchus or bronchiole, $r_{2,4}^{(j)}$ is the radius of the j bronchi or bronchiole, $G_{(j)}^{(2,4)}(r, \psi, r_{2,4}^{(j)}\psi_{(i)})$ is the Green's function.

We shall omit the analytical part of the solution and corresponding algorithms associated with this model, as they are described in detail in Section 6.4.

The implementation of the above model consists of two stages. During the first stage we choose bronchi and bronchioles to be included in the calculations and using the algorithm described in Section 6.4 we estimate the oscillating velocities of their walls. Then, using formula (6.48), we calculate corresponding sound pressures. After that, during the second stage, with the help of algorithms presented in Section 6.6, we calculate sound fields inside the thorax and on its outer surface taking into account oscillations of the trachea walls and of the chosen bronchi and bronchioles. Of course, the number of bronchi and bronchioles in the bronchial tree is huge and it is not possible to account for all of them in the calculations. However, it is also not needed to take all of them into account given that the calculations in Section 6.4 showed that the major portion of the sound energy that passed from the bronchial tree through the bronchi and bronchiole walls into the lung parenchyma, is determined by the energy transmitted through the walls of the first few of the bronchi and bronchioles ($2 \le z \le 10$).

6.7.2 Analysis of the Numerical Results

Physical dimensions and parameters of biological tissues in each domain of the model shown in Fig. 6.12 were taken the same as in Section 6.6.3. The frequency range, that we are interested in, lies between 50 Hz and 1500 Hz. It is determined by the sound phenomena occurring during breathing [1,2,6,7,17,22,23]. The dimensions and physical characteristics of the bronchi and bronchiole walls included in the calculations were taken from the tables in Section 6.4.

In our calculations, we used the same generalized (integral) estimates of α_1 and α_2 [see Eqs. (6.41) and (6.42)] for oscillating speed distribution $|\upsilon(r_7)|$ over the surface of the thorax depending on frequency as we adopted in Section 6.6. Recall that parameter α_1 is the integral estimate of the oscillating speed on the surface of the thorax over the lungs and characterizes acoustic permeability of the lungs by speed. Parameter α_2 gives an integral estimate of the oscillating speed over the inter-pulmonary area and also describes its acoustic permeability speed. In addition, we estimated relative oscillating speeds on the surface of the thorax, υ_7 and υ_7^1, when $r = r_7$ at $\psi = 0$ and $\psi = \pi/2$, respectively.

We assumed that axially symmetric radial oscillations of the tracheal wall with the oscillating speed $\upsilon_0 = 1$ m/s were the main source of the sound inside the thorax. These fluctuations should undoubtedly excite sound waves both in the mediastinum and in the airspace filling the bronchial tree.

We first consider the extreme case when the tracheal wall vibrations excite sound only in the mediastinum, and there is no sound in the bronchial tree (this corresponds to the model of only "structural" way of sound propagation). This case is depicted in Fig. 6.21 by charts showing dependencies of values α_1, α_2, υ_7, and υ_7^1 on frequency. Features of the curves 1 and 2 in this figure were analysed in detail in Section 6.6.3.2. We would like, however, to remind that the peak at about 70 Hz is caused by resonance between the lungs and rib-muscle layer, while the rapid decline of the curve 1 as the frequency increases is determined by large losses of sound energy in the lung parenchyma. The behaviour of curves 3 and 4 on the whole resembles the behaviour of curves 1 and 2, respectively.

We now turn to a more complicated case when the oscillations of the tracheal wall excite sound not only in the mediastinum, but also in the air that fills the bronchial tree. This, as mentioned before, leads to oscillations of the bronchiole and bronchi walls and subsequent excitation of the sound directly in the lung parenchyma. We first consider a simplified problem assuming that only the walls of the two lobar bronchi vibrate, which are placed at coordinates $r = 0.09$ m, $\psi = 25°$, and $\psi = 335°$. The results of the calculations for this case are presented in Fig. 6.22. As can be seen, the situation has changed significantly compared to the case where the sound propagation in the bronchial tree was not taken into account. First, the level of values α_1 and α_2 at around 70 Hz, fell by 2.5

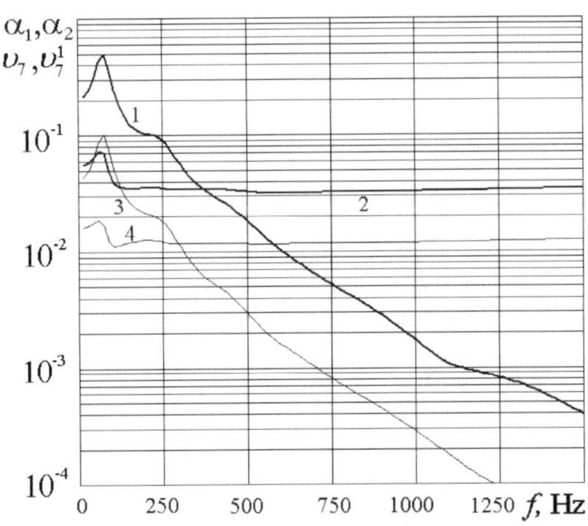

FIG. 6.21: The trachea walls generate sound only in the mediastinum (there is no sound inside the bronchial tree); 1, 2—dependencies of α_1 and α_2 on frequency; 3, 4—dependencies of relative oscillating velocities υ_7 and υ_7^1 on the thorax surface on frequency in points $\psi = 0$ and $\psi = \pi/2$, respectively

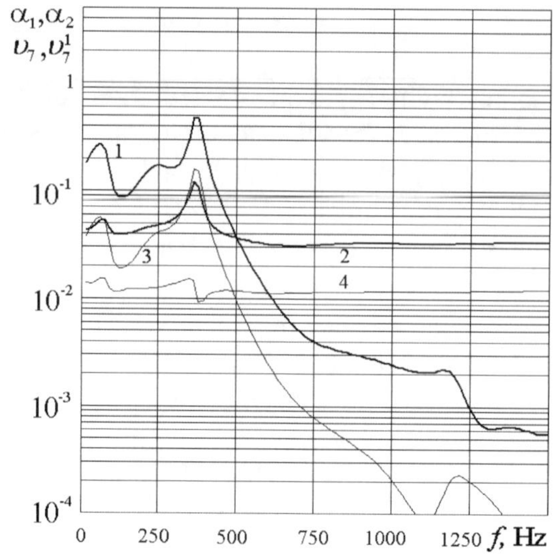

FIG. 6.22: The trachea walls generate sound in the mediastinum and inside the bronchial tree, and only wall oscillations of the two lobar bronchi in the lung parenchyma are taken into account; 1, 2—dependencies of α_1 and α_2 on frequency; 3, 4—dependencies of relative, oscillating velocities υ_7 and υ_7^1 on the thorax surface on frequency in points $\psi = 0$ and $\psi = \pi/2$, respectively

and 1.5 times, respectively, which can be explained by the interference between the sound waves that travelled through the mediastinum (the "structural" way), and the waves that passed through the bronchial tree (the "air" way). Secondly, the frequency characteristic of α_1 now shows two new peaks at frequencies around 375 Hz and 1140 Hz, while that of α_2, one new peak at 375 Hz. These new peaks are explained by the resonance phenomena of the air column in the bronchial tree, which in essence is a branched waveguide of finite length. In Section 6.5.1 we established that the input impedance of the bronchial tree significantly dependent on frequency and is characterized by successive resonances and anti-resonances. In particular, it is shown that the first and second anti-resonance frequencies of the input impedance of the bronchial tree are at about 400 Hz and 1150 Hz, which virtually coincides with frequencies for peak values of α_1 in Fig. 6.22.

We now complicate the task and take into account fluctuations in the walls not only of the lobar bronchi, but also of the next two generations of bronchi. We leave the coordinates of two lobar bronchi unchanged, while the coordinates of the next generation of bronchi (with the total of four bronchi) we set as $r = 0.11$ m, $\psi = 15°$, $\psi = 35°$, $\psi = 335°$, $\psi = 345°$, and, finally, the coordinates for the following generation, with the total of eight bronchi, we set at $r = 0.13$ m, $\psi = 10°$, $\psi = 20°$, $\psi = 30°$, $\psi = 40°$, $\psi = 350°$, $\psi = 340°$, $\psi = 330°$, and $\psi = 320°$.

Thus, we now have fourteen bronchi, whose walls can radiate sound in the lung parenchyma. We shall now examine how this impact on the characteristics that we study and turn to Fig. 6.23. As can be observed, the nature of dependencies of all the studied parameters on frequency changed relatively little, however their level increased significantly

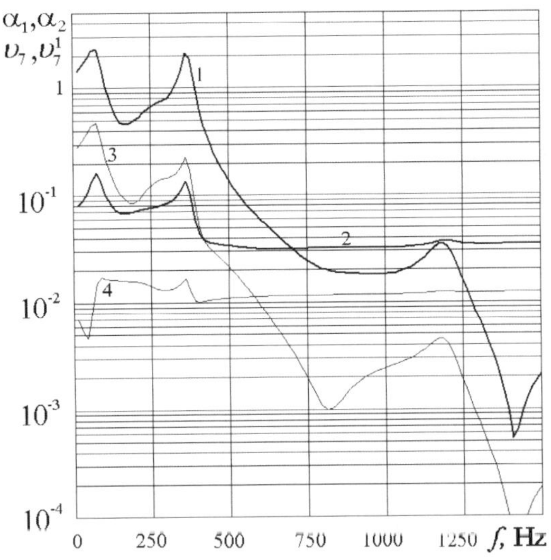

FIG. 6.23: The trachea walls generate sound in the mediastinum and inside the bronchial tree, and wall oscillations of the 14 bronchi in the lung parenchyma are taken into account; 1, 2—dependencies of α_1 and α_2 on frequency; 3, 4—dependencies of relative oscillating velocities υ_7 and υ_7^1 on the thorax surface on frequency in points $\psi = 0$ and $\psi = \pi/2$, respectively

(almost by an order of magnitude) (compare Figs. 6.22 and 6.23). This fact indicates that most of the sound energy generated by the walls of the trachea, reaches the surface of the thorax by passing through the air medium filling the bronchial tree, i.e., by the so-called "air" way. A fraction of energy penetrating through the mediastinum, i.e. via the "structural" way, is relatively small. This phenomenon can give the following simple physical explanation. Since the characteristic impedance of tissues of the mediastinum is much larger than the characteristic impedance of the lung parenchyma (approximately by two orders of magnitude), the proportion of energy that can penetrate from the mediastinum into a lung, is less than 0.1% [33,36,37]. The situation is different for the sound energy, excited by the trachea wall, which emanates into the air medium of the bronchial tree. Obviously, as the sound propagates along the bronchial tree it attenuates, however not as much as on the mediastinum-lung border. More importantly, sound energy immediately gets into the lung parenchyma due to oscillations of the bronchi and bronchiole walls that are inside the lungs. We have made additional calculations for the larger amount of bronchi and also using other coordinates for their locations in the lung parenchyma. The calculation results showed almost no significant differences from the graphs shown in Fig. 6.23.

This confirms the conclusion reached in Section 6.4, that most of the sound energy transmitted through the bronchi walls into the lungs is determined by only the first few generations of them. Thus, the results of numerical analysis have shown that the sound levels on the surface of the thorax of a healthy person is mainly determined by the part of the sound energy that has passed through the bronchial tree, i.e., by "air" means.

6.8 THE NATURE OF RESPIRATION NOISE

From the above analysis (see Section 6.2) it becomes clear that tracheal and bronchial noise occurs due to nonstationary pressure pulsations in glottis during act respiration.

Let us now consider the physical causes for the occurrence of vesicular noise. Unfortunately, in contrast to tracheal noise, the nature of the occurrence of vesicular noise is still under discussion [50]. For instance, in [6,51], it was shown that the nature of this noise has still not been decisively established, and in [52] it is stated directly that both the location and the mechanism of the occurrence of vesicular noise are unknown. However, a number of other researchers believe that vesicular noise occurs in the lungs directly in the alveoli (vesicles). So, Laennec himself [1], who introduced this term, believed that vesicular noise occurred owing to air friction along alveoli walls. The famous clinicians of the former Soviet Union Kovalevskiy [1], Gubergritz [2], the author of the article "Respiratory Noise" in the *Large Medical Encyclopedia* [4], and the well known Swiss clinician Hermann Sahli [53] also believe that vesicular noise is generated directly in the alveoli, but mainly owing to oscillations of the alveoli walls occurring as a result of elastic tension of the alveoli during their expansion during inspiration and decreasing tension during expiration. There are also researchers who hypothesize that the lung parenchyma of a healthy person is "silent" [54] and vesicular noise is nothing other than tracheal noise penetrating through the biological tissues of the chest wall and through the air in the bronchial tree to the surface of the chest wall (see, e.g., [16,17,55]).

We are impressed by the hypothesis in [1,2,4,53], according to which vesicular noise is generated directly in the alveoli owing to oscillations of the alveoli walls occurring as a result of elastic tension of the alveoli walls during their expansion during inspiration and weakening tension during expiration. However, it cannot be ruled out that in actuality, vesicular noise is a mixture occurring owing both to oscillations of the alveoli walls and a certain fraction of tracheal noise reaching the surface of the chest wall.

The aim of this research is a comprehensive study of normal tracheal and vesicular noise, which on this basis attempts to answer the question: is vesicular noise tracheal noise penetrating through the biological tissues of the chest wall and through the air in the bronchial tree to the surface of the chest wall, or are the natures of vesicular and tracheal noise completely different?

6.8.1 Technique for Recording Human Respiratory Noise

Tracheal and vesicular noise was recorded by the worldwide conventionally practiced technique [51,56–59] (see Fig. 6.24). As sensors, we used piezoceramic accelerometers *1* and *2* with a diameter of 25 mm with built in preamplifiers of in house design [17,60,61]. They were attached to the body surface by two sided tape. Accelerometer *1* was attached to the neck slightly beneath the glottis, and accelerometer *2*, on the chest wall under the right clavicle. The signals recorded by the accelerometers were fed to the input of a dual channel analog amplifier with filters *3* and, after corresponding amplification, were fed to the audio input of computer *4*. The obtained signals were processed by computer *4*

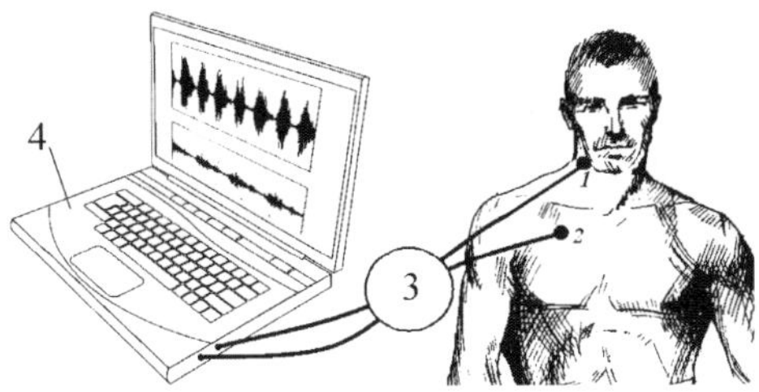

FIG. 6.24: Diagram of setup for synchronous recording of tracheal and vesicular noise: *1*, accelerometer for recording tracheal noise; *2*, accelerometer for recording vesicular noise; *3*, dual channel analog amplifier with filters; *4*, computer

with Matlab and Adobe Audition 2.0 in the frequency range from approximately 70 to 1600 Hz. Noise was recorded on male volunteers aged from 40–70 years with healthy respiratory systems. The respiratory intensity in all cases remained moderate with an air expenditure of around 0.7 L/s.

6.8.2 Analysis of Results

First, let us consider the initial unprocessed signals. Figure 6.25 shows the typical oscillograms of tracheal and vesicular noise, which are recorded using the setup shown in Fig. 6.24. One can see that inherent to these signals are all the characteristic features that have been frequently studied and described in the world literature (see, e.g., [9,22,54,57,58,63,64] and the vast references therein). Indeed, tracheal noise is a sequence of characteristic alternating sound pulses occurring upon inspiration and expiration [see, e.g., Fig. 6.25(a)]. It is also noteworthy that pulses upon inspiration and expiration differ quite insignificantly from each other. In contrast, for vesicular noise, the most characteristic feature is precisely the fact that pulses occurring upon inspiration and expiration are completely different [Fig. 6.25(b)]. First, the pulse level upon inspiration is significantly larger than the pulse level upon expiration, and second, their shapes differ noticeably from each other. We now examine how the spectra of tracheal and vesicular noise differ (Fig. 6.26). One can see that the spectrum of tracheal noise is comparatively

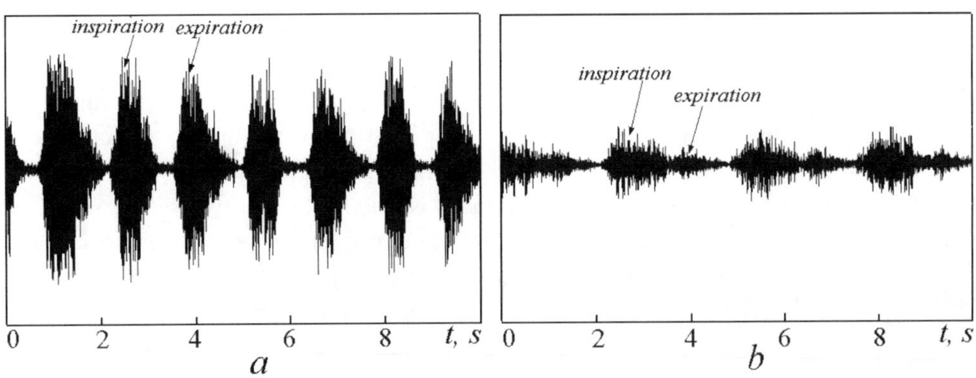

FIG. 6.25: Oscillograms of respiratory noise: (a) tracheal noise, (b) vesicular noise

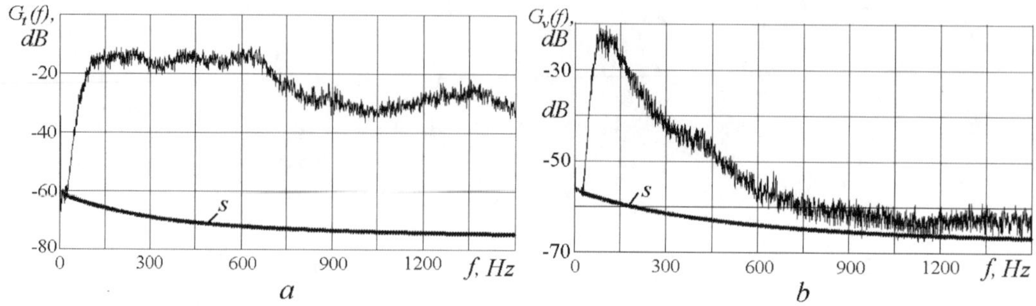

FIG. 6.26: Respiratory noise spectra: (a) tracheal noise; (b) vesicular noise; s—intrinsic noise of setup

uniform, although in the frequency range from 70 to 700 Hz, the spectrum level is approximately 8–10 dB higher than in the range from 700 to 1500 Hz. On the whole, this agrees well with the data available in the literature (see, e.g., [57,63]). In contrast, the spectrum level of vesicular noise in the frequency range from 70 to 700 Hz drops very rapidly (at a rate of approximately 10–15 dB per octave), and higher than 700 Hz it is almost completely comparable to the intrinsic noise level (curves denoted s) of our measurement setup [see Fig. 6.26(b)]. Such a characteristic dependence of the spectrum level of vesicular noise on frequency also agrees well with the literature data [57,63].

Let us analyze tracheal and vesicular noise spectrograms.*** We note immediately that McKusick et al. are the pioneers of using spectrography to analyze respiratory noises, who as far back as 1955 published [65] on this subject matter. However, way back then, the authors had only analog spectrographic equipment, which was quite cumbersome and lacked the necessary performance. Therefore, this method was not widespread then. However, with the advent of high speed computers, it became possible to use it effectively to analyze respiratory noise. The initiators of the wide use of spectrography related to the study of respiratory noise and in daily medical practice were Hans Pastercamp and his colleagues from the University of Manitoba in Winnipeg, Canada [22,58,64]. They even introduced the new term "digital respirosonography" to underscore that the spectrograms pertain precisely to respiratory noise. We will also use this terminology.

However, let us return to Fig. 6.27, which shows the respiratory sonograms for tracheal and vesicular noise.

As expected, for tracheal noise [Fig. 6.27(a)], the respiratory sonogram time regions corresponding to the inspiration and expiration phases hardly differ from each other and their spectra extend up to 1500 Hz or higher. In contrast, the plots respiratory sonogram vesicular noise adequate phases of inspiration and expiration, differ significantly from each other [see Fig. 6.27(b)]. During inspiration, the noise spectrum is wider (it goes up to approximately 1100 Hz, while during expiration, approximately to 700 Hz) and its level is significantly higher than the spectrum level during expiration, especially at relatively low frequencies. These features of the respiratory sonograms of

***The spectrograms are the time dependences of the instantaneous spectrum.

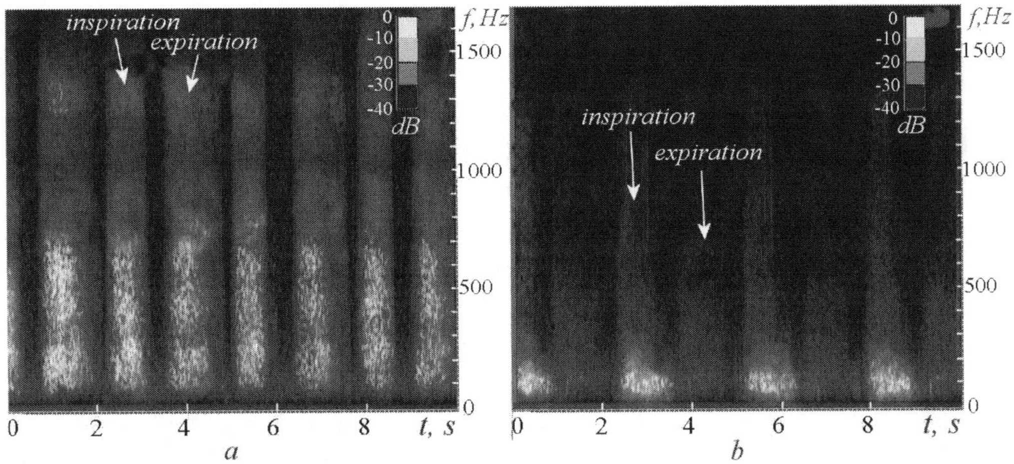

FIG. 6.27: Respiratory sonograms of respiratory noise: (a) tracheal noise, (b) vesicular noise

normal tracheal and vesicular noise are also well known from literature sources (see, e.g., [22,58,64]).

The aim of the above analysis was not so much to reveal the features of tracheal noise characteristics and differences from those of vesicular noise (they are quite well known), but to remind the reader of their essence and to certify that our setup for recording and processing noise obtains quite correct results that agree well with the known data. In addition, it is quite clear that these noise characteristics cannot effectively answer the following questions: is vesicular noise tracheal noise that penetrated through the bronchial tree, attenuated, changed its spectrum, and finally reached the surface of the chest wall, or are tracheal and vesicular noise and the physical causes of their occurrence fundamentally different?

In order to proceed, it is necessary, based on the noise characteristics obtained above, to use additional characteristics that would characterize the degree of the relationship between tracheal and vesicular noise. As it seems to us, the coherency function can play the role of such a characteristic [66–68]; it is determined in our case as

$$\gamma_{tv}^2(f) = \frac{|G_{tv}(f)|^2}{G_t(f)G_v(f)}, \tag{6.49}$$

where $G_t(f)$, $G_v(f)$, and $G_{tv}(f)$ are the frequency dependences of the tracheal noise spectrum, the vesicular noise spectrum, and their mutual spectrum, respectively.

Figure 6.28 shows the coherency function calculated using our experimentally recorded tracheal and vesicular noise (see Fig. 6.25).

One can see that in the entire frequency range, the coherency function on average is lower than 0.2 and only in a relatively narrow frequency region, from approximately 400 to 600 Hz, does it lie on average at a level of around 0.45–0.5. By interpreting the coherency function as the relative contribution of tracheal noise (inspiration process) in the vesicular noise spectrum (expiration process) [66,67], it is possible to conclude that

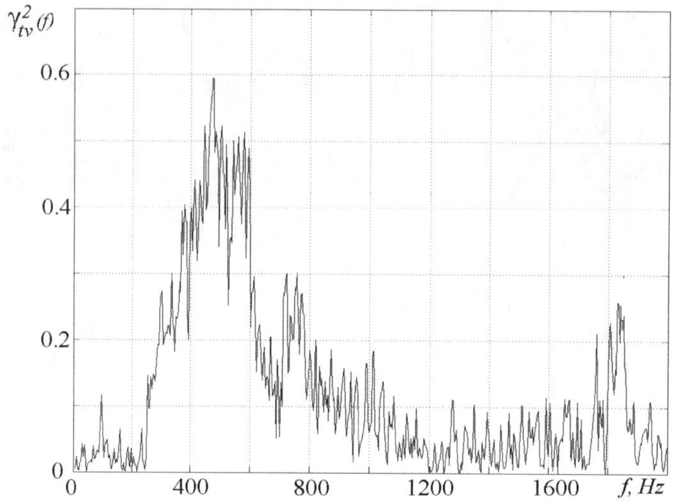

FIG. 6.28: Coherency function between tracheal and vesicular noise

only in the frequency range from 400 to 600 Hz does this contribution become noticeable. At the remaining frequencies, there is hardly any noticeable contribution. Based on this analysis, we can conclude that the physical causes for the occurrence of tracheal and vesicular noise are most likely different and vesicular noise is generated in the lungs them-selves. The anomalous increase in the coherency function near 500 Hz is easily explained if we recall that precisely in this frequency zone is a longitudinal resonance of the air column observed, which exists in the bronchial tree as an acoustic waveguide [27,68,69]. Naturally, at this resonance, the sound energy generated by the nonstationary airflow in the region of the glottis sharply increases and becomes comparable to the vesicular noise energy.

Unfortunately, such an optimistic conclusion can be overshadowed by the following circumstances. As is known [66], if the coherency function is larger than zero, but less than unity, then in practice the following situations are possible: (a) when recording signals, there is noticeable intrinsic noise and (b) the properties of biological tissues of the internal regions of the chest wall through which tracheal noise passes from the trachea to the surface of the chest wall are not strictly linear from the acoustical standpoint. The intrinsic noise from our setup is quite low (cf. the levels of curves s in Fig. 6.26 with the levels of tracheal and vesicular noise). As concerns the linearity of the properties of biological tissues, this question here remains open. Of course, the initial region of the airway along which tracheal noise passes is air filling the bronchial tree. And at sufficiently low acoustic signal levels characteristic of respiratory noise in general and tracheal noise in particular, here nonlinear distortions are impossible. However, it seems impossible to answer the question whether nonlinear distortions of tracheal noise signals can occur as they pass through the lung parenchyma, since the acoustic properties of the parenchyma have also been insufficiently studied.

Therefore, to reliably ascertain that vesicular noise is a type of noise separate, independent, and differing from tracheal noise that occurs only in the lung parenchyma, an *in vivo* experiment is necessary, where during the act of respiration there would be no tracheal noise, and vesicular noise, on the contrary, would be present. At first glance, the idea of such an experiment seems absurd, since respiratory noise occurs during the respiration process. Nevertheless, if we except as a working hypothesis that vesicular noise occurs only owing to contraction and expansion of the parenchyma, then the experiment is quite realistic. Indeed, if we close off the mouth and the nose so that air does not penetrate into the bronchial tree, and owing to the efforts of the chest muscles to alternately expand and contract the chest wall (just like in the act of respiration), the lung parenchyma will be forced to expand and contract. If our hypothesis is true, then we should detect noise similar to vesicular. As well, tracheal noise will not be detected, since there will be no airflow in the larynx and bronchial tree.

We conducted experiments according to this scenario (our volunteers mastered this exercise literally within a few minutes) and recorded noise signals at those points where respiratory noise had been recorded previously (see Fig. 6.24).

Let us now turn to Fig. 6.29, which shows the oscillograms of noise signals recorded in the course of this experiment. As expected, there was hardly any signal from the accelerometer fixed to the neck [see Fig. 6.29(a)]. But there was a signal from the accelerometer fixed beneath the clavicle [Fig. 6.29(b)], and it had a distinctly pronounced pulse character. In the signal, the areas of its increase and decrease are well distinguished, corresponding to expansion (increase in volume) and contraction (decrease in volume) of the chest wall. Figure 6.30 also shows the spectrum of this signal. One can see that in the frequency region from 70 to 600 Hz, the level of the spectrum drops rapidly, and higher than 600 Hz, it is virtually comparable to the level of intrinsic noise (curve *s*).

Simple comparison of the oscillograms in Figs. 6.25(b) and 6.29(b), as well as the spectra in Figs. 6.26(b) and 6.30, allows us to ascertain that the time and spectral characteristics of vesicular noise and noise caused only by expansion–contraction deformation of the lung parenchyma (in the absence of airflow in the bronchial tree) are quite close. This quite convincingly demonstrates that vesicular noise is a separate independent type of noise occurring in the lung parenchyma owing to conversion of part of the parenchyma

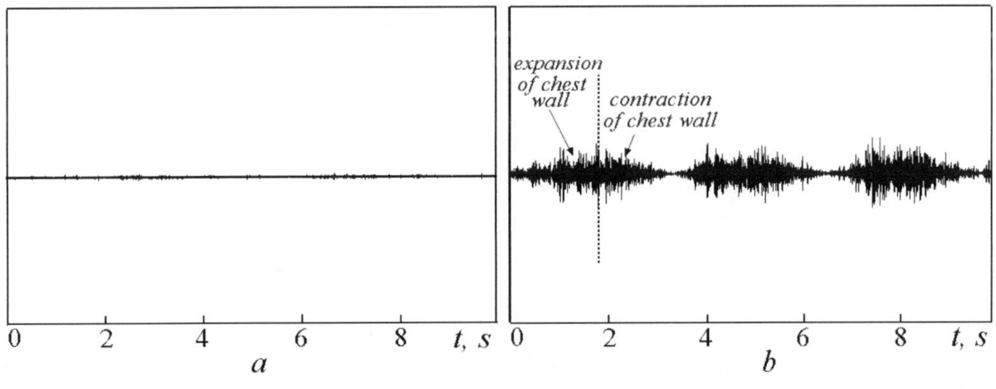

FIG. 6.29: Oscillograms of noise signals during contraction—expansion of parenchyma in absence of airflow in the glottis and bronchial tree: (a) at the neck slightly beneath the glottis, (b) on the chest wall under the clavicle

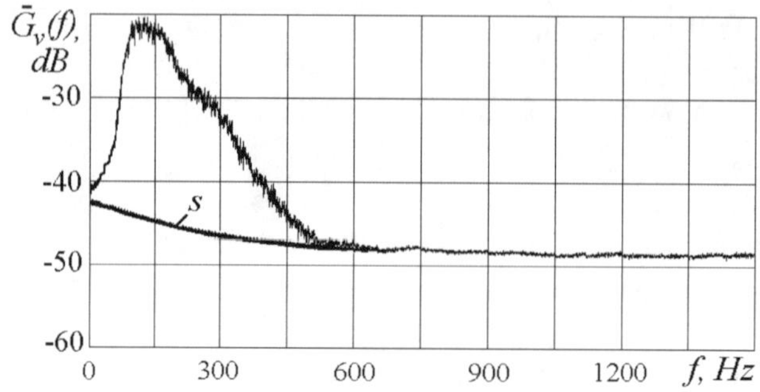

FIG. 6.30: Spectrum of noise signal depicted in Fig. 6.29(b); *s*—intrinsic noise of setup

deformation energy into sound energy. Therefore, the known clinician Gubergritz was profoundly correct when he intuitively presumed in his book [2] that "...vesicular respiration occurs from oscillations of the alveoli walls as they expand during inspiration..." Below, using a very simple mechanical model, we show how such a physical process can be achieved.

In the above analysis, we only compared the general integral characteristics of noise signals. It seems important to also compare the finer (local) structure of signals. Keeping this in mind, let us first recall certain features of lung parenchyma morphology. As is known [2,9], the main purpose of the parenchyma is to ensure passage of oxygen from air into the blood and carbon dioxide in the inverse direction. This process occurs in the alveoli, of which in both lungs of an adult person there are approximately 600–700 mln. The shape of alveoli is like an incorrect polyhedron of arbitrary orientation (see Fig. 6.31). The mean lateral dimension of alveoli is around 250–300 μm. There are no statistical data at present on the number of faces (alveoli walls), their dimensions, and shape. The thickness of alveoli walls is around 2–8 μm. Thus, it is obvious that the lung parenchyma represents a biological material of cellular structure for which the

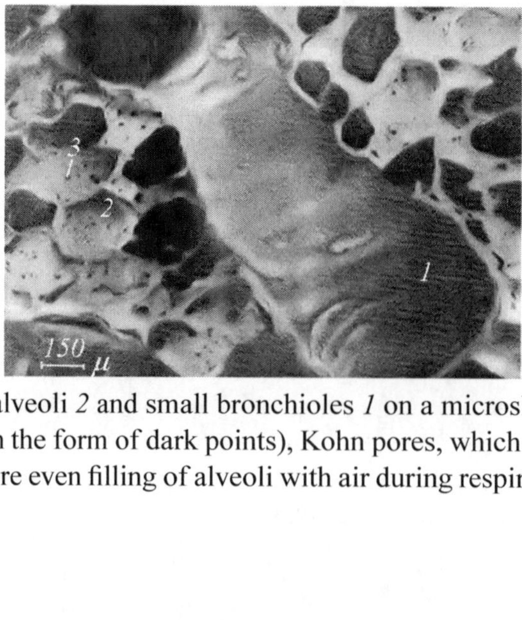

FIG. 6.31: Multiple alveoli *2* and small bronchioles *1* on a microslice of lung; opening *3* in the alveoli walls (in the form of dark points), Kohn pores, which join neighboring alveoli, which ensures more even filling of alveoli with air during respiration (figure borrowed from [73]).

shape of cells (alveoli) and their orientation in space are unordered. A material with such characteristics can properly be categorized as fractal materials [71,72].

If we accept the hypothesis that the noise generated by the parenchyma during its expansion–contraction deformations are caused by oscillations of a large number of differently shaped and differently sized alveoli walls, then the fractal structure of the parenchyma as a fractal material should also manifest itself in the structure of the noise generated by it. Taking into account these considerations, we will attempt to reveal the fractal (multifractal) properties of respiratory noise, which should shine a light on their origin [71].

At present, the ideas of multifractal theory are applied to studying objects of various nature. Important representatives of such objects are signals recorded in field experiments, including signals from biological systems. Therefore, the involvement of finer mathematical tools for studying signals, in comparison to conventional statistical processing, makes it possible to study the finer local structure of the complex signals received.

Over the last several decades, signals obtained in experiments, as a rule, are represented in digital (discrete) form, i.e., as time series. To calculate the multifractal characteristics of random time series, there are two main practical methods: the wavelet transform modulus maxima (WTMM) method, proposed at the beginning of the 1990s and based on an estimate of the so-called skeleton of the modulus maxima of continuous wavelet transformations [74]; and the later multifractal detrended fluctuation analysis (MF_DFA) [75]. Both methods analyze the remainder after the trend is eliminated or, in other words, "pure noise." However, elimination of trends itself and calculation of the singularity spectrum $f(\alpha)$ are conducted by different methods.

Below, MF_DFA is used, the algorithm for which reduces to the following steps [75].

1. Let the analyzed signal be represented as a finite sampling of a time series $\xi(i)$, $i = 1, 2, ..., N$. Here, integer-valued samples i correspond to certain time instants at which the signal value is recorded, which are separated by equal time intervals. The fluctuation profile is formed from this series,

$$Y(i) = \sum_{k=1}^{i} (\xi(k) - \bar{\xi}), \quad i = 1, 2, ..., N, \tag{6.50}$$

   with the mean value $\bar{\xi}$. The obtained series $Y(i)$ is divided into nonintersecting intervals $Y(v, i)$ containing a different number of points s; the ν value indicates a specific series interval. The number of such intervals is equal to the total value $N_s = [N/s]$.

2. For each interval N_s, it is necessary to find the local trend $y(v, i)$, e.g., in the form of a polynomial to a certain power, and estimate the deviations from the local trend:

$$\Delta Y(v, i) = Y(v, i) - y(v, i). \tag{6.51}$$

3. Then we determine the dispersion $\mu(v, s)$ of sampling $\Delta Y(v, i)$ for each interval,

$$\mu(v, s) = \frac{1}{s} \sum_{i=1}^{s} (\Delta Y(v, i))^2, \tag{6.52}$$

and, averaging these values to a power of q, we determine the fluctuation function,

$$F(q, s) = \left\{ \frac{1}{N_s} \sum_{v=1}^{N_s} [\mu(v, s)]^{q/2} \right\}^{1/q}. \tag{6.53}$$

4. At a fixed value q, we construct the dependence $F(q, s)$ in dual logarithmic coordinates. If this series reduces to a self similar set exhibiting long_range correlations, then fluctuation function $F(q, s)$ is represented as a power dependence,

$$F(q, s) \propto s^{h(q)}, \tag{6.54}$$

and it plots as a straight line with a slope equal to the generalized Hurst exponential $h(q)$. We calculate $h(q)$ for different values of q. For multifractal series with positive q, the main contribution to function $F(q, s)$ comes from intervals ν exhibiting large deviations $\mu(v, s)$, and for negative q, intervals with small $\mu(v, s)$ values dominate.

The standard representation of scaling properties assumes a transition from the generalized Hurst exponent $h(q)$ to the mass exponent $\tau(q)$ and singularity spectrum $f(\alpha)$ [75,76]:

$$\tau(q) = qh(q) - 1, \quad \alpha(q) = d\tau(q)/dq, \quad f(\alpha) = q\alpha(q) - \tau(q). \tag{6.55}$$

The typical graphic representation of function $f(\alpha)$ is an "inverted parabola." The integral estimate of the singularity spectrum is given by two of its parameters, which determine the position of the maximum and the width of the spectrum $f(\alpha)$:

$$f(\alpha^*) = \max f(\alpha) \text{ and } \Delta\alpha = \alpha_{\max} - \alpha_{\min}.$$

Usually, $f(\alpha^*) = 1$; i.e., the α^* value corresponds to the most typical singularity index, which is encountered so frequently that the dimension of the corresponding set A_{α^*} simply coincides with the dimension of the containing space—a one-dimensional time axis equal to 1. For a monofractal signal, theoretically the $\Delta\alpha$ value should be equal to zero, but in practice, as a result of the finiteness of the sampling, such a condition is actually not fulfilled and the $\Delta\alpha$ value, although very small, does not stay equal to zero.

It is possible to distinguish two types of multifractality in discrete time series [75]:

1. Multifractality resulting from the scatterer of the time series values, i.e., a time series property like a nonuniform set. Obviously, for a random rearrangement of these values, the probability density function for a random variable does not change its form.

2. Multifracticality resulting from the presence of a correlation sequence of alternating terms of the time series. Randomization of the time series eliminates this type of multifractality.

If two types of multifractality exist, then randomization weakens the multifractal properties. As well, the spectrum usually narrows and is displaced toward small fractal dimensions.

Note that different sets related to a signal can be subjected to multifractal analysis. For instance, this can be the very set of terms of a discrete time series or set of points intersected by a signal of a certain level [71]. We studied the moduli of values for a time series of respiration noise.

Taking into account this calculation procedure, we now construct the singularity spectrum obtained from our recordings of respiration noise signals.[†††] The duration of a recording was 17.162 s, and the sampling frequency was 3496 Hz. This corresponded to construction of a discrete time series with 60,000 samples. Note that there were approximately 8500 samples per inspiration–expiration cycle.

First of all, it is necessary to be certain there is scaling of fluctuation function $F(q, s)$. Figure 6.32 shows the dependence $\ln[F(q, s)]$ on $\ln(s)$ for different values of parameter q for respiratory noise at the trachea and under the clavicle, as well as for noise under the clavicle owing to expansion–contraction of the parenchyma in the absence of respiration. The number of time samples in interval s changed from 25 to 2000 (30 intervals in total). One can see that the points in the plot are arranged along the corresponding straight lines, which speaks to the presence of scaling in the signal structure and, accordingly, the validity of using the singularity spectrum to analyze local structural features of the signals studied by us.

We turn to Fig. 6.33, which shows the singularity spectra for all noise considered by us. One can see that vesicular noise spectra (curve *2*) and noise occurring due to the expansion–contraction of the parenchyma in the absence of respiration (curve *3*) practically coincide. Their integral characteristics are very close: for curve *2* $\alpha^* = 0.856$, $\Delta\alpha = 0.636$ and for curve *3* $\alpha^* = 0.890$, $\Delta\alpha = 0.669$. In contrast to the spectra, the noise spectrum at the trachea (curve *1*) shifted significantly toward smaller values $\alpha^* = 0.660$, and the spectral width increased up to $\Delta\alpha = 0.882$. Such results are a weighty confirmation in favor of the hypothesis on the different nature of vesicular and tracheal noise.

In Fig. 6.33, there is yet one more singularity spectrum (curve *4*), which was obtained after randomization of the terms of the series describing the signal of vesicular respiration noise beneath the clavicle. The randomization procedure is a test that confirms the presence or absence of long-range correlations in the set of values of the time series. As one can see from Fig. 6.33, randomization substantially reduced the multifractal properties, transforming the initial series into a signal close in properties to white noise $\alpha^* \approx 0.5$.

This speaks to the significant influence on the formation of multifractal properties of the initial series of long-range correlations.

[†††] All calculations were performed in Mathcad 14.0 based on programs developed by the authors.

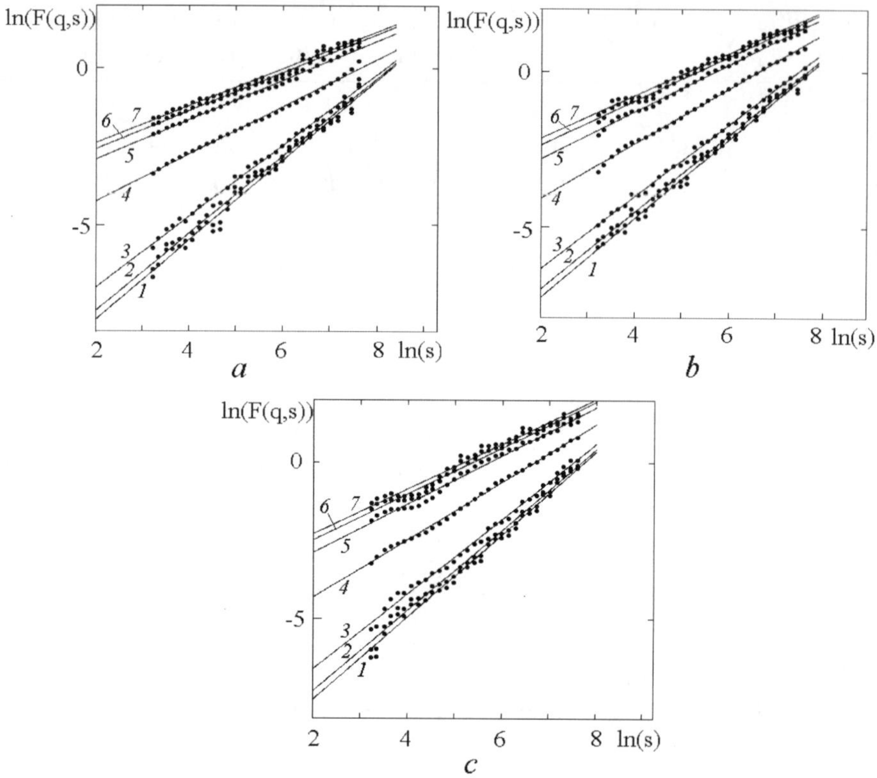

FIG. 6.32: Dependences of quantity $\ln[F(q, s)]$ on $\ln(s)$ for different q values: (a) for respiratory noise at the trachea; (b) for respiratory noise under the clavicle; (c) for noise under the clavicle occurring due to expansion–contraction of the parenchyma in the absence of respiration; the number of straight lines corresponds to the following: 1—$q = -15$, 2—$q = -10$, 3—$q = -5$, 4—$q = 0$, 5—$q = 5$, 6—$q = 10$, 7—$q = 15$, $s = 25...2000$

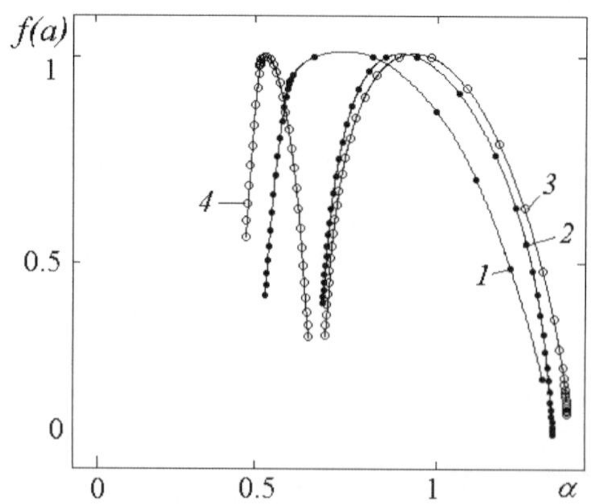

FIG. 6.33: Singularity spectrum: *1*, respiratory noise at the trachea; *2*, respiratory noise under the clavicle; *3*, noise under the clavicle occurring due to expansion–contraction of the parenchyma in the absence of respiration; *4*, for the case where terms of the series describing respiratory noise under the clavicle are randomized

Thus, it is possible to conclude that vesicular and tracheal noise, as well as noise occurring owing to expansion–contraction of the parenchyma (in the absence of respiration), possess multifractal properties resulting from the presence of a correlated sequence of alternating terms of their time series.

Based on the results obtained, it is possible to state with certainty that the physical causes for the occurrence of vesicular and tracheal noise are completely different. Vesicular noise occurs owing to expansion–contraction deformation of lung parenchyma during respiration, and tracheal noise, as is known, owing to pressure pulsations on the innersurface of the trachea, caused bynonstationarity of the airflow in the glottal zone). These deformations are transformed into lateral oscillations of the alveoli walls, which generate sound waves at eigenfrequencies. Since alveoli walls number in hundreds of millions and the shape and dimensions of walls are very diverse, the total sound signal generated by wall oscillations is sufficiently wideband; i.e., it is in essence a noise signal.

6.9 MODELING THE MECHANISM OF THE VESICULAR SOUND GENERATION

Analysis performed in the previous section has shown that the nature of vesicular and tracheal noises to be different. It is shown that vesicular sound is likely to be generated as a result of periodic stretching of parenchyma during respiration, whereas tracheal noise, as already well-known, is generated due to pressure pulsationson the inner surface of the trachea due to unsteady air flow in the glottis. However, the question about the mechanism of the energy transformation, stored in the lung parenchyma due to its deformation, into the sound energy continues to be controversial.

The purpose of this section is to show, at least in theory, that the proposed by clinicians mechanism of vesicular breath sound generation in the lungs is quite believable.

6.9.1 Theoretical Model

First of all, let us recall some of the physical and geometric characteristics of the parenchyma. Parenchyma is the biological tissue of the lungs, having mainly a honeycomb structure [73 and see Fig. 6.31] (in the first approximation resembling foam). It is these cells which are randomly oriented air-filled irregular polyhedrons with the maximum size of the order of $L \approx 300 \times 10^{-6}$ m that are called alveoli. The amount alveoli in human lungs is approximately 600 million. Alveolar walls are quite thin (about $h \approx (6...10) \times 10^{-6}$ m) and contain tiny blood vessels (capillaries). The density of the walls of the biological tissue is close to the density of water, i.e., $\rho \approx 10^3$ kg/m^3. The surface of the alveolar walls is wetted with a thin layer of liquid which tends to flatten alveoli as a result of surface tension. However, the layer of bio cells lining the wall surface excretes a particular substance, surfactant, which reduces the surface tension of the liquid and thus ensures stability of the alveolar shape [73]. Thus, the alveolar walls, in the absence of respiration, have some initial tension, which, however, cannot disturb the shape of the alveoli. During inhalation, parenchyma increases in volume due to workings of the intercostals and the diaphragm muscles, which increase the volume of the thorax.

Naturally, the volume of the alveoli increases due to stretching of their walls. It is understood that during exhalation, this process is reversed. With this in mind, it is in the process of respiration, as stated in [1,2], that periodic stretching of the walls transforms into transverse vibrations of the alveolar walls, that in turn excite acoustic oscillations recorded on the surface of the chest as noise vesicular breathing.

We now show that the described mechanism of transformation of periodic deformation of the alveolar wall stretching into their transverse vibrations is possible. To do this, we consider a simple two-dimensional physical model that can illustrate the mechanism that leads to the excitation of transverse vibrations of the alveolar walls under their periodic extension. Since $h \ll L$, then it is perfectly acceptable to use a rectangular membrane, fixed on a contour (Fig. 6.34), as the simplest model of an alveolar wall.

Suppose there are a constant (time-independent) membrane tension F_0 and a small, compared with the dimensions of the membrane, initial transverse deviation of the centre of the membrane $A(0)$ at time $t = 0$. We now apply periodic tension $F(t)$ to the membrane (which will simulate the tension resulting in periodic stretching of an alveolar wall during respiration) and would like to see if this leads to the transverse vibrations of the membrane.

To solve this problem, we use a well-known differential equation of membrane oscillations [37],

$$\frac{\partial^2 w(x,y,t)}{\partial t^2} = c^2(t)\left[\frac{\partial^2 w(x,y,t)}{\partial x^2} + \frac{\partial^2 w(x,y,t)}{\partial y^2}\right], \tag{6.56}$$

where $w(x, y, t)$ is the transverse deflection of the membrane from the equilibrium position, $c^2(t) = [F_0 + F(t)]/\bar{\rho}$, $c(t)$ is the velocity of propagation of perturbations in the membrane, $\bar{\rho} = \rho h$ is the surface density of the membrane's bio tissue.

Given the above assumptions, we can write the boundary conditions as follows:

$$\begin{aligned} &w(x,y,t) = 0, \quad 0 \le x \le L_x, \quad y = 0, \quad y = L_y, \\ &w(x,y,t) = 0, \quad 0 \le y \le L_y, \quad x = 0, \quad x = L_x. \end{aligned} \tag{6.57}$$

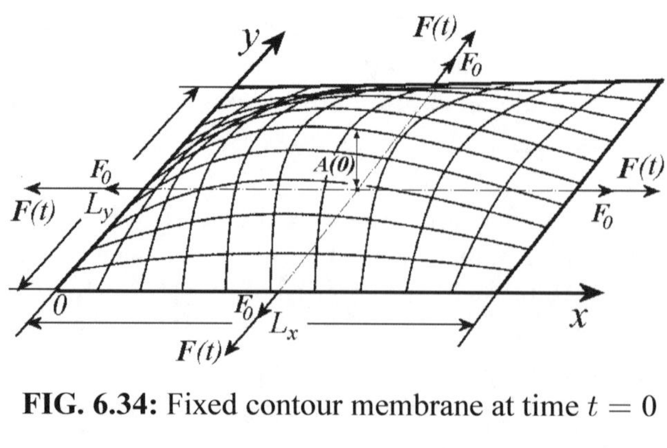

FIG. 6.34: Fixed contour membrane at time $t = 0$

We shall focus on the case when the membrane oscillates only in its first mode, when there are no sections on the surface of the membrane oscillating in antiphase. Then, taking into account (6.57), we get

$$w\left(x,y,t\right)=A\left(t\right)\sin\left(\frac{\pi x}{L_x}\right)\sin\left(\frac{\pi x}{L_y}\right). \tag{6.58}$$

Substituting (6.58) into equation (6.56) we obtain the equation for the amplitude of the oscillations $A(t)$. Assuming for simplicity that $L_x = L_y = L$, we have

$$\frac{d^2A\left(t\right)}{dt^2}+2\left(\frac{\pi}{L}\right)^2\frac{F_0+F\left(t\right)}{\bar{\rho}}A\left(t\right)=0;\quad x=L/2;\quad y=L/2. \tag{6.59}$$

In order to proceed with solving equation (6.59), let us define the initial conditions as

$$A\left(0\right)=0,1L;\quad \frac{dA\left(0\right)}{dt}=0. \tag{6.60}$$

Here the value of the initial deviation $A(0)$ is chosen taking into account the inequality $A(0) \ll L$. Now we need to set the value of the constant tension of the membrane F_0. Unfortunately, unlike the linear dimensions of the alveolarwalls and their density, there is no specific information about the magnitude of tension F_0 that we could find in the literature available to us. Therefore, in order to estimate it, even roughly, we used the equation relating the first Eigen frequency f_1 of the square membrane with its geometric and mechanical characteristics [37]:

$$f_1=\frac{1}{\sqrt{2}L}\sqrt{\frac{F_0}{\bar{\rho}}}, \tag{6.61}$$

from where

$$F_0=2f_1^2L^2\bar{\rho}. \tag{6.62}$$

We know from experimental data [1,2,4,6,22] that the main energy of vesicular sound lies approximately in the 50 Hz to 250 Hz range. Assuming, for example, an average frequency of $f_1 = 100$ Hz and using the above characteristics of the alveolar walls as well as formula (6.62), we can easily demonstrate that the tension F_0 can be of the order of $\sim 2\times10^{-5}$ N/m.

Next we need to set a variable tension $F(t)$ of the membrane. For simplicity, we choose a periodic function

$$F\left(t\right)=\frac{\bar{F}}{2}\left(1+\sin\left(2\pi\tilde{f}t-\pi/2+\phi\right)\right), \tag{6.63}$$

which will model the tension impacting an alveolar wall during respiration. Here fixed frequency $\tilde{f} = 0.345$ Hz (period $\tilde{T} = 1/\tilde{f}$ is thus 2.9 s) approximately corresponds to the real frequency of respiration, ϕ is the initial phase and $\bar{F}$ is the amplitude. Note that function $F(t)$ does not go negative for any value of t.

We now need to set specific values for parameters F_0, $\bar{\rho}$, L, and ϕ. It is apparent that if we assign some specific numbers to these parameters, we can simulate the mechanism of oscillation excitation only in one alveolar on one frequency. In reality, as we know from [1,2,4], walls of multiple alveoli vibrate simultaneously. These alveoli have different shapes, sizes, initial tensions and, consequently, different Eigen frequencies. In addition, the alvcolar walls in the parenchyma obviously cannot oscillate synchronously with the same phase. Moreover, the distance from different alveoli to the point where the sound is registered, is different, which causes a different phase delay. It is precisely due to the superposition of this multitude of oscillations with different frequencies and phases that a noise signal characteristic to vesicular breathing is generated.

In order to take this into account, at least to some extent, we proceed as follows. We assume that we have some fairly representative ensemble of membranes. To use some specific numbers, assume their quantity M to be 250. Each time we compute a numerical solution of the equation (6.59), we choose parameters F_0, $\bar{\rho}$, L, and ϕ randomly within the following ranges:‡‡‡ $3 \times 10^{-6} \leq F_0 \leq 60 \times 10^{-6}$ N/m, $4 \times 10^{-3} \leq \bar{\rho} \leq 16 \times 10^{-3}$ kg/m^2, $2.5 \times 10^{-4} \leq L \leq 3.5 \times 10^{-4}$ m, $0 \leq \phi \leq \pi$. Let $\bar{F}$ be 6×10^{-6} N/m. Then the total number of numerical solutions of the Eq. (6.59) will be $M = 250$ ($m = 1, 2, 3, ..., M$). As a result, we get 250 different values of amplitudes of the membrane centre deviations, which we average out and then assume that the sound pressure developed in the environment (i.e., in the parenchyma) at some time t_i is proportionate to this average amplitude.

We computed the numerical solution of the differential equation (6.59) over the observation time interval equal to 32 seconds. At a sampling frequency $f_d = 2048$ Hz the number of partitioning points of the observation interval was $N = 59{,}440$. Accordingly, the set of values $t_i = i/f_d$ ($i = 1, 2, ..., N$) defined the current discrete point of the observation time interval. Thus, the Eq. (6.59) was solved 250 time sat each point t_i. Thereafter, at each time interval point t_i we computed average amplitude of the membrane centre deviation.

$$\tilde{A}(t_i) = \frac{1}{M} \sum_{m=1}^{M} A_m(t_i). \tag{6.64}$$

Finally, let us point to one very important fact. As shown in [16], the lung parenchyma has the property of significantly absorbing sound vibrations propagating there. Moreover, the level of absorption increases as the frequency of sound oscillations increases. One can roughly assume that the presence of absorption leads to the decrease in the level of sound oscillations as the frequency increases proportionate to $\sim 1/f^2$. In order to account for this property of the parenchyma, the calculated signal $\tilde{A}(t)$ was passed through a low frequency filter $\Phi(f)$ with a cut-off frequency of about 50 Hz, and a slope of about $\sim$ 12 dB per octave. Therefore, the final form of the above simulated signal can be represented as

$$\tilde{A}(t, f) = \tilde{A}(t)\Phi(f). \tag{6.65}$$

‡‡‡Though the ranges were selected randomly, but so that, the first Eigen frequencies of the walls were to be in the range from $\sim$ 50 to $\sim$ 250 Hz. Any other information about real parameters range of the F_0, $\bar{\rho}$, L, and ϕ are absent today.

Thus, within the framework of the adopted model, we obtained the model signal (essentially a time series), imitating, to some extent, vesicular breath sound, recorded on the surface of the thorax. In this case, the signal source is the membranes whose transverse vibrations are caused solely by their periodic tension.

6.9.2 Analysis of the Numerical Results

First of all, let us see which frequency range Eigen frequencies of the membranes belong to. As an example, Fig. 6.35 shows the values of Eigen frequencies of the membranes obtained as a result of one random sampling of its parameters within abovementioned ranges. As is evident from the figure, the values of Eigen frequencies lie in the $\sim$ 40 Hz to $\sim$ 270 Hz range. This range of Eigen frequencies roughly corresponds to the frequency range where the major portion of the energy of vesicular sounds is concentrated.

Next, consider the model signal $\tilde{A}(t, f)$ obtained from calculations and simulating vesicular breath sound. We then compare it with the real signal $R(t)$ of vesicular sound recorded on the surface of the thorax of a healthy person at a point slightly below the right clavicle (see Fig. 6.36).

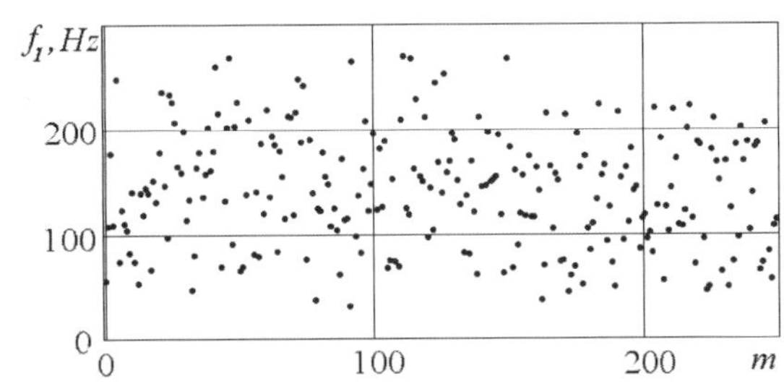

FIG. 6.35: One of the outcomes of random sampling of the Eigen frequencies of the membrane

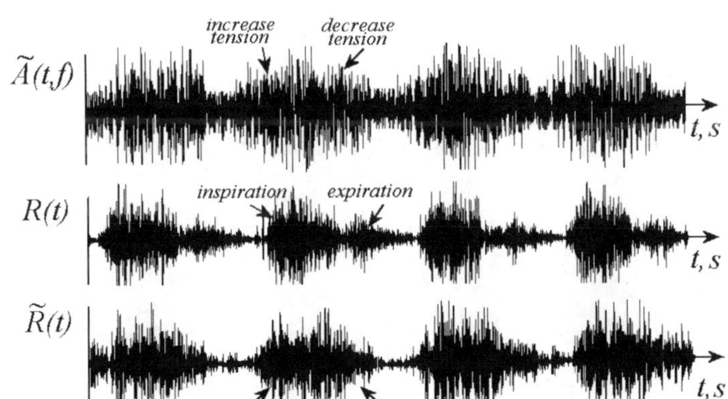

FIG. 6.36: $\tilde{A}(t, f)$—the model signal; $R(t)$—the real signal of the vesicular breath sound; $\tilde{R}(t)$—the real signal generated by periodically increasing the chest volume, in the absence of air flow in the bronchial tree (in accordance with the method proposed in Section 6.8)

As might be expected, the overall shape of the model signal does not exactly repeat the shape of the real vesicular sound signal. The model signal does not show pronounced signal separation into phases of growth and decay of the membrane tension force (corresponding to the phases of inhalation and exhalation, which are more pronounced in the real vesicular sound signal).

Now, referring to Fig. 6.37, it is interesting to compare spectra of these three signals. As can be seen, the behaviour of all three spectra is virtually identical, at least in the frequency range from about 60 Hz to 300 Hz (the model signal outside this frequency range simply does not exist, since the range of the Eigen frequencies of the membranes is limited by the adopted ranges of the physical parameters of the alveolar walls, see Fig. 6.35). This fact directly indicates that the real dispersion of geometric and mechanical parameters of the alveolar walls is much wider than the one we adopted for calculations. It is also possible that during respiration it is not only the alveolar walls that oscillate, but also the alveolar duct walls and the walls of small respiratory bronchioles.

Above, we have compared the obtained model signal and the real signal of the vesicular sound using their general enough characteristics such as waveform and spectrum. Now it is important to investigate and compare the finer structure of these signals, namely, the existence and nature of the correlated sequence of alternating members of their time series. The most expedient approach is to analyse the fractal properties of these signals. As in Section 6.8, we conduct the fractal analysis of signals based on MF-DFA (Multifractal Detrended Fluctuation Analysis) [75]. Therefore, we will not provide a detailed description of the algorithm for this method, and immediately turn to the analysis of the results.

We turn to Fig. 6.38, which shows all spectra of singularities for all considered signals. As can be seen, all singularity spectra are rather close to each other. This result leads to the conclusion that a rather representative ensemble of elementary membranes whose transverse vibrations are caused by their periodic stretching can generate multifractal noise-like signal sufficiently close to the vesicular sound.

Figure 6.38 also shows the three singularity spectra (curves 1′–3′), which were obtained after randomly mixing members of series describing the signals under consideration (curves 1–3). Mixing procedure is a standard test, which is designed to confirm the presence or absence of long-range correlations in the set of values of the time series.

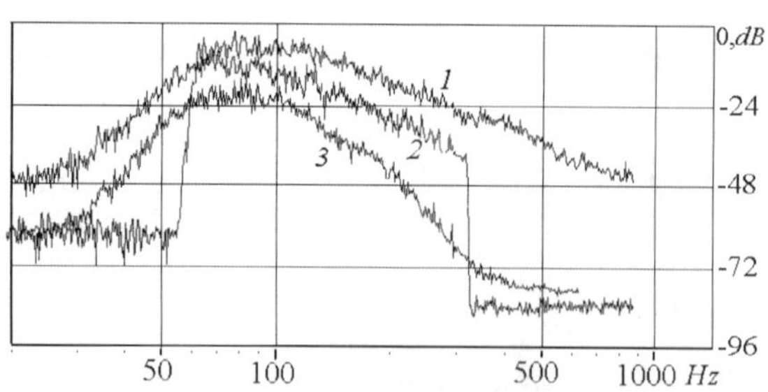

FIG. 6.37: Spectra of: 1—the vesicular breath sound signal, 2—the model signal, 3—the signal arising due to the periodic increase in the chest volume in the absence of air flow in the bronchial tree

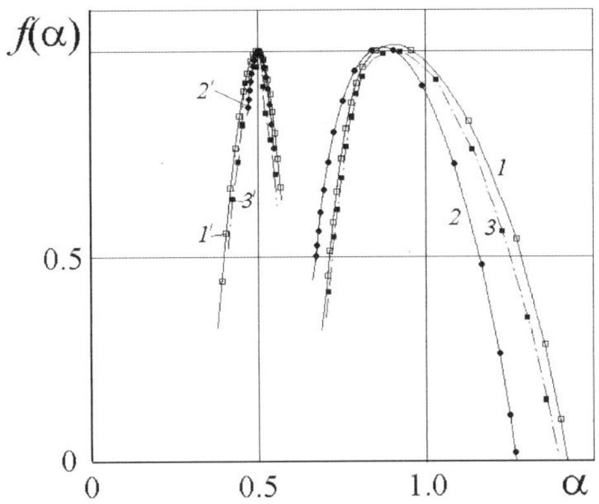

FIG. 6.38: Singularity spectra of: 1—the vesicular sound signal; 2—the model signal, 3—the signal arising due to the periodic increase in the chest volume in the absence of air flow in the bronchial tree; 1′–3′—respectively for cases where the terms of the series, describing these signals, are randomly mixed

The most expedient approach is to analyse the fractal properties of these signals. As in [1], we conduct the fractal analysis of signals based on MF-DFA (Multifractal Detrended Fluctuation Analysis) [75]. Therefore, we will not provide a detailed description of the algorithm for this method, and immediately turn to the analysis of the results.

As follows directly from Fig. 6.38, mixing led to the fact that the multifractal properties of the signal disappeared and the initial series almost turned into monofractal signals with properties similar to white noise. Recall that white noise is a uniform signal with the value of the index of singularity $\alpha = 0, 5$. This suggests significant influence of the original series of long-range correlations on the multifractal properties.

We proposed a model of the alveolar walls in the form of a membrane. It is shown that with its periodic tension transverse vibrations arise, which generate sound propagating into the environment (the parenchyma). A complex noise signal, which can be formed during simultaneous excitation of a sufficiently representative ensemble of membranes with different geometric and mechanical characteristics, was modelled based on the solution of the problem of parametric vibrations of the membrane. Analysis of such model signal showed that its shape, spectrum and fractal properties are close enough to the shape, spectrum and fractal properties of the real vesicular sound. This result allows us to conclude that that the vesicular sound mainly occurs due to mechanical vibrations of the alveolarwalls in their periodic tension during respiration.

In conclusion we note that the analysis of the acoustic properties of the respiratory system of the person, in this chapter, is based original papers [8,10,11,17,45,46,77], published by the authors at different times.

REFERENCES

1. Kovalevskiy, A.A., *Percussion and Auscultation*, Tomsk: Tomsk University Printing House, p. 170, 1961. [in Russian]

2. Gubergrits, A.Y., *Direct Examination of the Patient*, Izhevsk: Udmurtia Printing House, p. 332, 1996. [in Russian]
3. Laennec, R.T.H., De l'auscultation ou traite du diagnostic de maladies des poumons et du coeur, fonde principalement sur ce nouven moyen d'exploration, Paris: Brosson et Chaude, 1918.
4. Great Medical Encyclopedia, Moskow: Soviet Encyclopedia Printing House, vol. 2, pp. 366–368, 1975. [in Russian]
5. Chedd, G., *Sound*, Moscow: Mir, p. 205, 1975. [in Russian]
6. Gavriely, N., *Breath Sounds Methodology*, London–Tokyo: CRC Press, p. 203, 1995.
7. Vovk, I.V., Grinchenko, V.T., and Makarenkov, A.P., Acoustics of breathing and cardiac function, *Akustychnyi Visnyk*, vol. **14**, no. 1, pp. 3–9, 2011. [in Russian]
8. Vovk, I.V. and Vovk, O.I., On physical modeling of noises generated by airflow in the elements of respiratory system, *Int. J. Fluid Mech. Res.*, vol. **28**, pp. 116–134, 2001.
9. Weibel, E.R., *Morphometry of Human Lungs*, Moscow: Medicine, p. 175, 1970. [in Russian]
10. Vovk, I.V. and Vovk, O.I., Sound propagation in human bronchial tree. Part I. Theory, *Int. J. Fluid Mech. Res.*, vol. **28**, nos. 1-2, pp. 313–329, 2001.
11. Basovskiy, V.G., Vovk, I.V., and Vovk, O.I., Sound propagation in human bronchial tree. Part II. Numerical results analysis, *Int. J. Fluid Mech. Res.*, vol. **28**, no. 3, pp. 313–329, 2001.
12. Basovskiy, V.G., Vovk, I.V., and Vovk, O.I., Concerning the capability of the air flow in bronchi with stenosis to generate the tone acoustic vibrations, *Akustychnyi Visnyk*, vol. **6**, no. 1, pp. 14–23, 2003. [in Russian]
13. Vovk, I.V. and Grinchenko, V.T., *Sound Produced by Flow*, Kiev: Naukova Dumka, p. 221, 2010. [in Russian]
14. Harper, P., Kraman, S.S., Pasterkamp, H., and Wodicka, G.R., An acoustic model of respiratory tract, *IEEE Trans. Biomed. Eng.*, vol. **48**, pp. 543–550, 2001.
15. Beck, K., Rosenhouse, G., Mahagnah, M., and Chon, R.M., Measurements and theory of normal traheal breath sounds, *Ann. Biomed. Eng.*, vol. **33**, no. 10, pp. 1344–1351, 2005.
16. Wodicka, G.R., Stevens, K.N., Golub, H.L., Cravalno, E.G., and Shanon, D.C., Model of acoustic transmission in the respiratory system, *IEEE Trans. Biomed. Eng.*, vol. **36**, no. 9, pp. 925–934, 1989.
17. Vovk, I.V., Grinchenko, V.T., and Oleynik, V.N., The problems of simulation of acoustical properties of chest and breathing noise measurement, *J. Acoust.*, vol. **41**, no. 5, pp. 758–768, 1995. [in Russian]
18. Sapin, M.R. and Bilich, G.L., *Human Anatomy*, Moscow: Higher School, p. 544, 1989. [in Russian]
19. Sinelnikov, R.D., *Atlas of Human Anatomy*, vol. 2, Moscow: Medicine, p. 472, 1979. [in Russian]
20. Sorokin, V.N., *The Theory of Speech Formation*, Moscow: Radio i Svyaz, p. 312, 1985. [in Russian]
21. Borisyuk, A.O. and Grinchenko, V.T., Vibration and noise generation by elastic elements excited by a turbulent flow, *J. Sound Vib.*, vol. **204**, no. 2, pp. 213–237, 1997.

22. Pasterkamp, H., Carson, C., Daien, D., and Oh, Y., Digital respirosonography. New images of lung sounds, *Chest*, vol. **96**, no. 6, pp. 1405–1412, 1989.

23. Vovk, I.V., Dakhnov, S.L., Krizhanovskiy, V.V., and Oliynyik, V.N., The opportunities of diagnostic of pulmonary pathology by means of computer registration and handling of noise breathing, *Akustychnyi Visnyk*, vol. **1**, no. 2, pp. 3–18, 1998. [in Russian]

24. Horsfield, K., Dart, G., Olson, D.E., Filley, G.F., and Cumming, G., Models of the human bronchial tree, *J. Appl. Physiol.*, vol. **31**, no. 2, pp. 207–217, 1971.

25. Snyder, W.S., Cook, M.J., Nasset, E.S., Karhausen, L.R., Parry Howells, G., and Tipton, I.H., Report of the Task Group on Reference Man, International Commission on Radiological Protection, Pergamon Press, No. 23, p. 480, 1974.

26. Fredberg, J.J. and Hoenig, A., Mechanical response of lungs at high frequencies, *J. Biomed. Eng.*, vol. **100**, pp. 57–66, 1978.

27. Habib, R.H., Chalker, R.B., Suki, B., and Jackson, A., Airway geometry and wall mechanical properties estimated from subglottal input impedance in humans, *J. Appl. Physiol.*, vol. **77**, no. 1, pp. 441–451, 1994.

28. Cochran, J.A. and Pecina, R.G., Mode propagation in continuously curved waveguides, *Radio Sci.*, vol. **1**, no. 6, pp. 679–696, 1966.

29. Lapin, A.D., Sound propagation in non-uniform waveguides, *J. Acoust.*, vol. **13**, no. 2, pp. 231–234, 1967. [in Russian]

30. Levin, L., On the inadequacy of discrete mode-matching techniques in some waveguide discontinuity problems, *IEEE Trans. Microw. Theory Techniques*, vol. **MTT-18**, no. 7, pp. 364–372, 1970.

31. Vovk, I.V., Grinchenko, V.T., and Leyko, A.G., Analysis of acoustic properties of plane waveguide with non-uniform walls, Report of Ukrainian Academy of Science USSR, *Series A, Phys.-Mat. Techn. Sci.*, vol. **9**, pp. 795–798, 1975. [in Russian]

32. Goncharov, I.Y. and Matsypura, V.T., Sound propagation in the waveguide with the bends, *Akustychnyi Visnyk*, vol. **1**, no. 2, pp. 57–64, 1998. [in Russian]

33. Isakovich, M.A., *General Acoustics*, Moscow: Nauka, p. 495, 1973. [in Russian]

34. Bulanaya, M.A., Vovk, I.V., Grinchenko, V.T., and Matsypura, V.T., Sound signal propagation in the waveguide with inhomogeneity in the form of a chamber, *Akustychnyi Visnyk*, vol. **12**, no. 3, pp. 3–19, 2009. [in Russian]

35. Lighthill, J., *Waves in Fluids*, Cambridge, UK: Cambridge University Press, p. 504, 1978.

36. Rzhevkin, S.N., *A Course of Lectures on Sound Theory*, Moscow: MSU Printing House, p. 335, 1960. [in Russian]

37. Grinchenko, V.T., Vovk, I.V., and Matsypura, V.T., *Basic Acoustics*, Kiev: Naukova Dumka, p. 640, 2007. [in Russian]

38. Grinchenko, V.T. and Vovk, I.V., *Wave Problems of Sound Scattering by Elastic Shells*, Kiev: Naukova Dumka, p. 240, 1986. [in Russian]

39. Grinchenko, V.T., The evolution of method of solving of the problems of sound radiation and scattering in non-canonical regions, *Hydromechanics*, vol. **70**, pp. 27–40, 1996. [in Russian]

40. Ishizaka, K., Matsudaira, M., and Kaneko, T., Input acoustic-impedance measurement of the subglotal system, *J. Acoust. Soc. Am.*, vol. **60**, no. 1, pp. 190–197, 1976.

41. Skudrzyk, E., *Simple and Complex Vibratory Systems*, University Park: Penn State University Press, p. 514, 1968.

42. Nemerovskiy, L.I., Concerning acoustical vibrations propagation in the acoustic pipes in lungs, *J. Acoust.*, vol. **22b** no. 3, pp. 416–421, 1976. [in Russian]

43. Wei, Q., Hu, Y., Gelfand, G., and MacGregor, J.H., Segmentation of lung lobes in high-resolution isotropic CT images, *IEEE Trans. Biomed. Eng.*, vol. **56**, no. 5, pp. 1383–1393, 2009.

44. Nabih, A. and Said, N., Characterzation of the thorax via mechanical impedance, Tech. Suppl. Rep., Washington, DC: National Highway Traffic Safety Admin., Dept. Transportation, p. 72, Nov. 1986.

45. Vovk, I.V., Kosovets, L.I., Matsypura, V.T., and Oliynyk, V.N., Simulation of the process of sound propagation in human chest. Part 1. Theory, *Akustychnyi Visnyk*, vol. **14**, no. 2, pp. 16–25, 2011. [in Russian]

46. Vovk, I.V., Kosovets, L.I., Matsypura, V.T., and Oliynyk, V.N., Simulation of the process of sound propagation in human chest. Part II. The analysis of normal acoustical properties, *Akustychnyi Visnyk*, vol. **14**, no. 3, pp. 3–11, 2011. [in Russian]

47. Buller, A.J. and Dornhorst, A.C., The physics of some pulmonary signs, *Lancet*, vol. **29**, no. 271, pp. 649–665, 1956.

48. Korenbaum, V.I., Nuzhdenko, A.V., Tagiltsev, A.A., and Kostiv, A.E., The investigation of complex acoustical signals transmission in human respiratory system, *Akustychnyi Visnyk*, vol. **56**, no. 4, pp. 537–544, 2010. [in Russian]

49. Shenderov, E.L., *Sound Radiation and Scattering*, Leningrad: Sudostroenie, p. 301, 1989. [in Russian]

50. Korenbaum, V.I., Kulakov, Yu.V., and Tagil'tsev, A.A., Origin of vesicular respiratory noises, *Fiziol. Chel.*, vol. **23**, no. 4, pp. 133–135, 1997.

51. Gavriely, N. and Mahagna, M., Breath sounds spectra during nose and mouth breathing, *18th Int. Lung Sounds Conf.*, Lake Louse, Canada: University of Calgary, 1993.

52. Playsongsang, Y., Iyer, V.K., and Ramamoorthy, P.A., Reproducibility of the vesicular breath sounds in normal subjects, *Respiration*, vol. **58**, nos. 3-4, pp. 158–162, 1991.

53. Sahli, H., *Lehrbuch der klinuscher Untersuchungsmethoden*, Leipzig/Wien: Deuticke, 1920.

54. Forgacs P., The functional basis of pulmonary sounds, *Chest*, vol. **73**, pp. 57–86, 1978.

55. Vovk, I.V., Zalutskii, K.E., and Krasnyi, L.G., Acoustic model of the human respiratory system, *Acoust. Phys.*, vol. **40**, pp. 676–680, 1994.

56. Mussel, M.J., The need for standards in recording and analysing respiratory sounds, *Med. Biol. Eng. Comput.*, vol. **30**, no. 2, pp. 129–139, 1992.

57. Chowdhury, S.K. and Majumder, A.K., Frequency analysis of adventitious lung sounds, *J. Biomed. Eng.*, vol. **4**, pp. 305–312, 1982.

58. Pasterkamp, H., Kraman, S.S., and Wodicka, G.R., Respiratory sounds, *Am. J. Respir. Crit. Care Med.*, vol. **156**, pp. 974–987, 1997.

59. Earis, J.E. and Cheetham, B.M.G., Current methods used for computerized respiratory sound analysis, *Eur. Respir. Rev.*, vol. **10**, pp. 586–590, 2000.

60. Adamov, A.I., Vovk, I.V., Grinchenko, V.T., Vinogradnyi, G.P., Makarenkov, A.P., Maidannik, V.G., and Yatsenko, V.P., Device for Auscultation, RF Patent No. 2062047, 1992.

61. Oliynik, V.N., Rational design of piezoaccelerometers for measurements on compliant surfaces, *Int. J. Fluid Mech. Res.*, vol. **29**, no. 5, pp. 613–627, 2002.

62. Vovk, I.V. and Vovk, O.I., On physical modeling of noises generated by airflow in the elements of respiratory system, *Int. J. Fluid Mech. Res.*, vol. **28**, nos. 1-2, pp. 116–134, 2001.

63. Gavriely, N., Palti, Y., and Alroy, G., Spectral characteristics of normal breath sounds, *J. Appl. Physiol.*, vol. **50**, pp. 307–314, 1981.

64. Pastercamp, H., New developments and knowledge in research of respiratory sounds, *Pneumologie*, vol. **46**, pp. 602–611, 1992.

65. McKusick, V.A., Jenkins, J.T., and Webb, G.N., The acoustic basis of the chest examination, *Am. Rev. Tuberc.*, vol. **72**, pp. 12–34, 1955.

66. Bendat, J.S. and Piersol, A.G., *Random Data: Analysis and Measurement Procedures*, New York: Wiley, 1971.

67. Bendat, J.S. and Piersol, A.G., *Engineering Applications of Correlation and Spectral Analysis*, Canada: Wiley, 1980.

68. Chung, J.Y., Rejection of flow noise using a coherence function method, *J. Acoust. Soc. Am.*, vol. **62**, pp. 388–395, 1977.

69. Vovk, I.V. and Vovk, O.I., Sound propagation in human bronchial tree. Part I: Theory, *Int. J. Fluid Mech. Res.*, vol. **28**, no. 3, pp. 313–329, 2001.

70. Basovsky, V.G., Vovk, I.V., and Vovk, O.I., Sound propagation in human bronchial tree. Part II. Analysis of numerical results, *Int. J. Fluid Mech. Res.*, vol. **28**, pp. 330–340, 2001.

71. Zosimov, V.V. and Lyamshev, L.M., Fractals and scaling in acoustics, *Acoust. Zh.*, vol. **40**, no. 5, pp. 709–737, 1994. [In Russian]

72. Pietronero, L. and Tosatti, E., Surface theory of melting, *Solid State Commun.*, vol. **32**, no. 3, pp. 255–259, 1979.

73. West, J.B., *Pulmonary Pathophysiology: The Essentials, 7th ed.*, Lippincott Williams: Wilkins, Philadelphia, 2007.

74. Muzy, J.F., Bacry, E., and Arneodo, A., Wavelets and multifractal formalism for singular signals: Application to turbulence data, *Phys. Rev. Lett.*, vol. **57**, no. 25, pp. 3515–3518, 1991.

75. Kantelhardt, J.W., Zschiegner, S.A., Koscielny-Brude, E., Bande, A., Halvin, S., and Stanley, E., Multifractal detrended fluctuation analysis of nonstationary time series, *Physica A Stat. Mech. Appl.*, vol. **316**, pp. 87–114, 2002.

76. Feder, E., *Fractals*, Plenum: New York, 1988.

77. Vovk, I.V., Grinchenko, V.T., and Matsypura, V.T., The nature of breath sounds and their multifractal properties, *Acoust. Phys.*, vol. **59**, no. 5, pp. 590–600, 2013.

CHAPTER 7

Sound Propagation in the Irregular Waveguides

The walls indeed had ears. Or rather one ear:
a round hole with a tube—a kind of secret phone—to transfer
all that was said in the dungeon right to the room of Signore Tomato.

Gianni Rodari, "Adventures of Cipollino"

Wave propagation systems with irregularities are widely used in the acoustic radio-electronic devices (see, for example, [1–3] and bibliography in these works). We will pay special attention to the irregularities relating to the waveguide geometry change. The environment settings are unchanged. The theory is well developed for the single-mode waveguides [4]. If the ratio of waveguide dimension to wave length is rather large (for example, more than five), then the methods of geometric diffraction theory can be used. The theory of waveguides with irregularities is not so well developed for the case, when characteristic dimension is comparable with the length of wave. The reason for this is that there are some difficulties when we are constructing the rigorous solutions of the boundary problems. It should be noted, that there are a lot of waveguide phenomena in this range of the wave dimensions. These phenomena are sometimes unexpected and practically useful. The experts in electrodynamics pay great attention to such problems, using mainly the integral methods. Work [5] should be particularly noted. The statement of such boundary problems and the difficulties of their solution are considered in this work. Monograph [1] should also be noted.

A number of problems of sound propagation in the irregular waveguides are considered in chapter seven.

7.1 SOUND PROPAGATION IN A WAVEGUIDE WITH A BEND

7.1.1 The Problem Statement and the Solution Construction

Figure 7.1 shows a parallel-plate waveguide with a bend, filled with an ideal liquid with density ρ and sound velocity [6]. The dimensions of the waveguide before and after the bend are specified by values h_1 and h_2. The bend is characterized by angle θ, which can be acute, right, and obtuse angle [Figs. 7.1(a)–7.1(c)]. The complex amplitude of pressure in the waveguide satisfies Helmholtz equation

$$\frac{\partial^2 p}{\partial x^2} + \frac{\partial^2 p}{\partial y^2} + k^2 p = 0, \quad k = \frac{\omega}{c}, \tag{7.1}$$

ISBN: 978–1–56700–459–5 © 2018 by Begell House Inc.

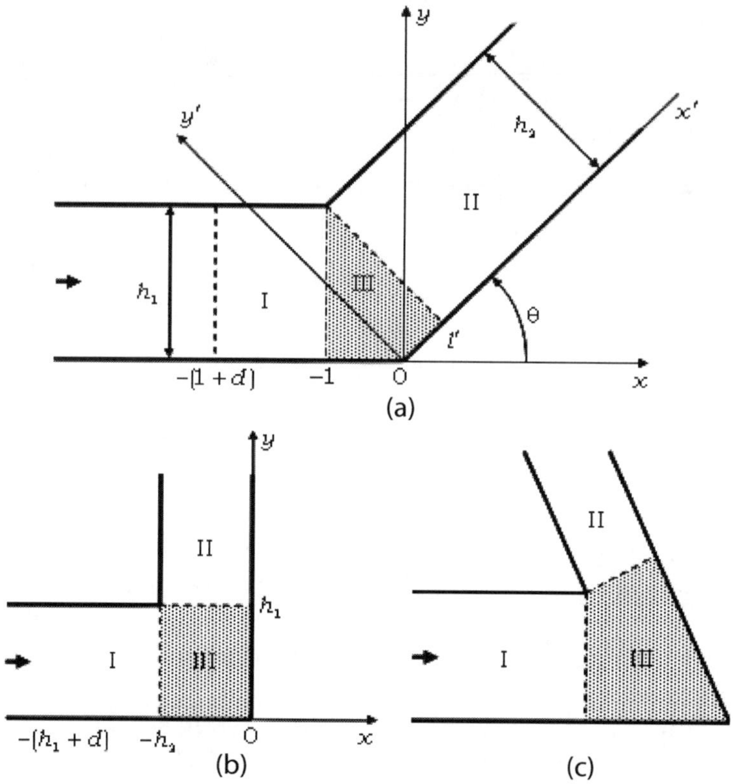

FIG. 7.1: Waveguide with a bend geometry: (a) bend angle $\theta < 90°$, (b) $\theta = 90°$, and (c) $\theta > 90°$

here ω—harmonic wave frequency. The surfaces of the waveguide are rigid. In this case, the normal derivative of pressure at these surfaces is equal to zero:

$$\left.\frac{\partial p}{\partial n}\right|_S = 0. \tag{7.2}$$

The perpendiculars are dropped from the angular point to the lower boundaries. The waveguide is divided into three domains. Two Cartesian systems of coordinates, XOY and $X'OY'$, were introduced to construct the problem solution. These coordinate systems have the common center O [see Fig. 7.1(a)].

Let one of the propagating normal waves (modes) of domain I run on the bend on the left hand side

$$p_0^{(q)}(x, y) = \cos(\alpha_q y)\exp[ik_q(x + l)], \tag{7.3}$$

where q—enumerates the normal waves running on the bend. The reflected wave and the wave propagated in area II appear in this situation. The expressions for the sound fields in domains I, II, III should be written in accordance with partial domains method.

The reflected wave is represented as superposition of normal waves in domain I. The pressure in domain I, in the coordinate system XOY, is then written as

$$p_I(x, y) = \cos(\alpha_q y) \exp[ik_q (x + l)] + \sum_{n=0}^{\infty} A_n \cos(\alpha_n y) \exp[-ik_n(x + l)], \quad (7.4)$$

and, using condition (7.2) on the acoustically rigid surfaces of domain I, we find

$$\alpha_n = \frac{n\pi}{h_1}, \quad k_n = \begin{cases} \sqrt{k^2 - a_n^2}, & k > \alpha_n, \\ i\sqrt{a_n^2 - k^2}, & k < \alpha_n. \end{cases} \quad (7.5)$$

Expression (7.4) is the general solution for Eq. (7.1) for domain I. General solution means that the form of the pressure representation does not change, no matter what are the conditions at the boundary of domains I and III.

The sound field in domain II, which determines the wave propagated through the bend, is written in a similar way, but only in the coordinate system $X'OY'$

$$p_{II}(x', y') = \sum_{n=0}^{\infty} B_n \cos(\beta_n y') \exp[i\gamma_n(x' - l')], \quad (7.6)$$

where

$$\beta_n = \frac{n\pi}{h_2}, \qquad \gamma_n = \begin{cases} \sqrt{k^2 - \beta_n^2}, & k > \beta_n, \\ i\sqrt{\beta_n^2 - k^2}, & k < \beta_n. \end{cases} \quad (7.7)$$

The field in domain III is represented in the form of the superposition of normal waves of domains I and II:

$$p_{III} = \sum_{n=0}^{\infty} C_n \cos(\alpha_n y) \exp[ik_n(x + l)] + \sum_{n=0}^{\infty} D_n \cos(\beta_n y') \exp[-i\gamma_n(x' - l')]$$
$$+ \sum_{n=0}^{\infty} E_n \cos(\xi_n x) \exp[i\eta_n y] + \sum_{n=0}^{\infty} F_n \cos(\tau_n x') \exp[i\zeta_n y'], \quad (7.8)$$

where

$$\xi_n = \frac{n\pi}{l}, \quad \eta_n = \begin{cases} \sqrt{k^2 - \xi_n^2}, & k > \xi_n, \\ i\sqrt{\xi_n^2 - k^2}, & k < \xi_n, \end{cases}$$
$$\tau_n = \frac{n\pi}{l'}, \quad \zeta_n = \begin{cases} \sqrt{k^2 - \tau_n^2}, & k > \tau_n, \\ i\sqrt{\tau_n^2 - k^2}, & k < \tau_n. \end{cases} \quad (7.9)$$

The solution for domain III is general. Expression (7.8) contains four Fourier series with arbitrary coefficients. It allows us to fulfill the matching conditions of the fields at the boundary of domains I, III and II, III and boundary conditions at the rigid surfaces in domain III. The first sum in expression (7.8), allows us to fulfill matching conditions of the sound fields at the boundary of domains I and III: $x = -l$, $y = [0, h_1]$, due to the completeness of the system of functions $\cos(\alpha_n y)$ on the interval $y = [0, h_1]$. The second sum, allows us to fulfill the matching conditions of the sound fields at the boundary of domains II and III: $x' = l'$, $y' = [0, h_2]$, due to the completeness of the system of functions $\cos(\beta_n y')$ on the interval $y' = [0, h_2]$. The third sum allows us to fulfill boundary condition at the rigid surface: $x = [-l, 0]$, $y = 0$, and the fourth—at the rigid surface $x' = [0, l']$, $y' = 0$.

The matching conditions between the named domains and boundary conditions at the rigid surfaces of domain III have the following form:

$$\left.\begin{array}{l} p_I = p_{III} \\ \dfrac{\partial p_I}{\partial x} = \dfrac{\partial p_{III}}{\partial x} \end{array}\right\} \quad x = -l, \quad y = [0, h_1],$$

$$\left.\begin{array}{l} p_{II} = p_{III} \\ \dfrac{\partial p_{II}}{\partial x'} = \dfrac{\partial p_{III}}{\partial x'} \end{array}\right\} \quad x' = l', \quad y' = [0, h_2], \tag{7.10}$$

$$\frac{\partial p_{III}}{\partial y} = 0, \quad x = [-l, 0], \quad y = 0,$$

$$\frac{\partial p_{III}}{\partial y'} = 0, \quad x' = [0, l'], \quad y' = 0.$$

These formulas show the relation between the coordinate systems XOY and $X'OY'$

$$\begin{cases} x = x'\cos\theta - y'\sin\theta, \\ y = x'\sin\theta + y'\cos\theta, \end{cases} \qquad \begin{cases} x' = x\cos\theta + y\sin\theta, \\ y' = -x\sin\theta + y\cos\theta, \end{cases}$$

If we substitute (7.4), (7.6), (7.8) in conditions (7.10), we obtain the system of functional equations. The functional system can be transformed into algebraic system, using the property of orthogonality of the system of functions $\cos(\alpha_n y)$ on the interval $y = [0, h_1]$, of the function system $\cos(\beta_n y')$ on the interval $y' = [0, h_2]$, of the function system $\cos(\xi_n x)$ on the interval $x = [-l, 0]$ and function system $\cos(\tau_n x')$ on the interval $x' = [0, l']$. As a result we obtain the infinite system of linear algebraic equations of second kind in the unknown coefficients A_n, B_n, C_n, D_n, E_n, F_n. The system is not given here and bellow in the chapter, because of awkwardness.

The obtained solution allows us to analyze the process of the wave propagation in the waveguide. Energetic characteristics of the waves coming through the bend of the waveguide are of interest. Transmission coefficient $W^{(q)}$ is defined as the ratio of the average flow of the wave power in domain II to the average flow of power of the qth normal wave in domain I:

$$W^{(q)} = \left[\int_0^{h_2} \left(p_{II}\left(\frac{1}{i\omega\rho}\frac{\partial p_{II}}{\partial x'}\right)^* + p_{II}^*\frac{1}{i\omega\rho}\frac{\partial p_{II}}{\partial x'}\right) dx'\right] \Bigg/ \left[\int_0^{h_1} \left(p_0^{(q)}\left(\frac{1}{i\omega\rho}\frac{\partial p_0^{(q)}}{\partial x}\right)^* + p_0^{(q)*}\frac{1}{i\omega\rho}\frac{\partial p_0^{(q)}}{\partial x}\right) dx\right], \tag{7.11}$$

where *—complex conjugation sign. After a set of transformations has been done, we obtain

$$W^{(q)} = \sum_{n=0}^{N_2} W_n^{(q)}, \tag{7.12}$$

where

$$W_n^{(q)} = |B_n|^2 \cdot \frac{\varepsilon_n \mathrm{Re}(\gamma_n)}{\varepsilon_q \mathrm{Re}(k_q)} \cdot \frac{h_2}{h_1}, \quad \varepsilon_0 = 1, \quad \varepsilon_n = 0.5, \quad n > 0. \tag{7.13}$$

According the expressions (7.12) and (7.13), transmission coefficient $W^{(q)}$ is represented in the form of a sum of the energetic excitation coefficients $W_n^{(q)}$ of the normal waves in domain II. In other words, we can say that expression (7.12) is the sum of the transformation coefficients of the qth normal wave in domain I into normal waves in domain II. As it follows from expression (7.13), the number of summands N_2 in Eq. (7.12) is equal to the number of the propagating waves in domain II [$\mathrm{Re}(\gamma_n) \neq 0$].

Analogously, reflection coefficient $V^{(q)}$, determined as ratio of the average flow of power in the reflected wave to the average flow of power in the incident wave, is represented as follows:

$$V^{(q)} = \sum_{n=0}^{N_1} V_n^{(q)}, \tag{7.14}$$

where

$$V_n^{(q)} = |A_n|^2 \cdot \frac{\varepsilon_n \mathrm{Re}(k_n)}{\varepsilon_q \mathrm{Re}(k_q)}. \tag{7.15}$$

The necessity to fulfill the condition $V^{(q)} + W^{(q)} = 1$ follows from the law of conservation of energy.

7.1.2 Flat Piston Radiation in a Waveguide with a Bend

The constructed solution allows us to consider the problem of the rigid piston vibration in a waveguide with a bend. Let the piston be located in domain I in the cross-section $x = -(l + d)$ [Fig. 7.1(a)]. The piston executes harmonic vibrations with frequency ω. The following condition on the surface of vibrating piston is then added to the system of functional Eq. (7.10)

$$\left.\frac{1}{i\omega\rho}\frac{\partial p_I}{\partial x}\right|_{x=-(l+d)} = \upsilon_0, \tag{7.16}$$

where υ_0—the amplitude of the piston particle velocity. The field in domain I, in this case, is written as follows:

$$\begin{aligned} p_I = &\sum_{n=0}^{\infty} A_n \cos(\alpha_n y) \exp\left[-ik_n(x + l)\right] \\ &+ \sum_{n=0}^{\infty} G_n \cos(\alpha_n y) \exp\left[ik_n(x + l + d')\right]. \end{aligned} \tag{7.17}$$

Piston radiation resistance Z should be analyzed in this problem. Let us specify the value of Z with regard to the piston radiation resistance in an infinite flat waveguide $(h_1\rho c)$:

$$\varsigma = \frac{Z}{(h_1\rho c)} = \frac{1}{(h_1\rho c)\upsilon_0}\int_0^{h_1} p_I\big|_{x=-(l+d)}\,dy, \tag{7.18}$$

where $p_I\ (x = -(l+d), y)$—pressure amplitude on the piston surface. Solving the system of Eqs. (7.10) and (7.16) and writing the Eq. (7.18), we obtain the expression for the piston radiation resistivity normalized to ρc:

$$\zeta = -\frac{ik}{\nu_0}\left[A_0 \exp\left(ikd\right) + G_0 \exp\left(-ikd\right)\right].$$

Setting the value $\upsilon_0/k = 1$, we obtain

$$\zeta = -i\left[A_0 \exp\left(ikd\right) + G_0 \exp\left(-ikd\right)\right]. \tag{7.19}$$

7.1.3 Numerical Results Analysis

The infinite system of linear algebraic equations was solved by reduction method. The number of equations in the reduced system was determined by the wave dimensions of the waveguide domains: h_1, h_2, l, l' [Fig. 7.1(a)]. The normal waves, which determine the order of equation, contain the number of all the propagating normal waves and several non-propagating waves (three or four waves).

First of all, we should check the fulfillment of matching conditions of the fields at the boundary of partial domains and the fulfillment of the boundary conditions at the rigid surfaces in domain III. These conditions are fulfilled with accuracy up to three decimal places. Some degradation of accuracy can by observed when approaching thc corner points. Such results are wholly satisfactory. The low of conservation of energy is fulfilled with accuracy no worse than 0.1%. It allows us to calculate the reflection and transmission energy coefficients.

Figure 7.2 shows frequency dependencies of transmission coefficient $W^{(0)}$ of zero mode ($q = 0$) with different bend angles (curve 1—$\theta = 5°$, 2—$\theta = 45°$, 3—$\theta = 65°$, 4—$\theta = 90°$). Characteristic dimensions of domains I and II were taken equal to $h_1 = h_2$; their wave values h_1/λ and h_2/λ were determined with regard to the length of sound wave $\lambda = 2\pi/k$. As we can see, the decrease of transmission coefficient is observed with an increase of bend angle θ.

When $h_1 = h_2 \ll \lambda$, no matter what is the bend angle, transmission coefficient $W^{(0)}$ is equal to unity. It corresponds to sound propagation in "...narrow tube. The wave propagates in such a tube, neglecting the bends, as if the tube is straight" [7, p. 168]. There are the dips in characteristics for h_1 and h_2 divisible by half-wave length $\lambda/2$. When the bend angle $\theta = 90°$, we obtain zero transmission coefficients $W^{(0)}$. When $h_1/\lambda = 0.5$ we have $W^{(0)} = 0$ for all bend angles.

We turn now to the dependencies of normal waves excitation coefficients $W_n^{(q)}$ in domain II to understand the nature of the curves represented in Fig. 7.2. Figures 7.3(a)

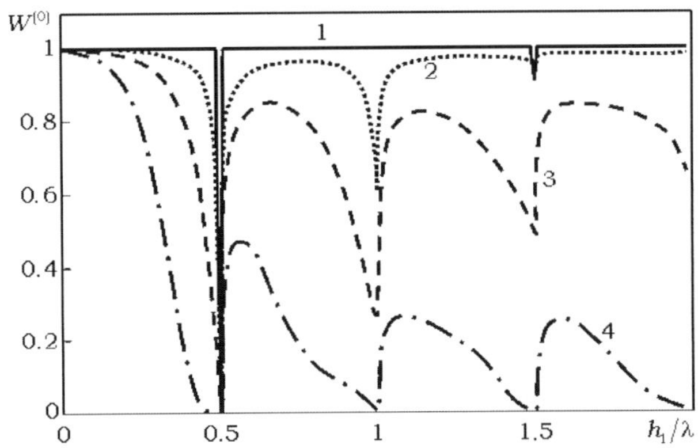

FIG. 7.2: Frequency characteristics of transmission coefficient $W^{(0)}$ for $h_1 = h_2$: 1—$\theta = 5°$, 2—$\theta = 45°$, 3—$\theta = 65°$, 4—$\theta = 90°$

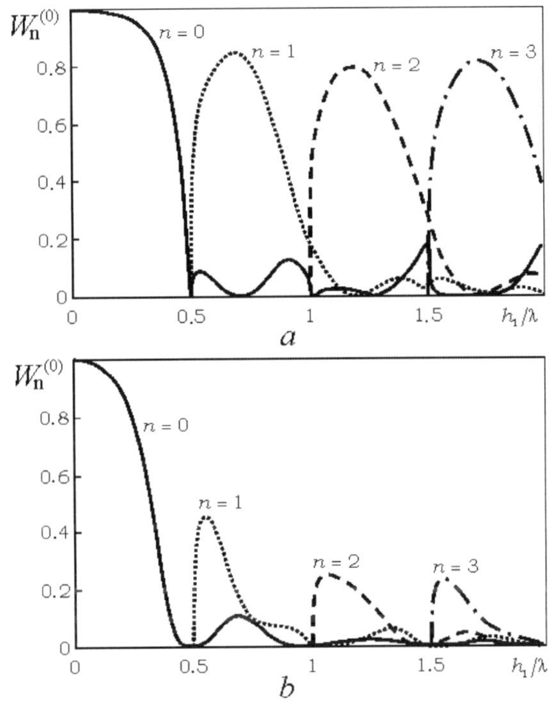

FIG. 7.3: Frequency characteristics of the normal waves excitation coefficients in domain II $W_n^{(0)}$ for $h_1 = h_2$: a – $\theta = 65°$, b—$\theta = 90°$

and 7.3(b) illustrates the frequency dependencies of the excitation coefficients of the first normal waves in domain II (the number of the curve specifies the corresponding mode) when zero normal wave ($q = 0$) is propagating in domain I.

A zero normal wave in the uniform waveguide with rigid boundaries is propagating whatever is the wave dimension of the waveguide. The normal wave with a number $n > 0$ is propagating in domain II when the condition $k > \beta_n$ is fulfilled or $h_2 > n\pi/k =$

$n\lambda/2$. In this case, the average flow of power in the given normal wave is not equal to zero. Figure 7.3(a) corresponds to the bend with an angle $\theta = 65°$, Fig. 7.3(b)—$\theta = 90°$ when $h_1 = h_2$.

The graphs show that if the wave size $h_2/\lambda < 0.5$, then only a zero normal wave is propagating in domain II (curve 0). The situation changes if $h_2/\lambda > 0.5$. It is illustrated in Fig. 7.3. If the bend angle $\theta = 65°$ [Fig. 7.3(a)] for $h_2/\lambda > 0.5$ the first normal wave is excited (curve 1). It's amplitude exceeds considerably the value of zero mode. When $h_2/\lambda > 1$, the second normal wave (curve 2) dominates and then higher modes dominate.

Thus, it is valid to say that the energy of zero normal wave in domain I transforms into normal waves in domain II with numbers $n > 0$. When we analyze the graphs in Figs. 7.3(a) and 7.3(b) for bend angles $\theta \leq 90°$, it is important to note the decreasing of excitation efficiency of the normal waves in domain II for h_2 divisible by $\lambda/2$. It leads to the decreasing of transmission coefficient of zero mode through the bend (see Fig. 7.2). If $\theta = 90°$ [Fig. 7.3(b)], the modes excitation coefficients are equal to zero in the frequency range points, where h_2 is divisible by half wave-length $\lambda/2$. Consequently, the transmission coefficient is equal to zero, i.e., there is a waveguide "blocking" effect. The same situation is observed in the waveguide when there are unequal but divisible by each other dimensions h_1 and h_2. Figure 7.4 shows frequency dependencies of excitation coefficients $W_n^{(0)}$ of the normal waves in domain II when $h_2 = h_1/2$; bend angle $\theta = 90°$. As we can see, the "blocking" of the waveguide for zero mode in domain I happens in the same way as in the case when $h_1 = h_2$, for value h_2 divisible by $\lambda/2$, and also for $h_2/\lambda = 0.25$, where $h_1/\lambda = 0.5$.

Providing that $h_1 \ll \lambda$ and $h_2 \ll \lambda$, the obtained results agree well with the theory of a single-mode waveguide. The energy transmission coefficient of zero normal wave for such a waveguide is determined as follows [4]:

$$W^{(0)} = \frac{4h_2/h_1}{(1 + h_2/h_1)^2}. \tag{7.20}$$

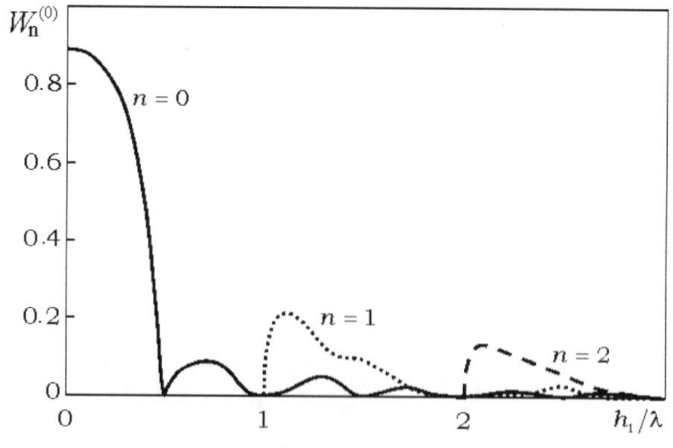

FIG. 7.4: Frequency dependencies of energy coefficients of the modes excitation in domain II $W_n^{(0)}$ when $h_2 = h_1/2$, $\theta = 90°$

When $h_2/h_1 = 0.5$, according to Eq. (7.20), we have $W^{(0)} \approx 0.89$. It corresponds to the data in Fig. 7.4.

The analysis of the given above data for the waveguide with equal (or divisible by each other) values h_1 and h_2 shows, that a decrease and "blocking" of the waveguide for the bend angle $\theta = 90°$ (Fig. 7.2) depends on the moment of the next normal wave generation in the waveguide. Let us consider, for example, the waveguide with a normal bend, ($\theta = 90°$) when $h_1 = h_2 = \lambda$, Fig. 7.3(b). The waveguide is "blocked" in this point of frequency range. Here $|A_0| = 1$, $|A_n| = 0$, for $n > 0$; $|B_0| = 0$, $|B_1| = 0$, $|B_2| = 2$, $|B_n| = 0$, for $n > 2$. The pressure distribution corresponds to the second normal wave in domain II at the surface $x = [-h_2, 0]$, $y = h_1$ [Fig. 7.1(b)]. But particle velocity is equal to zero and thus the flow of power is also equal to zero. Consequently, normal zero wave, running on the bend, is totally reflected.

The situation changes, if the waveguide typical dimensions are not divisible by half wave-length. Figure 7.5 shows the dependencies of the normal waves excitation coefficients in domain II versus h_1/λ when the dimension of the domain II is fixed $h_2/\lambda = 1.2$; $\theta = 90°$, $q = 0$. There are three propagating normal waves in domain II when we select the dimension $h_2/\lambda = 1.2$. As a consequence, the transmission coefficient of the zero mode through the bend is free of zeros. Although when h_1 is divisible by $\lambda/2$ the excitation coefficient of zero mode in domain II $W_n^{(0)}$ is equal to zero.

Figure 7.6 shows the dependencies of the excitation coefficients of the first normal waves in domain II when the second normal wave runs on the bend ($q = 2$) in domain I. Here the dimension $h_1/\lambda = 1.2$ is fixed, and the wave dimension in domain II changes h_2/λ; bend angle $\theta = 90°$. As we can see, if the value of h_2 in domain II is not sufficient to give raise to the second propagating normal wave, the second mode in domain I transforms into zero mode. Then the second mode transforms into the first mode and so on. With an increase of h_2/λ, higher propagating modes appear in domain II. It means that inverse transformation of a higher normal wave into a lower wave is possible.

Such dependency of the frequency response on transmission coefficient $W_n^{(0)}$ is not observed for the waveguides with a bend angle $\theta > 90°$, as it is in the case when bend angles $\theta = 90°$ (Fig. 7.2). For example, Fig. 7.7(a) shows the values of the zero mode transmission coefficient $W^{(0)}$ for the bend angles $\theta = 125°$ (curve 1) and $\theta = 145°$

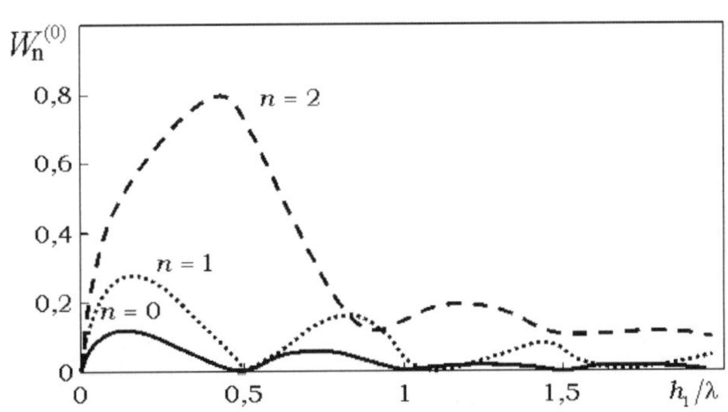

FIG. 7.5: Frequency dependencies of the excitation coefficients of the modes in domain II $W_n^{(0)}$ when $h_2/\lambda = 1.2$; $\theta = 90°$

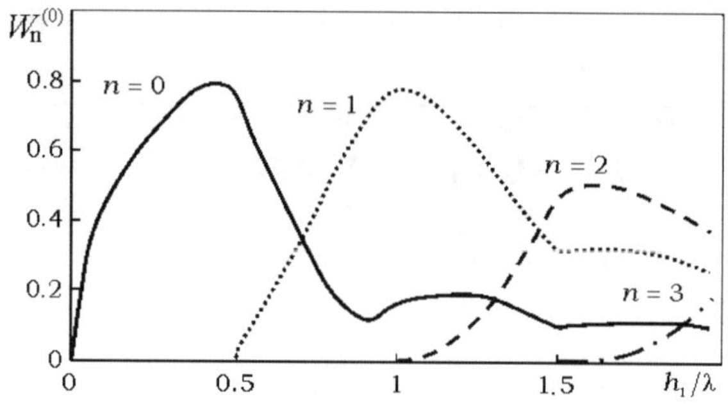

FIG. 7.6: Frequency dependencies of the excitation coefficients of the modes in domain II $W_n^{(2)}$ when $h_1/\lambda = 1.2$, $\theta = 90°$

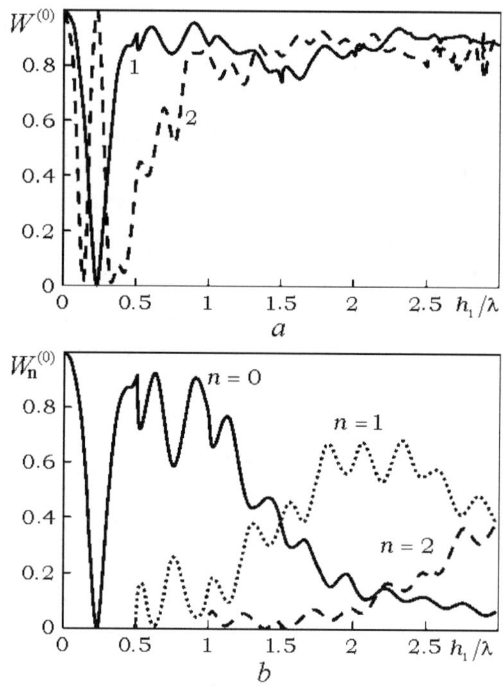

FIG. 7.7: Frequency dependencies: (a)—of transmission coefficient $W^{(0)}$ for $h_1 = h_2$: 1—$\theta = 125°$, 2—$\theta = 145°$; (b)—of excitation coefficients of the modes in domain II $W_n^{(0)}$, $\theta = 125°$

(curve 2) for $h_1 = h_2$. Figure 7.7(b) shows excitation coefficients of the normal waves $W_n^{(0)}$ in domain II for bend angle $\theta = 125°$.

Of special note are two singularities in frequency response of the transmission coefficient $W^{(0)}$. When $\theta = 125°$ for $h_1/\lambda > 0.5$ the bend is almost sound transparent. It is determined by excited modes superposition $W_n^{(0)}$. A deep dip in characteristics of $W^{(0)}$ is observed for $h_2/\lambda < 0.5$, when only zero mode is propagating in domain II.

We turn our attention now to the graphs, which characterize the waveguide when the bend angle θ changes. Figure 7.8(a) shows transmission coefficient $W^{(0)}$ dependencies

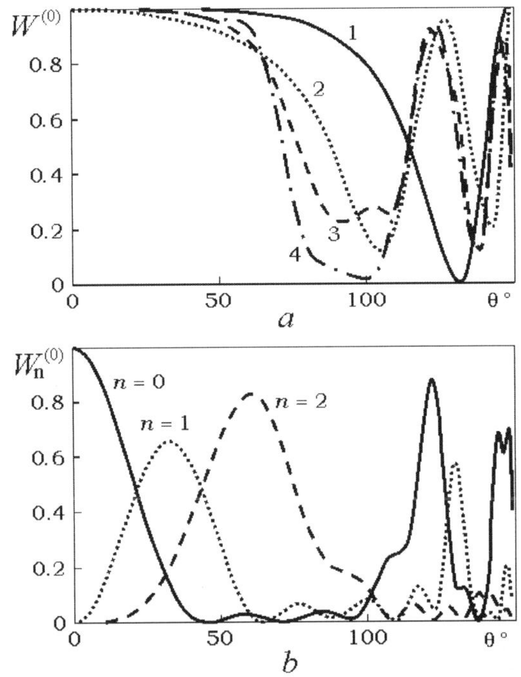

FIG. 7.8: Waveguide bend angle θ dependencies: (a)—transmission coefficient $W^{(0)}$ when $h_1 = h_2$: 1—$h_1 = 0.2\lambda$; 2—$h_1 = 0.6\lambda$; 3—$h_1 = 1.2\lambda$; 4—$h_1 = 2.4\lambda$; (b)—excitation coefficient of the normal waves in domain II $W_n^{(0)}$ when $h_1 = h_2 = 1.2\lambda$

on angle θ when the values h_1 and h_2 are fixed: curve 1—$h_1 = h_2 = 0.2\lambda$; 2—$h_1 = h_2 = 0.6\lambda$; 3—$h_1 = h_2 = 1.2\lambda$; 4—$h_1 = h_2 = 2.4\lambda$.

A zero mode runs on the bend in domain I ($q = 0$). One deep dip in the vicinity of angle $\theta \approx 133°$ is observed for a small wave dimension h_1 and h_2 (curve 1). With an increase of wave dimension h_1 and h_2 this dip moves in the domain of smaller angles and the second dip appears in the vicinity of angle $\theta = 140°$. The calculation is difficult for the angles $\theta > 150°$ because of the increase of domain III.

The nature of these dips can be explained when considering the nature of excitation of normal waves in domain II. Figure 7.8(b) shows the dependencies of excitation coefficients $W_n^{(0)}$ as a function of angle θ when $h_1 = h_2 = 1.2\lambda$. It corresponds to curve 3 in Fig. 7.8(a). Such wave dimension determines three propagating modes in domain II. As we can see, the dips on curve 3 [Fig. 7.8(a)] correspond to the domains of ineffective excitation of normal waves in domain II. Graphs in Fig. 7.8(b) allows us to determine the bend angles θ of waveguide (for specified values of h_1 and h_2). The largest transformation of one mode into the other is observed for these angles. In this case, when $\theta \approx 60°$ ($h_1 = h_2 = 1.2\lambda$) zero mode in domain I transforms into the second mode in domain II with efficiency more than 80%.

Figure 7.9 shows the graphs of transmission coefficients $W^{(q)}$ as the functions of bend angle θ when zero ($q = 0$), the first ($q = 1$) and the second ($q = 2$) modes in domain I run on the bend. Characteristic dimensions of domains are set $h_1 = h_2 = 1.2\lambda$.

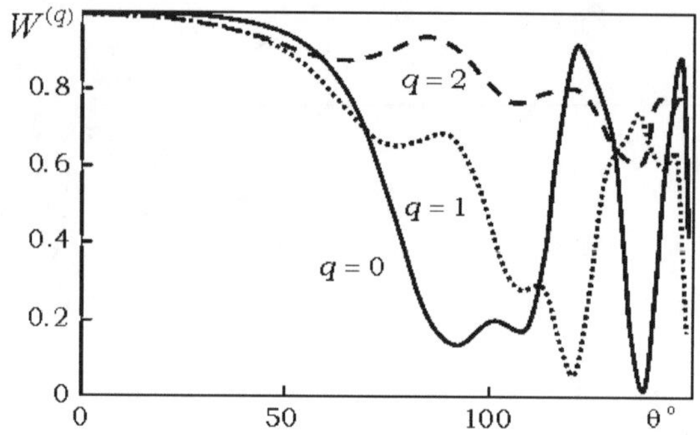

FIG. 7.9: Transmission coefficient $W^{(q)}$ dependency, $q = 0, 1, 2$ on the bend angle of the waveguide θ, $h_1 = h_2 = 1.2\lambda$

It is interesting to note that with an increase of q of the incident mode the irregularity of characteristic of $W^{(q)}$ decreases as function of bend angle θ. It is determined by the superposition of the excited modes in domain II.

We turn now to the analysis of the problem of vibrating piston [Fig. 7.1(a)], which is situated at a distance $l + d$ from the origin of coordinates O in domain I. Figure 7.10 shows frequency dependencies of real (curve 1) and imaginary (curve 2) parts of radiation resistivity (with regard to ρc) when $d = h_1 = h_2$ [Fig. 7.1(a)] for different bend angles θ: Fig. 7.10(a)—$\theta = 5°$, Fig. 7.10(b)—$\theta = 30°$, Fig. 7.10(c)—$\theta = 65°$, and Fig. 7.10(d)—$\theta = 90°$. Frequency dependencies of radiation resistance have the properties, typical for any complex vibrating system [8]. The normal bend graphs show it [Fig. 7.10(d)]. The radiation resistance is approximately equal to ρc for all the bend angles when the typical dimensions of the waveguide are less than 0.1λ, at low frequencies. It corresponds to almost complete transmission of sound energy through the bend of the waveguide (Fig. 7.2).

Here imaginary part has small positive values, i.e., it has the nature of elasticity. When $d = h_1 = h_2 \approx 0.1\lambda$, the graph of the imaginary part of radiation resistance crosses zero axis, and changes the sign, and takes on negative values. It means that Imζ has now the nature of mass. The frequency, at which Imζ crosses zero axis, is resonance frequency. With further increase of frequency, when Imζ crosses zero axis, it can be classified as antiresonance. Then we can observe the interchange of resonances and antiresonances. When $d = h_1 = h_2 = \lambda/2$ real part of radiation resistance is equal to zero for all the bend angles, and imaginary part takes negative values (for $\theta = 90°$ [Fig. 7.10(d)] we have Im$\zeta \to -\infty$). This situation corresponds to the case when there is no sound transmission through the bend of the waveguide, and it is in agreement with the graphs in Fig. 7.2.

Figure 7.11 shows frequency dependencies of real (curve 1) and imaginary (curve 2) parts of resistivity of piston radiation when $\theta = 90°$, $d = h_1$: Fig. 7.11(a) corresponds to $h_2 = h_1/2$, and Fig. 7.11(b) corresponds to $h_2 = 2h_1$. The behavior of real Reζ and imaginary Imζ parts of piston resistivity in the domain of low frequencies is of interest. As it follows from the graphs, Imζ has the nature of elasticity in the case of $h_2 = h_1/2$ [Fig. 7.11(a)], and it has the nature of mass in the case $h_2 = 2h_1$ [Fig. 7.11(b)]. Such

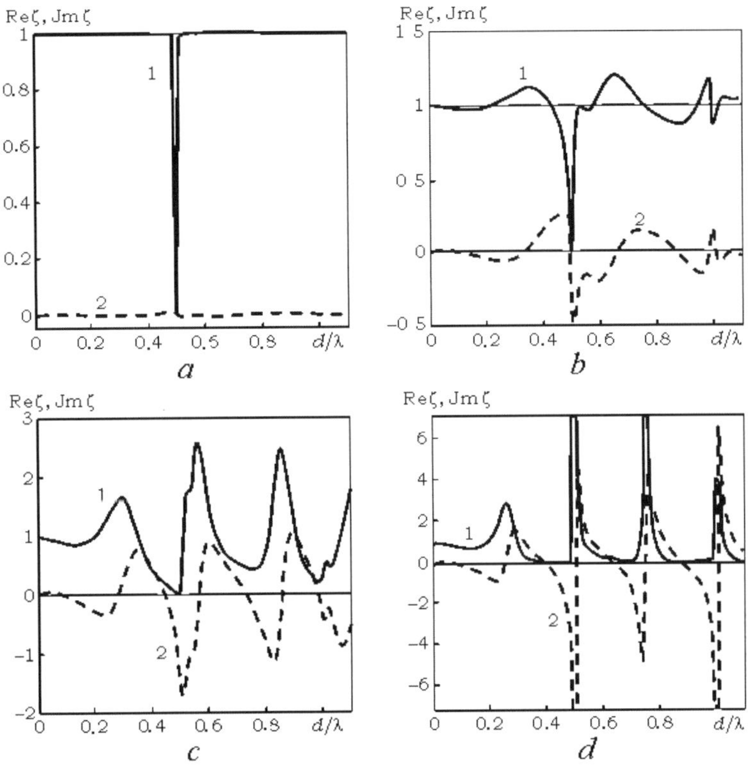

FIG. 7.10: Frequency dependencies of real (curve 1) and imaginary (curve 2) parts of piston radiation resistivity with regard to ρc when $d = h_1 = h_2$: (a)—$\theta = 5°$, (b)—$\theta = 30°$, (c)—$\theta = 65°$, (d)—$\theta = 90°$

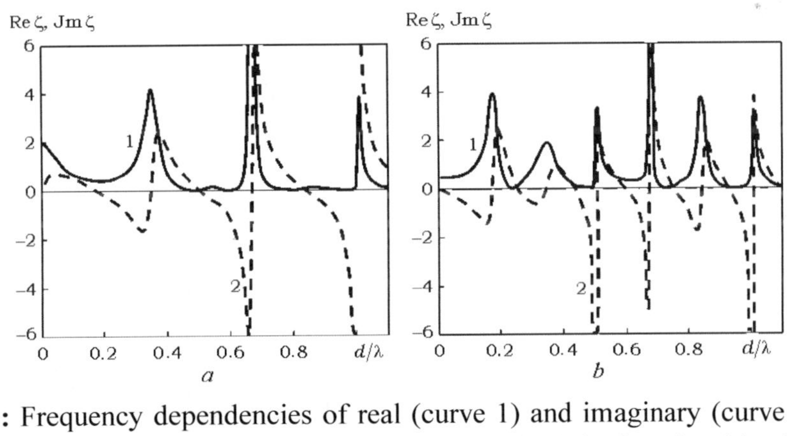

FIG. 7.11: Frequency dependencies of real (curve 1) and imaginary (curve 2) parts of piston radiation resistivity with regard to ρc when $\theta = 90°$, $d = h_1$: (a)—$h_2 = h_1/2$, (b)—$h_2 = 2h_1$

nature of the imaginary part of radiation resistance corresponds to the character of the phenomena, which occur in the waveguide with a jump of cross section in the domain of low frequencies.

Real part of radiation resistivity when $d/\lambda \ll 1$ tends to the value: in the case when $h_2 = h_1/2$, Re$\zeta = 2$, and when $h_2 = 2h_1$, Re$\zeta = 0.5$. These results are in agreement with a single-mode waveguide theory. The piston radiation resistivity for such a waveguide with the cross section jumping with regard to ρc is determined by ratio [4]: $\zeta = h_2/h_1$.

We can observe the interchange of resonances and antiresonances with further increase of frequency. They are spaced more closely at the frequency axis for the case when $h_2 = 2h_1$ [Fig. 7.11(b)], because the range of variation in domain II is larger than in the case when $h_2 = h_1/2$ [Fig. 7.11(a)].

7.2 SOUND PROPAGATION IN A BRANCHING WAVEGUIDE

7.2.1 The Problem Statement and the Solution Construction

Figure 7.12 shows parallel-plate waveguide, having a vertical branch [9]. The angle between the branch and the horizontal waveguide is 90°. The waveguide is filled with ideal liquid with density ρ and sound velocity c. Characteristic dimensions of the waveguide and the branch are specified by values h_1 and h_2, respectively. The dimension of the waveguide on the left hand and on the right hand of the branch remains constant. The boundary surfaces of the waveguide are perfectly rigid.

As it is shown in Fig. 7.12, Cartesian coordinate system XOY is introduced. The dashed lines partition the whole domain of the waveguide with the branch into the following four domains: domain I—$x = [-\infty, 0]$, $y = [0, h_1]$; domain II—$x = [h_2, \infty]$, $y = [0, h_1]$; domain III—$x = [0, h_2]$, $y = [h_1, \infty]$; domain IV—$x = [0, h_2]$, $y = [0, h_1]$.

Let one of the propagating normal waves of domain I be incident on the branch on the left hand side:

$$p_0^{(q)}(x, y) = \cos(\alpha_q y) \exp(ik_q x). \tag{7.21}$$

As a result, we obtain reflected wave and the waves propagated into domains II and III (Fig. 7.12). Field in domain I is represented as follows:

$$p_I(x, y) = \cos(\alpha_q y) \exp(ik_q x) + \sum_{n=0}^{\infty} A_n \cos(\alpha_n y) \exp(-ik_n x), \tag{7.22}$$

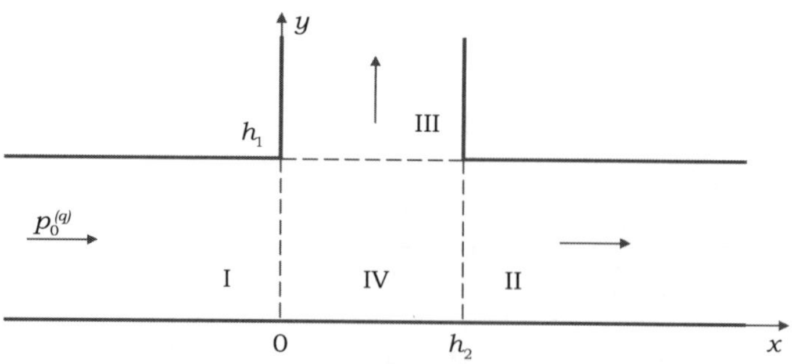

FIG. 7.12: Geometry of the waveguide with a branch

where α_n and k_n are determined from the formulas

$$\alpha_n = \frac{n\pi}{h_1}, \quad k_n = \begin{cases} \sqrt{k^2 - a_n^2}, & k > \alpha_n, \\ i\sqrt{a_n^2 - k^2}, & k < \alpha_n. \end{cases}$$

Field in domain II

$$p_{II}(x, y) = \sum_{n=0}^{\infty} B_n \cos(\alpha_n y) \exp\left[ik_n(x - h_2)\right], \tag{7.23}$$

in domain III

$$p_{III}(x, y) = \sum_{n=0}^{\infty} C_n \cos(\beta_n x) \exp\left[i\gamma_n(y - h_1)\right], \tag{7.24}$$

where β_n and γ_n are determined from the formulas

$$\beta_n = \frac{n\pi}{h_2}, \quad \gamma_n = \begin{cases} \sqrt{k^2 - \beta_n^2}, & k > \beta_n, \\ i\sqrt{\beta_n^2 - k^2}, & k < \beta_n. \end{cases}$$

Field in domain IV is represented as follows:

$$\begin{aligned} p_{IV}(x, y) = &\sum_{n=0}^{\infty} \left[D_n f_1^{(n)}(x) + E_n f_2^{(n)}(x)\right] \cos(\alpha_n y) \\ &+ \sum_{n=0}^{\infty} \left[F_n g_1^{(n)}(y) + H_n g_2^{(n)}(y)\right] \cos(\beta_n x), \end{aligned} \tag{7.25}$$

where

$$f_1^{(n)}(x) = \begin{cases} \cos\left(\sqrt{k^2 - \alpha_n^2} \cdot x\right), & k > \alpha_n, \\ \operatorname{ch}\left(\sqrt{\alpha_n^2 - k^2} \cdot x\right), & k < \alpha_n, \end{cases}$$

$$f_2^{(n)}(x) = \begin{cases} \sin\left(\sqrt{k^2 - \alpha_n^2} \cdot x\right), & k > \alpha_n, \\ \operatorname{sh}\left(\sqrt{\alpha_n^2 - k^2} \cdot x\right), & k < \alpha_n, \end{cases}$$

$$g_1^{(n)}(y) = \begin{cases} \cos\left(\sqrt{k^2 - \beta_n^2} \cdot y\right), & k > \beta_n, \\ \operatorname{ch}\left(\sqrt{\beta_n^2 - k^2} \cdot y\right), & k < \beta_n, \end{cases}$$

$$g_2^{(n)}(y) = \begin{cases} \sin\left(\sqrt{k^2 - \beta_n^2} \cdot y\right), & k > \beta_n, \\ \operatorname{sh}\left(\sqrt{\beta_n^2 - k^2} \cdot y\right), & k < \beta_n. \end{cases}$$

Presentation of sound field (7.25) is the general solution of Helmholtz equation for domain IV. Expression (7.25) contains four Fourier series with arbitrary coefficients. It allows us to fulfill the matching conditions of the fields at the boundaries of domains I, IV; II, IV; III, IV and the boundary condition on the rigid surface of domain IV.

Following the condition at the acoustically rigid surface in domain IV, namely

$$\frac{\partial p_{IV}}{\partial y} = 0 \ \text{ for } \ y = 0, \ \ x = [0, h_2],$$

we obtain the coefficients $H_n = 0$, $n = 0, 1, 2, \ldots$. Then the pressure in domain IV is as follows:

$$p_{IV}(x, y) = \sum_{n=0}^{\infty} \left[D_n f_1^{(n)}(x) + E_n f_2^{(n)}(x) \right] \cos(\alpha_n y) + \sum_{n=0}^{\infty} F_n g_1^{(n)}(y) \cos(\beta_n x). \quad (7.26)$$

Let us write the matching conditions of the sound fields at the boundaries of the named domains:

$$\left.\begin{aligned} p_I &= p_{IV} \\ \frac{\partial p_I}{\partial x} &= \frac{\partial p_{IV}}{\partial x} \end{aligned}\right\} \quad x = 0, \qquad y = [0, h_1],$$

$$\left.\begin{aligned} p_{II} &= p_{IV} \\ \frac{\partial p_{II}}{\partial x} &= \frac{\partial p_{IV}}{\partial x} \end{aligned}\right\} \quad x = h_2, \qquad y = [0, h_1], \qquad (7.27)$$

$$\left.\begin{aligned} p_{III} &= p_{IV} \\ \frac{\partial p_{III}}{\partial y} &= \frac{\partial p_{IV}}{\partial y} \end{aligned}\right\} \quad x = [0, h_2], \quad y = h_1.$$

Substituting the expressions (7.22)–(7.24), (7.26) into conditions (7.27) we obtain the system of functional equations. Using the property of orthogonality of the function system $\cos(\alpha_n y)$ at the interval $y = [0, h_1]$ and the function system $\cos(\beta_n x)$ at the interval $x = [0, h_2]$ we can perform the conversion of the system of functions into algebraic system. As a consequence, we obtain infinite system of linear algebraic equations of the second kind in unknown coefficients A_n, B_n, C_n, D_n, E_n, F_n.

The energetic characteristics of sound transmission through the branch are of interest, as it was in the problem of waveguide with a bend. The energetic coefficients of sound transmission in domain II and III must be determined. By analogy with (7.12), the sound transmission coefficient in domain II is as follows:

$$W^{(q)} = \sum_{n=0}^{N_2} W_n^{(q)}, \qquad (7.28)$$

where the excitation coefficients of normal waves in domain II

$$W_n^{(q)} = |B_n|^2 \frac{\varepsilon_n \mathrm{Re}(k_n)}{\varepsilon_q \mathrm{Re}(k_q)}. \qquad (7.29)$$

Sound transmission coefficient in the branch (domain III) is written as follows:

$$W_y^{(q)} = \sum_{n=0}^{N_3} W_{yn}^{(q)}, \qquad (7.30)$$

where the excitation coefficients of the normal waves in domain III

$$W_{yn}^{(q)} = |C_n|^2 \frac{\varepsilon_n \mathrm{Re}\,(\gamma_n)}{\varepsilon_q \mathrm{Re}\,(k_q)} \cdot \frac{h_2}{h_1}. \tag{7.31}$$

The branch $V^{(q)}$ reflection coefficient is determined by expression (7.14). The values N_2 and N_3 in Eqs. (7.28) and (7.30) determine the number of propagating modes in corresponding areas. According to the low of conservation of energy

$$W^{(q)} + W_y^{(q)} + V^{(q)} = 1. \tag{7.32}$$

7.2.2 Numerical Results Analysis

The solution of infinite system of linear equations and the control of solution convergence is performed in a similar way as in the problem of the waveguide with a bend. The law of conservation of energy (ratio (7.32)) is fulfilled with accuracy not less than 0.1%.

Figure 7.13 shows the frequency dependencies of the transmission coefficients of zero normal wave ($q = 0$) in domain II of waveguide (curve 1) and in the branch (curve 2) for different ratios h_1 and h_2.

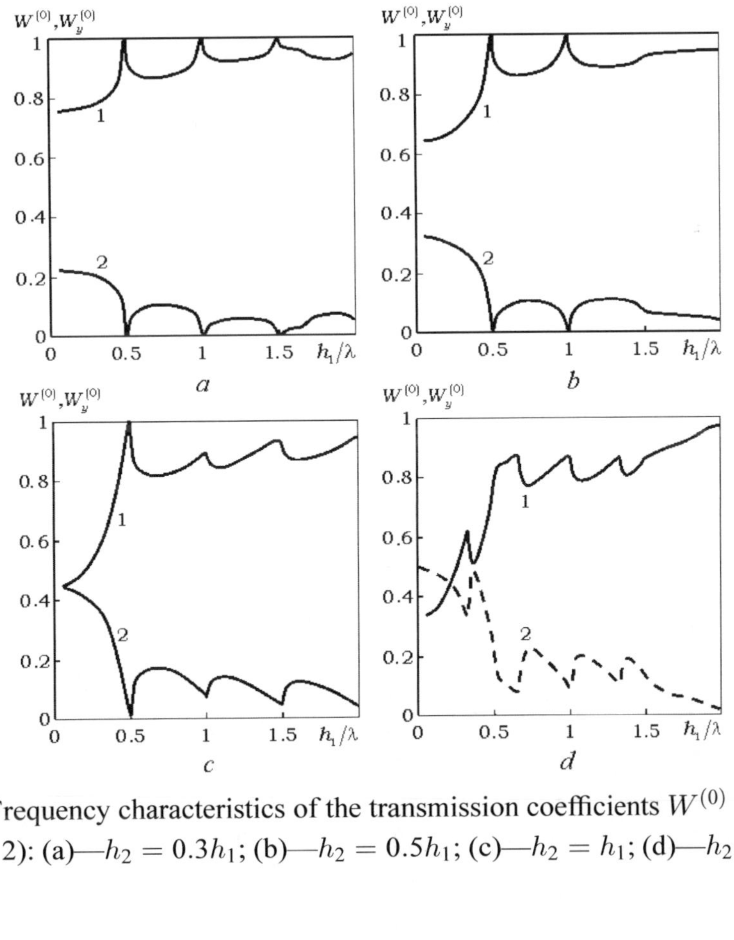

FIG. 7.13: Frequency characteristics of the transmission coefficients $W^{(0)}$ (curve 1) and $W_y^{(0)}$ (curve 2): (a)—$h_2 = 0.3h_1$; (b)—$h_2 = 0.5h_1$; (c)—$h_2 = h_1$; (d)—$h_2 = 1.5h_1$

First of all, let us verify that values $W^{(0)}$ and $W_y^{(0)}$ for $h_1 \ll \lambda$ and $h_2 \ll \lambda$ correspond to the data of the single-mode waveguide theory. Following work [4], we can write the following formulas for the energy transmission coefficients of plane wave (zero mode) through the branch:

$$W^{(0)} = \left(\frac{2}{2 + h_2/h_1}\right)^2, \quad W_y^{(0)} = \left(\frac{2}{2 + h_2/h_1}\right)^2 \cdot \frac{h_2}{h_1}. \tag{7.33}$$

Performing the calculations by Eq. (7.33), we obtain the following results: when $h_2/h_1 = 0.3$ we have $W^{(0)} \approx 0.756$, $W_y^{(0)} \approx 0.227$; when $h_2/h_1 = 1$ we have $W^{(0)} = W_y^{(0)} \approx 0.44$. If we compare the following data with the corresponding graphs in Figs. 7.13(a) and 7.13(c), we can see good agreement between the results.

The analysis of the graphs in Fig. 7.13 allows us to draw a conclusion, that the coefficient $W^{(0)}$ increases and $W_y^{(0)}$ decreases with a rise in frequency of zero mode of domain I. Zero mode does not go into the side branch when $h_2/\lambda < 0.5$, and h_1 is divisible by half-wave length λ. It goes into domain II without reflection. The similar phenomena were observed when the waveguide with a bend was investigated. The situation is different, if $h_2/\lambda > 0.5$. It is related to the fact that normal waves with numbers $n > 0$ appear. Figure 7.14 illustrates the mode structure of the field in domain II [Fig. 7.14(a)] and the side branch [Fig. 7.14(b)] for $h_1 = h_2$. As it follows from the graphs, if zero mode ($n = 0$) is dominant in domain II, the first mode is dominant in the branch when $h_2/\lambda > 0.5$, and the second mode is dominant in the side branch when $h_2/\lambda > 1$. Thus, we can observe the transformation of zero mode energy of domain I into the higher modes of domain III. The mode structure of domain II is determined by zero normal wave.

We suppose that the first mode ($q = 1$) is incident in domain I. Figure 7.15 shows the frequency dependencies of coefficients $W^{(1)}$ (curve 1) and $W_y^{(1)}$ (curve 2) for $h_1 = h_2$. Figures 7.16(a) and 7.16(b) shows the mode structure of the field in domains II and III. It is interesting to note that, when the dimensions of the waveguide h_1 and h_2 are less than 0.9λ, the most part of energy penetrates into the branch ($W_y^{(1)} > W^{(1)}$). With an increase

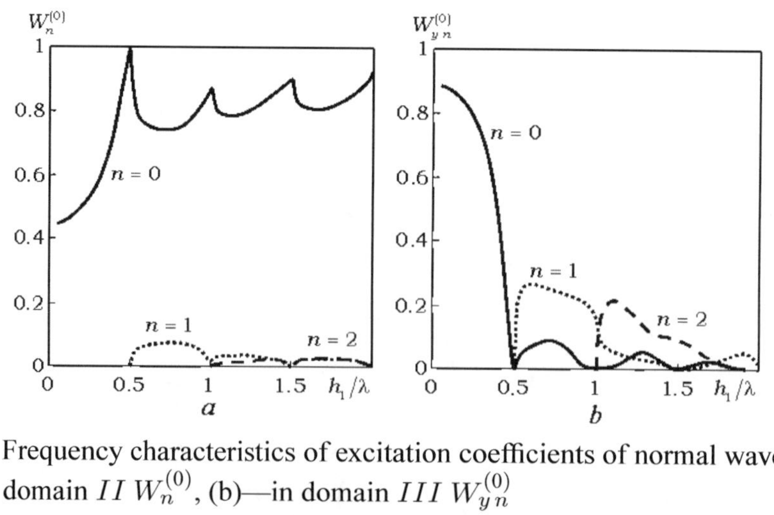

FIG. 7.14: Frequency characteristics of excitation coefficients of normal waves for $h_1 = h_2$: (a)—in domain II $W_n^{(0)}$, (b)—in domain III $W_{yn}^{(0)}$

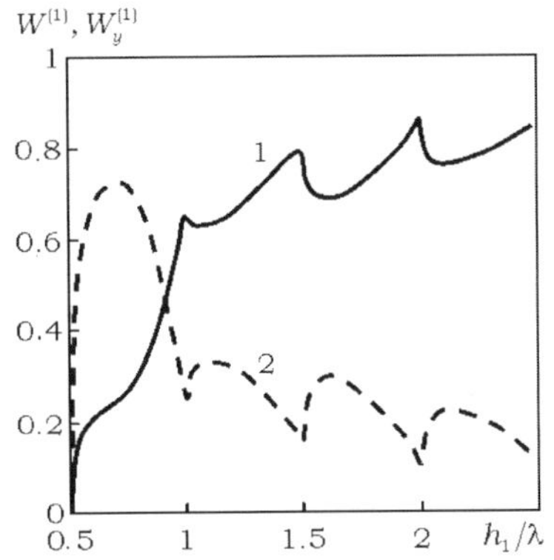

FIG. 7.15: Frequency characteristics of the transmission coefficients $W^{(1)}$ (curve 1) and $W_y^{(1)}$ (curve 2) when $h_1 = h_2$

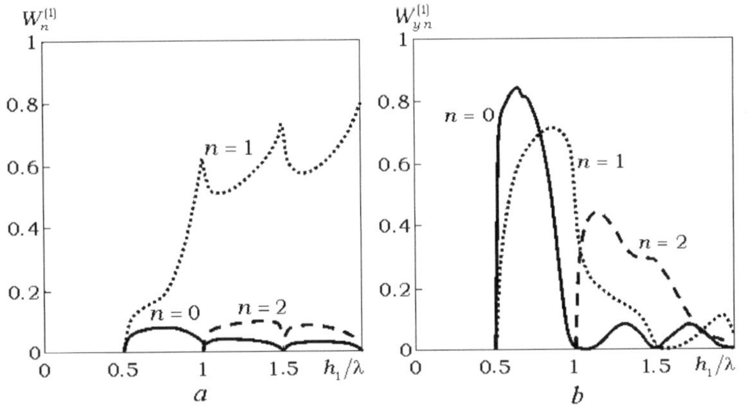

FIG. 7.16: Frequency characteristics of excitation coefficients of normal waves for $h_1 = h_2$: (a)—in domain II $W_n^{(1)}$, (b)—in domain III $W_{y\,n}^{(1)}$

of frequency, we can observe an increase of $W^{(1)}$coefficient and a decrease of $W_y^{(1)}$ coefficient. These curves have the rises and dips in the vicinity of "critical" frequencies. These frequencies correspond to the generation of the next propagating mode in the waveguide. The next propagating mode in domain III dominates with an increase of wave dimension of the branch, on a certain frequency interval. Mode structure of domain II is determined by the first normal wave.

7.3 SOUND TRANSMISSION THROUGH THE AREA OF FLAT AND WEDGE SHAPED WAVEGUIDES MATCHING

The problem of the waves passing through the matching area of the flat and wedge shaped waveguides is considered in this paragraph (Fig. 7.17). This problem attracted a lot of attention of the experts in acoustics and electrodynamics. Most of the works are concerned with a particular case: radiation from the waveguide with a flat flange [Fig. 7.17(b)]. The analysis of this problem, for the case of zero mode propagation in a flat waveguide, was

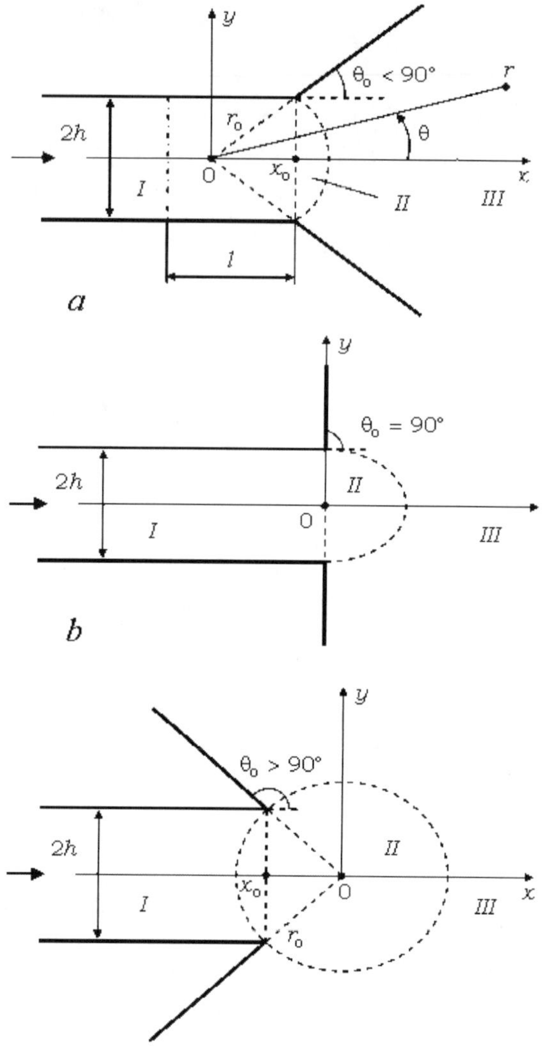

FIG. 7.17: Waveguide structure geometry, formed by the matching of parallel-plate waveguide and wedge shaped waveguide

carried out in works [10,11]. The problem of sound field excitation in a waveguide with a rigid flange when sound is incident from a half-space was considered in work [12]. The plane wave radiation from a waveguide with a flare [Fig. 7.17(a)], was considered in work [13]. An approximate solution, which yields an infinite system of the linear algebraic equations of the first kind, is proposed in this work. The system of equations analysis allows us to obtain the numerical results for small wave dimensions of the waveguide and limited flare angle of the flare. Work [5] is concerned with the problems, occurring when solving the problems of the waves propagation in the nonuniform waveguides [Fig. 7.17(a)]. The author gives attention to the necessity of the correct representation of the wave field characteristics at the different sections of nonuniform waveguide. There are the examples of the cases when the representation is not correct and, consequently, it is not possible to obtain reliable quantitative estimates.

A strict solution of the problem of sound transmission through the matching area of flat and wedge shaped waveguides is constructed bellow. The flare angle of the wedge shaped waveguide is arbitrary [14]. Solution analysis allows us to estimate the energy of the wave getting through the matching area and investigate the sound field structure. The problem of piston radiating in the waveguide with a flare is also slowed.

7.3.1 Problem Statement and Solution Construction

Let us consider the problem of sound propagation in the waveguide shown in Fig. 7.17. The flat semi-infinite waveguide is matching with the wedge shaped waveguide with flare angle $2\theta_0$. The dimension of the flat waveguide is $2h$. As it is shown in Fig. 7.17, the flare angle of the wedge shaped waveguide θ_0 can be acute, right or obtuse.

The waveguide structure is filled with an ideal liquid with density ρ and sound velocity c. The harmonic field of sound pressure in the waveguide satisfies Helmholtz equation (7.1). The surfaces of the waveguide are rigid. The normal derivative on these surfaces is equal to zero, i.e., condition (7.2) is fulfilled.

As it is shown in Fig. 7.17(a) Cartesian coordinate system (x, y) and polar coordinate system (r, θ) are introduced with common coordinate origin O. All the sound field existence domain is partitioned into three domains, which are shown in Fig. 7.17: domain I—plane semi-infinite waveguide $x \leq x_0$, $|y| \leq h$; domain III—wedge-shaped waveguide $r \geq r_0$, $|\theta| \leq \theta_0$; domain II is represented as a part of the circle of radius r_0, separating domains I and III.

Let one of the normal modes in domain I run on the matching area of the plane and wedge shaped waveguides. The reflected and penetrated into domain III waves appear as a consequence of interaction between the mode and the matching area. Following the method of partial domains, we write the expressions for the sound fields in domains I, II, III. For domain I, the expression includes an incident wave and the totality of propagating and non-uniform reflected waves:

$$p_I(x, y) = \cos(\alpha_q y)\exp(ik_q(x - x_0)) + \sum_{n=0}^{\infty} A_n \cos(\alpha_n y)\exp(-ik_n(x - x_0)), \quad (7.34)$$

where q—enumerates the normal wave, which is incident on the matching area of the waveguides; value α_n is determined according condition (7.2) on the rigid surfaces of the waveguide, namely

$$\alpha_n = \frac{n\pi}{h}, \quad k_n = \begin{cases} \sqrt{k^2 - a_n^2}, & k > \alpha_n, \\ i\sqrt{a_n^2 - k^2}, & k < \alpha_n. \end{cases} \quad (7.35)$$

The set of arbitrary coefficients A_n allows us to fulfill the matching conditions at the boundary of domain II. The wave propagated in domain III is represented as the superposition of normal waves of the wedge shaped waveguide. This expression can be represented in polar coordinates:

$$p_{III}(r,\theta) = \sum_{n=0}^{\infty} B_n \cos(\beta_n \theta) \frac{H^{(I)}_{\beta_n}(kr)}{H'^{(I)}_{\beta_n}(kr_0)}. \qquad (7.36)$$

Here $\beta_n = n\pi/\theta_0$, $H^{(I)}_{\beta_n}(kr)$—Hankel function of the first kind. The angle function choice in solution (7.36) provides for the fulfillment of boundary conditions on the sides of the wedge shaped waveguide. The radial functions are selected according to the radiation conditions. The sequence of arbitrary coefficients B_n must be determined from the matching conditions with domain II. The specific solutions of Helmholtz equation should be used in Cartesian and polar coordinate systems to construct the general solution of boundary problem in domain II [15]. The general solution differs for the case of flare angle of the wedge shaped waveguide $\theta_0 \leq 90°$ [Figs. 7.17(a) and 7.17(b)] and for $\theta_0 > 90°$ [Fig. 7.17(c)].

7.3.2 Solution Construction for Flare Angle $\theta_0 \leq 90°$

The system of functions $\cos(\alpha_n y)$, $n = 0, 1, 2, ...$ is complete and orthogonal at the boundary between domains I and II for $x \leq x_0$, $|y| \leq h$ [Figs. 7.17(a) and 7.17(b)]. The system of functions $\cos(\beta_n \theta)$, $n = 0, 1, 2, ...$ has the same properties at the boundary between domains II and III for $r \geq r_0$, $|\theta| \leq \theta_0$. Consequently, pressure in domain II can be represented as follows:

$$p_{II} = p^{(1)}_{II} + p^{(2)}_{II}, \qquad (7.37)$$

where

$$p^{(1)}_{II}(x,y) = \sum_{n=0}^{\infty} D_n \cos(\alpha_n y) \exp(ik_n(x - x_0)), \qquad (7.38)$$

$$p^{(2)}_{II}(r,\theta) = \sum_{n=0}^{\infty} C_n \cos(\beta_n \theta) \frac{J_{\beta_n}(kr)}{J_{\beta_n}(kr_0)}, \qquad (7.39)$$

where $J_{\beta_n}(kr)$—Bessel function of the first kind.

According to Figs. 7.17(a) and 7.17(b), matching conditions between the mentioned domains are as follows:

$$\left.\begin{aligned} p_I &= p_{II} \\ \frac{\partial p_I}{\partial x} &= \frac{\partial p_{II}}{\partial x} \end{aligned}\right\} \quad x = x_0, \quad |y| \leq h,$$
$$\left.\begin{aligned} p_{II} &= p_{III} \\ \frac{\partial p_{II}}{\partial r} &= \frac{\partial p_{III}}{\partial r} \end{aligned}\right\} \quad r = r_0, \quad |\theta| \leq \theta_0. \qquad (7.40)$$

We substitute solutions (7.34), (7.36), (7.37) into condition (7.40), including formulas determining relation between Cartesian and polar coordinate systems. And we obtain the system of functional equations. We transform the functional system into an algebraic system using the property of orthogonality of the function system $\cos(\alpha_n y)$ for the first two equations of system (7.40) and system of functions $\cos(\beta_n \theta)$ for the third and fourth equations of system (7.40). We obtain the infinite system of linear algebraic equations of the second kind in unknown coefficients A_n, B_n, C_n, D_n.

7.3.3 Solution Construction for Flare Angle $\theta_0 > 90°$

The coordinate origin (point O), is inside domain II, according to Fig. 7.17(c). Consequently, the angle functions, which determine the sound field in domain II should be periodical, with a period equal 2π. It corresponds to the round-trip around point O, when the distance r is fixed. The system of functions $\cos(\beta_n\theta)$, $n = 0, 1, 2, \ldots$ in expression (7.39) for solution $p_{II}^{(2)}$ does not satisfy the periodicity condition with a period 2π. It should be replaced with the system of functions $\cos(n\theta)$, $n = 0, 1, 2, \ldots$, which has the required periodicity and it is complete and orthogonal at the interval $0 \le \theta \le 2\pi$. Thus, the field in domain II, is determined by relation $p_{II} = p_{II}^{(1)} + p_{II}^{(2)}$. Here $p_{II}^{(1)}$ is determined by expression (7.38). We should write a new expression for $p_{II}^{(2)}$:

$$p_{II}^{(2)}(r,\theta) = \sum_{n=0}^{\infty} C_n \cos(n\theta)\frac{J_n(kr)}{J_n(kr_0)}. \tag{7.41}$$

But it is not possible to use the completeness and orthogonality of the function system $\cos(n\theta)$, $n = 0, 1, 2, \ldots$ because of the configuration of domain II boundary [Fig. 7.17(c)]. This difficulty can be overcome, if we complete the boundary to a circle with radius r_0 for solution $p_{II}^{(2)}$. It is shown in Fig. 7.17(c) as dashed line in domain I: $r = r_0$, $|\theta| \ge \theta_0$. And we obtain the possibility to use the completeness and orthogonality of the function system $\cos(n\theta)$ when meeting boundary condition. But the difficulty is that the conditions are not given on the additional sector of the boundary. The key moment is that the arbitrary value of particle velocity or pressure can be given on the additional sector of the boundary. Let it be determined by some function $f(r_0, \theta)$.

Two variants of the conditions of the sound fields matching at the boundaries of the domains appear for flare angles $\theta_0 > 90°$ [Fig. 7.17(c)]. If the pressure is given at the additional sector, we have the first variant:

$$\begin{aligned}
&p_I = p_{II}, && x = x_0, \quad |y| \le h,\\
&\frac{\partial p_I}{\partial x} = \frac{\partial p_{II}}{\partial x}, && x = x_0, \quad |y| \le h,\\
&\frac{\partial p_{II}}{\partial r} = \frac{\partial p_{III}}{\partial r}, && r = r_0, \quad |\theta| \le \theta_0,\\
&\begin{cases} p_{II} = p_{III}, \\ p_{II}^{(2)} = f(r_0,\theta), \end{cases} && \begin{matrix} r = r_0, \quad |\theta| \le \theta_0, \\ r = r_0, \quad \theta_0 \le |\theta| \le \pi. \end{matrix}
\end{aligned} \tag{7.42}$$

If the particle velocity is given, we have the second variant:

$$p_I = p_{II}, \quad x = x_0, \quad |y| \le h,$$

$$\frac{\partial p_I}{\partial x} = \frac{\partial p_{II}}{\partial x}, \qquad x = x_0, \quad |y| \le h,$$
$$p_{II} = p_{III}, \qquad r = r_0, \quad |\theta| \le \theta_0,$$
$$\begin{cases} \dfrac{\partial p_{II}}{\partial r} = \dfrac{\partial p_{III}}{\partial r}, & r = r_0, \quad |\theta| \le \theta_0, \\ \dfrac{1}{i\omega\rho}\dfrac{\partial p_{II}^{(2)}}{\partial r} = f(r_0, \theta), & r = r_0, \quad \theta_0 \le |\theta| \le \pi. \end{cases} \tag{7.43}$$

The first two equations in systems (7.42) and (7.43) are transformed into algebraic equations due to the function system $\cos(\alpha_n y)$, which is orthogonal at the interval $|y| \le h$. The third equation is transformed due to the function system $\cos(\beta_n \theta)$, which is orthogonal at the interval $|\theta| \le \theta_0$. The fourth equation, which contains two equalities within the brace brackets, due to the function system $\cos(n\theta)$. The function system $\cos(n\theta)$ is orthogonal at the interval $\theta = [0, 2\pi]$.

When we solve the problem, we use the finite system of equations, obtained after reduction of infinite system. Thus, the choice of the function $f(r_0, \theta)$ is noteworthy. The problem analysis can show the preferable function setting $f(r_0, \theta)$, to decrease the pressure jump or particle velocity in the vicinity of corner points $r = r_0$, $\theta = \pm\,\theta_0$ for solution $p_{II}^{(2)}$.

The obtained solution allows us to analyze the sound field in the waveguide structure, Fig. 7.17. The analysis of angle distribution of intensity in the wedge shaped waveguide is of interest. It is the calculation of energy characteristic of directivity:

$$R_J^{(q)}(r, \theta) = \frac{|p_{III}(r, \theta)|^2}{|p_{III}(r, \theta^*)|^2_{\max}}, \tag{7.44}$$

where q—enumerates the incident mode in domain I; θ^*—angle, which determines the direction of sound maximum intensity in domain III. We take into account the asymptotics of Hankel functions $H_{\beta_n}^{(1)}(kr)$ for $kr \gg \beta_n$ [11]

$$H_{\beta_n}^{(1)}(kr) \approx \sqrt{\frac{2}{\pi kr}} \cdot \exp\left[i\left(kr - \frac{\pi}{2}\beta_n - \frac{\pi}{4}\right)\right],$$

according to representation (7.36), Eq. (7.44) can be written as follows:

$$R_j^{(q)}(\theta) = \left|\sum_{n=0}^{\infty} \frac{B_n \cos(\beta_n\theta)\exp(-i\beta_n\pi/2)}{H_{\beta_n}^{\prime(1)}(kr_0)}\right|^2 \Bigg/ \left|\sum_{n=0}^{\infty} \frac{B_n \cos(\beta_n\theta^*)\exp(-i\beta_n\pi/2)}{H_{\beta_n}^{\prime(1)}(kr_0)}\right|^2. \tag{7.45}$$

The energetic characteristics of sound wave propagation through the region of matching of the flat and wedge shaped waveguides are of great importance. Transmission coefficient $W^{(q)}$ is determined as the ratio of the average power flow of the wave in domain

III to the average power flow of qth normal wave in domain I. After the transformations we obtain the following formula for transmission coefficient:

$$W^{(q)} = \frac{4\theta_0}{\pi h \mathrm{Re}\,(k_q)\,\varepsilon_q} \sum_{n=0}^{\infty} \varepsilon_n \left| \frac{B_n}{H'^{(1)}_{\beta_n}(kr_0)} \right|^2, \tag{7.46}$$

where $\varepsilon_0 = 1$, $\varepsilon_n = 0, 5$, $n > 0$. According to expression (7.46), the transmission coefficient $W^{(q)}$ is represented as the sum of energetic coefficients of normal waves excitation of domain III. Otherwise speaking, expression (7.46) is the sum of transformation coefficients of qth normal wave in domain I into normal waves in domain III. Reflection coefficient $V^{(q)}$ is represented as follows:

$$V^{(q)} = \frac{1}{\mathrm{Re}\,(k_q)\,\varepsilon_q} \sum_{n=0}^{N} \varepsilon_n \mathrm{Re}\,(k_n)\,|A_n|^2\,. \tag{7.47}$$

Here N determines the number of the modes propagating in domain I. It follows from the law of conservation of energy, that the following condition must be satisfied

$$V^{(q)} + W^{(q)} = 1.$$

7.3.4 Flat Piston Radiation in a Waveguide with a Flare

The constructed solution allows us to consider the problem of rigid piston vibration. This piston is situated in domain I at a distance l from the region of matching of plane and wedge shaped waveguides, Fig. 7.17(a). Let us consider the situation, when the flare angle of the wedge shaped waveguide is less than 90°. Let the piston perform the harmonic vibrations with frequency ω. The condition on the surface of the vibrating piston is added to the system of Eq. (7.40)

$$\frac{1}{i\omega\rho}\frac{\partial p_I}{\partial x} = \upsilon_0, \quad x = x_0 - l, \quad |y| < h, \tag{7.48}$$

where υ_0—is the amplitude of piston particle velocity. The field in domain I can be written as follows:

$$p_I(x, y) = \sum_{n=0}^{\infty} \left[A_n \cos\left(k_n(x - x_0)\right) + E_n \sin\left(k_n(x - x_0)\right)\right] \cos\left(\alpha_n y\right). \tag{7.49}$$

We write system (7.40), (7.48) including expression (7.49) and transform the system into algebraic system. We obtain the infinite system of linear algebraic equations of the second kind in unknown coefficients A_n, B_n, C_n, D_n, E_n.

The resistance value of piston radiation Z should be analyzed in this problem. Let us specify the value of Z with regard to the value of $2h\rho c$—piston radiation resistance in an infinite flat waveguide:

$$z = \frac{Z}{2h\rho c} = \frac{1}{2h\rho c \cdot \upsilon_0} \int_{-h}^{h} p\,(x = x_0 - l, y)\,dy, \tag{7.50}$$

where $p\,(x = x_0 - l, y)$—pressure amplitude on the piston surface. We write Eq. (7.50), according to expression (7.49), and we obtain the expression for resistivity of piston radiation in the waveguide with a flare:

$$z = -\frac{ik}{\upsilon_0}\,[A_0 \cos{(kl)} - E_0 \sin{(kl)}].$$

Setting the value of $\upsilon_0/k = 1$, we obtain

$$z = -i\,[A_0 \cos{(kl)} - E_0 \sin{(kl)}]. \tag{7.51}$$

7.3.5 Numerical Results Analysis

The obtained infinite systems of equations were solved by reduction method. Let the N_1, N_2, N_3 determine the number of included modes in domains I, II, III when performing reduction. The fulfillment of the matching conditions at the boundaries of partial domains should be controlled in the first place.

Let us consider the situation, when the flare angle $\theta_0 \leq 90°$. Figure 7.18 shows the pressure discrepancy graphs (curve 1) and particle velocity (curve 2) along the boundary separating the domains in Figs. 7.17(a) and 7.17(b).

Figure 7.18(a) shows the boundary between domains I and II, Fig. 7.18(b)—between II and III, $2h = 0.4\lambda$, $\theta_0 = 70°$, $N_1 = N_2 = N_3 = 7$. The zero mode in domain I $(q = 0)$ is incident on the area of matching. The discrepancy is determined as the ratio of pressure difference modulus or particle velocity on the left and on the right hand side of the boundary to the field amplitude in the wave incident on the area of matching of the flat and wedge shaped waveguides. The curves, shown in the figures, illustrate the high accuracy of the fulfillment of the sound fields matching conditions along almost the entire interface between the domains. Some matching deterioration can be observed only in the vicinity of angular points of the domain boundary. But the influence of such local errors on the field characteristics is not significant far form angular point. This is

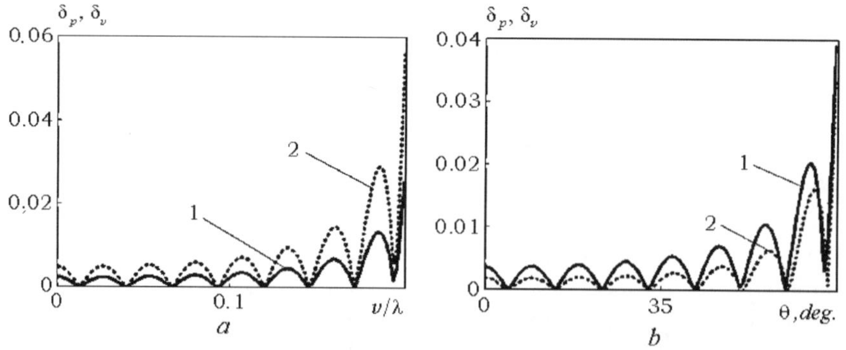

FIG. 7.18: Pressure discrepancy (curve 1) and velocity (curve 2): (a)—along the boundary of domains I and II; (b)—along the boundary of domains II and III; $2h = 0, 4\lambda$, $\theta_0 = 70°$, $N_1 = N_2 = N_3 = 7$, $q = 0$

evidenced by high precision of the law of conservation of energy fulfillment. This law is fulfilled with accuracy not less than 0.1% for the included summands in series (3.34), (3.36), (3.37).

When the flare angle of the waveguide $\theta_0 > 90°$, with the increase of θ_0, we can observe an increase of r_0, which determines the size of domain II. The growth of radius r_0 is unlimited in the limiting case, when $\theta_0 \to 180°$. When the angle θ_0 increases, it is necessary to increase the number of included modes to satisfy required accuracy of the calculations. And also it is necessary to answer the question: which variant of matching conditions (7.42) or (7.43), is preferable? Figure 7.19(a) shows the velocity discrepancy graphs, and Fig. 7.19(b)—pressure discrepancy graphs for the case $2h = 0.4\lambda$, $\theta_0 = 130°$, $q = 0$, $N_1 = 11$, $N_2 = N_3 = 15$. Curves 1 correspond to the system of Eq. (4.42), curves 2—to the system (4.43). Function $f(r_0, \theta)$ in (7.42) and (7.43) is set equal to zero. Angular region $130° \leq \theta \leq 180°$ determines the sound field along the straight-line boundary between domains I and II. As we can see, the values of discrepancies at this boundary are the same as for the case of flare angle $\theta_0 \leq 90°$. The situation is different for the curved boundary, separating domains II and III ($0° \leq \theta \leq 130°$). The discrepancy increases.

The difference in the discrepancy character along the boundaries between domains I, II and II, III [Fig. 7.17(c)], is related to the fact that: the solution $p_{II}^{(1)}$ was not changed when to flare angles $\theta_0 > 90°$, but solution $p_{II}^{(2)}$ was set on the full circle with radius r_0. And the system $\cos(n\theta)$ appeared instead of the function system $\cos(\beta_n\theta)$. Thus, the angular point $\theta = \theta_0$ is inside the interval of defining function system $\cos(n\theta)$. Thus, the deterioration in convergence of solution at the boundary $r = r_0$, $0° \leq \theta \leq 130°$ comes to be understood. As for the choice of system (7.42) or (7.43), as it is seen from Fig. 7.19, the first is preferable, where the pressure is set on the additional region.

We turn now to the choice of function $f(r_0, \theta)$, which is included in system (7.42). Let us compare the calculation results for two cases:

1. $f(r_0, \theta) = 0$,
2. $f(r_0, \theta) = p_{III}(r_0, \theta_0)$.

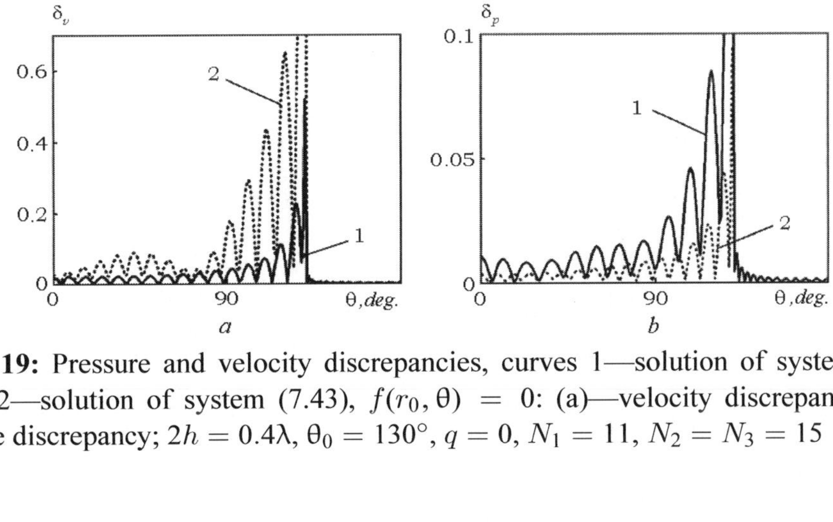

FIG. 7.19: Pressure and velocity discrepancies, curves 1—solution of system (7.42), curves 2—solution of system (7.43), $f(r_0, \theta) = 0$: (a)—velocity discrepancy; (b)—pressure discrepancy; $2h = 0.4\lambda$, $\theta_0 = 130°$, $q = 0$, $N_1 = 11$, $N_2 = N_3 = 15$

The first variant corresponds to function $f(r_0, \theta) = 0$. This leads to a pressure jump in the point (r_0, θ_0), that is an undesirable phenomenon. We attempt to decrease this jump in the second variant. The solution $p_{II}^{(2)}$ continues as a constant value equal to the value of pressure in domain III for $r = r_0$, $\theta = \theta_0$. The calculations show some improvement in the quality of fulfillment of boundary conditions for the second variant in comparison with the first. Of course, there is no possibility to determine the value of function $f(r_0, \theta)$, which continues solution $p_{II}^{(2)}$ continuously on the arc $r = r_0$, $130° \leq |\theta| \leq 180°$. We should analyze each specific problem and attempt to decrease the field jump at the boundary section under consideration.

In conclusion, the values of energy coefficients of the reflection and transmission of sound are given in Table 7.1 for three calculation variants.

Figure 7.20 shows frequency characteristics of transmission coefficient $W^{(0)}$ of zero mode ($q = 0$) in domain I for different flare angles of wedge shaped waveguide θ_0. Curve 7 determines the sound radiation from the open end of the plane semi-infinite waveguide and is calculated according to work [24], using the formula

$$W^{(0)} = 1 - \left[\exp\left(-2\pi h/\lambda\right)\right]^2, \tag{7.52}$$

where $\lambda = 2\pi/k$—wave length. As we can see, with an increase of flare angle θ_0 a decrease of transmission coefficient is observed. When the size of the waveguide $h \ll \lambda$,

TABLE 7.1: Energy coefficients of sound reflection and sound transmission

$\theta_0 = 130°, 2h = 0.4\lambda, q = 0,$ $N_1 = 11, N_2 = N_3 = 15$		$V^{(0)}$	$W^{(0)}$	$V^{(0)} + W^{(0)}$
System of Eq. (4.42)	$f(r_0, \theta) = 0$	0.06373	0.093404	0.99777
System of Eq. (4.42)	$f(r_0, \theta) = p_{III}(r_0, \theta_0)$	0.06405	0.93477	0.99882
System of Eq. (4.43)	$f(r_0, \theta) = 0$	0.05248	1.01241	1.06489

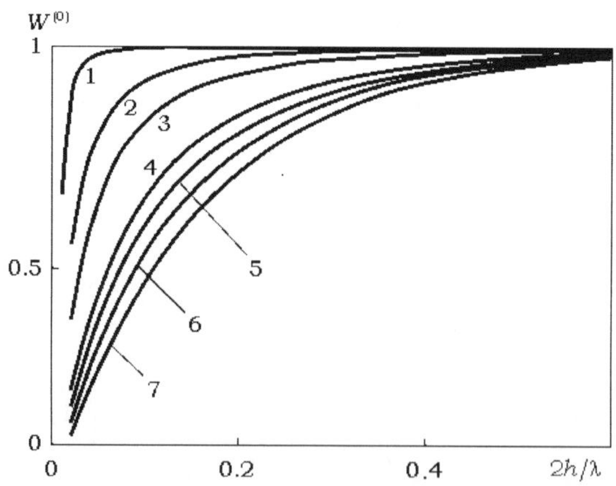

FIG. 7.20: Frequency characteristics of transmission coefficient $W^{(0)}$ for different flare angles θ_0: 1—$\theta_0 = 5°$, 2—$\theta_0 = 25°$, 3—$\theta_0 = 45°$, 4—$\theta_0 = 90°$, 5—$\theta_0 = 110°$, 6—$\theta_0 = 140°$, 7—$\theta_0 = 180°$

transmission coefficient tends to zero, whatever is the flare angle of the wedge shaped waveguide. When the value $2h > 0.6\lambda$ the transmission coefficient tends to a unity.

Summarizing the data shown in graphs Fig. 7.20, we can say, that waveguide structure shown in Fig. 7.17 behaves like high-pass filter. The cutoff frequency and the frequency response slope of such a filter depend on flare angle θ_0.

We turn now to the investigation of angular distribution of sound intensity in the wedge shaped waveguide. We use the calculation Eq. (7.45). Note that when flare angle $\theta_0 < 90°$ we can say that we have a flat waveguide with a flare, and when $\theta_0 \geq 90°$—the sound is radiated from the flat waveguide in outer space.

Figure 7.21 shows dependencies of energy characteristic of directivity $R_J^{(0)}(\theta)$ for different flare angles of wedge shaped waveguide θ_0: Fig. 7.21(a)—$\theta_0 = 90°$, (b)—$\theta_0 = 130°$, (c)—$\theta_0 = 25°$. Zero mode ($q = 0$) is the incident wave in domain I. The curves parameter is the value of ratio of flat waveguide dimension to wave length.

Figure 7.21(a) shows dashed lines which correspond to the directivity patterns of plane piston in an infinite rigid screen; they are calculated by formula $(\sin \xi/\xi)^2$, where $\xi = kh \sin\theta$ [16]. As it follows from the graphs, with an increase of the value of $2h/\lambda$, directivity pattern is becoming sharp. Dashed lines in Fig. 7.21(a) coincide with continuous lines when $2h/\lambda = 1$ and $2h/\lambda = 1.5$. It means that when the wave value $2h/\lambda$ is large enough, the running plane wave (zero mode) is almost not reflected from the plane end of the waveguide. All the wave energy goes to the outer domain, and directivity

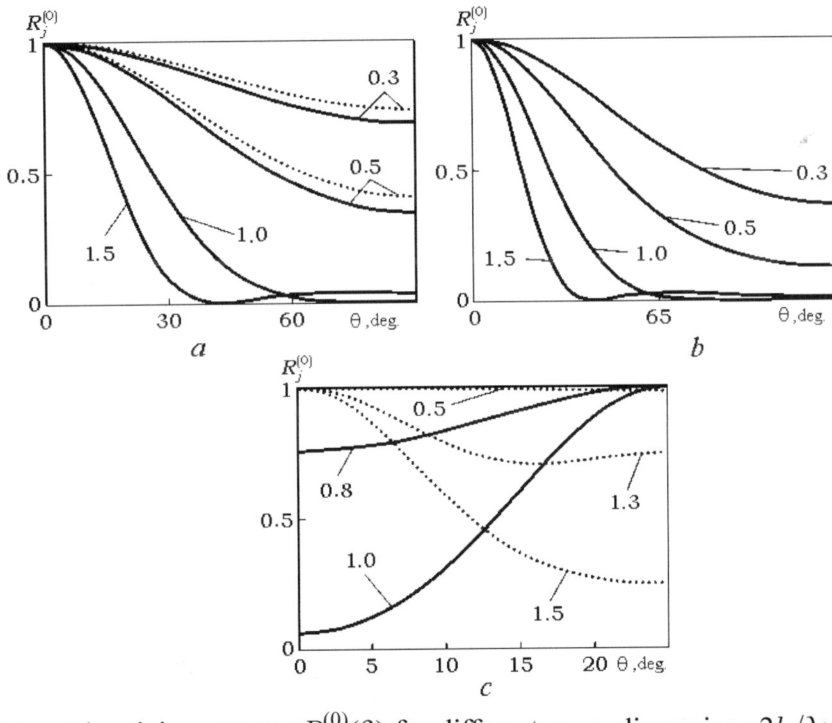

FIG. 7.21: Directivity patterns $R_J^{(0)}(\theta)$ for different wave dimensions $2h/\lambda$: (a)—$\theta_0 = 90°$, (b)—$\theta_0 = 130°$, (c)—$\theta_0 = 25°$

pattern is determined by expression $(\sin\xi/\xi)^2$, where $\xi = kh\sin\theta$. This result is consistent with the data given in Fig. 7.20.

Comparing the graphs in Fig. 7.21(a) and Fig. 7.21(b) it should be noted, that an increase of flare angle ($\theta_0 > 90°$) almost does not change the main lobe of directivity pattern when the wave dimension of the flate waveguide is $2h/\lambda \geq 1$. When $2h/\lambda < 1$ the directivity patterns is becoming sharp, when the angle changes from $\theta_0 = 90°$ to $\theta_0 = 130°$. When the flare angles of the flare are relatively small [$\theta_0 = 25°$, Fig. 7.21(c)] we can observe a decrease of sound energy along the axis of the waveguide ($\theta = 0$). This phenomenon is explained by amplitude-phase ratio of excited zero wave to the first symmetrical normal wave in the wedge shaped waveguide.

Peak radiation displacement from the axis of the waveguide can be observed for the other flare angles θ_0 for definite value of $2h/\lambda$. Figure 7.22 shows computed data.

The value of ratio $2h/\lambda$ was calculated when the flare angle is fixed, and the value of directivity pattern is minimum along the acoustic axis $R_J^{(0)}(\theta = 0)$. The values of function $R_J^{(0)}(\theta = 0)$ determine the numbers in the vicinity of the points on the plane $(\theta_0,\ 2h/\lambda)$, Fig. 7.22. The minimum value $R_J^{(0)}(\theta = 0)$ is observed when the parameters of the waveguide are: $\theta_0 \approx 25°$, $2h/\lambda \approx 1$. When $\theta_0 < 25°$ we observe the slow growth of minimum value $R_J^{(0)}(\theta = 0)$, when $2h/\lambda$ increases. We observe the rapid growth of minimum $R_J^{(0)}(\theta = 0)$ with a decrease of $2h/\lambda$, and when $\theta_0 > 25°$. When $\theta_0 > 45°$ we have $R_J^{(0)}(\theta = 0) = 1$.

Figure 7.23(a) illustrates the energetic characteristics of directivity in the wedge shaped waveguide, when the field of incident wave is determined by the first symmetric normal wave in domain I ($q = 1$). Here curve 1 corresponds to $\theta_0 = 15°$, 2—$\theta_0 = 30°$, 3—$\theta_0 = 40°$, 4—$\theta_0 = 50°$, 5—$\theta_0 = 90°$, 6—$\theta_0 = 130°$, $2h/\lambda = 1.1$. For small flare angles ($\theta_0 < 30°$), the distribution of the field in the cross section is almost the same as in an incident wave. When the flare angle increases the characteristic changes. The sound is almost not radiated along the waveguide axis. Most of the sound energy is near the walls of the waveguide.

The situation does not change, if we increase the wave dimension $2h/\lambda$ when the flare angle θ_0 is fixed. Figure 7.23(b) corresponds to the flare angle $\theta_0 = 90°$. The value

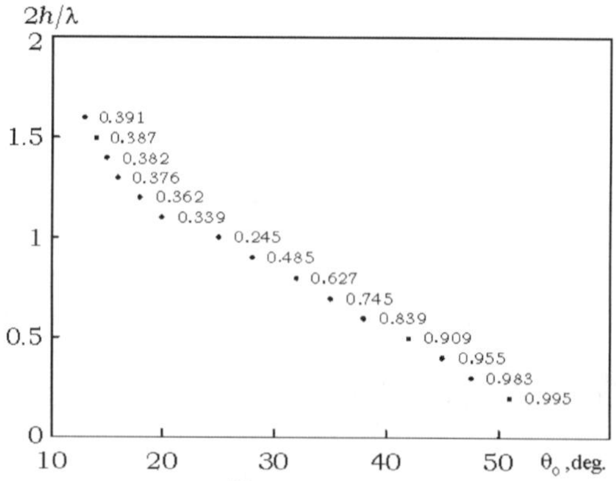

FIG. 7.22: Minimum value of $R_J^{(0)}(\theta = 0)$ for different waveguide parameters

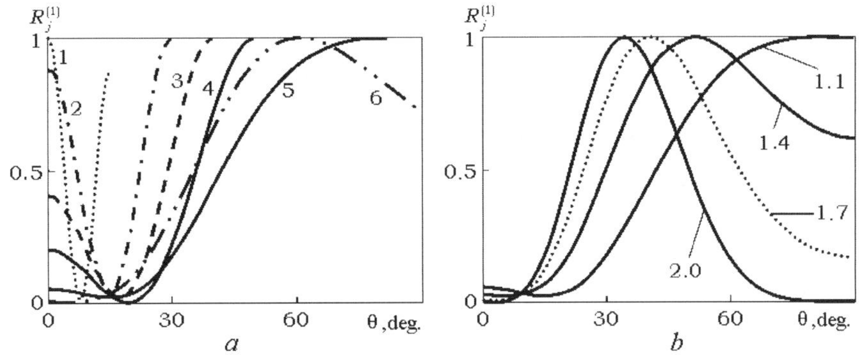

FIG. 7.23: Directivity patterns $R_J^{(1)}(\theta)$: (a)—for different flare angles θ_0 ($2h/\lambda = 1.1$): 1—$\theta_0 = 15°$, 2—$\theta_0 = 30°$, 3—$\theta_0 = 40°$, 4—$\theta_0 = 50°$, 5—$\theta_0 = 90°$, 6—$\theta_0 = 130°$; (b)—for different wave dimensions $2h/\lambda$ ($\theta_0 = 90°$)

$2h/\lambda$ is the parameter of the curves. As we can see, in the angular range $0 \leq \theta \leq 10°$ the flow of energy is almost equal to zero. We can observe the displacement and width decrease of the main lobe of directivity pattern with an increase of $2h/\lambda$.

The described situation generates a question: how fast the structure of the flows of energy in the wedge shaped waveguide changes? To answer this question it is necessary to investigate the field of intensity vector in close proximity to the matching zone of the plane and wedge shaped waveguides. Intensity vector projections $\vec{I} = (I_x, I_y)$ on the axis of Cartesian coordinate system (Fig. 7.17) are determined by formulas [4]:

$$I_x = 0.5\text{Re}\,(p\upsilon_x^*), \quad I_y = 0.5\text{Re}\,(p\upsilon_y^*).$$

Here p—sound pressure, υ_x and υ_y—particle velocity projections on the axis of Cartesian coordinate system x and y. Symbol $*$ means complex conjugation. When the projections I_x and I_y, are defined we can construct intensity vector $\vec{I}$ in some point of the sound field existence domain.

Figure 7.24 shows the field of intensity vector near the area of matching of the plane and wedge shaped waveguides. Figures 7.24(a)–7.24(c) corresponds to the case of incident plane wave (zero modes, $q = 0$), and Fig. 7.24(d)—corresponds to case of the first symmetric mode ($q = 1$). Cartesian coordinates on the axis are indicated in fractions of the wave length λ.

The length of each arrow defines the intensity modulus. The direction of arrow defines the direction of the flow of energy. The beginning of the arrow defines the point of sound field, in which the intensity vector is defined. Figures 7.24(a)–7.24(c) illustrates the process of space modification of the flows of energy in the wedge shaped waveguide. We can see that the flow of energy is concentrated along the waveguide axis in the vicinity of matching area [Fig. 7.24(a)]. It means that the flow of energy behaves almost in the same way as in the flat waveguide. At a distance about 3λ [Fig. 7.24(b)] the intensity modulus becomes constant along the cross section of the waveguide. When the x coordinate increases [Fig. 7.24(c)], the flow of energy decreases along the waveguide axis,

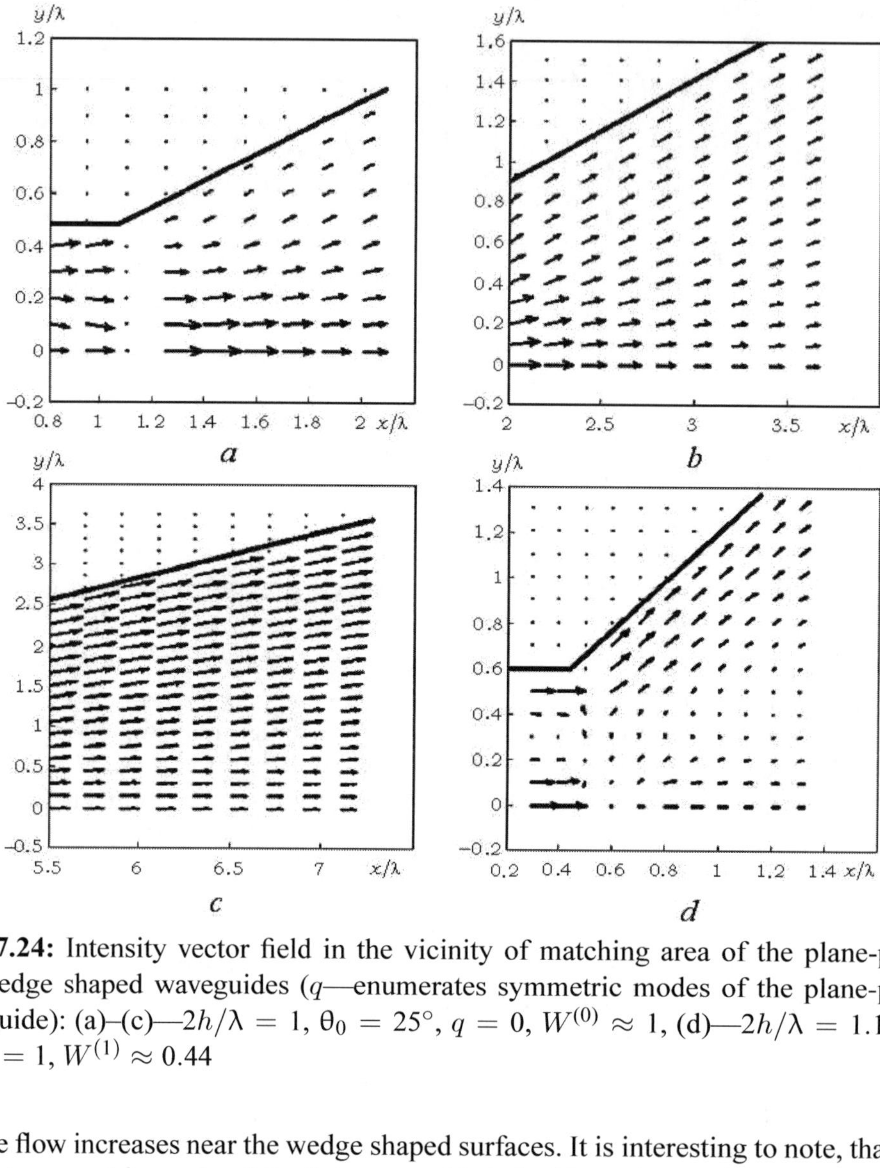

FIG. 7.24: Intensity vector field in the vicinity of matching area of the plane-parallel and wedge shaped waveguides (q—enumerates symmetric modes of the plane-parallel waveguide): (a)–(c)—$2h/\lambda = 1$, $\theta_0 = 25°$, $q = 0$, $W^{(0)} \approx 1$, (d)—$2h/\lambda = 1.1$, $\theta_0 = 50°$, $q = 1$, $W^{(1)} \approx 0.44$

and the flow increases near the wedge shaped surfaces. It is interesting to note, that when the first symmetric mode of the flat waveguide is incident, the spatial reconstruction of energy in the wedge shaped waveguide happens faster [Fig. 7.24(c)].

The plane piston radiation conditions in the waveguide with a flare [Fig. 7.17(a)] are illustrated in graphs in Fig. 7.25. This figure shows the frequency characteristics of the real and imaginary parts of radiation resistivity, calculated by formula (7.51). Continuous lines define real parts, and dashed lines define imaginary parts of radiation resistivity. The curves parameter is the flare angle: curve 1—$\theta_0 = 10°$, curve 2—$\theta_0 = 50°$, curve 3—$\theta_0 = 90°$. We assume that $2h/\lambda = l$, Fig. 7.17(a). When $2h/\lambda < 0.3$, we observe the nonuniformity of frequency characteristic of resistivity of piston radiation. The nonuniformity increases with an increase of flare angle, and it has the resonance characteristics. When the real part of radiation impedance is a maximum, the imaginary part changes

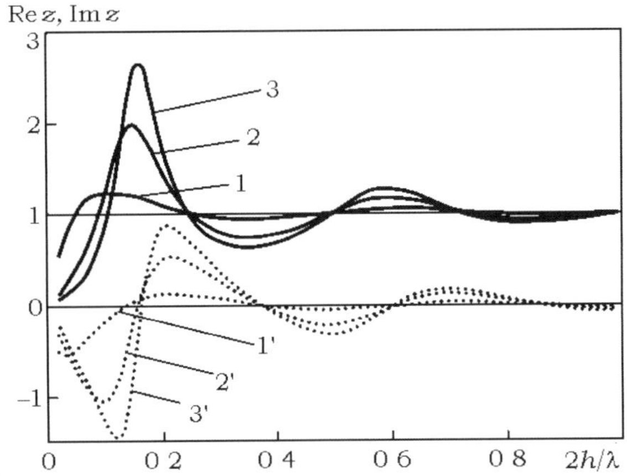

FIG. 7.25: Frequency dependencies of piston radiation resistivity in fractions of ρc for different flare angles θ_0 ($2h/\lambda = l$): continuous lines—real parts, dashed lines—imaginary parts; 1—$\theta_0 = 10°$, 2—$\theta_0 = 50°$, 3—$\theta_0 = 90°$

its nature from mass to elastic. With an increase of value of $2h/\lambda$, the value of radiation impedance is oscillating and tends to the radiation impedance of the piston in the endless flat waveguide.

Figure 7.26 shows real (continuous lines) and imaginary (dashed lines) parts of piston radiation impedance, when the flare angle is fixed $\theta_0 = 90°$, and the niche depth is variable l/λ. The curves parameter is the piston wave size: curve 1—$2h/\lambda = 0.3$, curve 2—$2h/\lambda = 0.5$, curve 3—$2h/\lambda = 0.7$, curve 4—$2h/\lambda = 1.5$. With an increase of l/λ, we observe the periodical change of the behavior of piston radiation impedance. The spatial period is the half of the wave length. The value of the oscillations of imaginary and real parts of radiation impedance depends on the wave size of the piston. With an

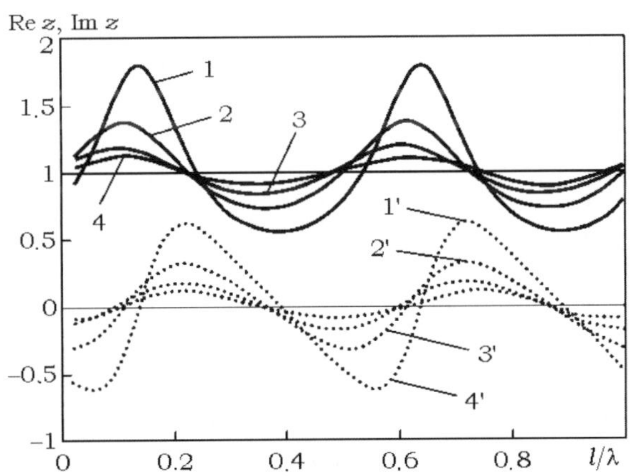

FIG. 7.26: The dependencies of piston radiation resistivity in fractions of ρc on niche depth l/λ ($\theta_0 = 90°$): continuous lines—real parts, dashed lines—imaginary parts; 1—$2h/\lambda = 0.3$, 2—$2h/\lambda = 0.5$, 3—$2h/\lambda = 0.7$, 4—$2h/\lambda = 1.5$

increase of $2h/\lambda$, the oscillations decrease and piston radiation impedance approaches the piston radiation impedance in an infinite flat waveguide.

If the wave size of the niche tends to zero, the situation approaches the problem of piston radiation in an infinite screen. The graphs in Fig. 7.26 are in accordance with the results of work ([11], p. 96, Fig. 27, curve $n = 0$) when $l/\lambda \ll 1$. Piston radiation impedance in an infinite screen for different piston vibration modes is calculated in this work.

7.4 ACOUSTIC SIGNAL PROPAGATION IN THE WAVEGUIDE WITH STEPWISE VARIATION OF CROSS SECTION

The processes of tone and impulse signals propagation in the flat waveguide with discontinuous variation of the cross section are investigated in the given paragraph [25]. This discontinuity is called "a step." The step discontinuity, together with the other discontinuities characterized by the sharp change in the waveguide parameters, is widely used in the acoustic [4] and radioelectronic [1] devices. It should be noted that the impulse signal propagation in the irregular waveguides with discontinuities, similar to the discontinuity called "a step," is not as well studied, as harmonic wave propagation. Impulse propagation in the waveguide is accompanied by the series of effects, which arouse great interest [26].

7.4.1 Problem Statement and Solution Construction

Figure 7.27 illustrates the waveguide, formed in the result joining of two semi-infinite parallel-plate waveguides with different cross sections. The cross sectional dimensions of the waveguide vary from h_1 to h_2 in cross section $x = 0$. The waveguide is filled with an ideal compressible liquid with density ρ and sound velocity c. All the boundaries of the waveguide are acoustically rigid.

Let us consider the harmonic wave propagation in the waveguide with frequency ω_0. We assume that plane pressure wave is incident from the left in the waveguide. This wave is zero mode of the waveguide, with unit amplitude

$$p_0(x,t) = \exp\left[-i\left(\omega_0 t - k_0 x\right)\right], \tag{7.53}$$

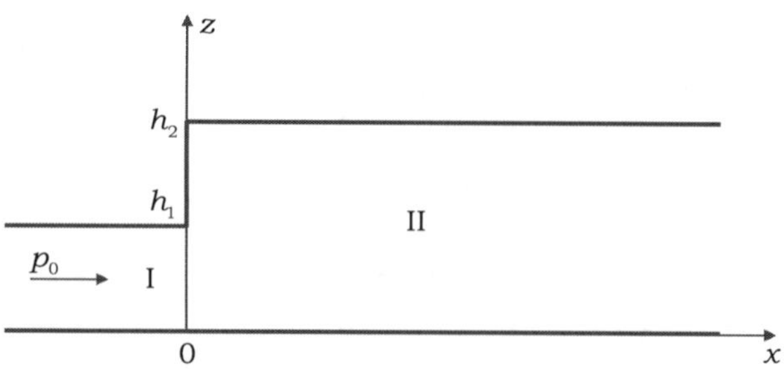

FIG. 7.27: Flat waveguide with stepwise variation

where $k_0 = \omega_0/c$. According to the partial domains method, all the sound field existence domain is partitioned into two domains (Fig. 7.27): I—$x \leq 0$, $0 \leq z \leq h_1$, II—$x \geq 0$, $0 \leq z \leq h_2$. When the flat wave p_0 is incident at the interface of domains I and II, the reflected wave p_1 and propagated wave p_2 appear. The field in domain II is represented as superposition of normal waves of the flat waveguide with dimension h_2, namely

$$p_{II}(x, z, t) = \sum_{n=0}^{\infty} B_n \cos\left(\frac{n\pi z}{h_2}\right) \exp\left[-i\left(\omega_0 t - \gamma_n x\right)\right], \tag{7.54}$$

where

$$\gamma_n = \begin{cases} \sqrt{k_0^2 - (n\pi/h_2)^2}, & k_0 > n\pi/h_2, \\ i\sqrt{(n\pi/h_2)^2 - k_0^2}, & k_0 < n\pi/h_2. \end{cases} \tag{7.55}$$

Analogous expression can be constructed for the pressure field in domain I of the waveguide. Incident wave (7.53) is given, and reflected wave can be represented as follows

$$p_I(x, z, t) = \sum_{n=0}^{\infty} A_n \cos\left(\frac{n\pi z}{h_1}\right) \exp\left[-i\left(\omega_0 t + \eta_n x\right)\right], \tag{7.56}$$

where

$$\eta_n = \begin{cases} \sqrt{k_0^2 - (n\pi/h_1)^2}, & k_0 > n\pi/h_1, \\ i\sqrt{(n\pi/h_1)^2 - k_0^2}, & k_0 < n\pi/h_1. \end{cases} \tag{7.57}$$

The expressions contain two sequences of arbitrary values A_n and B_n, $n = 0, 1, 2, \ldots$. The matching conditions at the interface of domains I and II can be fulfilled due to the selection of these coefficients:

$$p_0 + p_1 = p_2, \quad x = 0, \quad z = [0, h_1], \tag{7.58}$$

$$\frac{\partial p_{II}}{\partial x} = \begin{cases} \dfrac{\partial (p_0 + p_I)}{\partial x}, & x = 0, \quad z = [0, h_1], \\ 0, & x = 0, \quad z = [h_1, h_2]. \end{cases} \tag{7.59}$$

This statement follows form the Fourier series properties. The set of functions $\cos(n\pi z/h_2)$ and $\cos(n\pi z/h_1)$, $n = 0, 1, 2, \ldots$ forms full and orthogonal system of functions on the intervals $z = [0, h_2]$ and $z = [0, h_1]$, respectively. The infinite system of linear algebraic equations of the second kind can be obtained in the same way, as it was shown above.

7.4.2 Numerical Results Analysis: Tonal Signal

The energetic characteristics of plane wave penetration through the stepwise variation zone of the waveguide cross section are of interest. The energy coefficients of transmission W and reflection V of incident plane wave are defined by Eqs. (7.12)–(7.14) where $q = 0$. We now turn attention to the step widening of the waveguide, where $h_1 < h_2$.

We use the following notation: $h'_1 = h_1/\lambda_0$, $h'_2 = h_2/\lambda_0$, $x' = x/\lambda_0$, $z' = z/\lambda_0$, where $\lambda_0 = 2\pi/k_0$—wave length.

Figure 7.28 shows the examples of the frequency characteristics of transmission coefficient W for different values of ratio h_1/h_2. The wave size h'_2 is plotted along abscissa axis. If the wave dimensions h_1 and h_2 are small ($h'_1 \ll 1$ and $h'_2 \ll 1$), the results shown in Fig. 7.28 are in good agreement with single-mode waveguide theory. According to this theory, the energy coefficient of zero normal wave transmission through the stepwise discontinuity of the waveguide is determined as follows [4]:

$$W^{(0)} = \frac{4h_1/h_2}{(1 + h_1/h_2)^2}. \tag{7.60}$$

For example, when $h_1/h_2 = 0.2$ we obtain $W^{(0)} \approx 0.556$, and when $h_1/h_2 = 0.3$—$W^{(0)} \approx 0.71$ (see, Fig. 7.28, $h'_2 \ll 1$).

The characteristic feature of the curves in Fig. 7.28 is the presence of dramatic dips, which appear with an increase of frequency. The less is value of ratio h_1/h_2 the more is quantity of these dips. Waveguide becomes almost sound transparent with further increase of sound wave frequency. When $h_1/h_2 = 0.3$ [Fig. 7.28(b)] for the values $h'_2 > 1.05$ the transmission coefficient is close or equal to unity.

Such character of the curves in Fig. 7.28 is connected with mode structure of the sound field in domain II. Let us discuss this moment in more detail. We assume that domain I (Fig. 7.27) is a single-mode waveguide ($h_1 < 0.5\lambda_0$). The value h_1 is constant. Wave dimension of domain II h_2/λ_0 is variable.

Figure 7.29 shows the dependency of energy transmission coefficient W (curve 1) on wave size of domain II h'_2. Figure 7.29(a) corresponds to the wave size of domain I $h'_1 = 0.1$, and Fig. 7.29(b)—$h'_1 = 0.3$. As we can see, the value change of energy

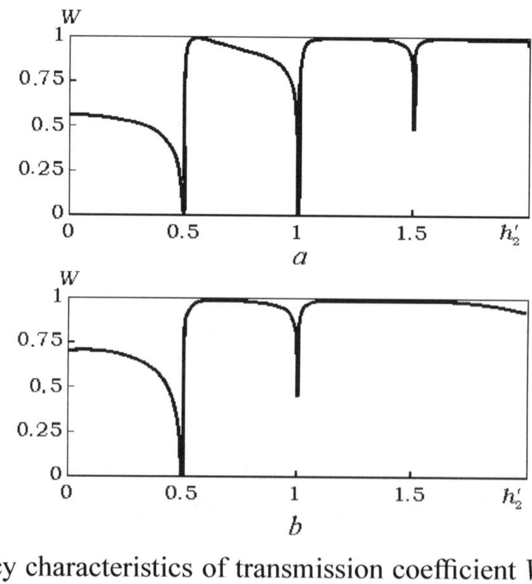

FIG. 7.28: Frequency characteristics of transmission coefficient W: (a)—$h_1/h_2 = 0.2$, (b)—$h_1/h_2 = 0.3$

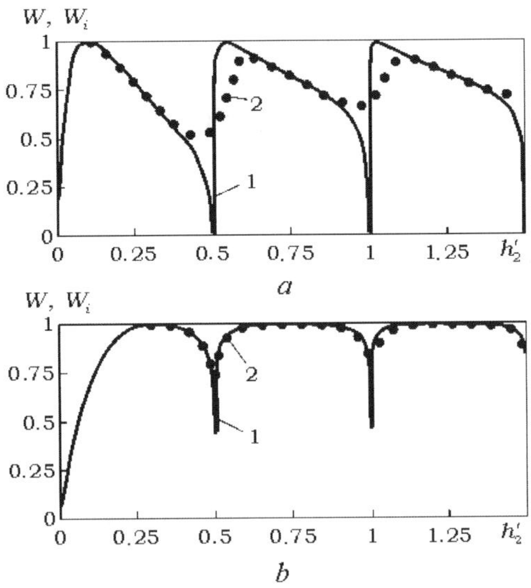

FIG. 7.29: Energy transmission coefficient dependency on wave size h_2' of domain II, curve 1—W, curve 2—W_i impulse signal: (a)—$h_1' = 0.1$, (b)—$h_1' = 0.3$

coefficient W is of periodic nature. Period corresponds to the half wave length $\lambda_0/2$. It may be noted that there are dramatic dips of curve 1 when h_2 is divisible by $\lambda_0/2$. The dips depth depends on wave value of the cross section h_1' of domain I. When $h_1' \leq 0.2$, the transmission coefficient W is almost equal to zero in the indicated above points of characteristics. When the value h_1' increases, these dips decrease and when $h_1' > 0.4$, ($h_2 > h_1$) the stepwise variation becomes almost transparent.

Let us consider Fig. 7.30 to understand the run of the curves in Figs. 7.28 and 7.29. This figure illustrates the dependencies of energy excitation coefficients of the modes W_n, $n = 0, 1, 2$ in domain II on wave dimension h_2' when a zero mode is incident on the step. Figure 7.30(a) corresponds to the wave size of domain I $h_1' = 0.1$, and Fig. 7.30(b)—$h_1' = 0.3$. According to Fig. 7.30 [see, Eq. (7.55)], when values h_2 are divisible by $\lambda_0/2$, these values correspond to the next mode generation in domain II. If the wave size in domain I $h_1' \leq 0.2$, the excitation coefficients of the other modes W_n are almost equal to zero [Fig. 7.30(a)] when the next mode appear in domain II. As we can see in Fig. 7.29(a), when h_2 is divisible by $\lambda_0/2$, the incident wave is almost completely reflected from the step of the waveguide.

7.4.3 Numerical Results Analysis: Pulse Signal

Let time dependence of initial signal be an infinite sequence of the intervals of sinusoid [27], Fig. 7.31:

$$p_0(t) = \begin{cases} \sin(\omega_0 t), & 0 \leq t \leq \tau_i, \\ 0, & \tau_i \leq t \leq T_i, \end{cases} \tag{7.61}$$

where the carrier frequency ω_0 at the time interval of pulse duration τ_i is a constant value, T_i—pulse period. Let us specify the parameters, which are widely used in pulse

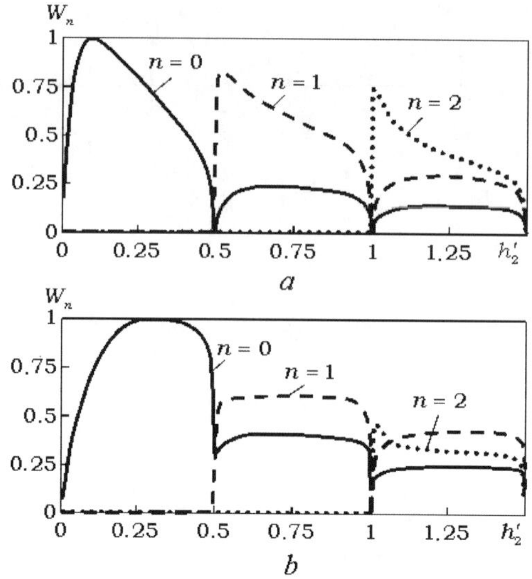

FIG. 7.30: Dependency of energy excitation coefficients $W_n, n = 0, 1, 2$ of normal waves in domain II on wave size h'_2: (a)—$h'_1 = 0.1$, (b)—$h'_1 = 0.3$

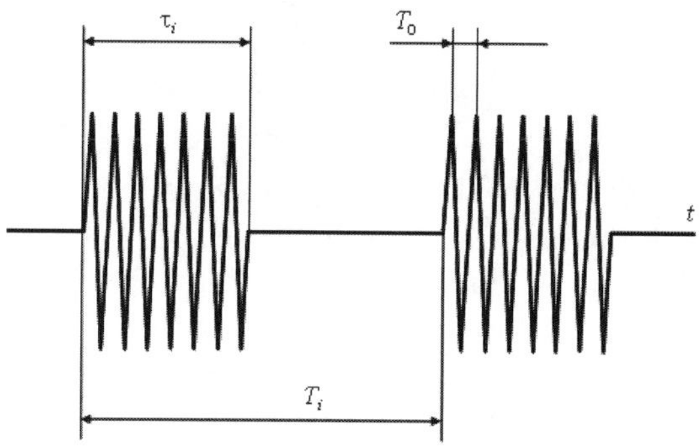

FIG. 7.31: Impulse signal model

equipment, namely ratio $q = T_i/\tau_i$ and quantity $N = \tau_i/T_0$ of periods $T_0 = 2\pi/\omega_0$ of the carrier in impulse with duration τ_i.

It is convenient to use dimensionless values N, q when performing numerical calculations. Let us define the dimensionless time as normalized to impulse duration $\tau_i = NT_0$, that is $t' = t/\tau_i$. We will normalize space values to the length of sound wave $\lambda_0 = cT_0$ at carrier frequency ω_0. Than Eq. (7.61) can be written as

$$p_0(t') = \begin{cases} \sin(2\pi N t'), & 0 \le t' \le 1, \\ 0, & 1 \le t' \le q. \end{cases} \tag{7.62}$$

Let us represent the initial signal (7.61) as Fourier series

$$p_0(t) = \sum_{s=1}^{\infty} [a_s \cos(\omega_s t) + b_s \sin(\omega_s t)], \tag{7.63}$$

or, if we use dimensionless values N, q, we have

$$p_0\left(t'\right)=\sum_{s=1}^{\infty}\left[a_s\cos\left(\frac{2\pi}{q}st'\right)+b_s\sin\left(\frac{2\pi}{q}st'\right)\right],\tag{7.64}$$

where coefficients $\omega_s = 2\pi f_s = s\omega_1 = s\Omega_i$ and b_s are defined by known formulas, value $p_s = \sqrt{a_s^2+b_s^2}$ is the amplitudes of harmonic components. The frequencies of harmonics $\omega_s = 2\pi f_s = s\omega_1 = s\Omega_i$, $s = 1, 2, 3, \ldots$ are divisible by impulses repetition frequency $\Omega_i = 2\pi/T_i$. According to Eq. (7.61) the constant component ($s = 0$) is absent in series (7.63).

Each component of sum (7.63) form zero mode with frequency $\omega_s = s\omega_1 = s\Omega_i$ in domain I. Consequently, the field of incident wave p_0 will have the form of the superposition of zero modes with corresponding frequencies ω_s, $s = 1, 2, 3, \ldots$ in domain I:

$$p_0\left(x,t\right)=\sum_{s=1}^{\infty}\left(a_s+ib_s\right)\exp\left[-i\left(\omega_s t-k_s x\right)\right],\tag{7.65}$$

where $k_s = \omega_s/c$. Let us solve the posed problem for each harmonic component of series (7.65), as it was done in Section 7.4.2. If we use superposition principle, we obtain the following expression for pressure field in domain II:

$$p_{II}\left(x,z,t\right)=\sum_{s=1}^{\infty}\left\{\left(a_s+ib_s\right)\exp\left(-i\omega_s t\right)\times\sum_{n=0}^{\infty}\left[B_{sn}\cos\left(\frac{n\pi z}{h_2}\right)\exp\left(i\frac{\omega_s}{c}x\sqrt{1-\left(\frac{n\pi c}{\omega_s h_2}\right)^2}\right)\right]\right\},\tag{7.66}$$

or, using dimensionless values, we have

$$p_{II}\left(x',z',t'\right)=\sum_{s=1}^{\infty}\left\{\left(a_s+ib_s\right)\exp\left(-i\frac{2\pi s}{q}t'\right)\times\sum_{n=0}^{\infty}\left[B_{sn}\cos\left(\frac{n\pi z'}{h_2'}\right)\exp\left(i\frac{2\pi s}{Nq}x'\sqrt{1-\left(\frac{nNq}{s2h_2'}\right)^2}\right)\right]\right\}.\tag{7.67}$$

The field of reflected wave pressure in domain I is of the similar form

$$p_{I}\left(x,z,t\right)=\sum_{s=1}^{\infty}\left\{\left(a_s+ib_s\right)\exp\left(-i\omega_s t\right)\times\sum_{n=0}^{\infty}\left[A_{sn}\cos\left(\frac{n\pi z}{h_2}\right)\exp\left(i\frac{\omega_s}{c}x\sqrt{1-\left(\frac{n\pi c}{\omega_s h_1}\right)^2}\right)\right]\right\}.\tag{7.68}$$

Let us define the energy coefficient W_i of pulse signal transmission (7.65) through the waveguide step as ratio of energy, passing through the arbitrary cross section of the waveguide in domain II during time interval $[0, T_i]$ to energy E_i of incident wave (7.65) which passes through arbitrary cross section of the waveguide in domain I during time interval $[0, \tau_i]$, that is

$$W_i = \frac{1}{E_i} \int\limits_0^{h_2} \int\limits_0^{\tau_i} \mathrm{Re} p_{II} \mathrm{Re} \upsilon_{xII} dt dz, \tag{7.69}$$

where

$$E_i = \frac{1}{E_i} \int\limits_0^{h_1} \int\limits_0^{\tau_i} \mathrm{Re} p_0 \mathrm{Re} \upsilon_{x0} dt dz.$$

Consequently, reflection coefficient is determined by ratio

$$V_i = \frac{1}{E_i} \int\limits_0^{h_2} \int\limits_0^{\tau_i} \mathrm{Re} p_I \mathrm{Re} \upsilon_{xI} dt dz. \tag{7.70}$$

Let us set the following parameters values in numerical calculations: $N = 10$, $q = 10$. Amplitude spectrum $p_s = \sqrt{a_s^2 + b_s^2}$, $s = 1, 2, ..., 200$ of signal (7.64) for the first two hundred harmonic components is represented in Fig. 7.32(a). The spectrum of given signal can be regarded as narrow-band, because frequency band $[\omega_{90}, \omega_{110}]$ contains 90% of signal energy. Note, that the frequency of hundredth harmonic is equal to carrier frequency, that is $\omega_{100} = \omega_0$.

We turn our attention now to numerical results. Curves 2, shown in Fig. 7.29, determine the transmission coefficient of impulse signal W_i through the waveguide step. Signal (7.64) is narrow-band for the selected parameters. The shape of curve 1 is close to curve 2 for tone signal with frequency ω_0, with the exception of some zones. These zones appear around values h_2, which are divisible by $\lambda_0/2$. The dips of curves 2 of transmission coefficient W_i are less than the dips of curves 1 of transmission coefficient W of tone signal with frequency ω_0 in these zones due to the presence of spectral components.

We turn now to the time dependencies of pressure in the waveguide. Let z coordinate of observation point be equal to zero ($z = 0$), and wave dimension of domain I $h_1' = 0.3$. Thus domain I is a single mode waveguide, with respect to carrier frequency ω_0. Let the wave size of domain II be equal $h_2' = 0.4$, that is domain II is a single-mode waveguide relative to carrier frequency ω_0.

Figure 7.33 shows time dependency of pressure in observation point with coordinate $x' = 40$. We exclude the time of impulse propagation to the point of observation with velocity c. As we can see, impulse signal 1 almost holds its shape, because it is formed basically by zero modes with frequencies $\omega_s = s\omega_1 = s\Omega_i$, $s = 1, 2, 3, ...$. It is caused by the fact that, for the given narrow-band signal, the first mode of domain II becomes uniform for harmonic components with frequencies $\omega > \omega_{125}$ [see Fig. 7.32(a), curve 3]. These components have relatively small amplitudes [Fig. 7.32(a)] and, consequently,

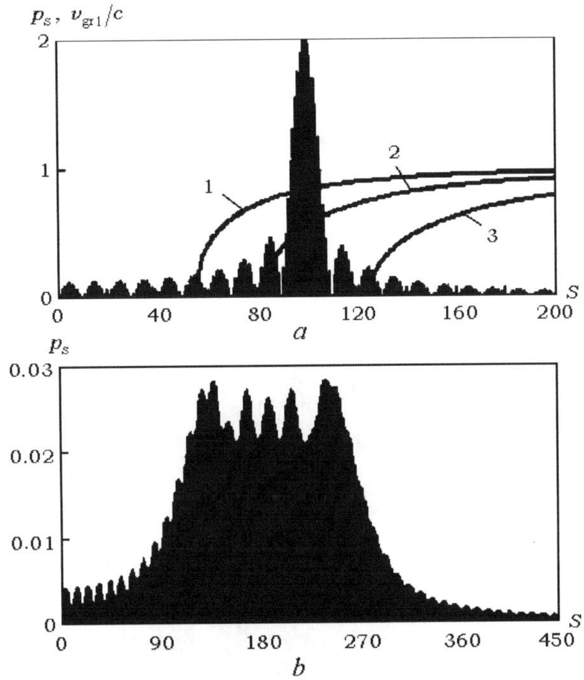

FIG. 7.32: Amplitude spectrums of signals (7.64)—(a) and (7.72)—(b); The values of normalized group velocities υ_{gp1}/c of the first mode of domain II are also given on the graph: 1—$h_2' = 0.9$, 2—$h_2' = 0.6$, 3—$h_2' = 0.4$

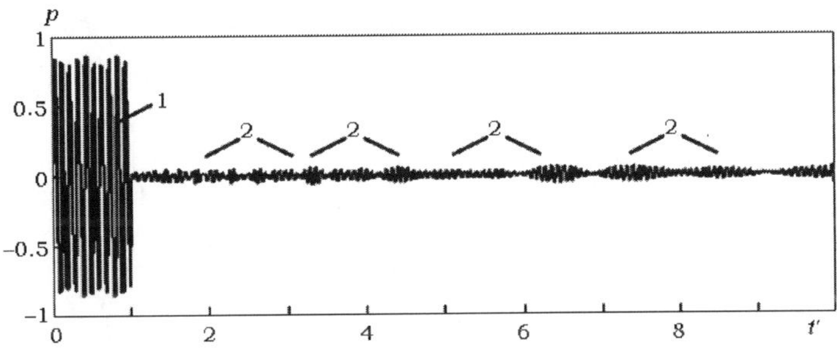

FIG. 7.33: Time dependency of pressure in the point with coordinates $x' = 40$, $z = 0$; $h_1' = 0.3$, $h_2' = 0.4$

they form small perturbations 2, which correspond to the first mode. They come to observation point after the main signal 1 with some delay (Fig. 7.33).

We increase the wave dimension of domain II to the value $h_2' = 0.6$ (Fig. 7.34). It causes the appearance of two uniform modes in domain II (zero and the first). The harmonic components of impulse signal for these modes are situated in energy-carrying subspectrum of initial signal. The harmonic components of the first mode with frequencies ω_s, $s > 83$ are the uniform waves [see Fig. 7.32(a), curve 2]; the second mode is uniform for harmonics with frequencies ω_s, $s > 116$. As it follows from the graphs in

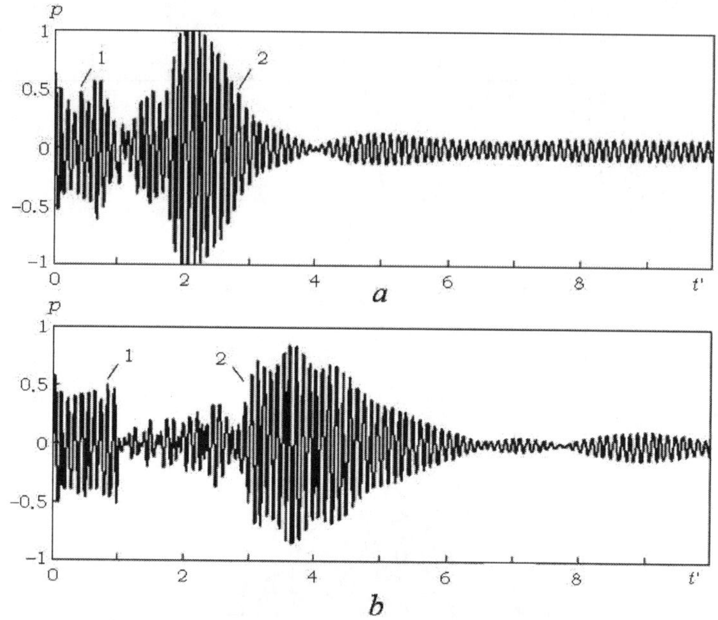

FIG. 7.34: Time dependencies of pressure in different points of observation, $h_1' = 0.3$, $h_2' = 0.6$, $z = 0$: (a)—$x' = 20$, (b)—$x' = 40$

Fig. 7.34, initial impulse divides almost into two impulses during its propagation, because of dispersive properties of the first mode (these impulses are indicated as 1 and 2, respectively, in Fig. 7.34). The first impulse is formed by dispersionless zero modes with frequencies ω_s, $s \geq 1$, and the second impulse, which changes its shape, is formed by the dispersive first modes with frequencies ω_s, $s > 83$. We can say that signal 2 comes into the point of observation with a delay relative to signal 1. For example, the time interval between impulse 1 and impulse 2 is about $2\tau_i$ at the distance $x' = 40$ [Fig. 7.34(b)].

The time dependencies of pressure in Fig. 7.34 correspond to the observation points with coordinate $z = 0$. The surface $z = 0$, in this problem, is acoustically rigid. Consequently, the value of pressure amplitude on the surface $z = 0$—is a maximum for any mode of the waveguide. If recording device moves along z coordinate, the pressure amplitude constancy remains only for zero mode. Consequently, almost undisturbed output signal will be observed in the points of observation with coordinate $z = h_2/2$, which correspond to zero values of pressure in the first mode. This signal is formed by zero modes of domain II with frequencies $\omega_s = s\omega_1 = s\Omega_i$, $s = 1, 2, 3, \ldots$. These data are illustrated in Fig. 7.35, where the time dependency of pressure in the point with coordinate $x' = 40$, $z = h_2/2$; $h_1' = 0.3$, $h_2' = 0.6$ is shown.

Let us increase the characteristic dimension of domain II to the value $h_2' = 0.9$ (zero and the first uniform modes remain in energy-carrying subspectrum of the signal). Figure 7.36 shows the time dependency of pressure in the point with coordinates $x' = 40$, $z = 0$.

Comparing Figs. 7.36 and 7.34(b) we note that the influence of dispersion properties of the first mode on initial impulse, when it is propagating in domain II is less, than in

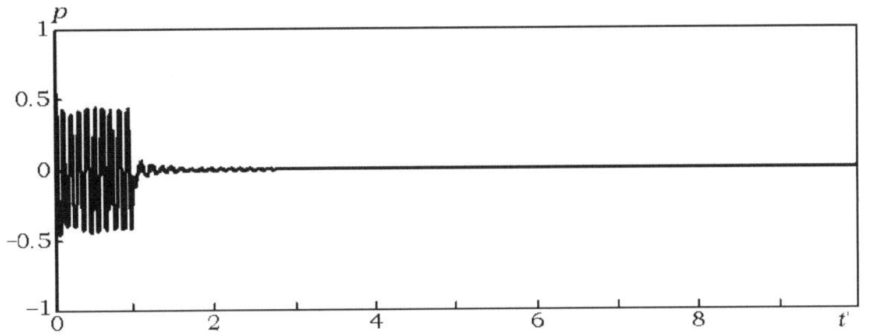

FIG. 7.35: Time dependency of pressure in the point with coordinate $x' = 40$, $z = h_2/2$; $h_1' = 0.3$, $h_2' = 0.6$

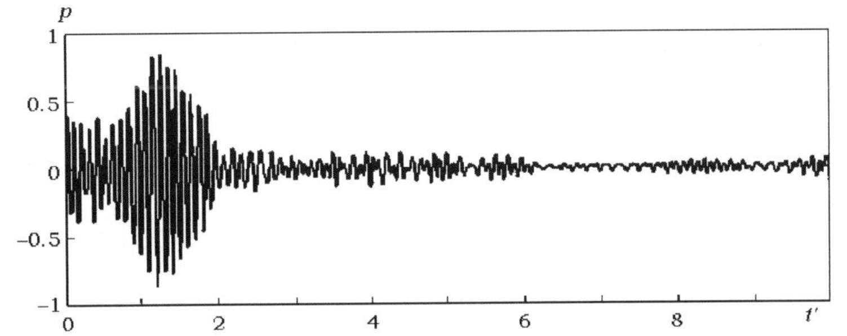

FIG. 7.36: Time dependency of pressure in the point with coordinate $x' = 40$, $z = 0$; $h_1' = 0.3$, $h_2' = 0.9$

Fig. 7.34(b). To explain this fact, we turn to the graphs in Fig. 7.32(a). The frequency dependencies of the group velocities of the first mode of domain II are shown here, together with the graph of amplitude spectrum of initial signal, for three variants of wave dimension h_2': curve 1—$h_2' = 0.9$, curve 2—$h_2' = 0.6$, curve 3—$h_2' = 0.4$. As we can see, numerical values of the group velocity of the first mode $[\omega_{90}, \omega_{110}]$, differ from zero mode velocity to a lesser extent, than corresponding numerical values for curve 2 ($h_2' = 0.6$) when the value of $h_2' = 0.9$ (curve 1) in the energy-carrying subspectrum of the signal. It explains the result mentioned above.

Let us now consider the frequency-modulated signal propagation in the waveguide. The carrier frequency increases from ω_0 to $\omega_0 (1 + \alpha\tau_i)$ in such signal on the time interval of impulse duration τ_i. In other words time dependence of initial signal is of the form [27]

$$p_0(t) = \begin{cases} \sin(\omega_0 t (1 + \alpha t)), & 0 \le t \le \tau_i, \\ 0, & \tau_i \le t \le T_i. \end{cases} \tag{7.71}$$

Using dimensionless parameters, we rewrite expression (7.71) as follows:

$$p_0(t') = \begin{cases} \sin(2\pi N t' (1 + \beta t')), & 0 \le t' \le 1, \\ 0, & 1 \le t' \le q. \end{cases} \tag{7.72}$$

Here $\beta = \alpha\tau_i$. The values of dimensionless parameters $N = 10$, $q = 10$ are unchanged, and the constant $\beta = 0.9$. When signal (7.72) has such parameters it is a broad band signal. Here 90% of signal energy is in frequency band $[\omega_{90}, \omega_{260}]$ [Fig. 7.32(b)]. It is to be recalled that 90% of signal (7.64) energy is in frequency band $[\omega_{90}, \omega_{110}]$ [Fig. 7.32(a)].

Figure 7.37 shows frequency dependencies of pressure in four observation points when frequency modulating signal is propagating in the waveguide with characteristic

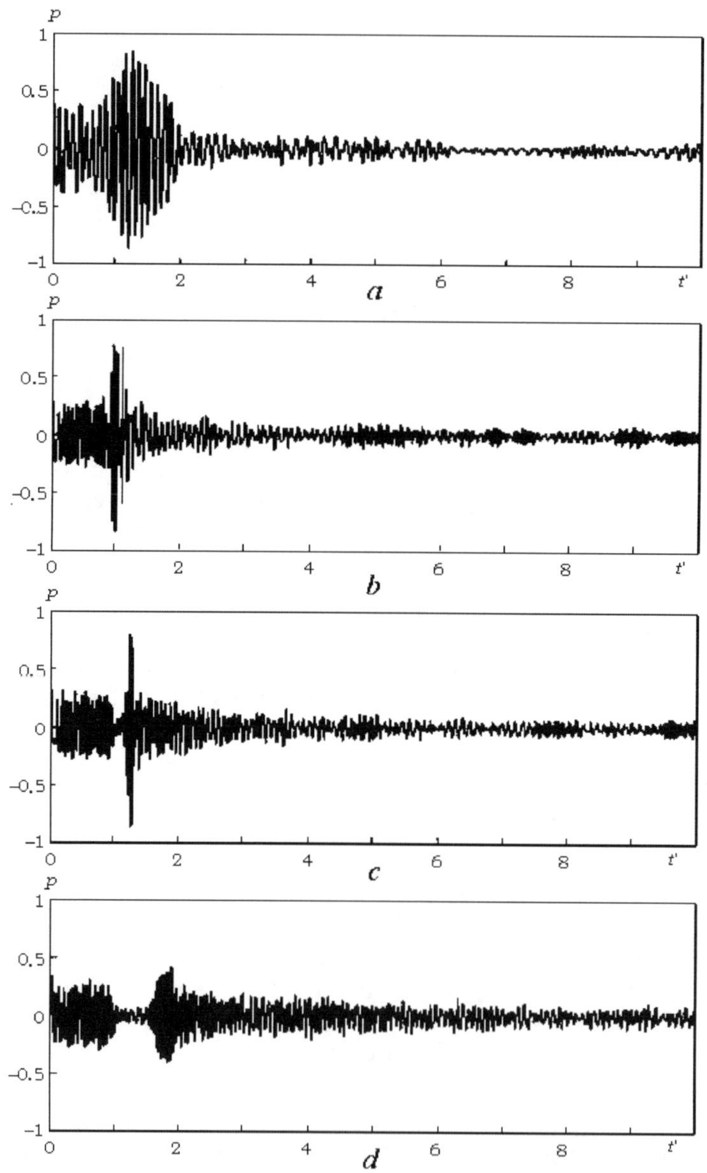

FIG. 7.37: Time dependencies of pressure in different observation points when frequency-modulating signal is propagating in the waveguide $h'_1 = 0.3$, $h'_2 = 0.6$, $z = 0$: (a)—$x' = 10$, (b)—$x' = 40$, (c)—$x' = 70$, (d)—$x' = 150$

dimensions $h_1' = 0.3$, $h_2' = 0.6$. The harmonic components of the first mode are uniform at frequencies $\omega > \omega_{83}$, of the second mode—at frequencies $\omega > \omega_{167}$, of the third mode—at frequencies $\omega > \omega_{250}$ for such dimensions of the waveguide.

Amplitude spectrum of the first three modes excitation of domain II is shown in Fig. 7.38, that is the modulus of coefficients $|(a_s + ib_s)\, B_{sn}|$ (here the mode number $n = 0, 1, 2$) at frequencies $\omega_s = s\omega_1 = s\Omega_i$. As we can see from the graphs in Fig. 7.38, when signal is propagating in domain II, the main part of signal energy is in zero and first modes.

When we compare the graphs in Fig. 7.37 with curves in Fig. 7.34 for nonmodulated impulse signal, it should be noted that there are the significant changes in time dependencies:

1. firstly, the time interval between the disturbances of zero and first modes declines considerably;

2. secondly, the dispersion focusing is observed for the first mode disturbance. Dispersion focusing was discussed in work [27]. This effect (the considerable increase of signal amplitude) appears near the zone of nonuniformity [Fig. 7.37(b)] and it remains unchanged when propagating at some distance [Fig. 7.37(c)]. The pressure jump fails at the distance $x' > 100$ for the chosen dimensions of the waveguide. Figure 7.37(d) shows time dependency in the point of observation with coordinate $x' = 150$. Pressure jump formation and its destruction are caused by interference procedures between harmonic components of impulse signal.

Let us place the observation point on the plane of zero pressure in the first mode, for example $x' = 40$, $z = h_2/2$ (Fig. 7.39). As we can see, there are signals 2 with the small amplitude in the pause of initial signal 1. We can address to the graphs in Fig. 7.38 to explain this fact. The harmonic components of the second mode become uniform at frequencies $\omega > \omega_{167}$. Consequently, these components determine whether there are signals 2 with the small amplitude (it is to be recalled that the observation point is in the plane $z = h_2/2$).

It might be supposed that the effect of dispersive focusing strengthens when zero mode is absent in domain II. The first mode in this case has more energy. Let the surface

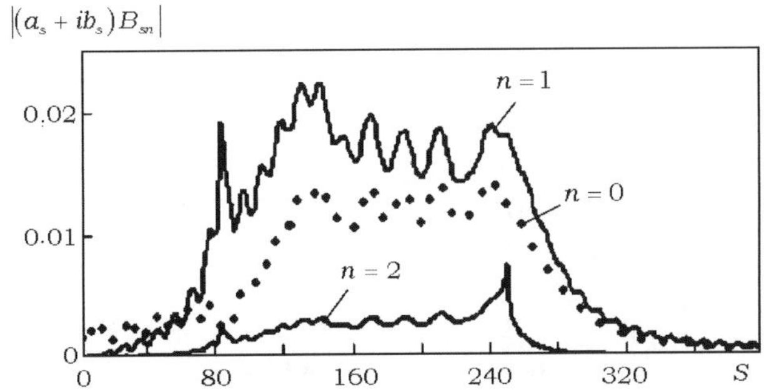

FIG. 7.38: Excitation amplitudes of zero (curve $n = 0$), first ($n = 1$), and second ($n = 2$) modes of domain II at frequencies $\omega_s = s\omega_1 = s\Omega_i$, $h_1' = 0.3$, $h_2' = 0.6$

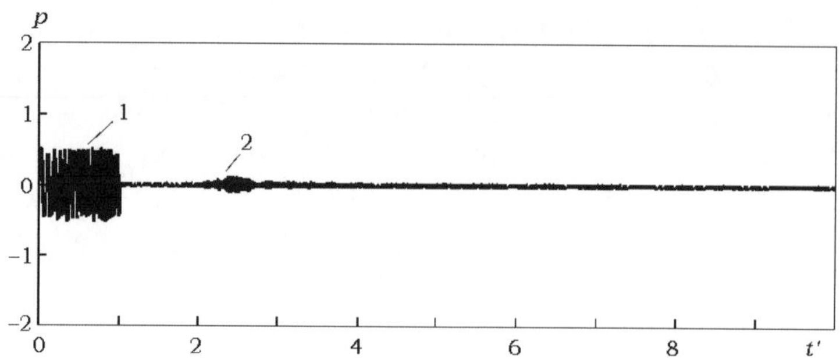

FIG. 7.39: Time dependence of pressure in the point with coordinates $x' = 40$, $z = h_2/2$; $h_1' = 0.3$, $h_2' = 0.6$

of the waveguide $x \geq 0$, $z = h_2$ be ideally soft, then zero mode is absent in domain II. Let $h_1' = 0.17$, $h_2' = 0.32$. The choice of such wave dimensions is conditioned by the fact that, the harmonic components of the first mode are uniform at frequencies $\omega > \omega_{78}$, the harmonic components of the second mode—at frequencies $\omega > \omega_{234}$. As a result, the main part of signal energy is in the first mode when it is propagating in domain II. Consequently, the pressure jump is observed at distances $x' \approx 20, ..., 100$. Its value on the surface $z = 0$ is twice as large as the amplitude of the incident wave pressure in domain I (Fig. 7.40 shows time dependency of pressure in the point with coordinates $x' = 70$, $z = 0$).

It may be noted that the received quantitative data can be useful in the design of air acoustic lines (for example, ventilating passages), and also when investigating sound propagation in a shallow sea.

7.5 SOUND PROPAGATION IN THE WAVEGUIDE, PARTLY PARTITIONED WITH RIBBON OF FINITE TRANSPARENCY

Let us consider the flat waveguide with acoustically rigid walls partially partitioned with thin ribbon ($\tau \to 0$) of finite transparency, Fig. 7.41. The waveguide is filled with ideal liquid with density ρ and sound velocity c. Let a plane harmonic wave propagate

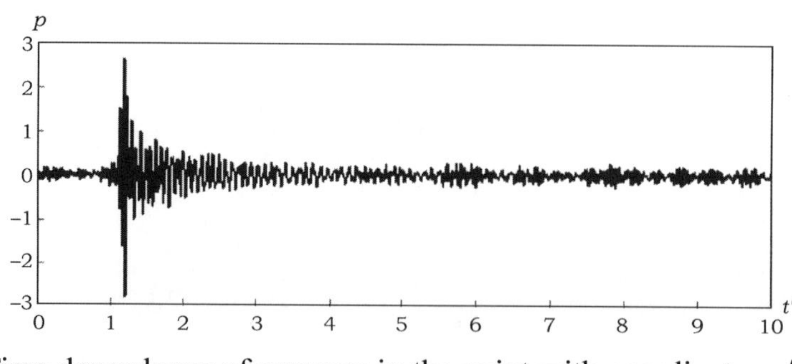

FIG. 7.40: Time dependency of pressure in the point with coordinates $x' = 70$, $z = 0$ (boundary $x \geq 0$, $z = h_2$ of domain II is ideally soft); $h_1' = 0.17$, $h_2' = 0.32$

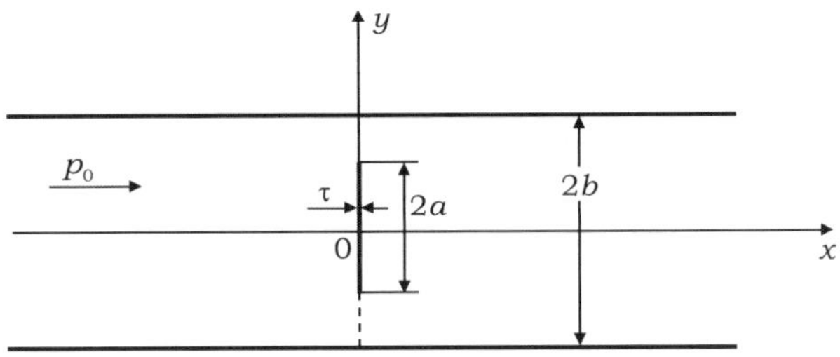

FIG. 7.41: Waveguide geometry

in the waveguide with pressure amplitude $p_0 = \exp(ikx)$. We omit time multiplier $\exp(-i\omega t)$.

Let us partition all the domain of field existence into two domains: the first domain $x \leq 0$ and the second domain $x \geq 0$. We write the pressure in each of the domains in the form of superposition of normal waves of the flat waveguide with acoustically rigid walls:

$$p_I = p_0 + \sum_{n=0}^{\infty} A_n \cos\left(\frac{n\pi y}{b}\right) \exp(-i\gamma_n x), \tag{7.73}$$

$$p_{II} = \sum_{n=0}^{\infty} B_n \cos\left(\frac{n\pi y}{b}\right) \exp(i\gamma_n x), \tag{7.74}$$

where $\gamma_n = \sqrt{k^2 - (n\pi/b)^2}$, $k = \omega/c$.

Let us investigate the sound transparency of the waveguide with different properties of the ribbon. In the first case, the ideally thin ribbon is characterized by surface mass M (kg/m^2). Then the normal speeds will be equal between each other at the boundary of domains

$$\frac{\partial p_I}{\partial x} = \frac{\partial p_{II}}{\partial x}, \quad x = 0, \quad 0 \leq |y| \leq b. \tag{7.75}$$

We substitute series (7.73), (7.74) in Eq. (7.75) and we use the properties of completeness and orthogonality of the functions $\cos(m\pi y/b)$, $m = 0, 1, ...$ Consequently, we find relation between the unknown complex coefficients

$$A_n = \delta_n - B_n, \quad \delta_0 = 1, \quad \delta_n = 0, \quad n > 0. \tag{7.76}$$

To find the unknowns B_n, it is necessary to formulate the second condition at the boundary of domains. We have the pressure balance on the interval $a \leq |y| \leq b$

$$p_I = p_{II}, \quad x = 0, \quad a \leq |y| \leq b. \tag{7.77}$$

Taking into account that particle velocity of each surface element of the ribbon is

$$\frac{1}{i\omega\rho}\frac{\partial p_I}{\partial x} = \frac{p_I - p_{II}}{Z_1}, \quad x = 0, \quad 0 \leq |y| \leq a, \tag{7.78}$$

where $Z_1 = -i\omega M$, and including (7.77), we obtain the second condition of fields conjugation

$$p_{II} = \begin{cases} p_I - \dfrac{Z_1}{i\omega\rho}\dfrac{\partial p_I}{\partial x}, & x = 0, \quad 0 \le |y| \le a, \\ p_I, & x = 0, \quad a \le |y| \le b. \end{cases} \tag{7.79}$$

Then we obtain the infinite system of linear algebraic equations of the second kind.

Let us consider some calculations results. Figure 7.42 shows us frequency dependencies of sound transmission coefficient W through the obstacle in the form of a ribbon with different values of M and perforation coefficient $\varepsilon = (b - a)/b$. The dependency of value W on frequency (curve 5) is also given here for the case of acoustically rigid ribbon (curve 5 is taken from [12]). Data analysis in Fig. 7.42 shows that when $M = 1000$ kg/m^2, curves 4 and 5 coincide in all the considered range of values $2b/\lambda$ with the exception of narrow interval in the vicinity of $2b/\lambda = 1$. Sound transparency of the waveguide increases with a decrease of surface density of the ribbon. The bigger is perforation coefficient ε the faster sound transparency increases.

Let us consider the second case, when the ribbon is characterized by surface elasticity S (N/m^3). The field matching conditions are as follows [2]:

$$p_I = p_{II}, \quad x = 0, \quad a \le |y| \le b, \tag{7.80}$$

$$\frac{1}{i\omega\rho}\frac{\partial p_{II}}{\partial x} = \begin{cases} \dfrac{1}{i\omega\rho}\dfrac{\partial p_I}{\partial x} - \dfrac{p_I}{Z_2}, & x = 0, \quad 0 \le |y| \le a, \\ \dfrac{1}{i\omega\rho}\dfrac{\partial p_I}{\partial x}, & x = 0, \quad a \le |y| \le b. \end{cases} \tag{7.81}$$

where $Z_2 = iS/\omega$. The results of calculation of transmission coefficients W are represented in Fig. 7.43. Curve 4, taken from [33], is given here for comparison. This curve corresponds to the case of acoustically soft ribbon. We can see that when $S = 1.12 \times 10^9$ N/m^3 in the range $2b/\lambda \ge 0.3$ curves 3 and 4 are very close to each other. But the curves 3 and 4 behavior differ from each other in the range of small values $2b/\lambda$. The

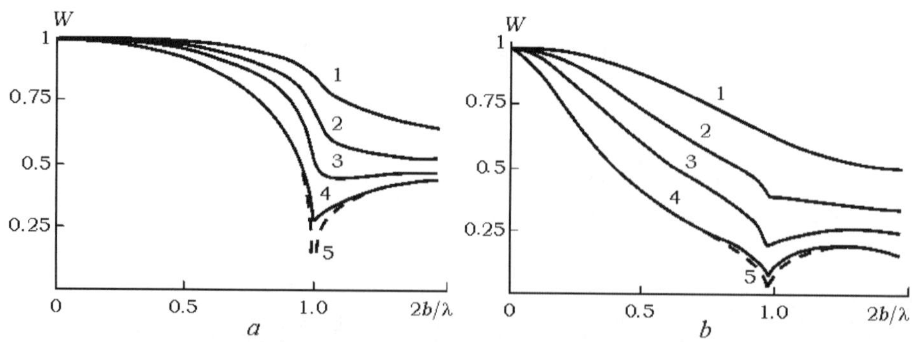

FIG. 7.42: Frequency dependencies of transmission coefficient W through the obstacle in the form of the ribbon with surface density M: (a)—$\varepsilon = 0.5$, (b)—$\varepsilon = 0.1$; curves 1–4 correspond to $M = 50, 100, 200, 1000$ kg/m^2, 5—acoustically rigid ribbon

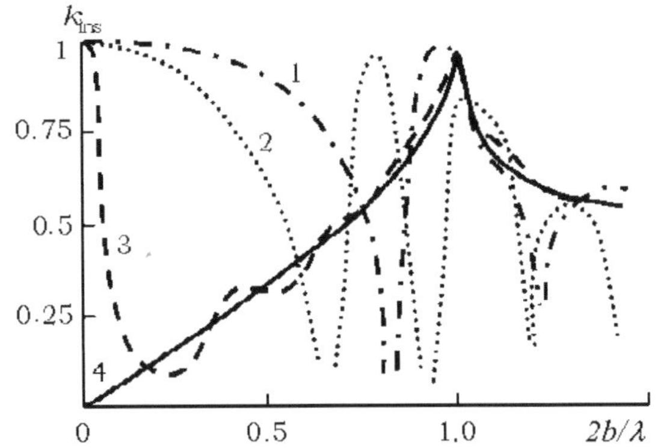

FIG. 7.43: Frequency dependencies of transmission coefficient W through the obstacle in the form of a ribbon with surface elasticity S, $\varepsilon = 0.5$: curves 1–3 correspond to $S = 4.5 \times 10^{10}$, 2.25×10^{10}, 1.125×10^{9} N/m^3, 4—acoustically soft ribbon

waveguide with an ideal ribbon ($S = 0$) becomes almost non sound transparent when $2b/\lambda$ decreases. The waveguide with a ribbon having finite surface elasticity ($S \neq 0$) remains sound transparent.

7.6 SOUND PROPAGATION IN THE FLAT WAVEGUIDE WITH ACOUSTICALLY SOFT INSERTIONS OF FINITE LENGTH

In this paragraph, let us consider the situation when the waveguide irregularity is caused by the inhomogeneity of the waveguide walls. Such design peculiarities of the waveguide are necessary to decrease the level of noise, which accompanies the processes in the waveguide structure. In other words, we consider the problem of creation the obstacle for the sound waves propagating in the pipeline without a change in the cross section of the waveguide.

Let us consider the flat waveguide with the rigid walls (Fig. 7.44). The walls of this waveguide are acoustically soft on the interval L. The waveguide is filled with an ideal liquid with density ρ and sound velocity c. The harmonic plane pressure wave $p_0 = \exp(ikx)$, $k = \omega/c$ with frequency ω is propagating in the waveguide in the direction of increase of Ox axis.

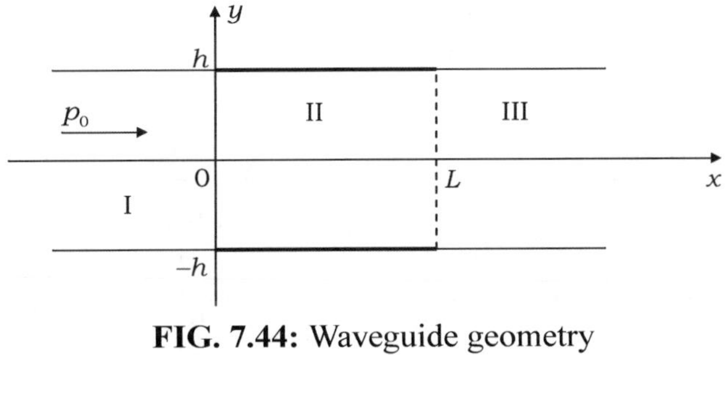

FIG. 7.44: Waveguide geometry

Let us partition the whole waveguide domain into three domains, Fig. 7.44. We represent the pressure fields in partial domains as follows:

$$p_I = p_0 + \sum_{n=0}^{\infty} A_n \cos\left(\frac{n\pi y}{h}\right) \exp\left(-i\gamma_n x\right), \tag{7.82}$$

$$p_{II} = \sum_{n=0}^{\infty} \cos\left(\frac{(2n+1)\pi y}{2h}\right) \left[C_n \exp\left(i\eta_n x\right) + D_n \exp\left(-i\eta_n \left(x - L\right)\right)\right], \tag{7.83}$$

$$p_{III} = \sum_{n=0}^{\infty} B_n \cos\left(\frac{n\pi y}{h}\right) \exp\left(-i\gamma_n \left(x - L\right)\right), \tag{7.84}$$

where $\gamma_n = \sqrt{k^2 - (n\pi/h)^2}$, $\eta_n = \sqrt{k^2 - ((2n+1)\pi/(2h))^2}$, $n = 0, 1, 2, \ldots$.

The matching conditions of the sound fields: the equality of pressure and normal velocities should be fulfilled at the boundaries of partial domains $x = 0$ and $x = L$, $|y| \leq h$.

Figure 7.45 shows frequency dependency of energy transmission coefficient W of the plane wave in the waveguide with acoustically soft insertions with length L: curve 1—$L/h = 0.25$, 2—$L/h = 0.5$, 3—$L/h = 1$. As we can see, the whole variation range of the waveguide width h/λ (λ—wave length) can be partitioned into two variants: the first—from $h/\lambda \ll 1$ to $h/\lambda \approx 0.25$, the second—$h/\lambda > 0.25$. There is no plane wave in the waveguide with the soft walls, and the lowest mode becomes uniform when $h/\lambda > 0.25$. As a result, when $h/\lambda > 0.25$ the transmission coefficient almost does not depend on the soft insertion length L.

The shape of the curves on the interval where the wave dimension $h/\lambda \leq 0.25$ is rather different. An increase of the soft insertion length L here leads to a considerable decrease of transmission coefficient W. It is caused by the fact that there are only nonuniform waves in domain II. Thus, a soft insertion is a significant obstruction for sound propagation in the waveguide.

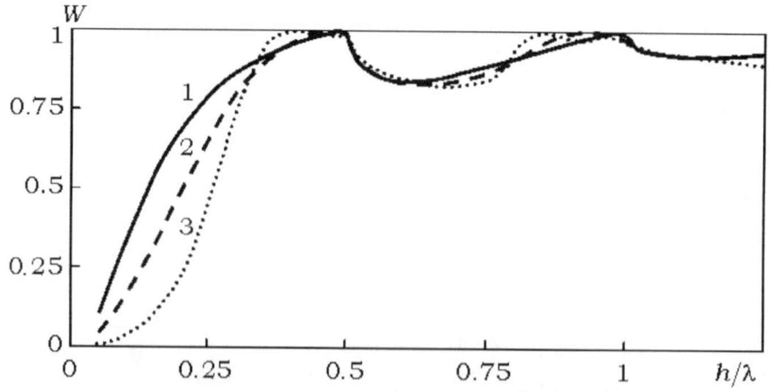

FIG. 7.45: Frequency dependency of energy transmission coefficient W of a plane wave in the waveguide with acoustically soft insertions with length L: 1—$L/h = 0.25$, 2—$L/h = 0.5$, 3—$L/h = 1$

7.7 ACOUSTIC PROPERTIES OF THE STETHOSCOPES

Auscultation (listening) of respiratory sounds, which appear in the respiratory channel of a person, is one of the most important diagnostic techniques of respiratory diseases. Auscultation is realized with a help of stethoscopes, phonendoscopes and microstethoscopes [34]. There are three main functional parts in the mentioned devices (monaural and binaural):

1. a head (socket), perceiving sound oscillations from the area of the body to which it is applied;
2. acoustic pipe—rigid or flexible tubes, connecting the head with the olives;
3. the olives, which are tightly embedded in the auditory passages of a doctor. The air in the acoustic pipe is a sound conductor.

Most of the modern devices are binaural. They have the devices, improving the obturation of auditory passages. In particular, the headband, consisting of comfortably bent metal tubes (which are the continuation of flexible acoustic pipe) with the olives at the ends [34]. According to [35], all the devices for auscultation can be broken down into four groups according to their constructional features: double-tube with flared end; double-tube with cup end; single-tube with flared end; single-tube with cup end.

Currently, a lot of companies produce the different models of stethoscopes, which differ from each other in sizes and main elements construction. Unfortunately, these differences cause the models differences in acoustic properties. This fact was proved repeatedly by experimental studies [35]. Consequently, a doctor uses one and the same model of stethoscope in the course of the work (he accommodates oneself to the acoustic properties of this model) he almost never changes the model of stethoscope, because it may lead to diagnostics errors.

From our point of view, there are considerable differences in acoustic properties of existing models of stethoscopes because of the fact that the stethoscopes development is not based on the rigorous analytical methods. It is based on practical experience of some company. This conclusion is confirmed by the analysis of the sources of literature of the recent 30 years. Although, there were some attempts to develop the approximate methods of calculation of the stethoscopes acoustic parameters (see, for example [36], where the transmission coefficient of the microstethoscope is based on the using of equivalent electric diagrams).

The purpose of this paragraph is to develop the acoustic and mathematic models of stethoscopes and to obtain the computational scheme of their acoustic properties [37].

7.7.1 Problem Statement and Solution Construction

The cross dimensions of the whole elements of the stethoscope are small in comparison with the wave length in the frequency range, used for auscultation. Thus, when modeling double-tube devices two tubes can be replaced with one. The cross-section area of this tube should be equal to the total area of the cross-section of the both tubes. The flared end can be replaced with a cylindrical chamber of equivalent volume. Its volume should correspond to a real flare design. The stethoscopes are schematized on Fig. 7.46.

Figure 7.46(a) shows acoustic model of a stethoscope with one sound pipe with bifurcate metal headband on the end. Figure 7.46(b)—shows acoustic model of a stethoscope with two sound pipes, each of which has metal headband. Two tubes are replaced with one, the cross-section area of which is equal to the total cross-section area of the two tubes. When the device has one sound pipe with bifurcate headband, the radius of equivalent tube section is equal $r_3 = r_3'\sqrt{2}$. When the device has two sound pipes $r_2 = r_2'\sqrt{2}$, $r_3 = r_3'\sqrt{2}$, where r_2' and r_3'—are the radiuses of the tubes of the sound pipe and headband respectively. Let the interior walls of all the stethoscope elements be absolutely rigid. It is acceptable, because the wave impedance of air is much less than the wave impedance of the material of the structure walls. We assume also that the sound field in the device is excited due to the harmonic vibrations of the surface $x = 0, 0 \le r \le r_1$ (imitating the skin vibrations) with the pressure amplitude $p_0 = 1$ Pa, uniformly distributed over the surface (piston radiation). We also assume that the device has some load with impedance Z_H from the opposite side.

When auscultating, the ends of the sound pipe with olives are placed in acoustic meatus. In this situation, the input impedance of the ear is the load for the stethoscope. There are a lot of works dedicated to the problem of the ear input impedance estimation (for example, [38]). There are a lot of differences in the obtained data in these works. It depends on the measurement conditions and different characteristics depending on

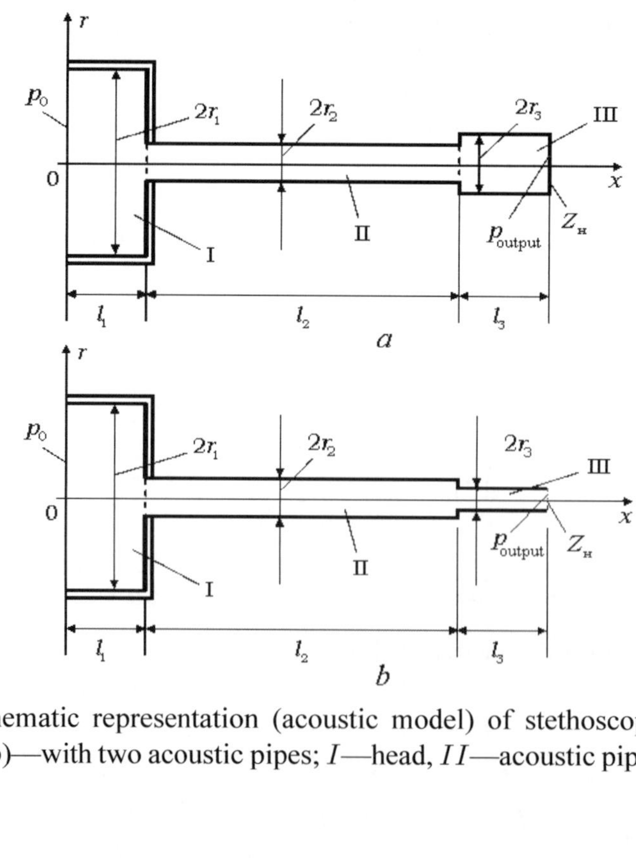

FIG. 7.46: Schematic representation (acoustic model) of stethoscope: (a)—with one acoustic pipe, (b)—with two acoustic pipes; *I*—head, *II*—acoustic pipe, *III*—headband

the edge, gender and other individual peculiarities. But there is a general tendency in the behavior of the frequency characteristics of the ear impedance, obtained by different researchers. Consequently, it is possible to use some average data for ear impedance, when constructing the mathematical model of auscultation. The results obtained in [38] are used below. The average readings in rather wide frequency range (125, ..., 10000 Hz) are given here. Low frequency range is especially important because the main frequency range of the most important auscultatory features is in the limits 180, ..., 1400 Hz [34].

Let us examine the harmonic sound field in the constructions shown in Fig. 7.46. All the field existence domain inside the stethoscope should be partitioned into three domains: domain I, domain II and domain III (see Fig. 7.46). Then the pressure field for each domain can be written as follows:

$$p_I = \sum_{n=0}^{\infty} A_n J_0(\alpha_n r) \exp(ik_n x) + \sum_{n=0}^{\infty} B_n J_0(\alpha_n r) \exp(-ik_n(x - l_1)), \tag{7.85}$$

$$\begin{aligned} p_{II} = &\sum_{n=0}^{\infty} C_n J_0(\beta_n r) \exp(i\gamma_n(x - l_1)) \\ &+ \sum_{n=0}^{\infty} D_n J_0(\beta_n r) \exp(-i\gamma_n(x - l_1 - l_2)), \end{aligned} \tag{7.86}$$

$$\begin{aligned} p_{III} = &\sum_{n=0}^{\infty} E_n J_0(\eta_n r) \exp(i\tau_n(x - l_1 - l_2)) \\ &+ \sum_{n=0}^{\infty} F_n J_0(\eta_n r) \exp(-i\tau_n(x - l_1 - l_2 - l_3)), \end{aligned} \tag{7.87}$$

where $k_n = \sqrt{k^2 - \alpha_n^2}$. Side walls are absolutely rigid, consequently the values α_n, β_n, and η_n are found from the following equations:

$$J_1(\alpha_n r_1) = 0, \quad J_1(\beta_n r_2) = 0, \quad J_1(\eta_n r_3) = 0. \tag{7.88}$$

The radius of acoustic pipe is small in comparison to the wave length in the considered frequency band. We should take into account viscous friction losses to construct adequate mathematical model of sound propagation process in a narrow acoustic pipe [16]:

$$\gamma_n = \sqrt{\delta_1^2 - \beta_n^2} - i\xi_1, \quad \tau_n = \sqrt{\delta_2^2 - \eta_n^2} - i\xi_2. \tag{7.89}$$

Here

$$\delta_1 = k\left(1 - \frac{1}{r_2'}\sqrt{\frac{\mu_1}{2\omega\rho}}\right)^{-1}, \quad \delta_2 = k\left(1 - \frac{1}{r_3'}\sqrt{\frac{\mu_2}{2\omega\rho}}\right)^{-1}, \tag{7.90}$$

$$\xi_1 = \frac{1}{r_2' c}\sqrt{\frac{\omega\mu_1}{2\rho}}, \quad \xi_2 = \frac{1}{r_3' c}\sqrt{\frac{\omega\mu_2}{2\rho}}. \tag{7.91}$$

In these relations μ_j, $j = 1, 2$—dynamic viscosity coefficient, which is $\mu_1 = 2 \times 10^{-3}$ Pa·s for air, and it is $\mu_2 = 4 \times 10^{-3}$ Pa·s for the walls with higher thermal conductivity (metal) [4]. Allowing for the losses, we should substitute the value of radius of one of the tubes of acoustic pipe or headband in Eqs. (7.90) and (7.91). It allows us to model the losses in the acoustic pipe more precisely. If these losses are not taken into account, then $\delta_j = k$, $\xi_j = 0$, $j = 1, 2$.

Sound fields matching conditions on the boundaries of partial domains are as follows:

$$p_I = p_{II}, \quad x = l_1, \quad 0 \le r \le r_1, \tag{7.92}$$

$$\frac{\partial p_I}{\partial x} = \begin{cases} \dfrac{\partial p_{II}}{\partial x}, & x = 0, \quad 0 \le r \le r_2, \\ 0, & x = l_1, \quad r_2 \le r \le r_1, \end{cases} \tag{7.93}$$

$$p_{II} = p_{III}, \quad x = l_1 + l_2, \quad \begin{cases} 0 \le r \le r_2, & r_2 < r_3, \\ 0 \le r \le r_3, & r_2 > r_3. \end{cases} \tag{7.94}$$

When $r_2 > r_3$

$$\frac{\partial p_{II}}{\partial x} = \begin{cases} \dfrac{\partial p_{III}}{\partial x}, & x = l_1 + l_2, \quad 0 \le r \le r_3, \\ 0, & x = l_1 + l_2, \quad r_3 \le r \le r_2. \end{cases} \tag{7.95}$$

When $r_2 < r_3$

$$\frac{\partial p_{III}}{\partial x} = \begin{cases} \dfrac{\partial p_{II}}{\partial x}, & x = l_1 + l_2, \quad 0 \le r \le r_2, \\ 0, & x = l_1 + l_2, \quad r_2 \le r \le r_3. \end{cases} \tag{7.96}$$

Bearing in mind the assumed above assumptions, the boundary conditions for the considered problem are as follows:

$$p_I = p_0, \quad x = 0, \quad 0 \le r \le r_1, \tag{7.97}$$

$$\frac{p_{III}}{Z_H} = \frac{1}{i\omega\rho}\frac{\partial p_{III}}{\partial x}, \quad x = l_1 + l_2 + l_3, \quad 0 \le r \le r_3. \tag{7.98}$$

Matching conditions (7.92)–(7.96) and boundary conditions (7.97), (7.98) are the system of six functional equations. Using the property of orthogonality of functions $J_0(\alpha_n r)$ on the interval $0 \le r \le r_1$, $J_0(\beta_n r)$ on the interval $0 \le r \le r_2$ and $J_0(\eta_n r)$ on the interval $0 \le r \le r_3$ ($n = 0, 1, 2, \ldots$), this system can be transformed into an infinite system of linear algebraic equations of the second kind with regard to the unknown coefficients A_n, B_n, C_n, D_n, E_n, and F_n. The solution of this system allows us to define all the characteristics of the sound field inside the stethoscope.

Let us examine the transmission characteristic of pressure of stethoscope. It is the relation of area mean pressure p_{output} in the cross section $x = l_1 + l_2 + l_3$ to area mean pressure p_{input} in the cross section $x = 0$:

$$H(f) = 20\lg \frac{p_{\text{output}}}{p_{\text{input}}}, \tag{7.99}$$

where

$$p_{\text{output}} = \frac{1}{\pi r_3^2}\int_0^{r_3} p_{III}\big|_{x=l_1+l_2+l_3} r dr, \quad p_{\text{input}} = \frac{1}{\pi r_1^2}\int_0^{r_1} p_0 r dr.$$

7.7.2 Numerical Results Analysis

We will use the data from work [39] and our own measurements, performed on real constructions, when choosing geometrical sizes for our models.

Figure 7.47 shows the graphs of transmission function $H(f)$ for single-tube [Fig. 7.47(a)] and double tube [Fig. 7.47(b)] stethoscopes. Curve 1 corresponds to the transmission function for some conditional base model for the flare dimensions $r_1 = 15$ mm, $l_1 = 7$ mm, the acoustic pipe dimensions $r_2 = 2\sqrt{2}$ mm for the case of double-tube device and $r_2 = 2$ mm for the case of single-tube device, $l_2 = 485$ mm and the headband dimensions $r_3 = 1{,}75\sqrt{2}$ mm, $l_3 = 200$ mm. Curve 2 shows the transmission function

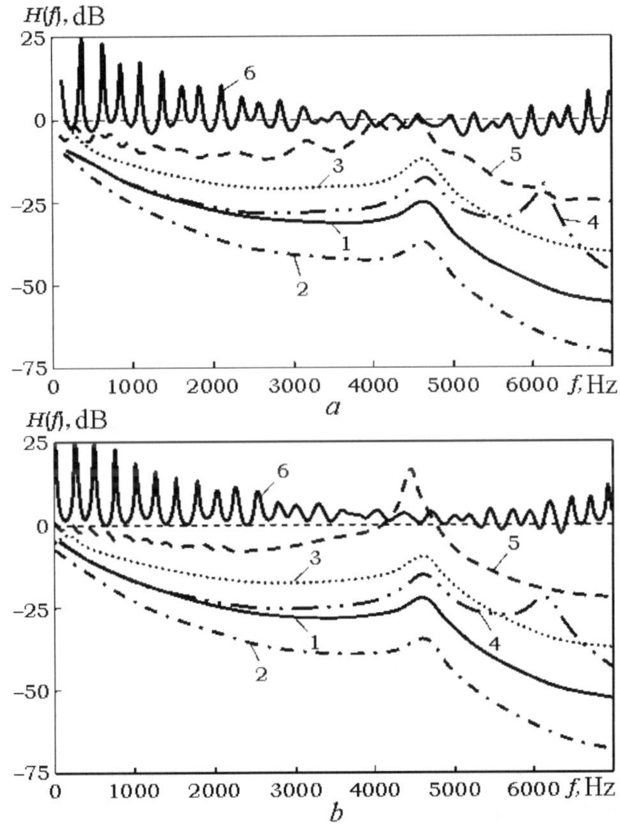

FIG. 7.47: Transmission function of stethoscope for different geometric dimensions of its constructional parts: (a)—for single-tube stethoscope, (b)—for double-tube stethoscope

when the acoustic pipe is 20 cm ($l_2 = 685$ mm) longer than the base model. Curve 3 shows the change of transmission function when the acoustic pipe is 20 cm ($l_2 = 285$ mm) shorter than the base model. Curve 4 corresponds to the doubled length of the socket ($l_1 = 14$ mm). Curve 5 describes the case when the radius of the acoustic pipe is doubled ($r_2 = 4\sqrt{2}$ mm for double-tube device and $r_2 = 4$ mm for single-tube device). Curve 6 describes the transmission function for the base model, but ignoring the viscous friction losses.

As we can see from the graphs in Fig. 7.47, the major role in transmission function construction plays losses accounting in the acoustic pipe, which appear because of viscosity. Numerous air column resonances in the construction are almost smoothed, except for the lowest frequencies only.

The socket geometrical dimensions are rather small. Thus, the socket length change influences the transmission characteristic only in the domain of high frequencies. The frequency, at which function has the maximums can be calculated by formula $f_p = (2n+1)c/(4l_1)$, $n = 0, 1, 2, \ldots$. This fact can explain the appearance of maximum on curve 4 at frequency 6127 Hz. Thus, we can increase the sound transfer efficiency at high frequencies, when increasing the length of the flare. The length of the socket should be less than 25 mm, to avoid the distortion in the frequency band of the main auscultatory characteristics.

As we can see, the dimensions of the acoustic pipe have the main impact on the transmission characteristic. An increase of the acoustic pipe length leads to the decreases of efficiency of sound transmission because of attenuation. It means that the acoustic pipe should be as short as possible. But it should be remembered that the device should be easy to use and it is undesirable to make it too short. Consequently, we should choose the device with the total length of the acoustic pipe and the headband not more than 40, ..., 70 cm. But the major role in the formation of transmission characteristic plays the acoustic pipe radius.

Figure 7.48 shows the graphs, illustrating the change of transmission characteristic $H(f)$ versus the change of radius of acoustic pipe and the headband. The radius of acoustic pipe in the region of headband changes together with the radius of acoustic pipe by formula $r_3 = r_2 - 0.25$ (mm). The other dimensions are equal to the base model.

For curve 1—$r_2 = 1$ mm, 2—$r_2 = 1.5$ mm, 3—$r_2 = 2$ mm (base model), 4—$r_2 = 3$ mm, 5—$r_2 = 4$ mm. The graphs show, that there is a loss of efficiency of sound transmission and even dips at low frequencies for rather narrow acoustic pipes. When the acoustic pipe radius is enlarged, sound transmission efficiency increases, because the viscous losses decrease and transmission characteristic becomes more wavy (especially at low frequencies).

The tube diameter being the same, the double-tube device is more efficient than single-tube device, because its total area of acoustic pipe is twice as large as the area of single-tube device.

One of the key features of the represented graphs is that all the curves have maximum in the vicinity of 4500, ..., 4700 Hz. The appearance of this maximum is caused by the resonance properties of ear membrane. It is the load for our model [38]. Thus, the transmission characteristic of stethoscope depends on individual characteristics of person ear.

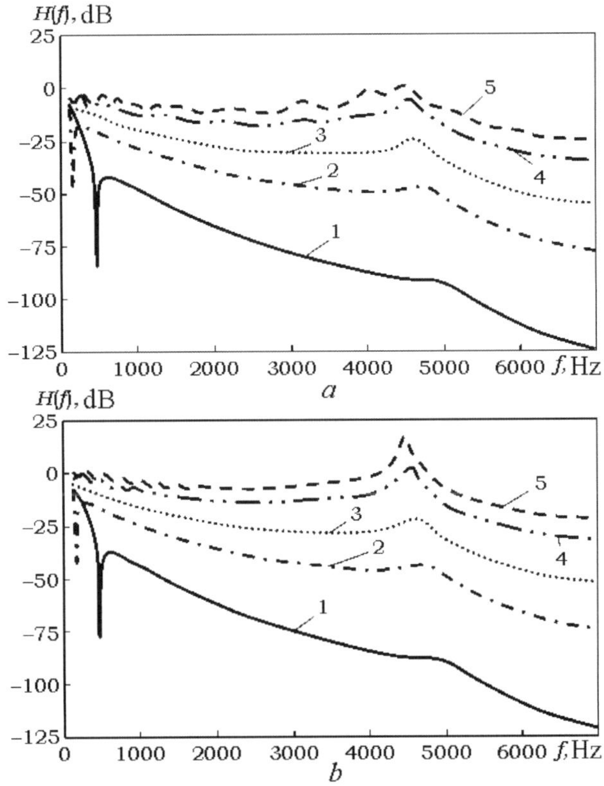

FIG. 7.48: The transmission function of stethoscope versus the change of acoustic pipe radius: (a)—for a single-tube stethoscope, (b)—for a double-tube stethoscope

The quantitative results obtained above are in agreement with experimental data, represented in article [40]. The dimensions of components of stethoscopes, which were investigated, are not given in this article (as well as in other similar articles) unfortunately. Consequently, the results of calculated and experimental data comparison are not sufficient. The graphs for single-tube [Fig. 7.49(a)] and double-tube [Fig. 7.49(b)] devices, obtained in the result of mathematic simulation are shown in Fig. 7.49 with curves 1, 2, 3. The averaged experimental data from [40] are shown with dots. Curve 1 describes the basic model with the dimensions indicated above, curve 2—shows the situation when the acoustic pipe length is 20 cm less than basic model ($l_2 = 285$ mm), curve 3—shows the situation when the acoustic pipe radius is longer than the radius of the basic model ($r_2 = 3$ mm).

As we can see, the levels of theoretical and experimental transmission functions almost coincide. We can observe the tendency for loss of efficiency of sound transmission with an increase of frequency, considering theoretical and experimental data. But the decreasing of efficiency shown by experimental data is not so smooth as it is shown by theoretical data (characteristic is more wavy). The last fact is probably connected with an inaccurate choice of the value of dynamic viscosity coefficient when calculating the losses in the acoustic pipe in the course of mathematic modeling.

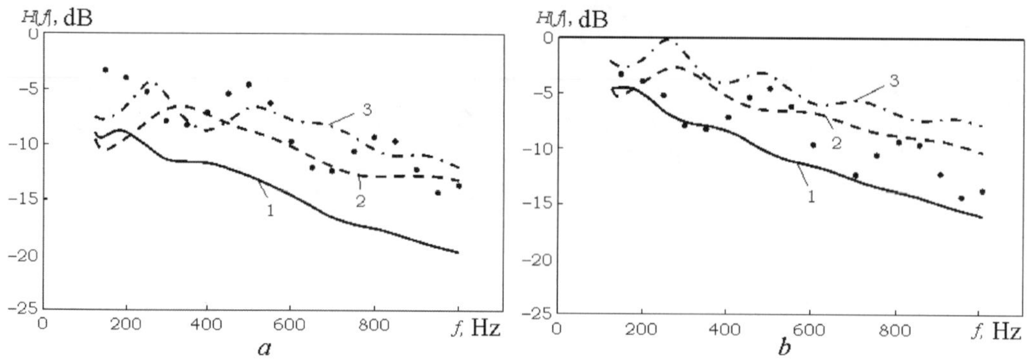

FIG. 7.49: The comparison of calculated transmission functions of stethoscope with experimental data: (a)—for single-tube stethoscope, (b)—for double-tube stethoscope

7.8 SOUND PROPAGATION IN CURVED WAVEGUIDE

Let us consider a flat waveguide, the surfaces of which are the arcs of circle on some interval, Fig. 7.50. A parallel-plate waveguide are adjacent to curved interval. Such constructions are often used in the models of acoustic and electromagnetic waveguides. That is why there is a considerable interest in the investigation of wave field in such waveguides (see, for example, [41,42]).

Let us introduce three coordinate systems to construct the problem solution: two Cartesian systems (x, y), (x', y') and polar coordinate system (r, ϕ) with the common origin at the point O. The radiuses of the surfaces of the waveguide curved segment are denoted by r_1 and r_2. The width of the waveguide is constant along the whole its length and is equal $h = r_2 - r_1$. Flare angle of the waveguide of its curved segment is denoted by ϕ_0. All the surfaces of the waveguide are supposed to be acoustically rigid or

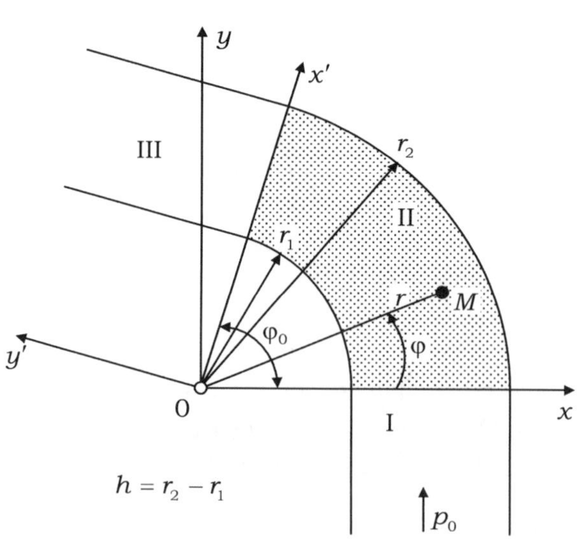

FIG. 7.50: Waveguide geometry

acoustically soft. The waveguide is filled with an ideal liquid with density ρ and sound velocity c. We do not write time factor $\exp(-i\omega t)$.

All the field existence domain is partitioned into three domains: domains I and III—parallel-plate waveguides, domain II—waveguide segment with curved surfaces. Let mth mode be propagating in domain I in the direction of increase of Oy. The fields in domains I and III then can be written as the superposition of corresponding modes of parallel-plate waveguide. The fields for the waveguide with acoustically rigid surfaces are written as

$$p_I = \cos(\alpha_m (x - r_1)) \exp(i\gamma_m y) + \sum_{n=0}^{\infty} A_n \cos(\alpha_n (x - r_1)) \exp(-i\gamma_n y), \quad (7.100)$$

$$p_{III} = \sum_{n=0}^{\infty} A_n \cos(\alpha_n (x' - r_1)) \exp(i\gamma_n y'), \quad (7.101)$$

where $\alpha_n = n\pi/h$, $\gamma_n = \sqrt{k^2 - \alpha_n^2}$, $k = \omega/c$.

The cosine functions are substituted with the functions $\sin(\alpha_n (x - r_1))$, where $\alpha_n = (n+1)\pi/h$, $n = 0, 1, 2, ...$ for the waveguide with acoustically soft surfaces in Eqs. (7.100) and (7.101).

7.8.1 Normal Waves in Curved Waveguide

Let us determine the expression for the normal waves of the curved waveguide segment. The particular solution of Helmholtz equation in the polar coordinates for domain II is written as [4]

$$p = [J_\nu (kr) + EY_\nu (kr)] \exp(i\nu\phi). \quad (7.102)$$

Using the boundary conditions at rigid surfaces of domain II ($\partial p/\partial r = 0$ when $r = r_1$ and $r = r_2$), we find the constant $E = -J'_\nu (kr_2) / Y'_\nu (kr_2)$. Consequently, we arrive at dispersion equation, which establishes a relation between frequency $\omega = kc$ and propagation constant ν:

$$J'_\nu (kr_2) Y'_\nu (kr_1) - J'_\nu (kr_1) Y'_\nu (kr_2) = 0. \quad (7.103)$$

The prime in Eq. (7.103) denotes the derivative of Bessel function with respect to argument. Likewise, we use the boundary conditions at the soft surfaces of domain II ($p = 0$ when $r = r_1$ and $r = r_2$), we determine the constants $E = -J_\nu (kr_2) / Y_\nu (kr_2)$ and we arrive at the following dispersion equation:

$$J_\nu (kr_2) Y_\nu (kr_1) - J_\nu (kr_1) Y_\nu (kr_2) = 0. \quad (7.104)$$

The dispersion curves for the acoustically soft surfaces of the waveguide are shown in Fig. 7.51(a), and the curves for the waveguide with the rigid walls are shown in Fig. 7.52(a). Non-dimensional frequency $\Omega = kr_2 = \omega r_2/c$ is plotted along the Y-axis, the propagation constant ν is plotted along X-axis. Radius $r_1 = r_2 q$, where constant $q < 1$ (in this case $q = 0.6$). The dashed curves correspond to the flat waveguide with

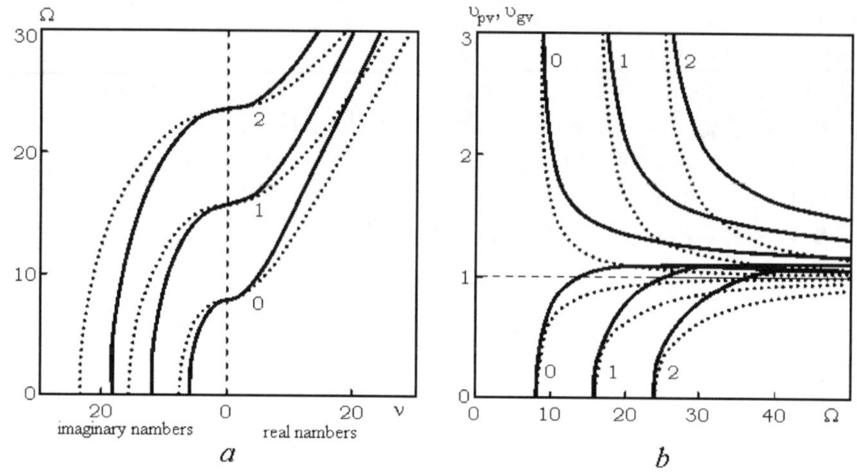

FIG. 7.51: Dispersion curves (a) and frequency dependencies of phase (upper curves group) and group velocities (lower curves group) (b) for curved waveguide with acoustically soft surfaces, $q = 0.6$

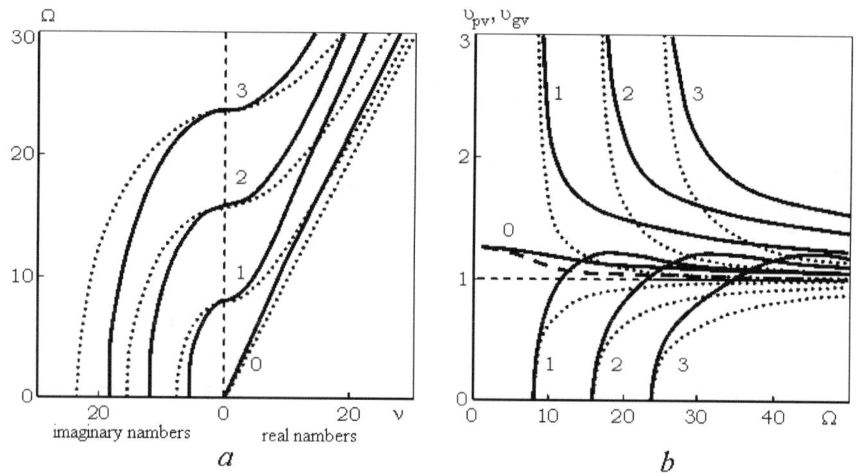

FIG. 7.52: Calculations for acoustically rigid waveguide. Notation corresponds to Fig. 7.51

the same cross dimension as the cross dimension of the curved waveguide. The figures near the curves enumerate the normal wave. As we can see, the critical frequencies of the modes of curved and plane-parallel waveguides coincide with graphical accuracy. But then the shapes of the curves are different from each other. This fact influences the construction of frequency dependencies of phase and group velocities of normal waves [Figs. 7.51(b) and 7.52(b)].

Thus, nth mode of the waveguide can be represented as follows:

$$p_n = R_n (kr) \exp (i\nu_n \phi), \tag{7.105}$$

where the eigenforms of the modes for acoustically rigid waveguide

$$R_n(kr) = J_{\nu_n}(kr) - \frac{J'_{\nu_n}(kr_2)}{Y'_{\nu_n}(kr_2)} Y_{\nu_n}(kr), \tag{7.106}$$

and for acoustically soft waveguide

$$R_n(kr) = J_{\nu_n}(kr) - \frac{J_{\nu_n}(kr_2)}{Y_{\nu_n}(kr_2)} Y_{\nu_n}(kr). \tag{7.107}$$

As an example, Figs. 7.53 (soft walls) and 7.54 (rigid walls) show the eigenforms of curved waveguide for the uniform and the first non-uniform modes. As we can see, the eigenforms of the first modes of indicated waveguides are very much different.

The propagation constant is indicated for each form ν_n, $n = 0, 1, ..., 5$. The points show the eigenforms of flat waveguide with the same wave width, as the wave width of the curved waveguide

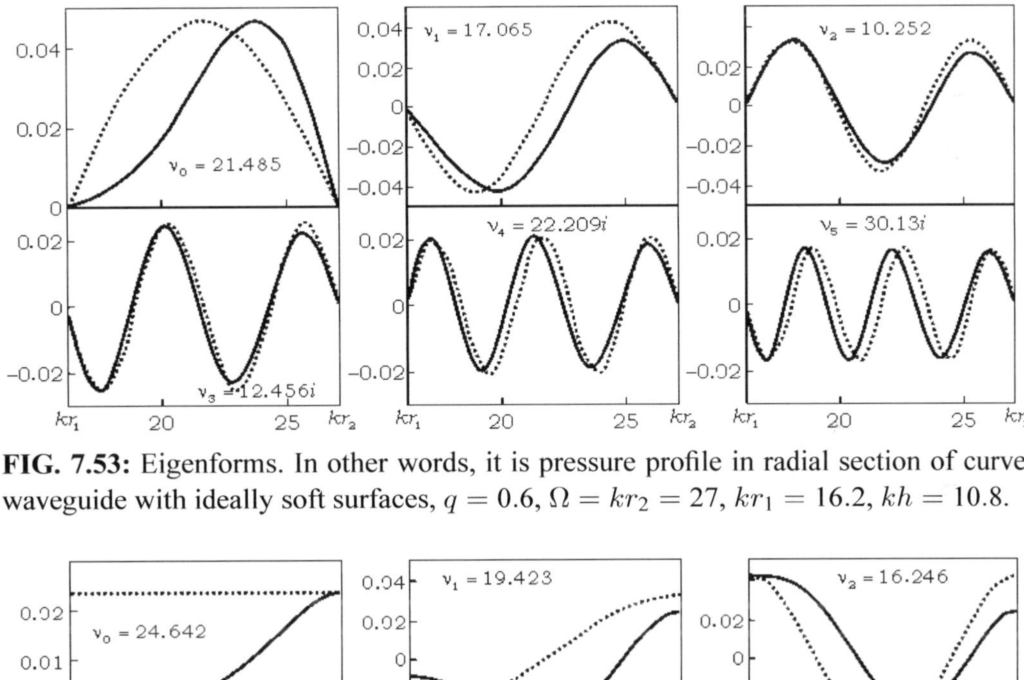

FIG. 7.53: Eigenforms. In other words, it is pressure profile in radial section of curved waveguide with ideally soft surfaces, $q = 0.6$, $\Omega = kr_2 = 27$, $kr_1 = 16.2$, $kh = 10.8$.

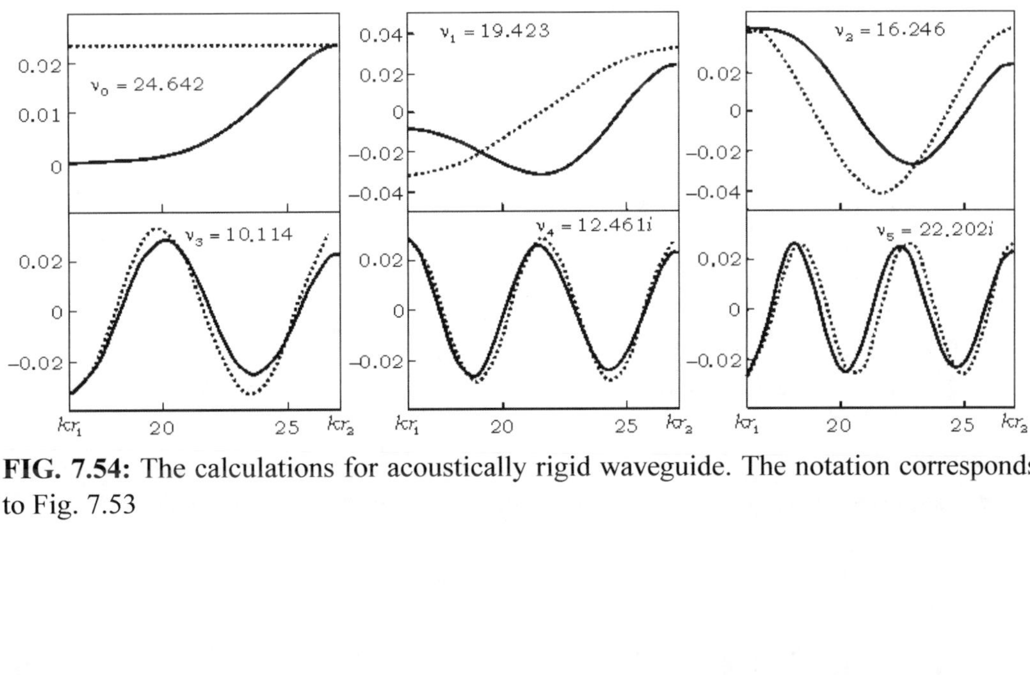

FIG. 7.54: The calculations for acoustically rigid waveguide. The notation corresponds to Fig. 7.53

We write the field in domain II as follows:

$$p_{II} = \sum_{n=0}^{\infty} R_n(kr)\left[B_n \exp(i\nu_n\phi) + C_n \exp(i\nu_n(\phi - \phi_0))\right]. \tag{7.108}$$

We should write the conditions of sound fields matching at the boundaries of partial domains and use the property of orthogonality of the set of functions $\cos(\alpha_n(x - r_1))$, $x = [r_1, r_2]$ and $R_n(kr)$, $r = [r_1, r_2]$ to solve the problem of wave propagation in the waveguide. Such procedure was more than once described in this book and that is why it is omitted here.

7.8.2 Sound Propagation in the Bent Waveguide for Different Variants of Bend Zone

7.8.2.1 Statement and Construction of Problem Solution

Figure 7.55 shows four variants of flat waveguide bent at right angle: waveguide in Fig. 7.55(a) is called "fracture waveguide," Fig. 7.55(b) shows waveguide with rounded bend zone. The corner radii in two bend points are the same. Such waveguide is called

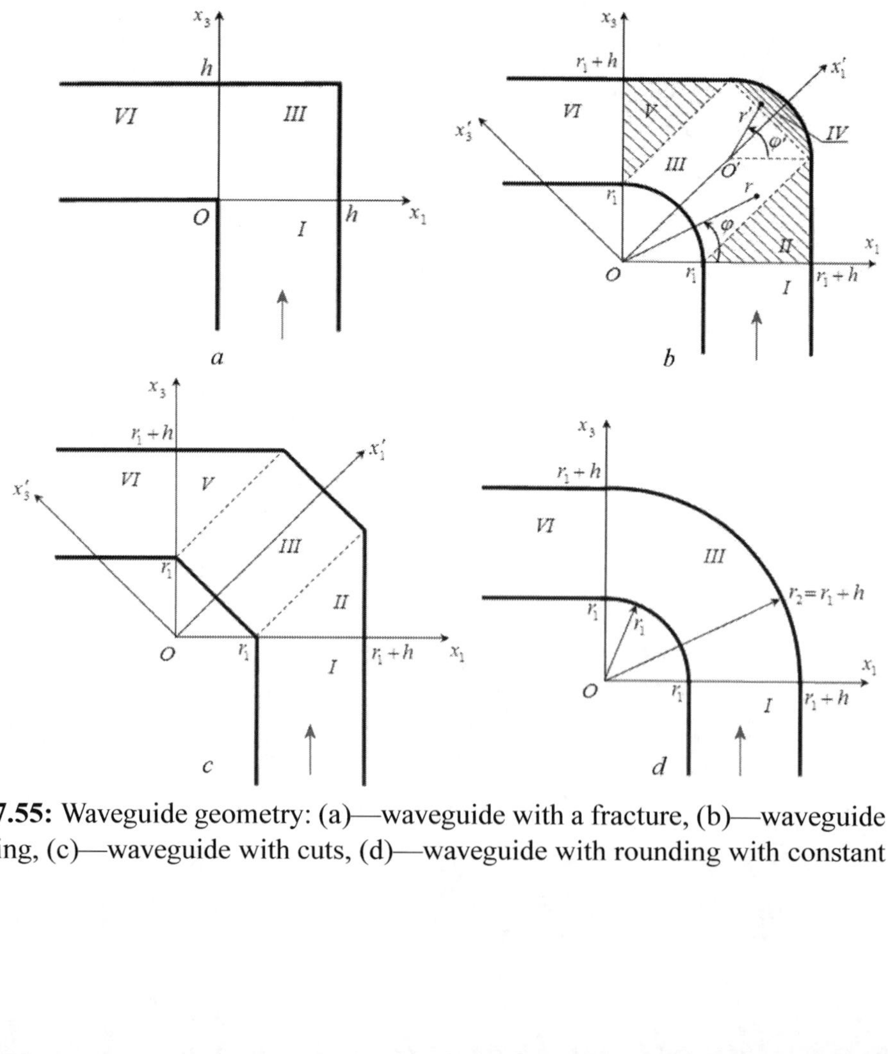

FIG. 7.55: Waveguide geometry: (a)—waveguide with a fracture, (b)—waveguide with a rounding, (c)—waveguide with cuts, (d)—waveguide with rounding with constant width

"waveguide with rounding." In Fig. 7.55(c) the rounded edges are cut off, such waveguide can be defined as "waveguide with cuts." The characteristic feature of the indicated waveguides is the change of cross dimension of waveguide in the bend zone. Figure 7.55(d) shows the variant of the waveguide bend rounding. The waveguide width is unchanged along its entire length. Such waveguide is called "the waveguide with rounding with constant width." The surfaces of the waveguide are acoustically rigid.

Waveguide with a fracture [Fig. 7.55(a)] is considered in Section 7.1. Let us compare the acoustic characteristics of the second, third and fourth waveguides [Figs. 7.55(b)–7.55(d)], with the characteristic features of the waveguide with a fracture.

To describe the waveguide geometry and to construct the solution of the problem of a harmonic wave propagation in the waveguide shown in Fig. 7.55(b), we introduce four coordinate systems: two Cartesian coordinate systems (x, y), (x', y'), polar coordinate system (r, ϕ) with the common center in point O and polar coordinate system (r', ϕ') with the center in point O'. Radii of the rounding of internal (point O) and external (point O') points of bend are the same; let us denote their value as r_1. The width of flat waveguides, adjacent to the bend is h. Thus, the initial values of the waveguide geometry are h and r_1.

All the domain of sound field existence is partitioned into six domains: domains I and VI—parallel-flat waveguides, domains II, III, IV, V—are the bend zone. The geometry of boundaries of the indicated zones is shown in Fig. 7.55(b).

Let the sth mode of plane-parallel waveguide

$$p_0^{(s)} = \cos(\alpha_s (x - r_1)) \exp(i\gamma_s y)$$

be propagating in domain I in positive direction of Oy axis. We write the fields in domains I and VI as the superposition of plane-parallel waveguide modes. The pressure fields for a waveguide with acoustically rigid surfaces are as follows:

$$p_I = \cos(\alpha_s (x - r_1)) \exp(i\gamma_s y) + \sum_{n=0}^{\infty} A_n^{(1)} \cos(\alpha_n (x - r_1)) \exp(-i\gamma_n y), \quad (7.109)$$

$$p_{VI} = \sum_{n=0}^{\infty} A_n^{(6)} \cos(\alpha_n (y - r_1)) \exp(-i\gamma_n x), \quad (7.110)$$

where $\alpha_n = n\pi/h$, $\gamma_n = \sqrt{k^2 - \alpha_n^2}$, $k = \omega/c$, ω—frequency. We don't write time factor $\exp(-i\omega t)$.

We write the fields in domains II and V as follows:

$$\begin{aligned} p_{II} = &\sum_{n=0}^{\infty} A_n^{(21)} \cos(\alpha_n (x - r_1)) \exp(i\gamma_n y) \\ &+ \sum_{n=0}^{\infty} A_n^{(22)} \cos(\alpha_n y) \exp(-i\gamma_n (x - (r_1 + h))) \\ &+ \sum_{n=0}^{\infty} A_n^{(23)} \cos(\beta_n (x' - a)) \exp(-i\eta_n (y' + a)), \end{aligned} \quad (7.111)$$

$$\begin{aligned} p_V &= \sum_{n=0}^{\infty} A_n^{(51)} \cos\left(\alpha_n \left(y - r_1\right)\right) \exp\left(i\gamma_n x\right) \\ &+ \sum_{n=0}^{\infty} A_n^{(52)} \cos\left(\alpha_n x\right) \exp\left(-i\gamma_n \left(y - \left(r_1 + h\right)\right)\right) \\ &+ \sum_{n=0}^{\infty} A_n^{(53)} \cos\left(\beta_n \left(x' - a\right)\right) \exp\left(i\eta_n \left(y' - a\right)\right), \end{aligned} \tag{7.112}$$

where $\beta_n = n\pi/H, \eta_n = \sqrt{k^2 - \beta_n^2}$. Geometrical dimensions of boundary zones are determined by values $H = \sqrt{2}h$ and $a = r_1/\sqrt{2}$.

The boundaries of domains III and IV contain curvilinear sections. The pressure fields in these zones are as follows:

$$\begin{aligned} p_{III} &= \sum_{n=0}^{\infty} A_n^{(31)} \cos\left(\beta_n \left(x' - a\right)\right) \exp\left(i\eta_n \left(y' + a\right)\right) \\ &+ \sum_{n=0}^{\infty} A_n^{(32)} \cos\left(\beta_n \left(x' - a\right)\right) \exp\left(-i\eta_n \left(y' - a\right)\right) \\ &+ \sum_{n=0}^{\infty} A_n^{(33)} \cos\left(\chi_n \left(y' + a\right)\right) \exp\left(-i\mu_n \left(x' - \left(a + H\right)\right)\right) \\ &+ \sum_{n=0}^{\infty} A_n^{(34)} \cos\left(2n\phi\right) \frac{H_{2n}^{(1)}\left(kr\right)}{H_{2n}'^{(1)}\left(kr_1\right)}, \end{aligned} \tag{7.113}$$

$$\begin{aligned} p_{IV} &= \sum_{n=0}^{\infty} A_n^{(41)} \cos\left(\chi_n \left(y' + a\right)\right) \exp\left(i\mu_n \left(x' - \left(a + H\right)\right)\right) \\ &+ \sum_{n=0}^{\infty} A_n^{(42)} \cos\left(2n\phi'\right) \frac{J_{2n}\left(kr'\right)}{J_{2n}'\left(kr_1\right)}, \end{aligned} \tag{7.114}$$

where $\chi_n = n\pi/(2a)$, $\mu_n = \sqrt{k^2 - \chi_n^2}$, $J_{2n}(kr)$ and $H_{2n}^{(1)}(kr)$—Bessel and Hankel functions of the first kind.

Domain IV [Fig. 7.55(b)] is absent in the waveguide with cuts [Fig. 7.55(c)], and curvilinear boundary of domain III should be replaced with linear section. We write the superposition of modes of the corresponding flat-parallel waveguide instead of the fourth sum in Eq. (7.113).

For the rounded waveguide of constant width [Fig. 7.55(d)] the transition section between two flat-parallel waveguides is formed by domain III. Pressure field in domain III is the sum of four infinite series:

$$\begin{aligned} p_{III} &= \sum_{n=0}^{\infty} B_n \cos\left(\alpha_n \left(x - r_1\right)\right) \exp\left(i\gamma_n y\right) + \sum_{n=0}^{\infty} C_n \cos\left(\alpha_n \left(y - r_1\right)\right) \exp\left(i\gamma_n x\right) \\ &+ \sum_{n=0}^{\infty} E_n \cos\left(2n\phi\right) \frac{J_{2n}\left(kr\right)}{J_{2n}'\left(kr_2\right)} + \sum_{n=0}^{\infty} F_n \cos\left(2n\phi\right) \frac{Y_{2n}\left(kr\right)}{Y_{2n}'\left(kr_1\right)}. \end{aligned} \tag{7.115}$$

Here $r_2 = r_1 + h$, $Y_{2n}(kr)$—Neumann function. The first two sums in Eq. (7.115) enables us to fulfill the conditions of matching the fields on the boundaries of particle domains I, III and III, VI. Boundary conditions on rigid surfaces of curvilinear domain III ($r = r_1$, $\phi = [0, \phi_0]$ and $r = r_2$, $\phi = [0, \phi_0]$) enables us to fulfil the third and the fourth sums in Eq. (7.115).

Let us write the sound fields matching conditions on the boundaries of particle domains (Fig. 7.55). These conditions are the equality of pressure and normal velocities on these boundaries. Using the orthogonality properties of the corresponding sets of functions, we obtain the infinite system of linear algebraic equations of the second kind in unknown coefficients. The infinite system of equations is solved by reduction method. This procedure is described in our book. The energy coefficients of reflection and transmission of the wave which is incident on the waveguide bend is defined by Eqs. (7.11)–(7.15).

7.8.2.2 Numerical Results Analysis: Tone Signal

Let zero mode be incident on the waveguide bend ($s = 0$) in domain I. Figure 7.56 shows dependencies of transmission coefficient $W^{(0)}$ for the waveguide with fractures and for the waveguide with rounding. It should be noted that when the rounding radius is small ($r_1/h = 0.06$ for curve 1) the calculation curves coincide with graphical accuracy for the waveguide with fractures and for the waveguide with rounding. When we increase the rounding radius (curves 2, 3, 4) the medium value of transition coefficient increases. Irregularity remains the same for all the radii of the rounding.

When $h \ll \lambda$, for all the curves, transmission coefficient $W^{(0)}$ is equal to unity, and it corresponds to the situation when the sound is propagating in "very narrow tube." In the vicinity of value $h/\lambda = 0.5$ (Fig. 7.56) there is a dip to zero in the characteristics of transmission coefficients $W^{(0)}$. In the waveguide with fracture (curve 1), when the value

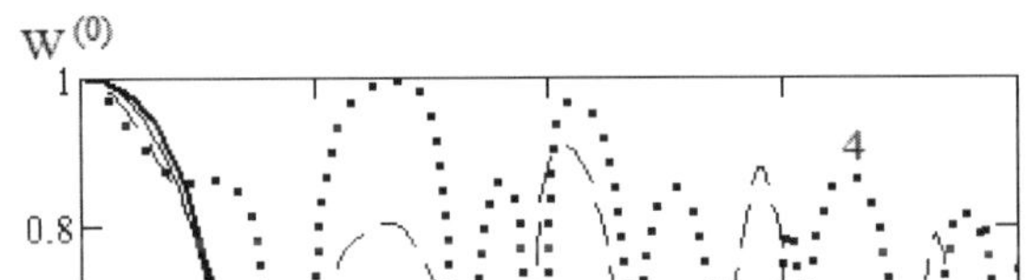

FIG. 7.56: Frequency dependencies of transmission coefficient $W^{(0)}$ for the waveguide with fracture and for the rounded waveguide: 1—$r_1/h = 0.06$ waveguide with fracture, 2—$r_1/h = 0.3$, 3—$r_1/h = 0.6$, 4—$r_1/h = 1$

h is divisible by the half wave length $\lambda/2$, the transmission coefficient is equal to zero $W^{(0)}$.

To understand the nature of curves represented in Fig. 7.56, let us consider similar dependencies of excitation coefficients of normal waves $W_n^{(0)}$ of domain VI. Figures 7.57(a) and 7.57(b) shows frequency dependencies of excitation coefficients of the first normal waves in domain VI (curve number determines the mode) when zero mode is incident on the bend in domain I. Zero mode of the waveguide with rigid boundaries is propagating for any wave dimension of the waveguide. Normal wave with number $n > 0$ is propagating when the condition $h > n\lambda/2$ is fulfilled. The graphs show that if the wave dimension $h/\lambda < 0.5$, than only zero normal wave is propagating in domain VI (curve 0). It determines the transmission of flat wave (zero mode) of domain I through the bent of the waveguide. The situation changes, if $h/\lambda > 0.5$. In the waveguide with a fracture [Fig. 7.57(a)] when $h/\lambda > 0.5$ only the first normal wave is excited (curve 1). This wave exceeds zero mode, and when $h/\lambda > 1$ the second normal wave dominates (curve 2) and than higher modes. Consequently, the energy of zero mode in domain I is transformed in normal waves in domain VI with numbers $n > 0$. If value h is divisible by half-wave length $\lambda/2$, the excitation coefficients of the modes are equal to zero. There is an effect of the waveguide "closing" in the waveguide with a fracture and in the waveguide with the small rounding radius [Fig. 7.57(a), curve 1]. It is closely connected with the generation of the next normal wave in the waveguide.

The nature of the curves changes when the rounding radius increases. Zero values of excitation coefficients disappear in the vicinity of the points where h/λ is divisible by half wave length starting with high frequencies. The effect of closing appears only in the vicinity of point $h/\lambda = 0.5$. Figure 7.57(b) shows the corresponding graphs for the case when $r_1/h = 1$. There is strong irregularity in frequency characteristics of excitation energy coefficients when $h/\lambda > 0.5$ in domain VI.

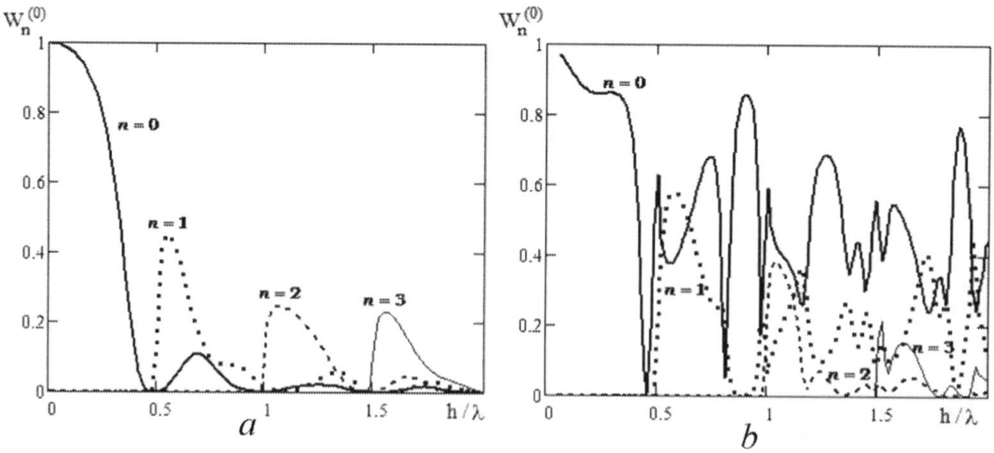

FIG. 7.57: Frequency characteristics excitation coefficient of normal waves $W_n^{(0)}$ in the domain VI for the waveguide with fracture and for rounded waveguide: (a)—$r_1/h = 0.06$ waveguide with fracture, (b)—$r_1/h = 1$

Figure 7.58 shows pressure amplitude field when $r_1/h = 1$, $h/\lambda = 0.453$. This situation corresponds to the waveguide closing [Fig. 7.57(b)]. Standing wave appears in domain I. Pressure distribution in domain VI corresponds to the first mode. For the given wave dimension h/λ the first mode is non-uniform and that is why the energy flow is equal to zero.

Figure 7.59 shows frequency characteristics of transmission coefficient $W^{(0)}$ for waveguide with cuts [Fig. 7.55(c)]. If we compare Figs. 7.56 and 7.59 we can find the common features of all the curves. The irregularity in frequency characteristics of the waveguide with cuts is stronger.

Let now turn to the rounded waveguide of constant width h [Fig. 7.55(d)]. Here r_1—internal radius of rounding, and external radius of rounding $r_2 = r_1 + h$. Figure 7.60 shows frequency characteristics of transmission coefficient $W^{(0)}$ (curve 1) and excitation coefficients of normal waves $W_n^{(0)}$ in domain VI. The waveguide of constant width

FIG. 7.58: Pressure amplitude field in the vicinity of the bend of the rounded waveguide: $r_1/h = 1$, $h/\lambda = 0.453$, zero mode is incident ($s = 0$)

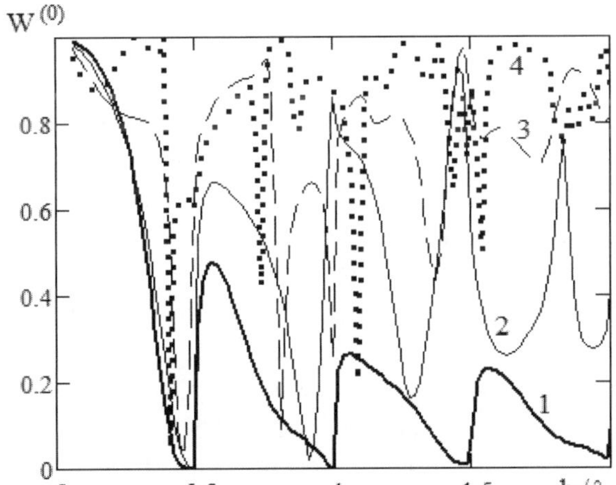

FIG. 7.59: Frequency characteristics of transmission coefficient $W^{(0)}$ for the waveguide with cuts [Fig. 7.55(c)]: 1—$r_1/h = 0.06$, 2—$r_1/h = 0.3$, 3—$r_1/h = 0.6$, 4—$r_1/h = 1$

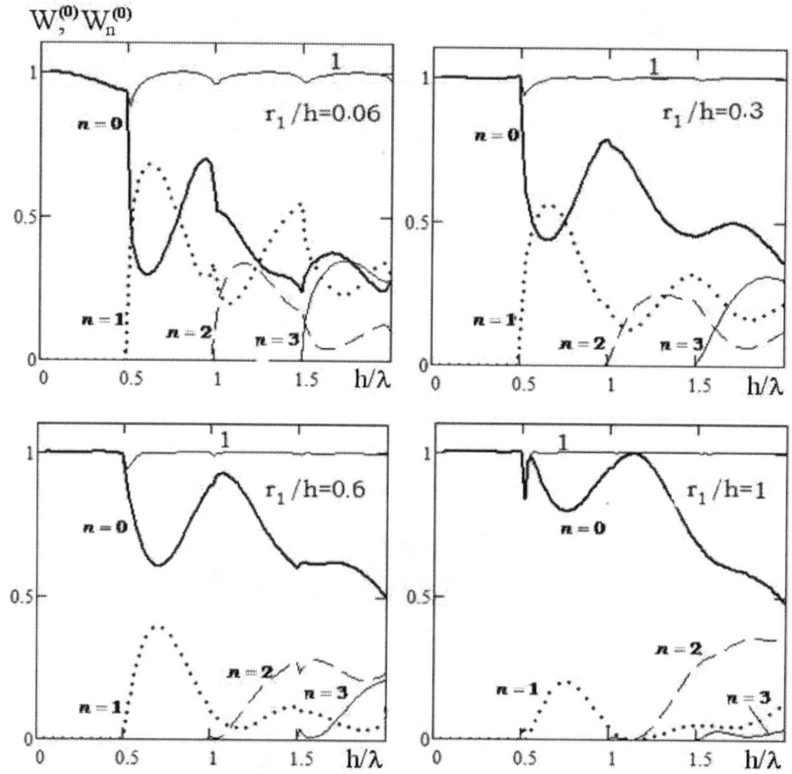

FIG. 7.60: Frequency characteristics of transmission coefficient $W^{(0)}$ (curve 1) and excitation coefficients of normal waves $W_n^{(0)}$ in domain VI for the rounded waveguide of constant width for different values r_1/h

is almost sound transparent in all the frequency range in contrast to the dependencies represented in Fig. 7.56. But, as it is seen in Fig. 7.60, mode field structure in domain VI depend on geometry parameters in waveguide r_1/h and h/λ. Comparing Figs. 7.56 and 7.60 we can see some similarities and differences. In the waveguide characteristics we can select frequency domains, where one of the waveguide modes dominates. There is no effect of waveguide "closing" in the rounded waveguide of constant width in these domains. As we can see in Fig. 7.60, with an increase of value of r_1/h, zero mode becomes more dominant.

Figure 7.61 shows the pressure amplitude fields in the vicinity of bend of waveguide of constant width. Zero mode is incident here, wave width $h/\lambda = 1.15$ (there are three uniform modes with such width). Figure 7.61(a) corresponds to radius $r_1/h = 1$, Fig. 7.61(b)—$r_1/h = 0.06$. As we can see in Fig. 7.61(b) field structure changes round the bend. It is due to the fact that all three uniform modes are energy-carrying (see Fig. 7.60). If we increase the radius of the rounding [Fig. 7.61(a)] the structure of the incident wave field round the bend almost doesn't change (zero mode dominates here). The flow of energy in the zone of rounding "is forced against" the surface with radius r_2, avoiding the low pressure zone.

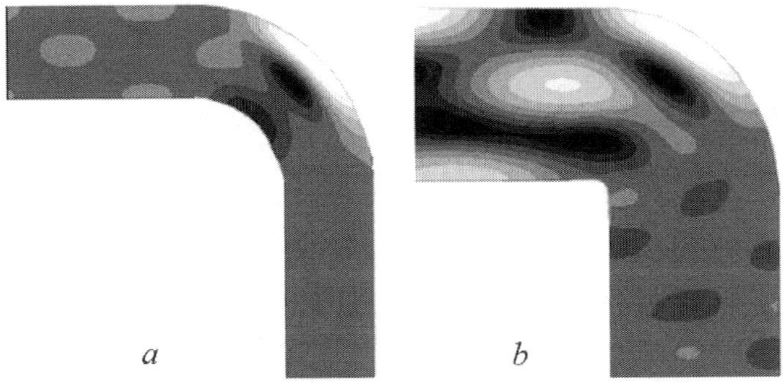

FIG. 7.61: Pressure amplitude fields in the vicinity of the waveguide bent with the rounding of constant width, $h/\lambda = 1.15$: (a)—$r_1/h = 1$, (b)—$r_1/h = 0.06$

7.8.2.3 Numerical Results Analysis: Impulse Signal

The change in the field structure round the bent should influence the impulse propagation in such waveguide. When impulse signal is propagating in the waveguide its space-time structure is influenced by two main mechanisms of the waveguide. One of them is caused by the fact that the waveguide is a filter (a normal wave is not uniform at the frequencies less than critical frequency). The other mechanism is determined by dispersion of waves propagated in the waveguide.

The solution for harmonic signal allows us to construct the solution of non-stationary problem. We use the model of impulse signal, which is described in Section 7.4.3. Let us take the following dimensionless parameters: $N = 10$, $q = 10$. Figure 7.62 shows the amplitude spectrum and time dependency of signal. The frequency of hundredth of harmonics ($s = 100$) is equal to carrier signal frequency. It should be noted that there is 90% of the whole signal energy in the frequency band $[\omega_{90}, \omega_{110}]$. Consequently, this impulse signal can be considered as narrow-band signal.

Let zero a mode be propagating in domain I of the waveguide with time dependency (see Section 7.4.3)

$$p_0 = \sum_{j=0}^{\infty} (a_j + ib_j) \exp\left(-i\omega_j t + i\gamma_{j,0} y\right). \tag{7.116}$$

The field in domain VI should be written as follows:

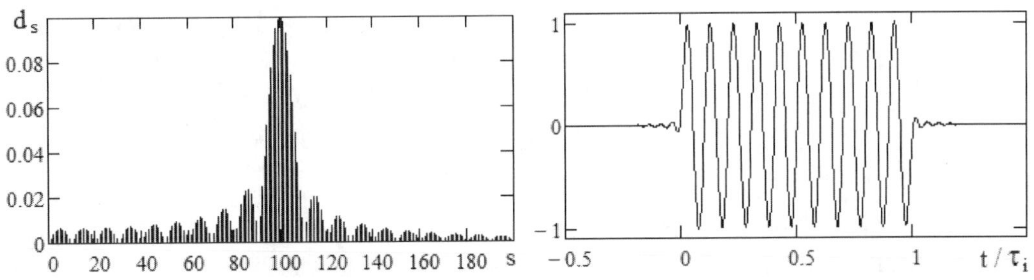

FIG. 7.62: Amplitude spectrum and time dependency of impulse signal

$$p_{VI} = \sum_{j=0}^{\infty}(a_j + ib_j)\sum_{n=0}^{\infty} D_{j,n}\cos(\alpha_n(y - r_1))\exp(-i\omega_j t - i\gamma_{j,n}x), \qquad (7.117)$$

where propagation constant $\gamma_{j,n} = (\omega_j/c)\sqrt{1-(\tilde{\omega}_n^2/\omega_j^2)}$, a $\tilde{\omega}_n = n\pi c/h$—critical frequency of nth mode; $d_j = \sqrt{a_j^2 + b_j^2}$—amplitude spectrum (Fig. 7.62).

Let us consider the situation, when one of the modes of the waveguide dominates in domain VI. For example, according to Fig. 7.60, zero mode is dominant, when $h/\lambda_0 = 1.15$, where λ_0—wave length at carrier frequency and $r_1/h = 1$. There is no dispersion in a zero mode and impulse holds its shape at any distance from the bend. This is shown in Fig. 7.63. Time dependency of pressure in the midpoint of the waveguide is shown here on the left, time dependency of pressure amplitude in the cross-section is shown on the right, which is at the distance $150\lambda_0$ from the bend. Non-dimensional time $t' = t/\tau_i$ is plotted along the abscise axis.

Let us decrease the radius of rounding r_1, the waveguide width remains the same, $h/\lambda_0 = 1.15$, $r_1/h = 0.06$. In this case, according to Fig. 7.60, there are three modes. Zero mode and the second mode are dominant. For this width of the waveguide [see Eq. (7.117)], the modes with the following numbers are propagating: zero mode ($n = 0$) for $j \geq 0$, first mode ($n = 1$) for $j > 44$, second mode ($n = 2$) for $j > 88$. All tree modes are in band of energy-carrying frequencies. Figure 7.64 show the change of impulse form when it is propagating in domain VI.

Figure 7.64(a) shows how impulse enters into domain VI. Impulse in the midpoint is almost undistorted and its amplitude is 2 times larger than the amplitude of incident impulse in domain I. As we can see in Fig. 7.64(a) (on the right) the impulse amplitude changes when the observation point moves long the waveguide cross-section. Figures 7.64(b) and 7.64(c) corresponds to the situation when the observation point moves away from the bend. Zero mode and the second mode are spatially separated here, because group velocity of the second mode is less than group velocity of zero mode. Initial impulse is deformed. It divides into two impulses with about the same level. The first impulse, formed by zero modes, has the same duration as initial impulse. The second impulse is formed by the second modes. It is time and space stretched because of dispersion.

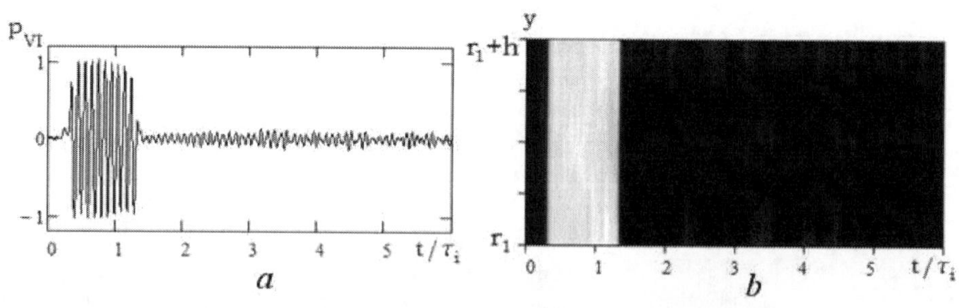

FIG. 7.63: Time dependencies of pressure in domain VI in the rounded waveguide of constant width in the points with coordinates $(y - r_1)/h = 0.5$, $x/\lambda_0 = -150$ (a) and pressure amplitude in cross-section $x/\lambda_0 = -150$ (b); $h/\lambda_0 = 1.15$, $r_1/h = 1$

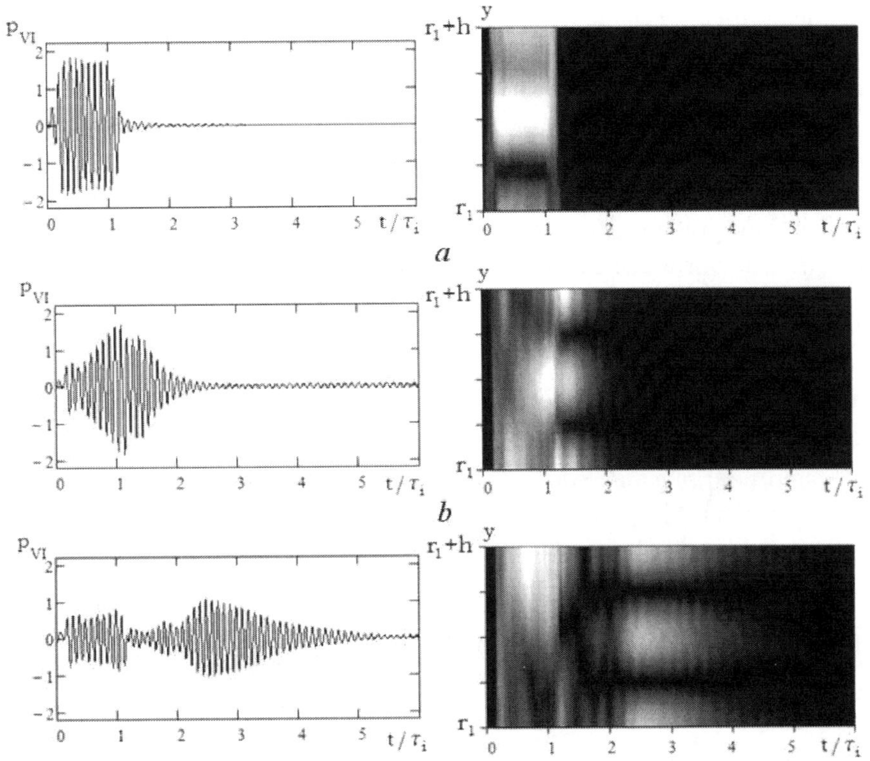

FIG. 7.64: Time dependencies of pressure in domain VI in the rounded waveguide of constant width in the point with coordinates $(y - r_1)/h = 0.5$, x/λ_0 and pressure amplitude in cross-section x/λ_0; $h/\lambda_0 = 1.15$, $r_1/h = 0.06$: (a)—$x/\lambda_0 = 0$, (b)—$x/\lambda_0 = -6$, (c)—$x/\lambda_0 = -20$

In conclusion we point out the relation of this investigation to music. Some wind instruments have complex system of bends sound-transmitting pipes. For example, a horn, a trombone and others. Music signal it's a temporal flow of impulses of different duration. We can say that the number of bends and their geometrical parameters, determine the tone color of the given instrument. Of course, the form of the modern musical instrument is the fruit of labor of many experts, which follow their perception of the sounding of a musical instrument in the process of its creation.

7.9 SOUND PROPAGATION IN A CYLINDRICAL WAVEGUIDE WITH SPHERICAL CAVITY

The main results, achieved using partial domains method, refer to the cases, when neighboring domains do not intersect (they have one common boundary). If the neighboring domains intersect, the traditional ways of application of partial domains methods can be ineffective [43]. Chapter 1 describes the approach, allowing us to use the partial domains method in the situation, when partial domains intersect. For illustrative purposes, let us consider the problem of plane wave propagation in the round cylindrical waveguide with spherical cavity.

Let us consider the waveguide, consisting of two semi-infinite cylindrical tubes with round cross-section, connected by spherical cavity, Figure 7.65 (the figure shows waveguide cross-section, the symmetry axis of the waveguide is in the cross-section's plane). Tube radius R_0, radius of spherical cavity r_0. All the surfaces of the waveguide are acoustically rigid. Waveguide is filled with ideal compressible liquid with density ρ and sound velocity c. Let us introduce two coordinate systems: cylindrical (x, R, ψ) and spherical (r, θ, ψ) with common center in point O. We direct the axis x of cylindrical coordinate system along the tube axis.

Let plane harmonic wave be incident on spherical cavity (zero mode of cylindrical waveguide) with frequency ω. In this case sound field has radial symmetry. It doesn't depend on angular coordinate ψ. We don't write temporal factor $\exp(-i\omega t)$.

The sound field existence domain is divided into three partial domains (Fig. 7.65). Domain I—semi-infinite tube $-\infty < x \leq -x_0$, $0 \leq R \leq R_0$; domain II—spherical cavity with radius r_0; domain III—semi-infinite tube $x_0 \leq x < \infty$, $0 \leq R \leq R_0$. Such partition is natural. But, as we can see, partial domains intersect. To solve this problem we use the approach described in Chapter 1.

We write the pressure in domain I as the superposition of modes of cylindrical waveguide:

$$p_I = \exp(ik(x + x_0)) + \sum_{n=0}^{\infty} A_n J_0(\beta_n R) \exp(-i\gamma_n (x + x_0)), \tag{7.118}$$

where β_n—roots of equation $J_0'(\beta R_0) = 0$ (the prime means the derivative with respect to argument of Bessel function of zero order), propagation constant $\gamma_n = \sqrt{k^2 - \beta_n^2}$, $k = \omega/c$. The first member of sum in Eq. (7.118) is incident plane wave with the amplitude equal to unity. The second member of sum determines the wave reflected from spherical cavity.

The field in domain III determines the wave, propagated through the spherical cavity

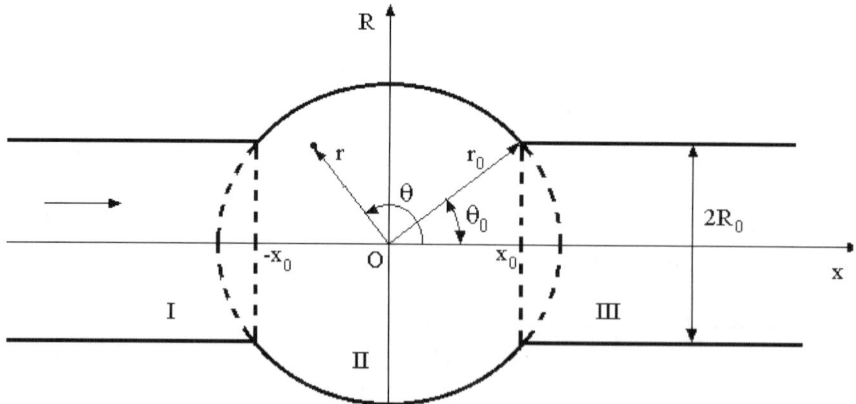

FIG. 7.65: Waveguide geometry

$$p_{III} = \sum_{n=0}^{\infty} B_n J_0 (\beta_n R) \exp (i\gamma_n (x - x_0)). \tag{7.119}$$

Let us write the pressure field in domain II as the sum of the modes of spherical cavity

$$p_{II} = \sum_{n=0}^{\infty} E_n P_n (\cos\theta) \frac{j_n (kr)}{j_n' (kr_0)}. \tag{7.120}$$

Here $P_n (\cos\theta)$—Legendre polinomials, $j_n (kr)$—spherical Bessel functions.

We write the matching conditions of the fields as follows:

1. $$p_I = p_{II}, \quad x = -x_0, \quad R = [0, R_0], \tag{7.121}$$

2. $$p_{II} = p_{III}, \quad x = x_0, \quad R = [0, R_0], \tag{7.122}$$

3. $$\frac{\partial p_{II}}{\partial r} = \begin{cases} \dfrac{\partial p_{III}}{\partial r}, & r = r_0, \quad \theta = [0, \theta_0], \\ 0, & r = r_0, \quad \theta = [\theta_0, \pi - \theta_0], \\ \dfrac{\partial p_I}{\partial r}, & r = r_0, \quad \theta = [\pi - \theta_0, \pi]. \end{cases} \tag{7.123}$$

The matching conditions of the fields are the pressure balance on the plane surfaces of the boundary of partial domains I, II and II, III [conditions (7.121) and (7.122)]. Condition (7.123) sets the matching of the pressure derivatives with respect to radial coordinate on spherical surface of the boundary of partial domains. We omit the further description of solution construction.

The calculations showed that the curves of pressure and normal velocity on the plane and spherical surfaces of boundaries of partial domains coincide with graphical accuracy. The small domain in the vicinity of conjunction line of cylindrical and spherical surfaces is an exception. The divergence of the curves of particle velocity is the natural in this domain.

Figure 7.66 shows the dependency of pressure amplitude along x coordinate for two values of radial coordinate R: Fig. 7.66(a)—$R = R_0/2$, Fig. 7.66(b)—$R = 0$. Waveguide parameters $r_0/R_0 = 1.6$, $R_0/\lambda = 1.1$. Thin lines correspond to the calculation of field in cylindrical domains I (on the left) and III (on the right). The bullet points determine the calculation in spherical domain II. Thin vertical lines form the overlap region of partial domains I, II and II, III. It is clearly seen that the calculation results of sound fields coincide in the intersection zones of partial domains I, II and II, III. This fact was discussed in Chapter 1.

Figure 7.67 shows energy transmission coefficient dependency W on the tube wave radius R_0/λ when the value $r_0/R_0 = 1.6$. As we can see, the spherical cavity in cylindrical waveguide can influence the frequency characteristic of energy transmission coefficient W. We can observe the dips to zero in frequency response of transmission coefficient W.

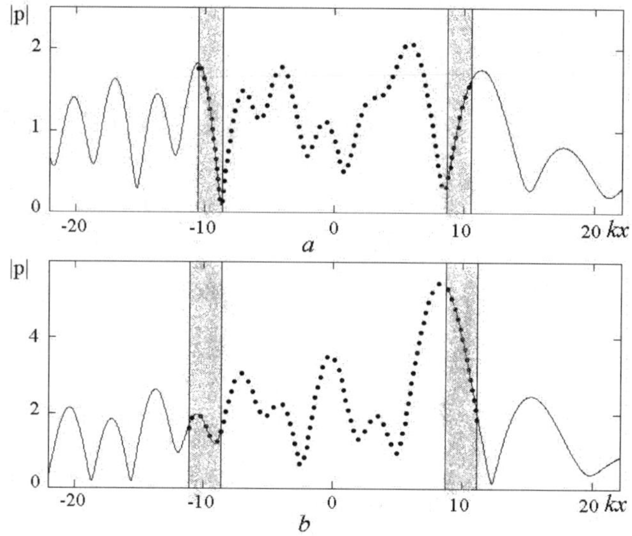

FIG. 7.66: Pressure amplitude dependency on coordinate x for different values of radial coordinate R, $r_0/R_0 = 1.6$, $R_0/\lambda = 1.1$, thin vertical lines form the overlap region of partial domains I, II (on the left) and II, III (on the right): (a)—$R = R_0/2$, (b)—$R = 0$

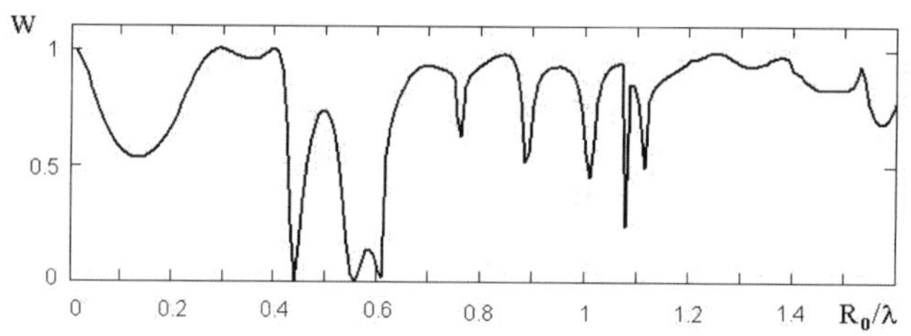

FIG. 7.67: Energy transmission coefficient dependency W on the wave radius of tube R_0/λ, $r_0/R_0 = 1.6$

7.10 PROPERTIES OF NORMAL WAVES IN ELASTIC FLUID-FILLED WAVEGUIDES

So far in this chapter properties of normal waves in the liquid filled waveguides with ideal (rigid or soft) walls were considered. The structure of the wave field in waveguide becomes more complicated when its walls are elastic and the part of wave energy can be transferred to them. Investigation of wave processes in the elastic cylinders with the liquid, started in the 19th century, stimulated by the desire to understand the patterns of pulse propagation in the blood vessels and pipelines. Study of the issues of blood flow in the human arteries, as specified in [46], originates in the work of T. Young (1808). The knowledge of the normal waves characteristics in compound waveguides consisting of cylindrical elastic pipes, filled by a liquid, are of interest for different application in instrument making, geoacoustics, seismology, ultrasonic fault detection, and medical acoustics. The known dependence of dispersion properties of the normal waves from physical characteristics of the waveguides gives way to diagnose of various devices and systems under the measured characteristics of waves, propagating in them.

The principal difference of the dispersion properties of elastic waveguides and acoustic waveguides (filled with ideal compressible fluid) becomes clear after the works of R.D. Mindlin [47]. The main important feature of the normal waves in elastic waveguides is strong frequency dependence of wave characteristics in waveguide cross section [48]. The physical reason of this dependence is interaction phenomena between shear and longitudinal waves in a bonded elastic body. Elastic-fluid interaction in composed waveguides causes yet more complicated dispersion properties of the normal waves. This is manifested, primarily, in analyzing the energy characteristics of normal waves.

Wave disturbances in a waveguide are a set of propagating and evanescent waves. Here we will only discuss the properties of propagating waves. The focus is on two aspects: the distribution of kinematic and energy characteristics of normal waves over cross-section of the waveguide and special features of the interaction between the wave motions in the fluid and elastic shell.

Figure 7.68 shows the axial section of cylindrical waveguide with geometrical characteristics of components and direction of coordinate axes. We are want to determine characteristics of harmonic wave, travelling in z-direction. Expressions for characteristics of waves are presented in dimensionless coordinates $r = R/R_0$, $z = Z/R_0$.

In this case, the dependence of all the characteristics of normal wave on the coordinate z and time t determines by the factor $\exp\left[i\left(\varsigma z - \omega t\right)\right]$ where ς is the dimensionless wave number. To describe the radial dependences of wave characteristics in fluid part of waveguide one has to use the solutions of corresponding Helmholtz equation. As a result, we have next expression for pressure in liquid

$$p\left(r, z, t\right) = D_0 J_0\left(\chi_0 r\right) \exp\left[i\left(\varsigma z - \omega t\right)\right]. \tag{7.124}$$

Construction the expressions for component of displacement vector $\mathbf{U}\left(U_r, U_z\right)$ carries out using partial solutions of the equations of infinitesimal element of isotropic elastic medium [49]

$$\left(\lambda + 2\mu\right) \text{grad}\,\text{div}\mathbf{U} - \mu\text{rotrot}\mathbf{U} = \rho_0 \frac{\partial^2 \mathbf{U}}{\partial t^2}, \tag{7.125}$$

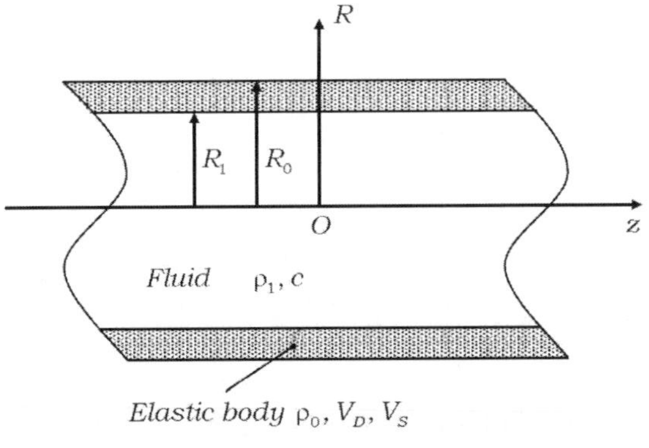

FIG. 7.68: Geometry of the waveguide

where λ and μ are Lame coefficients, ρ_0 is the density of elastic media. For the considered axisymmetric case the desired function take the forms

$$U_r(r,z,t) = -\alpha\left[A_1 J_1(\alpha r) + A_2 Y_1(\alpha r)\right] + \beta\left[B_1 J_1(\beta r) + B_2 Y_1(\beta r)\right]\exp\left[i(\varsigma z - \omega t)\right], \tag{7.126a}$$

$$U_z(r,z,t) = i\varsigma\left[A_1 J_0(\alpha r) + A_2 Y_0(\alpha r)\right] + i\frac{\beta^2}{\varsigma}\left[B_1 J_0(\beta r) + B_2 Y_0(\beta r)\right]\exp\left[i(\varsigma z - \omega t)\right]. \tag{7.126b}$$

In the presented expression (7.124) and (7.126a), (7.126b) of normal wave characteristics the next notations were introduced:

$$\chi_0^2 = \gamma_2^2\left(\frac{V_S}{c}\right)^2 - \varsigma^2, \quad \alpha^2 = \gamma_1^2 - \varsigma^2, \quad \beta^2 = \gamma_2^2 - \varsigma^2,$$

$$\gamma_1 = \omega R_0/V_D, \quad \gamma_2 = \omega R_0/V_S. \tag{7.127}$$

Here V_S is the shear wave velocity and V_D is the longitudinal wave velocity in elastic media, c is velocity of sound in fluid, J_n, Y_n are the Bessel functions. The general expressions (7.124) and (7.126a), (7.126b) contain five arbitrary coefficients D_0, A_1, A_2, B_1, B_2. To get relationship between their values it is necessary to use the boundary conditions on the free surface and the surface of contact between the elastic cylinder and fluid. These conditions produce next five equations

$$\sigma_{rr}(1,z,t) = \sigma_{rz}(1,z,t) = \sigma_{rz}(r_1,z,t) = 0,$$
$$\sigma_{rr}(r_1,z,t) = -p(r_1,z,t),$$
$$\frac{\partial U_r(r_1,z,t)}{\partial t} = -\frac{i}{\omega\rho_1 R_0}\frac{\partial p(r_1,z,t)}{\partial r}. \tag{7.128}$$

Here ρ_1 is density of fluid. The components of stress here are determined using displacement (7.126a), (7.126b) according to Hooke's law [49]. The Eq. (7.128) represent the conditions of the lack of stress on the outer surface of the waveguide, and coupling conditions on the contact surface of a liquid and an elastic shell.

Considering expressions, (7.124) and (7.126a), (7.126b) it is easy to see that the system (7.128) reduces to five homogeneous algebraic equations in the coefficients D_0, A_1, A_2, B_1, B_2. The coefficients of this system is rather cumbersome expressions. Their explicit expressions presented in [50]. Equating to zero the determinant of this algebraic system, that is necessary for the existence of a nonzero solution, forming the dispersion equation of the wave problem under consideration. The solutions of this equation relates values of dimensionless frequency γ_2 and wave number ς at given amounts the physical and geometrical parameters λ, μ, V_S/c, ρ_0/ρ_1, r_1.

There is no problem to get concrete value of a root of this dispersion equation. But, to give a complete description of the dispersion characteristics of normal waves it is necessary to construct dispersion spectra as a system of continues curves, dispersion branches

[47]. It looks promised to use notion of partial systems and construct a geometrical basis for qualitative analysis the wave interection effects in composite waveguide. As components (partial systems) of the considered compound waveguide will be used acoustical waveguide with rigid boundary and empty elastic cylinder. As a first concrete example thick-walled ($r_1 = 0.3$), still waveguide filled with water will be considered. Thus, we have the following values of the physical parameters: $V_S/c = 2.14$, $\rho_0/\rho_1 = 0.13$. The data concerning dispersion spectra of the waveguide is shown in Fig. 7.69.

According to the idea of partial systems, intensive interaction between component occurs for parameters corresponding the points of crossing dispersion branches [as in Fig. 7.69(a)]. Interaction between components in integrated system produces "repulsion" of dispersion branches. As a measure of intensity of the interaction it is possible to consider the distance between dispersion branches. We can see a illustration of the effects in Fig. 7.69(b). For this concrete values of geometrical and physical parameters the interaction between wave in liquid and elastic shell occurs in very narrow intervals of frequencies and wave numbers. It is interesting that lowest branch for "water" wave are practically dispersion less and corresponding branch is a little bellow the line C ($\chi_0 = 0$). General opinion about dashed dispersion branches is that corresponding normal waves on different parts are "successors" or water, or elastic waves. It is important for estimation of energy characteristics of waves, as that will be shown below.

The data on properties of normal wave for the case of thick-walled cylinder can be used to predict ones for the normal waves in waveguide with thin-walled cylinder in the low frequency domain. The second mode of the compound waveguide is connected "genetically" to the lowest normal wave in the empty elastic cylinder. The motion in this mode is practically longitudinal for small frequencies. Such type of motion do not produce a wave in ideal fluid. The phase velocity of this mode is equal to the velocity of rod mode and do not depend of thickness of cylinder. So, the second dispersion curve for compound waveguide is always the same for the small frequency domain.

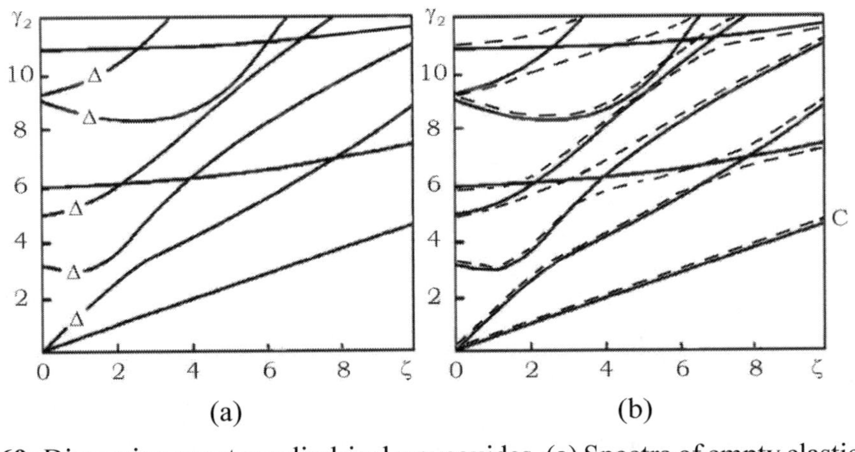

FIG. 7.69: Dispersion spectra cylindrical waveguides. (a) Spectra of empty elastic waveguide (marked by Δ) and fluid waveguide, (b) spectra of compound waveguide (dashed lines) against the curves of Fig. a

Specific features of normal waves shown in Fig. 7.69(b) for the pair still-water are preserved when thickness of cylinder decreases to value $r_1 \approx 0.9$. For thin walled cylinder manifestation of interaction effects has a specific feature for the lowest normal waves. Calculation results for the two different thickness of elastic shells.

In Fig. 7.70 the dispersion curves of the compound waveguides are shown by thickened dashed lines. To discuss the choice of corresponding partial subsystem the dispersion curves for empty elastic cylinder (thin continues lines, marked by triangles), acoustical waveguide with rigid wall (thin continues line), and acoustical waveguide with soft wall (dash-and-dot lines) are shown in Fig. 7.70. The wave interaction phenomena for such waveguides is more complicated. Strong interaction occurs not only near points of crossing the dispersion curves of partial subsystems. We can see sufficiently changes in dispersion brunches in low frequency domain. Physically it means that the important factors to occurrence the interaction effects are not only ratio of acoustical impedances of still and water but also bending rigidity elastic shells. In the same time many dispersion curves have the segments close to the dispersion curves of the partial subsystem, especially in high frequency domain.

As to waves of higher order (high frequency region), for both thickness of the cylinder, the general tendency is saved. Each dispersion curve has a segment, where the curve is very close (graphically coincides) to dispersion curves of the empty cylinder. For the corresponding values of frequency and wave number the normal wave transfers energy largely in the elastic cylinder. On other segments of curves, where the energy is transferred mainly in fluid, the branches of dispersion curves differ from appropriate branches of the acoustic waveguide as with rigid, and with soft walls. With the increasing of the normal wave number ς the specific tendency is tracked. When the frequency increases the "fluid" segments of the dispersion curves approaches to the dispersion curves of acoustical waveguide with rigid wall. For the cylinder with $r_1 = 0.9$ the tendency is more pronounced. Physical reason of that is the increase of mass component of an impedance with increase of frequency.

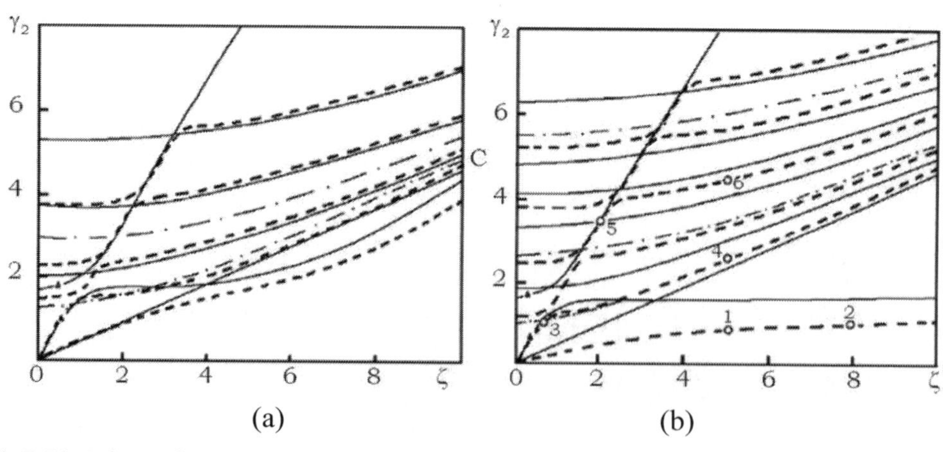

FIG. 7.70: Dispersion spectra for compound waveguide with still shell different thickness $d = 1 - r_1$: (a) $r_1 = 0.9$, (b) $r_1 = 0.99$

To illustrate manifestation of the interaction effects on energy characteristics of normal waves of the compound waveguide let us consider the average for the period energy flow in normal waves for several concrete values of frequencies and wave numbers. These values correspond the points on curves in Fig. 7.70(b) marked with circles. The common power flow in a normal wave is naturally presented in the sum of the flows in fluid core W_0 and in elastic shell W_1. In terms of the introduced function these values are given in following forms

$$W_0 = \frac{\omega\pi}{2} D_0^2 \varsigma \frac{\rho_1}{\rho_0} \left(\frac{c}{V_S}\right)^2 \int_0^{r_1} J_0^2 (\chi_0 r)\, r dr,$$

$$W_1 = \omega (R_0 - R_1) D_0^2 \int_{r_1}^{1} (-\sigma_{zz} U_z + \sigma_{rz} U_r)\, r dr.$$

These expressions contain only one amplitude parameter D_0. For given values of γ_2 and ς the constants A_i, B_i, $(i = 1, 2)$ in the Eqs. (7.124) and (7.126a), (7.126b) are expressed through value D_0 using the boundary and matching conditions. The results of calculations are presented in the Table 7.2.

These data give reason to talk about "liquid" and "elastic" parts of dispersion branches. A more complete analysis of the characteristics of normal waves in a waveguide thin walls are presented in [51]. The waveguides for which the sound speed in liquid is higher than the speed of shear waves in elastic shells can be determined as waveguides with complained walls. Specific feature of wave interaction I such waveguides presented in [52].

This section ends with a chapter and the entire book. Waveguide distribution is one of the most interesting phenomena in the physics of waves. Understanding structure and properties of normal waves in waveguides, it is important to build measurement systems and the development of signal processing algorithms in many sections of the engineering.

We conclude this chapter, as well as the entire book, with a section describing the properties of normal waves in a compound waveguide. Waveguide propagation is one of the most interesting phenomena in the physics of waves. Understanding the structure and properties of normal waves in waveguides is important to build measurement systems and the development of signal processing algorithms in many areas of engineering.

We note finally that the results given above are based on the following works: monograph [15], articles [6,9,14,21–23,25,27,31,37,43,50].

TABLE 7.2: Distribution of energy fluxes in normal waves by waveguide components

Number of point	W_0/W_1	Number of point	W_0/W_1
1	41.1	4	3.59×10^3
2	5.22	5	4.52×10^{-3}
3	3.07×10^{-2}	6	3.62×10^2

REFERENCES

1. Shestopalov, V.P., Kirilenko, A.A., and Rud, L.A., *Waves Resonance Scattering, Vol. 2, The Waveguides Discontinuities*, Kiev: Naukova dumka, p. 214, 1986. [in Russian]
2. Hwang, W.S., A boundary integral method for acoustic and scattering, *J. Acoust. Soc. Amer.*, vol. **101**, no. 6, pp. 3330–3335, 1997.
3. Sridhara, B.S. and Crocker, M.J., Review of theoretical and experimental aspects of acoustical modeling of engine exhaust systems, *J. Acoust. Soc. Amer.*, vol. **95**, no. 1, pp. 2363–2370, 1994.
4. Rzhevkin, S.N., *The Course of Lectures on Acoustic Theory*, Moscow: Lomonosov Moscow State University, p. 336, 1960. [in Russian]
5. Lewin, L., On the inadequacy of discrete mode-matching techniques in some wavequide discontinuity problems, *IEEE Trans. Microwave Theory Techniques*, vol. **18**, no. 7, pp. 364–372, 1970.
6. Goncharova, I.Y. and Matsypura, V.T., Sound propagation in the waveguide with a bend, *Akustychnyi Visnyk*, vol. **1**, no. 2, pp. 57–64, 1998. [in Russian]
7. Isakovich, M.A., *General Acoustics*, Moscow: Nauka, p. 495, 1973. [in Russian]
8. Skudrzyk, E., *Simple and Complex Vibratory Systems*, Pennsylvania State University Press, p. 514, 1968.
9. Matsypura, V.T., Wave propagation in the waveguide with a branch, *Elektronika i Svyaz*, vol. **5**, pp. 3–7, 1998. [in Russian]
10. Glazanov, V.E., *Hydroacoustic Systems Shielding*, Leningrad: Sudostroenie, p. 145, 1986. [in Russian]
11. Shenderov, E.L., *Wave Problems in Hydroacoustics*, Leningrad: Sudostroenie, p. 348, 1972. [in Russian]
12. Shenderov, E.L., Scattering of plane sound wave of a semi-infinite wavequide with a stiff flange, *Acoust. Phys.*, vol. **44**, no. 1, pp. 88–96, 1998.
13. Yampolskiy, V.G. and Suhnov, V.I., Sectored horn antenna analysis, *Radiotehnika*, vol. **7**, pp. 70–74, 1981. [in Russian]
14. Matsypura, V.T., Sound transmission through the matching area of flat and wedge shaped waveguides, *Akustychnyi Visnyk*, vol. **2**, no. 1, pp. 31–41, 1999. [in Russian]
15. Grinchenko, V.T. and Vovk, I.V., *Wave Problems of Sound Scattering by Elastic Shells*, Kiev: Naukova dumka, p. 238, 1986. [in Russian]
16. Tyulin, V.N., *Introduction to Sound Radiation and Scattering Theory*, Moscow: Nauka, p. 254, 1976. [in Russian]
17. Klim, S.Y., Diffraction coefficients and field patterns of obtuse angle dielectric wedge illuminate by E-polarized plane wave, *IEEE Trans. Antennal Propagation*, vol. **40**, no. 11, pp. 1427–1431, 1999.
18. Ufimtsev, P.Y., *Method of Edge Waves in Physical Theory of Diffraction*, Moscow: Soviet Radio, p. 243, 1962. [in Russian]
19. Abrahams, I.D. and Wicham, G.R., On the scattering of sound by two parallel staggered plates, I. Explicit matrix Winer-Hopf factorization, *Proc. Roy. Soc. London*, vol. **420**, pp. 131–156, 1988.

20. Lavrie, J.B., Abrachams, I.D., and Linton C.M., Acoustic radiation from two opposed semi-infinite coacsiial cylindrical waveguides, I. Overlapping edges, *Wave Motion*, vol. **18**, pp. 121–142, 1993.

21. Grinchenko, V.T. and Matsypura, V.T., Sound radiation from the open end of wedge shaped waveguide, I. Method of solution and computation algorithm, *Akustychnyi Visnyk*, vol. **2**, no. 4, pp. 32–41, 1999. [in Russian]

22. Grinchenko, V.T. and Matsypura, V.T., Sound radiation from the open end of wedge shaped waveguide, 2. Numerical results analysis, *Akustychnyi Visnyk*, vol. **3**, no. 2, pp. 63–71, 2000. [in Russian]

23. Matsypura, V.T., The energy structure of the wave scattering field by aperture of the wedge-shaped waveguide, *Electron. Commun.*, vol. **14**, pp. 16–19, 2002. [in Russian]

24. Vaynshteyn, L.A., *Diffraction Theory and Factorization Method*, Moscow: Soviet Radio, p. 431, 1966. [in Russian]

25. Bulanaya, M.A. and Matsypura, V.T., Acoustic signal propagation in the waveguide with the stepwise variation of crossection, *Akustychnyi Visnyk*, vol. **12**, no. 1, pp. 19–31, 2009. [in Russian]

26. Sirenko, Y.K., *Transition Processes Modeling and Analysis in the Open Periodic, Waveguide and Compact Resonators*, Kharkov: p. 363, 2003. [in Russian]

27. Bulanaya, M.A., Grinchenko, V.T., Vovk, I.V., and Matsypura, V.T., The singularities of acoustic impulse signal propagation in the flat regular waveguide, *Akustychnyii Visnyk*, vol. **11**, no. 4, pp. 9–23, 2008. [in Russian]

28. Lyapin, A.D., Sound attenuation in waveguides, *Soviet Phys.-Acoust.*, vol. **21**, no. 3, pp. 215–222, 1975.

29. Bogolepov, I.I., *Industrial Sound-Proofing*, Leningrad: Sudostroenie, p. 368, 1986. [in Russian]

30. Kutischev, M.A., The experimental studies of the baffles efficiency in the waveguides, *J. Acoust.*, vol. **26**, no. 1, pp. 99–103, 1980. [in Russian]

31. Bulanaya, M.A., Grinchenko, V.T., Vovk, I.V., and Matsypura, V.T., Acoustic signal propagation in the waveguide with an irregularity in the form of a chamber, *Akustychnyi Visnyk*, vol. **12**, no. 3, pp. 3–19, 2009. [in Russian]

32. Shenderov, E.L., *Sound Radiation and Scattering*, Leningrad: Sudostroenie, p. 304, 1989. [in Russian]

33. Shestopalov, V.P., Litvinenko, L.N., Maslov, S.A., and Sologub, V.G., *Waves Diffraction by Gratings*, Kharkov: KPU Publish ing House, 1973. [in Russian]

34. Soviet Encyclopedia Moscov, Small medical encyclopedia, vol. **1**, pp. 188–189, 1991. [in Russian]

35. Ertel, P.Y., Merle, L., Brown, R.K., and Stern, A.M., Stethoscope acoustic, Part 2. Transmission and filtration patterns, *Circulation*, vol. **34**, pp. 899–909, 1966.

36. Pashkevich, E.V. and Vorogushin, V.A., The algorithm of transmission factor of stethoscope, Theses All–S oviet Union Conference, The *Problems of Creating the Equipment for Medical Laboratory Research*, Leningrad, pp. 81–85, 1974. [in Russian]

37. Vovk, I.V. and Goncharova, I.Y., Analytical method for estimating the acoustical properties of stethoscopes, *Acustuchnyi Visnyk*, vol. **3**, no. 3, pp. 10–16, 2000. [in Russian]

38. Keefe, D.H., Bulen, J.C., Arehart, K.H., and Burns, E.M., Ear-canal impedance and reflection coefficient in human infants and adults, *J. Acoust. Soc. Amer.*, vol. **94**, no. 5, pp. 2617–2638, 1973.

39. Littmann, D., An approach to the ideal stethoscope, *J. Am. Med. Assoc.*, vol. **178**, no. 5, pp. 504–505, 1961.

40. Abella, M., Formolo, J., and Penney, D.G., Comparison of the acoustic properties of six popular stethoscopes, *J. Acoust. Soc. Am.*, vol. **91**, no. 4, Pt. 1, pp. 2224–2228, 1992.

41. Cochran, J.A. and Pecina, R.G., Mode propagation in continuously curved waveguides, *Radio Sci.*, vol. **34**, pp. 679–696, 1966.

42. Horvat, M. and Prosen, T., The bends on quantum waveguide and cross-products of Bessel functions, *J. Phys. A: Math. Gen.*, vol. **40**, no. 24, pp. 325–331, 2007.

43. Vovk, I.V., Gomilko, A.M., and Gorodetskaya, N.S., Concerning the application features of partial domains method in the wave problems, *J. Acoust.*, vol. **41**, no. 3, pp. 399–404, 1995. [in Russian]

44. Mizokhita, S., *Theory of Equations with Partial Derivatives*, Moscow: Mir, p. 504, 1977. [in Russian]

45. Mittra, R., (Eds.), *Computer Techniques for Electromagnetics*, UK: Pergamon Press, p. 403, 1973.

46. Rubinov, S.I. and Keller, J.B., Wave propagation in a fluid-filled tube, *J. Acoust. Soc. Am.*, vol. **50**, no. 1, Pt. 2, pp. 198–223, 1971.

47. Mindlin, R.D., *An Introduction to the Mathematical Theory of Vibration of Elastic Plate*, Singapore: World Scientific, p. 190, 2006.

48. Kumar, R., Dispersion of Axially Symmetric Waves in Empty and Fluid-Filled Cylindrical Shells, *Acustica*, vol. **27**, no. 6, pp. 317–329, 1972.

49. Landau, L.D. and Lifshitz, E.M., *Theory of Elastisit*, Vol. 7 of Course of Theoretical Physics, UK: Pergamon Press, p. 165, 1975.

50. Grinchenko, V.T. and Komissarova, G.L., Wave propagation in a hollow elastic cylinder with a fluid, *Int. Appl. Mech.*, vol. **20**, no. 1, pp. 18–23, 1984.

51. Komissarova, G.L., Propagation of normal waves through a fluid contained in thin-walled cylinder, *Int. Appl. Mech.*, vol. **38**, no. 1, pp. 103–112, 2002.

52. Komissarova, G.L., Features of wave propagation in liquid-filled cylinder with complain walls, *Int. J. Fluid Mech. Res.*, vol. **31**, no. 6, pp. 574–590, 2004.

Subject Index

A

B

C

D

E

F

G

H

I

K

L

M

N

P

Q

R

S